Student's Solutions Manual

THIRD EDITION
Elementary Algebra
CONCEPTS AND APPLICATIONS
Bittinger/Keedy

Judith A. Penna

ADDISON-WESLEY PUBLISHING COMPANY
Reading, Massachusetts • Menlo Park, California • New York
Don Mills, Ontario • Wokingham, England • Amsterdam • Bonn
Sydney • Singapore • Tokyo • Madrid • San Juan

Reproduced by Addison-Wesley from camera-ready copy supplied by the author.

ISBN 0-201-17394-8

Copyright © 1990 by Addison-Wesley Publishing Company, Inc.

All rights reserved. No part of this publication may be reproduced, stored in a retrieval system, or transmitted, in any form or by any means, electronic, mechanical, photocopying, recording, or otherwise, without the prior written permission of the publisher. Printed in the United States of America.

7 8 9 10 BA 959493

TABLE OF CONTENTS

Chapter 1	1
Chapter 2	23
Chapter 3	41
Chapter 4	63
Chapter 5	91
Chapter 6	125
Chapter 7	157
Chapter 8	195
Chapter 9	227
Chapter 10	251
Chapter 11	277
Answers for Exercises in the Appendixes	313

Special thanks are extended to Patsy Hammond for her typing. Her skill, hard work, and good humor made the author's work much easier.

CHAPTER 1 INTRODUCTION TO ALGEBRA AND PROBLEM SOLVING

Exercise Set 1.1

1. Substitute 29 for x: 29 - 6 = 23
   Substitute 34 for x: 34 - 6 = 28
   Substitute 47 for x: 47 - 6 = 41

2. 150 sec, 450 sec, 10 min

3. $bh = 6.5(15.4) = 100.1$ sq cm

4. 1935 sq m

5. $rt = 55(4) = 220$ mi

6. 0.52

7. $6x = 6 \cdot 7 = 42$

8. 49

9. $\frac{x}{y} = \frac{9}{3} = 3$

10. 7

11. $\frac{3p}{q} = \frac{3 \cdot 2}{6} = \frac{6}{6} = 1$

12. 3

13. $\frac{x+y}{5} = \frac{10+20}{5} = \frac{30}{5} = 6$

14. 9

15. $\frac{x-y}{8} = \frac{20-4}{8} = \frac{16}{8} = 2$

16. 2

17. $\frac{x}{y} = \frac{3}{6} = \frac{1}{2}$

18. $\frac{1}{4}$

19. $\frac{5z}{y} = \frac{5 \cdot 8}{2} = \frac{40}{2} = 20$

20. 2

21. $b + 6$, or $6 + b$

22. $t + 8$, or $8 + t$

23. $c - 9$

24. $d - 4$

25. $q + 6$, or $6 + q$

26. $z + 11$, or $11 + z$

27. $a + b$, or $b + a$

28. $d + c$, or $c + d$

29. $y - x$

30. $h - c$

31. $w + x$, or $x + w$

32. $t + s$, or $s + t$

33. $n - m$

34. $q - p$

35. $r + s$, or $s + r$

36. $d + f$, or $f + d$

37. $2x$

38. $3p$

39. $5t$

40. $9d$

41. $3b$

42. 97%x, or $0.97x$

43. Let x represent the number. Then we have 43%x, or $0.43x$.

44. $d - $29.95

45. The distance traveled is the product of the speed and the time, or $65t$ mi.

46. $x + 3y$

47. $y + 2x$

48. $a + 2 + b$

49. $2x - 3$

50. $a + 5$

51. $b - 2$

52. $x + x$, or $2x$

53. $s + s + s + s$, or $4s$

54. $\ell + \ell + w + w$, or $2\ell + 2w$

55. $y = 8$, $x = 2y = 2 \cdot 8 = 16$;
    $\frac{x+y}{4} = \frac{8+16}{4} = \frac{24}{4} = 6$

56. 5

Chapter 1   (1.2)

57. $x = 9$, $y = 3x = 3 \cdot 9 = 27$;
$$\frac{y - x}{3} = \frac{27 - 9}{3} = \frac{18}{3} = 6$$

58. 2

59. $\frac{y + x}{2} + \frac{3 \cdot y}{x} = \frac{4 + 2}{2} + \frac{3 \cdot 4}{2} = \frac{6}{2} + \frac{12}{2} = 3 + 6 = 9$

60. $w + 4$

61. The preceding odd number is 2 less than $d + 2$:
    $d + 2 - 2 = d$

62. $t - 3$, $t + 3$

63. $v + 2 + v - 2 = v + v$, or $2v$

64. 4

65. A year from now the number of dollars in the bank will be the sum of the original n dollars and the interest, 10% of n dollars:
    $n + 10\%n$, or $n + 0.1n$

Exercise Set 1.2

1. We write several factorizations of 56. There are other correct answers.
   $4 \cdot 14$, $7 \cdot 8$, $2 \cdot 4 \cdot 7$

2. $2 \cdot 51$, $6 \cdot 17$; there are other correct answers.

3. $1 \cdot 93$, $3 \cdot 31$

4. $4 \cdot 36$, $12 \cdot 12$; there are other correct answers.

5. $14 = 2 \cdot 7$

6. $3 \cdot 5$

7. $33 = 3 \cdot 11$

8. $5 \cdot 11$

9. $9 = 3 \cdot 3$

10. $5 \cdot 5$

11. $49 = 7 \cdot 7$

12. $11 \cdot 11$

13. We begin by factoring 18 in any way that we can and continue factoring until each factor is prime.
    $18 = 2 \cdot 9 = 2 \cdot 3 \cdot 3$

14. $2 \cdot 2 \cdot 3 \cdot 3$

15. We begin by factoring 40 in any way that we can and continue factoring until each factor is prime.
    $40 = 4 \cdot 10 = 2 \cdot 2 \cdot 2 \cdot 5$

16. $2 \cdot 2 \cdot 2 \cdot 7$

17. We begin by factoring 90 in any way that we can and continue factoring until each factor is prime.
    $90 = 2 \cdot 45 = 2 \cdot 9 \cdot 5 = 2 \cdot 3 \cdot 3 \cdot 5$

18. $2 \cdot 2 \cdot 2 \cdot 3 \cdot 5$

19. We go through the table of primes until we find a prime that divides 210. We continue dividing by that prime until it is not possible to do so any longer. We continue this process until each factor is prime.
    $210 = 2 \cdot 105 = 2 \cdot 3 \cdot 35 = 2 \cdot 3 \cdot 5 \cdot 7$

20. $2 \cdot 3 \cdot 5 \cdot 11$

21. We go through the table of primes until we find a prime that divides 91. The first such prime is 7.
    $91 = 7 \cdot 13$
    Both factors are prime, so this is the prime factorization.

22. $11 \cdot 13$

23. We go through the table of primes until we find a prime that divides 119. The first such prime is 7.
    $119 = 7 \cdot 17$
    Both factors are prime, so this is the prime factorization.

24. $13 \cdot 17$

25. The multiples of 3 are:
    3, 6, 9, 12, 15, 18, <u>21</u>, 24, 27, 30, 33, 36, 39, <u>42</u>, 45, 48, 51, 54, 57, 60, <u>63</u>, ...
    The multiples of 7 are:
    7, 14, <u>21</u>, 28, 35, <u>42</u>, 49, 56, <u>63</u>, ...
    The first three common multiples are 21, 42, and 63.

26. 54, 108, 162

27. The multiples of 20 are:
    20, 40, <u>60</u>, 80, 100, <u>120</u>, 140, 160, <u>180</u>, ...
    The multiples of 30 are:
    30, <u>60</u>, 90, <u>120</u>, 150, <u>180</u>, ...
    The first three common multiples are 60, 120, and 180.

28. 72, 144, 216

Chapter 1   (1.2)

29. $3 = 3$ (3 is prime)
    $15 = 3 \cdot 5$
    The LCM is $3 \cdot 5$, or 15.

30. 40

31. $30 = 2 \cdot 3 \cdot 5$
    $40 = 2 \cdot 2 \cdot 2 \cdot 5$
    The LCM is $2 \cdot 2 \cdot 2 \cdot 3 \cdot 5$, or 120.

32. 300

33. 13 and 23 are both prime. The LCM is $13 \cdot 23$, or 299.

34. 36

35. $18 = 2 \cdot 3 \cdot 3$
    $30 = 2 \cdot 3 \cdot 5$
    The LCM is $2 \cdot 3 \cdot 3 \cdot 5$, or 90.

36. 360

37. $30 = 2 \cdot 3 \cdot 5$
    $36 = 2 \cdot 2 \cdot 3 \cdot 3$
    The LCM is $2 \cdot 2 \cdot 3 \cdot 3 \cdot 5$, or 180.

38. 150

39. $24 = 2 \cdot 2 \cdot 2 \cdot 3$
    $36 = 2 \cdot 2 \cdot 3 \cdot 3$
    The LCM is $2 \cdot 2 \cdot 2 \cdot 3 \cdot 3$, or 72.

40. 420

41. 17 and 29 are both prime. The LCM is $17 \cdot 29$, or 493.

42. 72

43. $12 = 2 \cdot 2 \cdot 3$
    $28 = 2 \cdot 2 \cdot 7$
    The LCM is $2 \cdot 2 \cdot 3 \cdot 7$, or 84.

44. 315

45. 2, 3, and 5 are all prime. The LCM is $2 \cdot 3 \cdot 5$, or 30.

46. 105

47. $24 = 2 \cdot 2 \cdot 2 \cdot 3$
    $36 = 2 \cdot 2 \cdot 3 \cdot 3$
    $12 = 2 \cdot 2 \cdot 3$
    The LCM is $2 \cdot 2 \cdot 2 \cdot 3 \cdot 3$, or 72.

48. 176

49. $5 = 5$ (5 is prime)
    $12 = 2 \cdot 2 \cdot 3$
    $15 = 3 \cdot 5$
    The LCM is $2 \cdot 2 \cdot 3 \cdot 5$, or 60.

50. 360

51. $6 = 2 \cdot 3$
    $12 = 2 \cdot 2 \cdot 3$
    $18 = 2 \cdot 3 \cdot 3$
    The LCM is $2 \cdot 2 \cdot 3 \cdot 3$, or 36.

52. 360

53. $\frac{5}{6} = \frac{5}{6} \cdot 1$ Identity property of 1
    $= \frac{5}{6} \cdot \frac{8}{8}$ Using $\frac{8}{8}$ for 1
    $= \frac{40}{48}$ Multiplying numerators and denominators

54. $\frac{99}{110}$

55. $\frac{6}{7} = \frac{6}{7} \cdot 1$ Identity property of 1
    $= \frac{6}{7} \cdot \frac{100}{100}$ Using $\frac{100}{100}$ for 1
    $= \frac{600}{700}$

56. $\frac{yz}{10z}$

57. $\frac{s}{20} = \frac{s}{20} \cdot 1$ Identity property of 1
    $= \frac{s}{20} \cdot \frac{t}{t}$ Using $\frac{t}{t}$ for 1
    $= \frac{st}{20t}$

58. $\frac{mp}{3np}$

59. We will use $\frac{3}{3}$ for 1 since $24 = 8 \cdot 3$.
    $\frac{7}{8} = \frac{7}{8} \cdot 1 = \frac{7}{8} \cdot \frac{3}{3} = \frac{21}{24}$

60. $\frac{40}{48}$

61. We will use $\frac{y}{y}$ for 1 since $3y = 3 \cdot y$.
    $\frac{x}{3} = \frac{x}{3} \cdot 1 = \frac{x}{3} \cdot \frac{y}{y} = \frac{xy}{3y}$

62. $\frac{4b}{ab}$

63. $\frac{18}{45} = \frac{2 \cdot 9}{5 \cdot 9}$ Factoring numerator and denominator
    $= \frac{2}{5} \cdot \frac{9}{9}$ Factoring the fractional expression
    $= \frac{2}{5} \cdot 1 = \frac{2}{5}$

3

Chapter 1   (1.2)

64. $\dfrac{2}{7}$

65. $\dfrac{49}{14} = \dfrac{7 \cdot 7}{2 \cdot 7} = \dfrac{7}{2} \cdot \dfrac{7}{7} = \dfrac{7}{2} \cdot 1 = \dfrac{7}{2}$

66. $\dfrac{8}{3}$

67. $\dfrac{6}{42} = \dfrac{1 \cdot 6}{7 \cdot 6}$ Factoring and inserting a factor of 1 in the numerator

$= \dfrac{1}{7} \cdot \dfrac{6}{6}$

$= \dfrac{1}{7} \cdot 1 = \dfrac{1}{7}$

68. $\dfrac{1}{8}$

69. $\dfrac{56}{7} = \dfrac{8 \cdot 7}{1 \cdot 7} = \dfrac{8}{1} \cdot \dfrac{7}{7} = \dfrac{8}{1} \cdot 1 = 8$

70. 12

71. $\dfrac{5y}{5} = \dfrac{5 \cdot y}{5 \cdot 1} = \dfrac{y}{1} \cdot \dfrac{5}{5} = \dfrac{y}{1} \cdot 1 = y$

72. $\dfrac{a}{9}$

73. $\dfrac{x}{9xy} = \dfrac{1 \cdot x}{9 \cdot x \cdot y} = \dfrac{1}{9y} \cdot \dfrac{x}{x} = \dfrac{1}{9y}$

74. $\dfrac{1}{8p}$

75. $\dfrac{8a}{3ab} = \dfrac{8 \cdot a}{3 \cdot a \cdot b} = \dfrac{8}{3b} \cdot \dfrac{a}{a} = \dfrac{8}{3b}$

76. $\dfrac{9}{17q}$

77. $\dfrac{3pq}{6q} = \dfrac{3 \cdot p \cdot q}{2 \cdot 3 \cdot q} = \dfrac{p}{2} \cdot \dfrac{3q}{3q} = \dfrac{p}{2}$

78. $\dfrac{3}{s}$

79. $\dfrac{9nz}{19tn} = \dfrac{9 \cdot n \cdot z}{19 \cdot t \cdot n} = \dfrac{9z}{19t} \cdot \dfrac{n}{n} = \dfrac{9z}{19t}$

80. $\dfrac{13r}{3h}$

81. $\dfrac{1}{4} \cdot \dfrac{1}{2} = \dfrac{1 \cdot 1}{4 \cdot 2}$ Multiplying numerators and denominators

$= \dfrac{1}{8}$

82. $\dfrac{44}{25}$

83. $\dfrac{17}{2} \cdot \dfrac{3}{4} = \dfrac{17 \cdot 3}{2 \cdot 4} = \dfrac{51}{8}$

84. 1

85. $\dfrac{1}{2} + \dfrac{1}{2} = \dfrac{1 + 1}{2}$ Adding numerators; keeping the same denominator

$= \dfrac{2}{2} = 1$

86. $\dfrac{3}{4}$

87. $\dfrac{4}{9} + \dfrac{13}{18} = \dfrac{4}{9} \cdot \dfrac{2}{2} + \dfrac{13}{18}$ LCD is 18

$= \dfrac{8}{18} + \dfrac{13}{18}$

$= \dfrac{21}{18}$

$= \dfrac{7 \cdot 3}{6 \cdot 3} = \dfrac{7}{6}$ Simplifying

88. $\dfrac{4}{3}$

89. $\dfrac{3}{10} + \dfrac{8}{15} = \dfrac{3}{10} \cdot \dfrac{3}{3} + \dfrac{8}{15} \cdot \dfrac{2}{2}$ LCD is 30

$= \dfrac{9}{30} + \dfrac{16}{30}$

$= \dfrac{25}{30}$

$= \dfrac{5 \cdot 5}{6 \cdot 5} = \dfrac{5}{6}$ Simplifying

90. $\dfrac{41}{24}$

91. $\dfrac{5}{4} - \dfrac{3}{4} = \dfrac{2}{4}$

$= \dfrac{1 \cdot 2}{2 \cdot 2} = \dfrac{1}{2}$

92. 2

93. $\dfrac{13}{18} - \dfrac{4}{9} = \dfrac{13}{18} - \dfrac{4}{9} \cdot \dfrac{2}{2}$ LCD is 18

$= \dfrac{13}{18} - \dfrac{8}{18}$

$= \dfrac{5}{18}$

94. $\dfrac{31}{45}$

95. $\dfrac{11}{12} - \dfrac{2}{5} = \dfrac{11}{12} \cdot \dfrac{5}{5} - \dfrac{2}{5} \cdot \dfrac{12}{12}$ LCD is 60

$= \dfrac{55}{60} - \dfrac{24}{60}$

$= \dfrac{31}{60}$

96. $\dfrac{13}{48}$

97. $\dfrac{7}{6} \div \dfrac{3}{5} = \dfrac{7}{6} \cdot \dfrac{5}{3} = \dfrac{35}{18}$

98. $\dfrac{28}{15}$

99. $\dfrac{8}{9} \div \dfrac{4}{15} = \dfrac{8}{9} \cdot \dfrac{15}{4} = \dfrac{2 \cdot 4 \cdot 3 \cdot 5}{3 \cdot 3 \cdot 4} = \dfrac{2 \cdot 5}{3} \cdot \dfrac{4 \cdot 3}{4 \cdot 3} = \dfrac{10}{3}$

100. $\dfrac{7}{4}$

101. $\dfrac{1}{4} \div \dfrac{1}{2} = \dfrac{1}{4} \cdot \dfrac{2}{1} = \dfrac{1 \cdot 2}{2 \cdot 2 \cdot 1} = \dfrac{1}{2} \cdot \dfrac{2}{2} = \dfrac{1}{2}$

Chapter 1   (1.3)

102. $\frac{1}{2}$

103. $\frac{\frac{13}{12}}{\frac{39}{5}} = \frac{13}{12} \div \frac{39}{5} = \frac{13}{12} \cdot \frac{5}{39} = \frac{13 \cdot 5}{12 \cdot 3 \cdot 13} =$
$\frac{5}{12 \cdot 3} \cdot \frac{13}{13} = \frac{5}{36}$

104. $\frac{68}{9}$

105. $100 \div \frac{1}{5} = \frac{100}{1} \cdot \frac{5}{1} = \frac{500}{1} = 500$

106. 468

107. $\frac{3}{4} \div 10 = \frac{3}{4} \cdot \frac{1}{10} = \frac{3}{40}$

108. $\frac{1}{18}$

109. $xy = 7 \cdot 42 = 294$

110. 27%x, or 0.27x

111. $8 = 2 \cdot 2 \cdot 2$
$12 = 2 \cdot 2 \cdot 3$
The greatest number of times 2 occurs in the factorizations of 8 and 12 is three times. The greatest number of times 3 occurs is once. Thus, the LCM must contain exactly three factors of 2 and one factor of 3. LCM = $2 \cdot 2 \cdot 2 \cdot 3$.
a) No, b) No, c) No, d) Yes

112. 2592

113. Use a calculator to find multiples of 7800. Divide each one by 2700 to determine if it is also a multiple of 2700. The first such number, 70,200, is the LCM. (70,200 = 7800·9 = 2700·26)

114. $\frac{2}{3}$

115. $\frac{pqrs}{qrst} = \frac{p}{t} \cdot \frac{qrs}{qrs} = \frac{p}{t}$

116. $\frac{3sb}{2}$

117. $\frac{4 \cdot 9 \cdot 16}{2 \cdot 8 \cdot 15} = \frac{4 \cdot 3 \cdot 3 \cdot 2 \cdot 8}{2 \cdot 8 \cdot 3 \cdot 5} = \frac{4 \cdot 3}{5} \cdot \frac{3 \cdot 2 \cdot 8}{3 \cdot 2 \cdot 8} = \frac{12}{5}$

118. $\frac{r}{g}$

119. $\frac{3 \cdot (4xy) \cdot (5)}{2 \cdot (3x) \cdot (4y)} = \frac{5}{2} \cdot \frac{3 \cdot 4 \cdot x \cdot y}{3 \cdot 4 \cdot x \cdot y} = \frac{5}{2}$

120. 24 in.

121. Jupiter: $12 = 2 \cdot 2 \cdot 3$
Saturn: $30 = 2 \cdot 3 \cdot 5$
The LCM is $2 \cdot 2 \cdot 3 \cdot 5$, or 60.
Jupiter and Saturn will appear in the same direction every 60 years.

122. Every 420 years

123. Jupiter: $12 = 2 \cdot 2 \cdot 3$
Saturn: $30 = 2 \cdot 3 \cdot 5$
Uranus: $84 = 2 \cdot 2 \cdot 3 \cdot 7$
The LCM is $2 \cdot 2 \cdot 3 \cdot 5 \cdot 7$, or 420. The three planets will appear in the same direction every 420 years.

124.

Product	56	63	36	72	140	96	48	168	110	90	432	63
Factor	7	7	2	36	14	8	6	21	11	9	24	3
Factor	8	9	18	2	10	12	8	8	10	10	18	21
Sum	15	16	20	38	24	20	14	29	21	19	42	24

Exercise Set 1.3

1. 2^4 means $2 \cdot 2 \cdot 2 \cdot 2$

2. $5 \cdot 5 \cdot 5$

3. $(1.4)^5$ means $(1.4) \cdot (1.4) \cdot (1.4) \cdot (1.4) \cdot (1.4)$

4. 2.5

5. n^1 means n.

6. $m \cdot m \cdot m \cdot m \cdot m \cdot m$

7. $(7p)^2$ means $7p \cdot 7p$

8. $11c \cdot 11c \cdot 11c$

9. $(19k)^4$ means $19k \cdot 19k \cdot 19k \cdot 19k$

10. $104d \cdot 104d \cdot 104d \cdot 104d \cdot 104d$

11. $(10pq)^1$ means $10pq$.

12. $24ct \cdot 24ct \cdot 24ct$

13. $10 \times 10 \times 10 \times 10 \times 10 \times 10 = 10^6$

14. 6^4

15. $x \cdot x \cdot x \cdot x \cdot x \cdot x \cdot x = x^7$

16. y^3

17. $3y \cdot 3y \cdot 3y \cdot 3y = (3y)^4$

18. $(5m)^5$

19. $m^3 = 3^3 = 3 \cdot 3 \cdot 3 = 27$

20. 64

Chapter 1   (1.3)

21. $p^1 = 19^1 = 19$

22. 0

23. $x^4 = 4^4 = 4 \cdot 4 \cdot 4 \cdot 4 = 256$

24. 1

25. $y^2 - 7 = 10^2 - 7$
    $= 100 - 7$ Evaluating the power
    $= 93$ Subtracting

26. 37

27. $A = \pi r^2 \approx 3.14 \times (34 \text{ ft})^2$
    $= 3.14 \times 1156 \text{ ft}^2$ Evaluating the power
    $= 3629.84 \text{ ft}^2$

28. 576 m^2

29. $7 + 2 \times 6 = 7 + 12$ Multiplying
    $= 19$ Adding

30. 27

31. $8 \times 7 + 6 \times 5 = 56 + 30$ Multiplying
    $= 86$ Adding

32. 51

33. $19 - 5 \times 3 + 3 = 19 - 15 + 3$ Multiplying
    $= 4 + 3$ Subtracting and adding from left to right
    $= 7$

34. 9

35. $9 \div 3 + 16 \div 8 = 3 + 2$ Dividing
    $= 5$ Adding

36. 28

37. $7 + 10 - 10 \div 2 = 7 + 10 - 5$ Dividing
    $= 17 - 5$ Adding and subtracting from left to right
    $= 12$

38. 400

39. $(6 \cdot 3)^2 = 18^2$ Multiplying within parentheses
    $= 324$

40. 24

41. $4 \cdot 5^2 = 4 \cdot 25$ Evaluating the power
    $= 100$

42. 100

43. $(5 + 3)^3 = 8^3$ Adding within parentheses
    $= 512$

44. 11

45. $6 + 4^2 = 6 + 16$ Evaluating the power
    $= 22$

46. 9

47. $(3 - 2)^2 = 1^2$ Subtracting within parentheses
    $= 1$

48. 1

49. $12 - 2^3 = 12 - 8$ Evaluating the power
    $= 4$

50. 28

51. $2 \times 10^3 - 500 = 2 \times 1000 - 500$ Evaluating the power
    $= 2000 - 500$ Multiplying
    $= 1500$

52. 585

53. $6[9 + (3 + 4)] = 6[9 + 7]$ Adding inside the parentheses
    $= 6[16]$ Adding inside the brackets
    $= 96$

54. 64

55. $8 + (7 + 9) = 8 + 16$ Adding inside the parentheses
    $= 24$

56. 24

57. $15(4 + 2) = 15(6)$ Adding inside the parentheses
    $= 90$

58. 90

59. $12 - (8 - 4) = 12 - 4$ Subtracting inside the parentheses
    $= 8$

60. 0

61. $1000 \div 100 \div 10 = 10 \div 10$ Dividing in order from left to right
    $= 1$

62. 2

Chapter 1   (1.3)

63. We will do the calculations in the numerator and in the denominator and then divide the results.
$$\frac{80 - 6^2}{9^2 + 3^2} = \frac{80 - 36}{81 + 9}$$
$$= \frac{44}{90}$$
$$= \frac{22 \cdot 2}{45 \cdot 2} = \frac{22}{45} \cdot \frac{2}{2}$$
$$= \frac{22}{45}$$

64. $\frac{43}{39}$

65. $\frac{3(6 + 7) - 5 \cdot 4}{6 \cdot 7 + 8(4 - 1)} = \frac{3 \cdot 13 - 5 \cdot 4}{6 \cdot 7 + 8 \cdot 3}$
$$= \frac{39 - 20}{42 + 24}$$
$$= \frac{19}{66}$$

66. $\frac{36}{47}$

67. $3 \cdot (a + 10) = 3 \cdot (12 + 10)$
$\qquad = 3 \cdot 22$ Adding inside the parentheses
$\qquad = 66$

68. 60

69. $(t + 3)^3 = (4 + 3)^3$
$\qquad = 7^3$ Adding inside the parentheses
$\qquad = 343$

70. 125

71. $(x + 5) \cdot (12 - x) = (7 + 5) \cdot (12 - 7)$
$\qquad = 12 \cdot 5$ Working within parentheses
$\qquad = 60$

72. 96

73. $(5y)^3 - 75 = (5 \cdot 2)^3 - 75$
$\qquad = 10^3 - 75$ Working within parentheses
$\qquad = 1000 - 75$ Evaluating the power
$\qquad = 925$

74. 500

75. $\frac{y + 3}{2y} = \frac{5 + 3}{2 \cdot 5}$
$\qquad = \frac{8}{10}$ Calculating in the numerator and in the denominator
$\qquad = \frac{4}{5}$ Simplifying

76. $\frac{11}{5}$

77. $\frac{w^2 + 4}{5w} = \frac{4^2 + 4}{5 \cdot 4}$
$\qquad = \frac{16 + 4}{20}$ Calculating in the numerator and in the denominator
$\qquad = \frac{20}{20} = 1$

78. 3

79. $(x - 4) \cdot (8 + y) = (12 - 4) \cdot (8 + 2)$
$\qquad = 8 \cdot 10$ Working within parentheses
$\qquad = 80$

80. 32

81. $(4n)^3 = (4 \cdot 2)^3$
$\qquad = 8^3$ Multiplying inside the parentheses
$\qquad = 512$

$4n^3 = 4 \cdot 2^3$
$\qquad = 4 \cdot 8$ Evaluating the power
$\qquad = 32$

82. 1728; 108

83. $(4n)^3 = (4 \cdot 4)^3 = 16^3 = 4096$
$4n^3 = 4 \cdot 4^3 = 4 \cdot 64 = 256$

84. 8000; 500

85. $(3p)^2 = (3 \cdot 4)^2$
$\qquad = 12^2$ Multiplying inside the parentheses
$\qquad = 144$

$3p^2 = 3 \cdot 4^2$
$\qquad = 3 \cdot 16$ Evaluating the power
$\qquad = 48$

86. 441; 147

87. $(3p)^2 = (3 \cdot 11)^2 = 33^2 = 1089$
$3p^2 = 3 \cdot 11^2 = 3 \cdot 121 = 363$

88. 6084; 2028

89. $p + q$, or $q + p$

90. $\frac{1}{8}$

91. We begin by factoring 48 in any way that we can and continue factoring until each factor is prime.
$48 = 2 \cdot 24 = 2 \cdot 4 \cdot 6 = 2 \cdot 2 \cdot 2 \cdot 2 \cdot 3$

92. 168

7

Chapter 1   (1.4)

93. Choose any number except 0. For example, let
x = 1.
$3x^2 = 3 \cdot 1^2 = 3 \cdot 1 = 3$, but
$(3x)^2 = (3 \cdot 1)^2 = 3^2 = 9$.

94. Any number except 0 or -2

95. Choose any number except 2. For example, let x = 0.
$\frac{x+2}{2} = \frac{0+2}{2} = \frac{2}{2} = 1$, but x = 0.

96. Any number except 1; 0 is not acceptable

97. Let x represent the number.
$x^2 + 7$, or $7 + x^2$

98. $x + 7^2$

99. Let x represent the number.
$(x + 7)^2$, or $(7 + x)^2$

100. $x^2 + 7^2$

101. Let x represent the number.
$\frac{x+3}{(x+3)^2}$

102. $x(x + 5)$

103. A year ago Victor's age was x - 1. Thus, Carole's age is 2(x - 1).

104. a) $(x + y)^2$; b) $x^2 + y^2$

105. $\frac{10^5}{10^3} = \frac{10 \cdot 10 \cdot 10 \cdot 10 \cdot 10}{10 \cdot 10 \cdot 10} = 10 \cdot 10 = 10^2$

106. 10^5

107. $\frac{5^4}{5^2} = \frac{5 \cdot 5 \cdot 5 \cdot 5}{5 \cdot 5} = 5 \cdot 5 = 5^2$

108. 1

109. $x^3y^2 + zx = 2^3 \cdot 1^2 + 3 \cdot 2$
            = $8 \cdot 1 + 3 \cdot 2$ Evaluating the powers
            = 8 + 6   Multiplying
            = 14

110. 111

111. $x^2 + 2xy + y^2 = 7^2 + 2 \cdot 7 \cdot 8 + 8^2$
                         = $49 + 2 \cdot 7 \cdot 8 + 64$ Evaluating the powers
                         = 49 + 112 + 64   Multiplying
                         = 161 + 64   Adding in order from left to right
                         = 225

112. x^3y^3

113. $3a \cdot 3a \cdot 3a \cdot 2b \cdot 2b = (3a)^3(2b)^2$, or
$3a \cdot 3a \cdot 3a \cdot 2b \cdot 2b = 3 \cdot 3 \cdot 3 \cdot a \cdot a \cdot a \cdot 2 \cdot 2 \cdot b \cdot b = 27a^3 \cdot 4b^2$, or
$3a \cdot 3a \cdot 3a \cdot 2b \cdot 2b = 3 \cdot 3 \cdot 3 \cdot 2 \cdot 2 \cdot a \cdot a \cdot a \cdot b \cdot b = 108a^3b^2$

114. 0

115. $x^{410}y^2 = 1^{410} \cdot 3^2 = 1 \cdot 9 = 9$

116. 127

117. $(x^2)^2 = (3^2)^2$
        = 9^2 Working inside parentheses
        = 81

118. Answers may vary. For a 10 digit readout, 6^{13} and larger will be too large.

119. $\frac{5n^3}{n} = \frac{5 \cdot n \cdot n \cdot n}{n} = 5n^2$
The expressions are equivalent.

120. Yes

121. $xy^2 = x \cdot y \cdot y$. This is not equivalent to $x \cdot x \cdot y \cdot y$.

122. No

123. See the answer section in the text.

124. $1 + 2 + 3 + 4 + 5 + 6 + 7 + (8 \times 9) = 100$

Exercise Set 1.4

1. 8 + y, commutative law of addition

2. 3 + x

3. nm, commutative law of multiplication

4. ba

5. xy + 9, commutative law of addition
9 + yx, commutative law of multiplication
yx + 9, both commutative laws

6. ab + 11, 11 + ba, ba + 11

7. c + ab, commutative law of addition
ba + c, commutative law of multiplication
c + ba, both commutative laws

8. t + rs, sr + t, t + sr

9. $y^2 + x$, commutative law of addition

10. q^3p^2

Chapter 1   (1.4)

11. $t + xt^2$, commutative law of addition
    $t^2x + t$, commutative law of multiplication
    $t + t^2x$, both commutative laws

12. $(b + a)^2$

13. $(a + b) + 2$, associative law of addition

14. $(3 \cdot v) \cdot w$

15. $8 \cdot (x \cdot y)$, associative law of multiplication

16. $y + (z + 7)$

17. $a + (b + 3)$, associative law of addition

18. $5 + (x + y)$

19. $(3 \cdot a) \cdot b$, associative law of multiplication

20. $6 \cdot (x \cdot y)$

21. a) $(a + b) + 2 = a + (b + 2)$, associative law of addition
    b) $(a + b) + 2 = (b + a) + 2$, commutative law of addition
    c) $(a + b) + 2 = (b + a) + 2$
       $= b + (a + 2)$ Using the commutative law first, then the associative law

    There are other correct answers.

22. $3 + (x + y)$, $(x + 3) + y$, $x + (3 + y)$

23. a) $5 + (v + w) = (5 + v) + w$, associative law of addition
    b) $5 + (v + w) = 5 + (w + v)$, commutative law of addition
    c) $5 + (v + w) = 5 + (w + v)$
       $= (5 + w) + v$ Using the commutative law first, then the associative law

    There are other correct answers.

24. $(6 + x) + y$, $(x + 6) + y$, $x + (6 + y)$

25. a) $(x \cdot y) \cdot 3 = x \cdot (y \cdot 3)$, associative law of multiplication
    b) $(x \cdot y) \cdot 3 = (y \cdot x) \cdot 3$, commutative law of multiplication
    c) $(x \cdot y) \cdot 3 = (y \cdot x) \cdot 3$ Using the commutative law first, then the associative law
       $= y \cdot (x \cdot 3)$

    There are other correct answers.

26. $(b \cdot a) \cdot 5$, $b \cdot (a \cdot 5)$, $a \cdot (b \cdot 5)$

27. a) $7 \cdot (a \cdot b) = (7 \cdot a) \cdot b$
    b) $7 \cdot (a \cdot b) = (7 \cdot a) \cdot b = b \cdot (7 \cdot a)$
    c) $7 \cdot (a \cdot b) = 7 \cdot (b \cdot a) = (7 \cdot b) \cdot a$

    There are other correct answers.

28. $5 \cdot (y \cdot x)$, $(5 \cdot y) \cdot x$, $(5 \cdot x) \cdot y$

29. a) $2 \cdot c \cdot d = (2 \cdot c) \cdot d$
    b) $2 \cdot c \cdot d = 2 \cdot (c \cdot d)$
    c) $2 \cdot c \cdot d = 2 \cdot (c \cdot d) = 2 \cdot (d \cdot c)$

    There are other correct answers.

30. $3 \cdot (4 \cdot x)$, $3 \cdot (x \cdot 4)$, $(3 \cdot x) \cdot 4$

31. a) $(4a + 2) + b = (2 + 4a) + b$
    b) $(4a + 2) + b = 4a + (2 + b)$
    c) $(4a + 2) + b = 4a + (2 + b) = (2 + b) + 4a$

32. $(2 + 5x^2) + 6$, $(5x^2 + 2) + 6$, $2 + (6 + 5x^2)$

33. $2(b + 5) = 2 \cdot b + 2 \cdot 5 = 2b + 10$

34. $4x + 12$

35. $7(1 + t) = 7 \cdot 1 + 7 \cdot t = 7 + 7t$

36. $6v + 24$

37. $3(x + 1) = 3 \cdot x + 3 \cdot 1 = 3x + 3$

38. $7x + 56$

39. $4(1 + y) = 4 \cdot 1 + 4 \cdot y = 4 + 4y$

40. $9s + 9$

41. $6(5x + 2) = 6 \cdot 5x + 6 \cdot 2 = 30x + 12$

42. $54m + 63$

43. $7(x + 4 + 6y) = 7 \cdot x + 7 \cdot 4 + 7 \cdot 6y = 7x + 28 + 42y$

44. $20x + 32 + 12p$

45. $(a + b)2 = a(2) + b(2) = 2a + 2b$

46. $7x + 14$

47. $(x + y + 2)5 = x(5) + y(5) + 2(5) = 5x + 5y + 10$

48. $12 + 6a + 6b$

49. $2x + 4 = 2 \cdot x + 2 \cdot 2 = 2(x + 2)$

50. $5(y + 4)$

51. $30 + 5y = 5 \cdot 6 + 5 \cdot y = 5(6 + y)$

52. $7(x + 4)$

53. $14x + 21y = 7 \cdot 2x + 7 \cdot 3y = 7(2x + 3y)$

54. $6(3a + 4b)$

55. $5x + 10 + 15y = 5 \cdot x + 5 \cdot 2 + 5 \cdot 3y = 5(x + 2 + 3y)$

56. $9(a + 3b + 9)$

57. $9x + 27 = 9 \cdot x + 9 \cdot 3 = 9(x + 3)$
Check: $9(x + 3) = 9 \cdot x + 9 \cdot 3 = 9x + 27$

58. $6(x + 4)$

59. $9x + 3y = 3 \cdot 3x + 3 \cdot y = 3(3x + y)$
Check: $3(3x + y) = 3 \cdot 3x + 3 \cdot y = 9x + 3y$

60. $5(3x + y)$

61. $8a + 16b + 64 = 8 \cdot a + 8 \cdot 2b + 8 \cdot 8 = 8(a + 2b + 8)$

62. $5(1 + 4x + 7y)$

63. $11x + 44y + 121 = 11 \cdot x + 11 \cdot 4y + 11 \cdot 11 = 11(x + 4y + 11)$

64. $7(1 + 2b + 8w)$

65. $9a + 10a = (9 + 10)a = 19a$

66. $14x$

67. $10a + a = 10a + 1 \cdot a = (10 + 1)a = 11a$

68. $17x$

69. $\quad 2x + 9z + 6x$
$= 2x + 6x + 9z$
$= (2 + 6)x + 9z$
$= 8x + 9z$

70. $10a + 5b$

71. $\quad 7x + 6y^2 + 9y^2$
$= 7x + (6 + 9)y^2$
$= 7x + 15y^2$

72. $21m^2 + 6q$

73. $\quad 41a + 90 + 60a + 2$
$= 41a + 60a + 90 + 2$
$= (41 + 60)a + (90 + 2)$
$= 101a + 92$

74. $46x + 8$

75. $\quad 8a + 8b + 3a + 3b$
$= 8a + 3a + 8b + 3b$
$= (8 + 3)a + (8 + 3)b$
$= 11a + 11b$

76. $290y + 600z$

77. $\quad 8u^2 + 3t + 10t + 6u^2 + 2$
$= 8u^2 + 6u^2 + 3t + 10t + 2$
$= (8 + 6)u^2 + (3 + 10)t + 2$
$= 14u^2 + 13t + 2$

78. $13 + 15h + t$

79. $\quad 23 + 5t + 7y + t + y + 27$
$= 23 + 27 + 5t + 1 \cdot t + 7y + 1 \cdot y$
$= (23 + 27) + (5 + 1)t + (7 + 1)y$
$= 50 + 6t + 8y$

80. $135 + 106d$

81. $\quad \frac{1}{2}b + \frac{1}{2}b$
$= \left(\frac{1}{2} + \frac{1}{2}\right)b$
$= 1b$
$= b$

82. x

83. $\quad 2y + \frac{1}{4}y + y$
$= 2y + \frac{1}{4}y + 1 \cdot y$
$= \left(2 + \frac{1}{4} + 1\right)y$
$= 3\frac{1}{4}y$ or $\frac{13}{4}y$

84. $\frac{13}{2}a$ or $6\frac{1}{2}a$

85. $\frac{11}{12} + \frac{15}{16} = \frac{11}{12} \cdot \frac{4}{4} + \frac{15}{16} \cdot \frac{3}{3}$ LCD is 48
$= \frac{44}{48} + \frac{45}{48}$
$= \frac{89}{48}$

86. $\frac{5}{24}$

87. $(2x)^3 = (2 \cdot 2)^3 = 4^3 = 64$

88. 58

89. No

90. No

91. Yes; commutative law of addition

92. Yes; commutative law of addition

93. $bxy + bx = b(xy) + bx = (xy)b + bx = (yx)b + bx = yxb + bx$

The expressions are equivalent by the commutative and associative laws of multiplication.

Chapter 1   (1.5)

94. No

95. No

96. Yes; commutative and associative laws of multiplication

97. $a \div b = 64 \div 8 = 8$
$b \div a = 8 \div 64 = 1/8$
No

98. 20; 12; No

99. $a \div (b \div c) = 32 \div (8 \div 4) = 32 \div 2 = 16$
$(a \div b) \div c = (32 \div 8) \div 4 = 4 \div 4 = 1$
No

100. $P(1 + rt)$

101. a) $17x + 34 = 17(x + 2)$
$17(10) + 34 = 170 + 34 = 204$
$17(10 + 2) = 17 \cdot 12 = 204$
b) Yes, by the distributive law

102. $\frac{3}{2}$

Exercise Set 1.5

1. Let x = the number.

   Rewording: What number, added to, 60 is 112?
   Translating: x +   60   = 112

   $x + 60 = 112$

2. $7w = 2233$

3. Let y = the number.

   Rewording: 42 times what number, is 2352?
   Translating: 42 · y = 2352

   $42y = 2352$

4. $x + 345 = 987$

5. Let s = the number of squares your opponent gets.

   Rewording: What number, added to, 35 is 64?
   Translating: s +   35   = 64

   $s + 35 = 64$

6. $80y = \$53{,}400$

7. Let c = the cost of one box.

   Rewording: 4 times what number, is $7.96?
   Translating: 4 · c = $7.96

   $4c = \$7.96$

8. $d + \$0.2 = \$6.5$, where d is in billions of dollars.

9. $\begin{array}{c|c} x + 17 = 32 & \text{Writing the equation} \\ 15 + 17 \;|\; 32 & \text{Substituting 15 for } x \\ 32 & \end{array}$

   Since the left-hand and right-hand sides are the same, 15 is a solution.

10. No

11. $\begin{array}{c|c} x - 7 = 12 & \text{Writing the equation} \\ 21 - 7 \;|\; 12 & \text{Substituting 21 for } x \\ 14 & \end{array}$

    Since the left-hand and right-hand sides are not the same, 21 is not a solution.

12. Yes

13. $6x = 54$
    $6 \cdot 7 \;|\; 54$
    42

    7 is not a solution.

14. Yes

15. $\frac{x}{6} = 5$
    $\frac{30}{6} \;|\; 5$
    5

    5 is a solution.

16. No

17. $5x + 7 = 107$
    $5 \cdot 19 + 7 \;|\; 107$
    $95 + 7$
    102

    19 is not a solution.

18. Yes

19. $7(y - 1) = 63$
    $7(11 - 1) \;|\; 63$
    $7 \cdot 10$
    70

    11 is not a solution.

20. Yes

Chapter 1   (1.5)

21. $x + 7 = 24$
    $x = 24 - 7$ Subtracting 7 on both sides
    $x = 17$

    Check: $\begin{array}{c|c} x + 7 = 24 \\ \hline 17 + 7 & 24 \\ 24 & \end{array}$ Substituting 17 for x

22. 17

23. $x + 19 = 105$ Check: $\begin{array}{c|c} x + 19 = 105 \\ \hline 86 + 19 & 105 \\ 105 & \end{array}$
    $x = 105 - 19$
    $x = 86$

24. 175

25. $x + 99 = 476$ Check: $\begin{array}{c|c} x + 99 = 476 \\ \hline 377 + 99 & 476 \\ 476 & \end{array}$
    $x = 476 - 99$
    $x = 377$

26. 889

27. $x + 5064 = 7882$ Check: $\begin{array}{c|c} x + 5064 = 7882 \\ \hline 2818 + 5064 & 7882 \\ 7882 & \end{array}$
    $x = 7882 - 5064$
    $x = 2818$

28. 3895

29. $x + 2.78 = 8.44$ Check: $\begin{array}{c|c} x + 2.78 = 8.44 \\ \hline 5.66 + 2.78 & 8.44 \\ 8.44 & \end{array}$
    $x = 8.44 - 2.78$
    $x = 5.66$

30. 1.65

31. $x + \frac{1}{7} = \frac{6}{7}$ Check: $\begin{array}{c|c} x + \frac{1}{7} = \frac{6}{7} \\ \hline \frac{5}{7} + \frac{1}{7} & \frac{6}{7} \\ \frac{6}{7} & \end{array}$
    $x = \frac{6}{7} - \frac{1}{7}$
    $x = \frac{5}{7}$

32. $\frac{8}{13}$

33. $2.78 + x = 38.4$ Check: $\begin{array}{c|c} 2.78 + x = 38.4 \\ \hline 2.78 + 35.62 & 38.4 \\ 38.4 & \end{array}$
    $x = 38.4 - 2.78$
    $x = 35.62$

34. 65.7

35. $\frac{3}{4} + t = \frac{11}{12}$
    $t = \frac{11}{12} - \frac{3}{4}$
    $t = \frac{11}{12} - \frac{3}{4} \cdot \frac{3}{3}$ Multiplying to get the LCD, 12
    $t = \frac{11}{12} - \frac{9}{12}$
    $t = \frac{2}{12}$
    $t = \frac{1}{6}$ Simplifying

    Check: $\begin{array}{c|c} \frac{3}{4} + t = \frac{11}{12} \\ \hline \frac{3}{4} + \frac{1}{6} & \frac{11}{12} \\ \frac{9}{12} + \frac{2}{12} & \\ \frac{11}{12} & \end{array}$

36. $\frac{1}{18}$

37. $15x = 90$
    $x = \frac{90}{15}$ Dividing by 15 on both sides
    $x = 6$

    Check: $\begin{array}{c|c} 15x = 90 \\ \hline 15 \cdot 6 & 90 \\ 90 & \end{array}$

38. 7

39. $4x = 5$ Check: $\begin{array}{c|c} 4x = 5 \\ \hline 4 \cdot \frac{5}{4} & 5 \\ \frac{4 \cdot 5}{4 \cdot 1} & \\ 5 & \end{array}$
    $x = \frac{5}{4}$

40. $\frac{9}{2}$

41. $10x = 2.4$ Check: $\begin{array}{c|c} 10x = 2.4 \\ \hline 10(0.24) & 2.4 \\ 2.4 & \end{array}$
    $x = \frac{2.4}{10}$
    $x = 0.24$

42. 0.4

43. $2.9y = 8.99$ Check: $\begin{array}{c|c} 2.9y = 8.99 \\ \hline 2.9(3.1) & 8.99 \\ 8.99 & \end{array}$
    $y = \frac{8.99}{2.9}$
    $y = 3.1$

44. 6.2

Chapter 1   (1.6)

45. $6.2y = 52.7$ Check: $\begin{array}{c|c} 6.2y = 52.7 \\ 6.2(8.5) & 52.7 \\ & 52.7 \end{array}$
    $y = \frac{52.7}{6.2}$
    $y = 8.5$

46. 2.5

47. $117t = 2106$ Check: $\begin{array}{c|c} 117t = 2106 \\ 117 \cdot 18 & 2106 \\ 2106 \end{array}$
    $t = \frac{2106}{117}$
    $t = 18$

48. 21

49. $1.93t = 4053$ Check: $\begin{array}{c|c} 1.93t = 4053 \\ 1.93(2100) & 4053 \\ 4053 \end{array}$
    $t = \frac{4053}{1.93}$
    $t = 2100$

50. 6700

51. $\frac{1}{2}x = \frac{4}{5}$ Check: $\begin{array}{c|c} \frac{1}{2}x = \frac{4}{5} \\ \frac{1}{2} \cdot \frac{8}{5} & \frac{4}{5} \\ \frac{2 \cdot 4}{2 \cdot 5} \\ \frac{4}{5} \end{array}$
    $x = \frac{\frac{4}{5}}{\frac{1}{2}}$
    $x = \frac{4}{5} \cdot \frac{2}{1}$
    $x = \frac{8}{5}$

52. $\frac{1}{3}$

53. Area $= \frac{1}{2}bh = \frac{1}{2}(48)(17) = 408$ cm^2

54. $7(x + 2y + 3z)$

55. $45 = 3 \cdot 3 \cdot 5$
    $55 = 5 \cdot 11$
    $75 = 3 \cdot 5 \cdot 5$
    The LCM is $3 \cdot 3 \cdot 5 \cdot 5 \cdot 11$, or 2475.

56. $80x + 50y + 70z$

57. $5x + 3x = 10$
    $8x = 10$
    $x = \frac{10}{8} = \frac{2 \cdot 5}{2 \cdot 4}$
    $x = \frac{5}{4}$

58. 2

59. $225a = 27$
    $a = \frac{27}{225} = \frac{9 \cdot 3}{9 \cdot 25}$
    $a = \frac{3}{25}$

60. $\frac{3}{10}$

61. $0.0592y = 0.4736$
    $y = \frac{0.4736}{0.0592}$
    $y = 8$

62. 14

63. $0.125n = 1$
    $n = \frac{1}{0.125}$
    $n = 8$

64. 250

65. Answers may vary. $3x = 2$

66. Answers may vary. $x + 7 = 7 + x$

67. $x + 506,233 = 976,421$
    $x = 976,421 - 506,233$
    $x = 470,188$

68. 8424

Exercise Set 1.6

1.–16. These exercises are open-ended and, for the most part, do not have absolute answers.

17. A person has one day of birth.

18. 0

19. There is one groove on each side.

20. An electric train does not have smoke.

21. Familiarize. We want to know what should be added to 60 to get 112. We draw a picture of the situation. Let x = the unknown number.

60	x
112	

    Translate.
    What number, added to, 60 is 112?
    $x + 60 = 112$

    Carry out. We solve the equation.
    $x + 60 = 112$
    $x = 112 - 60$
    $x = 52$
    Check. $52 + 60 = 112$.
    State. The number is 52.

Chapter 1   (1.6)

22. 7.8

23. <u>Familiarize</u>. Let x = the unknown number. We draw a picture of the situation.

x	29
171	

<u>Translate</u>. We reword the problem.

A number, plus 29 is 171.
↓   ↓   ↓ ↓
x   +   29 = 171

<u>Carry out</u>. We solve the equation.

x + 29 = 171
x = 171 - 29
x = 142

<u>Check</u>. 142 + 29 = 171

<u>State</u>. The number is 142.

24. 864

25. <u>Familiarize</u>. Let y = the number. We draw a picture of the situation.

y	y	y	y	y	y	y
2233						

<u>Translate</u>.

Seven times what number, is 2233?
↓   ↓   ↓   ↓
7   ·   y   = 2233

<u>Carry out</u>. We solve the equation.

7y = 2233

$y = \frac{2233}{7}$

y = 319

<u>Check</u>. 7·319 = 2233

<u>State</u>. The number is 319.

26. 2236

27. <u>Familiarize</u>. Let x = the number. We draw a picture of the situation.

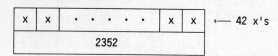

x	x	· · · · ·	x	x
2352				

← 42 x's

<u>Translate</u>. We reword the problem.

42 times a number, is 2352.
↓   ↓   ↓   ↓ ↓
42 ·   x   = 2352

<u>Carry out</u>. We solve the equation.

42x = 2352

$x = \frac{2352}{42}$

x = 56

<u>Check</u>. 42·56 = 2352

<u>State</u>. The number is 56.

28. 13

29. <u>Familiarize</u>. Let y = the number. We draw a picture of the situation.

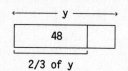

←―― y ――→
| 48 | |

2/3 of y

<u>Translate</u>.

Two-thirds of what number, is 48?
↓   ↓   ↓   ↓
$\frac{2}{3}$ ·   y   = 48

<u>Carry out</u>. We solve the equation.

$\frac{2}{3}y = 48$

$y = \frac{48}{\frac{2}{3}}$

$y = \frac{48}{1} \cdot \frac{3}{2} = \frac{2 \cdot 24 \cdot 3}{1 \cdot 2} = 72$

<u>Check</u>. We check to find if $\frac{2}{3}$ of 72 is 48.

$\frac{2}{3} \cdot 72 = \frac{2 \cdot 3 \cdot 24}{3 \cdot 1} = 48$

<u>State</u>. The number is 72.

30. 448

Chapter 1   (1.6)

31. Familiarize. We are trying to find what number added to 5 is 8. Let n = the number. This is the number of problems the student got right. We draw a picture of the situation.

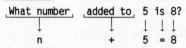

Translate. We reword the problem.

What number, added to, 5 is 8?
   ↓   ↓   ↓ ↓ ↓
   n   +   5 = 8

Carry out. We solve the equation.

$n + 5 = 8$

$n = 8 - 5$

$n = 3$

Check. $5 + 3 = 8$

State. The student got 3 problems right.

32. 24 yd

33. Familiarize. We are asking the question "37 added to what number is 101?" Let g = the number. This is the number of games the Twins won. We draw a picture of the situation.

37	g
101	

Translate. We reword the problem.

37, added to, what number, is 101?
↓   ↓   ↓   ↓ ↓
37   +   g   = 101

Carry out. We solve the equation.

$37 + g = 101$

$g = 101 - 37$

$g = 64$

Check. $37 + 64 = 101$

State. The Twins won 64 games.

34. 29

35. Familiarize. We are asking "What amount added to $48 is $115?" Let a = the unknown amount. This is the amount of cash in cash register A. We draw a picture.

$48	a
$115	

Translate. We reword the problem.

What number, added to, 48 is 115?
   ↓   ↓   ↓ ↓ ↓
   a   +   48 = 115

35. (continued)

Carry out. We solve the equation.

$a + 48 = 115$

$a = 115 - 48$

$a = 67$

Check. $67 + 48 = 115$

State. Cash register A contains $67.

36. 164,394

37. Familiarize. We are asking the question, "Twelve times what amount is $3.12?" Let p = the unknown amount. This is the price of one bagel. We draw a picture.

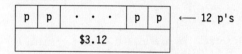

Translate. We reword the problem.

Twelve times, what number, is 3.12?
   ↓   ↓   ↓   ↓ ↓
   12   ·   p   = 3.12

Carry out. We solve the equation.

$12p = 3.12$

$p = \frac{3.12}{12}$

$p = 0.26$

Check. $12(\$0.26) = \$3.12$

State. Each bagel costs $0.26.

38. $3.75

39. Familiarize. We are asking "$80 times what number is $53,400?" Let n = the number. This is the number of hours the consultant worked. We draw a picture.

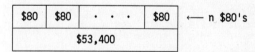

Translate. We reword the problem.

80 times, what number, is 53,400?
↓   ↓   ↓   ↓   ↓
80   ·   n   = 53,400

Carry out. We solve the equation.

$80n = 53,400$

$n = \frac{53,400}{80}$

$n = 667.5$

Check. $\$80(667.5) = \$53,400$

State. The consultant worked 667.5 hours.

40. 19,528.5 km²

Chapter 1   (1.6)

41. <u>Familiarize</u>. We are asking "2.5 times what number is 16.6?" Let t = the number. This is the length of time it takes a 150-watt bulb to use one kilowatt-hour of electricity.

    <u>Translate</u>.

    2.5 times what number is 16.6?
    ↓   ↓   ↓   ↓   ↓
    2.5 ·   t   = 16.6

    <u>Carry out</u>. We solve the equation.

    $2.5t = 16.6$

    $t = 6.64$

    <u>Check</u>. $2.5(6.64) = 16.6$

    <u>State</u>. It takes a 150-watt bulb 6.64 hr to use one kilowatt-hour of electricity.

42. About 3145 km²

43. <u>Familiarize</u>. We are asking "What temperature plus 13.5° C is 78.3° C?" Let t = the unknown temperature. This is the boiling point of methyl alcohol. We draw a picture.

t	13.5°C
78.3° C	

    <u>Translate</u>. We reword the problem.

    What number, plus 13.5 is 78.3?
    ↓   ↓   ↓   ↓   ↓
    t   + 13.5 = 78.3

    <u>Carry out</u>. We solve the equation.

    $t + 13.5 = 78.3$

    $t = 64.8$

    <u>Check</u>. $64.8 + 13.5 = 78.3$

    <u>State</u>. The boiling point of methyl alcohol is 64.8° C.

44. 209,367,000

45. <u>Familiarize</u>. We are asking "5.1 times what amount is $19,935.90?" Let a = the unknown amount. This is the original cost.

    <u>Translate</u>. We reword the problem.

    5.1 times what number, is 19,935.90
    ↓   ↓   ↓   ↓   ↓
    5.1 ·   a   = 19,935.90

    <u>Carry out</u>. We solve the equation.

    $5.1a = 19,935.90$

    $a = \dfrac{19,935.90}{5.1}$

    $a = 3909$

    <u>Check</u>. $5.1(\$3909) = \$19,935.90$

    <u>State</u>. The original cost was $3909.

46. 92 m

47. <u>Familiarize</u>. We are asking "391 times what distance is 150,000,000 km?" Let d = the unknown distance. This is the distance from the earth to the moon.

    <u>Translate</u>. We reword the problem.

    391 times what number, is 150,000,000?
    ↓   ↓   ↓   ↓   ↓
    391 ·   d   = 150,000,000

    <u>Carry out</u>. We solve the equation.

    $391d = 150,000,000$

    $d = \dfrac{150,000,000}{391}$

    $d = 383,631.7$ Rounding

    <u>Check</u>. $391(383,631.7) = 150,000,000$ (Rounding)

    <u>State</u>. The distance from the earth to the moon is about 383,631.7 km.

48. 0.288

49. <u>Familiarize and Translate</u>. We will use the formula given and substitute 15 for C.

    $F = 1.8(15) + 32$

    <u>Carry out</u>. We do the computation.

    $F = 1.8(15) + 32$

    $F = 27 + 32$

    $F = 59$

    <u>Check</u>. Substitute 59 for F in the formula and solve for C. The result should be 15.

    $59 = 1.8C + 32$

    $27 = 1.8C$ Subtracting 32 on both sides

    $15 = C$ Dividing by 1.8 on both sides

    <u>State</u>. 15° C corresponds to 59° F.

50. 120 watts

51. <u>Familiarize</u>. The total cost for the entire set of prints is $3.14 + $6.13, or $9.27. We are asking "36 times what amount is $9.27?" Let a = the unknown amount.

    <u>Translate</u>. We reword the problem.

    36 times what number, is 9.27?
    ↓   ↓   ↓   ↓   ↓
    36 ·   a   = 9.27

    <u>Carry out</u>. We solve the equation.

    $36a = 9.27$

    $a = 0.2575$

    <u>Check</u>. $36(\$0.2575) = \$9.27$

    <u>State</u>. The total cost for each print was $0.2575, or 25.75¢.

52. 7

Chapter 1 (1.7)

53. Familiarize. We are asking "2.54 times what number is 100?" Let n = the number.
    Translate.
    2.54 times what number is 100?
     ↓   ↓   ↓   ↓ ↓
    2.54 ·   n   = 100
    Carry out. We solve the equation.
    2.54n = 100
    $n = \frac{100}{2.54}$
    n = 39.37   Rounding
    Check. 2.54(39.37) = 100   (Rounding)
    State. There are about 39.37 inches in a meter.

54. 9.2 sec

55. You would need to know the person's salary now.

56. Marked price of the records

57. You would need to know the length of either the height or the base.

58. The length of time the car traveled

Exercise Set 1.7

1. 76% = 76 × 0.01   Replacing % by × 0.01
   = 0.76

2. 0.54

3. 54.7% = 54.7 × 0.01   Replacing % by × 0.01
   = 0.547

4. 0.962

5. 100% = 100 × 0.01 = 1

6. 0.01

7. 0.61% = 0.61 × 0.01 = 0.0061

8. 1.25

9. 240% = 240 × 0.01 = 2.4

10. 0.0073

11. 3.25% = 3.25 × 0.01 = 0.0325

12. 0.023

13. 20% = 20 × $\frac{1}{100}$ Replacing % by × $\frac{1}{100}$
    = $\frac{20}{100}$, or $\frac{1}{5}$

14. $\frac{80}{100}$, or $\frac{4}{5}$

15. 78.6% = 78.6 × $\frac{1}{100}$ Replacing % by $\frac{1}{100}$
    = $\frac{78.6}{100}$
    = $\frac{78.6}{100} \cdot \frac{10}{10}$ Multiplying by 1 to get a whole number in the numerator
    = $\frac{786}{1000}$, or $\frac{393}{500}$

16. $\frac{135}{1000}$, or $\frac{27}{200}$

17. $12\frac{1}{2}$% = $12\frac{1}{2}$ × $\frac{1}{100}$ = $\frac{25}{2}$ × $\frac{1}{100}$ = $\frac{25}{200}$, or $\frac{1}{8}$

18. $\frac{120}{100}$, or $\frac{6}{5}$

19. 0.042% = 0.042 × $\frac{1}{100}$ = $\frac{0.042}{100}$ = $\frac{0.042}{100}$ × $\frac{1000}{1000}$ =
    $\frac{42}{100,000}$, or $\frac{21}{50,000}$

20. $\frac{68}{10,000}$, or $\frac{17}{2500}$

21. 250% = 250 × $\frac{1}{100}$ = $\frac{250}{100}$, or $\frac{5}{2}$

22. $\frac{32}{1000}$, or $\frac{4}{125}$

23. 3.47% = 3.47 × $\frac{1}{100}$ = $\frac{3.47}{100}$ = $\frac{3.47}{100}$ × $\frac{100}{100}$ = $\frac{347}{10,000}$

24. $\frac{12,557}{100,000}$

25. 4.54 = 4.54 × 1
    = 4.54 × (100 × 0.01)   Replacing 1 by 100 × 0.01
    = (4.54 × 100) × 0.01   Using associativity
    = 454 × 0.01
    = 454%   Replacing × 0.01 by %

26. 100%

27. 0.998 = 0.998 × 1
    = 0.998 × (100 × 0.01)   Replacing 1 by 100 × 0.01
    = (0.998 × 100) × 0.01   Using associativity
    = 99.8 × 0.01
    = 99.8%   Replacing × 0.01 by %

28. 73%

29. 2 = 2 × 1 = 2 × (100 × 0.01) = (2 × 100) × 0.01 = 200 × 0.01 = 200%

30. 0.57%

17

Chapter 1   (1.7)

31. $0.072 = 0.072 \times 1 = 0.072 \times (100 \times 0.01) =$
    $(0.072 \times 100) \times 0.01 = 7.2 \times 0.01 = 7.2\%$

32. 134%

33. $9.2 = 9.2 \times 1 = 9.2 \times (100 \times 0.01) =$
    $(9.2 \times 100) \times 0.01 = 920 \times 0.01 = 920\%$

34. 1.3%

35. $0.0068 = 0.0068 \times 1 = 0.0068 \times (100 \times 0.01) =$
    $(0.0068 \times 100) \times 0.01 = 0.68 \times 0.01 = 0.68\%$

36. 67.5%

37. $\frac{1}{8} = \frac{1}{8} \times 1 = \frac{1}{8} \times \left(100 \times \frac{1}{100}\right) = \left(\frac{1}{8} \times 100\right) \times \frac{1}{100} =$
    $\frac{100}{8} \times \frac{1}{100} = \frac{100}{8}\%$, or 12.5%

38. $33\frac{1}{3}\%$

39. $\frac{17}{25} = \frac{17}{25} \times 1 = \frac{17}{25} \times \left(100 \times \frac{1}{100}\right) =$
    $\left(\frac{17}{25} \times 100\right) \times \frac{1}{100} = \frac{1700}{25} \times \frac{1}{100} = 68 \times \frac{1}{100} = 68\%$

40. 55%

41. $\frac{17}{100} = 17 \times \frac{1}{100} = 17\%$

42. 119%

43. $\frac{7}{10} = \frac{7}{10} \times \left(100 \times \frac{1}{100}\right) = \left(\frac{7}{10} \times 100\right) \times \frac{1}{100} =$
    $\frac{700}{10} \times \frac{1}{100} = 70 \times \frac{1}{100} = 70\%$

44. 80%

45. $\frac{3}{5} = \frac{3}{5} \times \left(100 \times \frac{1}{100}\right) = \left(\frac{3}{5} \times 100\right) \times \frac{1}{100} =$
    $\frac{300}{5} \times \frac{1}{100} = 60 \times \frac{1}{100} = 60\%$

46. 34%

47. $\frac{2}{3} = \frac{2}{3} \times \left(100 \times \frac{1}{100}\right) = \left(\frac{2}{3} \times 100\right) \times \frac{1}{100} =$
    $\frac{200}{3} \times \frac{1}{100} = \frac{200}{3}\%$, or $66\frac{2}{3}\%$

48. 37.5%

49. $\frac{7}{4} = \frac{7}{4} \times \left(100 \times \frac{1}{100}\right) = \left(\frac{7}{4} \times 100\right) \times \frac{1}{100} =$
    $\frac{700}{4} \times \frac{1}{100} = 175 \times \frac{1}{100} = 175\%$

50. 87.5%

51. $\frac{3}{4} = \frac{3}{4} \times \left(100 \times \frac{1}{100}\right) = \left(\frac{3}{4} \times 100\right) \times \frac{1}{100} =$
    $\frac{300}{4} \times \frac{1}{100} = 75 \times \frac{1}{100} = 75\%$

52. 99.4%

53. Familiarize. Let y = the percent.
    Translate.
    What percent of 68 is 17?
       y   %   ·   68   = 17
    Carry out. We solve the equation.
    $y\% \cdot 68 = 17$
    $y \times 0.01 \times 68 = 17$
    $y(0.68) = 17$
    $y = \frac{17}{0.68}$
    $y = 25$
    Check. We find 25% of 68:
    $25\% \cdot 68 = 0.25 \times 68 = 17$
    State. The answer is 25%.

54. 48%

55. Familiarize. Let y = the percent.
    Translate.
    What percent of 125 is 30?
       y   %   ·   125   = 30
    Carry out. We solve the equation.
    $y\% \cdot 125 = 30$
    $y \times 0.01 \times 125 = 30$
    $y(1.25) = 30$
    $y = \frac{30}{1.25}$
    $y = 24$
    Check. We find 24% of 125:
    $24\% \cdot 125 = 0.24 \times 125 = 30$
    State. The answer is 24%.

56. 19%

57. Familiarize. Let y = the number we are taking 30% of.
    Translate.
    45 is 30% of what number?
    45 = 30% · y
    Carry out. We solve the equation.
    $45 = 30\% \cdot y$
    $45 = 30 \times 0.01 \times y$
    $45 = 0.3y$
    $\frac{45}{0.3} = y$
    $150 = y$

18

Chapter 1   (1.7)

57. (continued)
    Check. We find 30% of 150.
        30%·150 = 0.3 × 150 = 45
    State. The answer is 150.

58. 85

59. Familiarize. Let y = the number we are taking 12% of.
    Translate.
        0.3 is 12% of what number?
        ↓   ↓   ↓   ↓
        0.3 = 12% ·   y

    Carry out. We solve the equation.
        0.3 = 12%·y
        0.3 = 12 × 0.01 × y
        0.3 = 0.12y
        $\frac{0.3}{0.12}$ = y
        2.5 = y

    Check. We find 12% of 2.5:
        12%·2.5 = 0.12(2.5) = 0.3
    State. The answer is 2.5.

60. 4

61. Familiarize. Let y = the unknown number.
    Translate.
        What number is 65% of 840?
        ↓   ↓   ↓   ↓
        y   = 65% · 840

    Carry out. We solve the equation.
        y = 65%·840
        y = 65 × 0.01 × 840
        y = 0.65 × 840
        y = 546

    Check. The check is the computation we used to solve the equation:
        65%·840 = 0.65 × 840 = 546
    State. The answer is 546.

62. 10,000

63. Familiarize. Let y = the percent.
    Translate.
        What percent of 80 is 100?
        y   %   · 80 = 100

    Carry out. We solve the equation.
        y%·80 = 100
        y × 0.01 × 80 = 100
        y(0.8) = 100
        y = $\frac{100}{0.8}$
        y = 125

63. (continued)
    Check. We find 125% of 80:
        125%·80 = 1.25 × 80 = 100
    State. The answer is 125%.

64. 2050%

65. Familiarize. We let x = the unknown number.
    Translate.
        What is 2% of 40?
        x   =   2% · 40

    Carry out. We solve the equation.
        x = 2%·40
        x = 2 × 0.01 × 40
        x = 0.02 × 40
        x = 0.8

    Check. The check is the computation we used to solve the equation:
        2%·40 = 0.02 × 40 = 0.8
    State. The answer is 0.8.

66. 0.8

67. Familiarize. Let y = the percent.
    Translate.
        2 is what percent of 40?
        2 =   y   %   · 40

    Carry out. We solve the equation.
        2 = y%·40
        2 = y × 0.01 × 40
        2 = y(0.4)
        $\frac{2}{0.4}$ = y
        5 = y

    Check. We find 5% of 40:
        5%·40 = 0.05 × 40 = 2
    State. The answer is 5%.

68. 2000

Chapter 1   (1.7)

69. <u>Familiarize</u>. Write down the information.
   Number of applicants: 16,000
   Number of applicants accepted: 600
Let y = the percent of applicants accepted.
<u>Translate</u>. We reword the problem.
   What percent of 16,000 is 600?
      y   %   · 16,000 = 600
<u>Carry out</u>. We solve the equation.
   y%·16,000 = 600
   y × 0.01 × 16,000 = 600
   y(160) = 600
   $y = \frac{600}{160}$
   y = 3.75
<u>Check</u>. We find 3.75% of 16,000:
   3.75% × 16,000 = 0.0375 × 16,000 = 600
<u>State</u>. The FBI accepts 3.75% of the applicants.

70. 7410

71. <u>Familiarize</u>. Write down the information.
   Population: 17% left-handed
   Bowlers: 160
Let y = the number of bowlers who are left-handed.
<u>Translate</u>. We reword the problem.
   What is 17% of 160?
      y = 17% · 160
<u>Carry out</u>. We solve the equation.
   y = 17%·160
   y = 17 × 0.01 × 160
   y = 0.17 × 160
   y = 27.2
<u>Check</u>. The check is the computation we used to solve the equation:
   17%·160 = 0.17 × 160 = 27.2
<u>State</u>. (We round to the nearest one.) You would expect 27 of the bowlers to be left-handed.

72. 7%

73. <u>Familiarize</u>. Write down the information.
   Number of items: 88
   Number correct: 76
We let y = the percent that were correct.
<u>Translate</u>. We reword the problem.
   What percent of 88 is 76?
      y   %   · 88 = 76
<u>Carry out</u>. We solve the equation.
   y%·88 = 76
   y × 0.01 × 88 = 76
   y(0.88) = 76
   $y = \frac{76}{0.88}$
   y = 86.4   Rounding
<u>Check</u>. We find 86.4% of 88:
   86.4%·88 = 0.864 × 88 = 76   (Rounding)
<u>State</u>. 86.4% were correct.

74. 52%

75. <u>Familiarize</u>. Write down the information.
   Original cost: $24 million
   Renovation cost: 2% of original cost
Let x = the renovation cost.
<u>Translate</u>. We reword the problem.
   What is 2% of $24 million?
      x   = 2% · 24,000,000
<u>Carry out</u>. We solve the equation.
   x = 2%·24,000,000
   x = 2 × 0.01 × 24,000,000
   x = 0.02 × 24,000,000
   x = 480,000
<u>Check</u>. The check is the computation we used to solve the equation:
   2%·24,000,000 = 0.02 × 24,000,000 = 480,000
<u>State</u>. The cost is $480,000.

76. $600,000, or $0.6 million

77. <u>Familiarize</u>. Write down the information.
   Amount spent on food: $208
   Percent of income: 26%
Let y = the monthly income.
<u>Translate</u>. We reword the problem.
   $208 is 26% of what?
      208 = 26% · y
<u>Carry out</u>. We solve the equation.
   208 = 26%·y
   208 = 26 × 0.01 × y
   208 = 0.26 × y
   $\frac{208}{0.26} = y$
   800 = y

Chapter 1   (1.7)

77. (continued)

   <u>Check</u>. We find 26% of 800:
   $$26\% \cdot 800 = 0.26 \times 800 = 208$$
   <u>State</u>. The monthly income was $800.

78. $34.31; $463.17

79. <u>Familiarize</u>. Write down the information.
   Percent of increase in volume: 9%
   Volume of water: 400 cm³

   Let x = the amount by which the volume will increase. Let t = the total volume of the ice.

   <u>Translate</u>. We reword the problem. We first translate to an equation to find the amount by which the volume will increase:

   What is 9% of 400?
   $$x = 9\% \cdot 400$$

   Then we write an equation to find the total volume of the ice:

   Total volume is original volume plus increase in volume.
   $$T = 400 + x$$

   <u>Carry out</u>. We solve the first equation.
   $$x = 9\% \cdot 400$$
   $$x = 9 \times 0.01 \times 400$$
   $$x = 0.09 \times 400$$
   $$x = 36$$

   Then we solve the second equation, substituting 36 for x.
   $$T = 400 + x$$
   $$T = 400 + 36$$
   $$T = 436$$

   <u>Check</u>. The check of each part of the problem is actually the computation we used to solve it:
   $$9\% \cdot 400 = 0.09 \times 400 = 36$$
   $$400 + 36 = 436$$

   <u>State</u>. The volume increases by 400 cm³. The total volume of the ice is 436 cm³.

80. $7800

81. <u>Familiarize</u>. Write down the information.
   Interest rate: 8%
   Amount after 1 year: $7776

   Let x = the amount originally invested (the principal).

   <u>Translate</u>. We reword the problem. (See Example 10.)

   Principal + Interest = Amount
   $$x + 8\%x = 7776$$

81. (continued)

   <u>Carry out</u>. We solve the equation.
   $$x + 8\%x = 7776$$
   $$x + 0.08x = 7776$$
   $$1x + 0.08x = 7776$$
   $$1.08x = 7776$$
   $$x = \frac{7776}{1.08}$$
   $$x = 7200$$

   <u>Check</u>. We take 8% of $7200 and add it to $7200:
   $$8\% \times \$7200 = \$576.$$
   Then $7200 + $576 = $7776.

   <u>State</u>. The original investment was $7200.

82. $1.50; $1.62

83. $12\% + 14\% = 0.12 + 0.14 = 0.26 = 26\%$

84. 68%

85. $1 - 10\% = 1 - 0.1 = 0.9 = 90\%$

86. 71%

87. $12 \times 100\% = 12 \times 1 = 12$

88. 0

89. $3(1 + 15\%) = 3(1 + 0.15) = 3(1.15) = 3.45$, or 345%

90. 98%

91. $\frac{100\%}{40} = \frac{1}{40} = 0.025$, or 2.5%

92. About 3.7%

93. First translate and carry out the solution to find the amount of the sales tax, s:

   What is 6% of $16.41?
   $$s = 6\% \cdot 16.41$$
   $$s = 0.06(16.41)$$
   $$s = 0.98 \quad \text{Rounding}$$

   The sales tax is $0.98.

   Find the sum of the meal price and the tax:
   $$\$16.41 + \$0.98 = \$17.39$$

   Next translate and carry out the solution to find the amount of the tip, t:

   What is 15% of $17.39?
   $$t = 15\% \cdot 17.39$$
   $$t = 0.15(17.39)$$
   $$t = 2.61 \quad \text{Rounding}$$

   The tip is $2.61.

Chapter 1 (1.7)

93. (continued)

Finally find the total paid:
$17.39 + $2.61 = $20
The check is left to the student.
The total paid is $20.

94. Rollies: $8.55; Warped: $8.65

95. <u>Familiarize</u>. Let x = the percent chance it won't rain during a particular time period.

<u>Translate</u>. We reword the problem.

	What percent,	added to	chance of showers	is 100%?
During the day:	x	+	60%	= 100%
Tonight:	x	+	30%	= 100%
Tomorrow morning:	x	+	5%	= 100%

<u>Carry out</u>. We solve the equations.

During the day: x + 60% = 100%
$\qquad$ x = 100% - 60%
$\qquad$ x = 40%

Tonight: x + 30% = 100%
$\qquad$ x = 100% - 30%
$\qquad$ x = 70%

Tomorrow morning: x + 5% = 100%
$\qquad$ x = 100% - 5%
$\qquad$ x = 95%

<u>Check</u>. In each case we add the chance of no rain to the chance of rain to determine if the sum is 100%.

During the day: 40% + 60% = 100%
Tonight: 70% + 30% = 100%
Tomorrow morning: 95% + 5% = 100%

<u>State</u>. The chances that it won't rain during the day, tonight, and tomorrow morning are 40%, 70%, and 95%, respectively.

96. 62.5%

97. First we translate and carry out the solution to find the amount of the price increase, p.

Price increase, is 25% of old price.
$\qquad$ p = 25% · 8800

p = 25%·8800
p = 0.25(8800)
p = 2200

The new price is $2200 higher than the old price, or $2200 + $8800, or $11,000.

Next we translate and carry out the solution to find what percent of the new price the price increase is. Let y = the percent.

$2200 is what percent of $11,000?
$\qquad$ 2200 = y % · 11,000

2200 = y%·11,000
2200 = y × 0.01 × 11,000
2200 = y × 110
$\frac{2200}{110}$ = y
20 = y

The check is left to the student.
The old price is 20% lower than the new price.

CHAPTER 2   THE REAL-NUMBER SYSTEM

Exercise Set 2.1

1. The integer 5 corresponds to winning 5 points, and the integer -12 corresponds to losing 12 points.

2. 18, -2

3. The integer -17 corresponds to owing $17, and the integer 12 corresponds to having $12 in a bank account.

4. 1200, -560

5. The integer -1286 corresponds to 1286 ft below sea level. The integer 29,028 corresponds to 29,028 ft above sea level.

6. Team A: -34, Team B: 34

7. The integer 750 corresponds to a $750 deposit, and the integer -125 corresponds to a $125 withdrawal.

8. -3,000,000

9. The integers 20, -150, and 300 correspond to the interception of the missile, the loss of the starship, and the capture of the base, respectively.

10. -3, 128

11. Since 5 is to the right of 0, we have 5 > 0.

12. 9 > 0

13. Since -9 is to the left of 5, we have -9 < 5.

14. 8 > -8

15. Since -6 is to the left of 6, we have -6 < 6.

16. 0 > -7

17. Since -8 is to the left of -5, we have -8 < -5.

18. -4 < -3

19. Since -5 is to the right of -11, we have -5 > -11.

20. -3 > -4

21. Since -6 is to the left of -5, we have -6 < -5.

22. -10 > -14

23. Since 2.14 is to the right of 1.24, we have 2.14 > 1.24.

24. -3.3 < -2.2

25. Since -14.5 is to the left of 0.011, we have -14.5 < 0.011.

26. 17.2 > -1.67

27. Since -12.88 is to the left of -6.45, we have -12.88 < -6.45.

28. -14.34 > -17.88

29. Convert to decimal notation: $\frac{5}{12} = 0.4166\ldots$ and $\frac{11}{25} = 0.44$. Since $0.4166\ldots$ is to the left of 0.44, we have $\frac{5}{12} < \frac{11}{25}$.

30. $-\frac{14}{17} < -\frac{27}{35}$

31. -5 > x

32. x < -1

33. 120 > -20

34. 20 > -25

35. -500,000 < 1,000,000

36. 60 > 20

37. -2 > -10

38. x ⩾ -9

39. p ⩽ 0

40. n ⩾ 0

41. -3 ⩾ -11 is true, since -3 > -11 is true.

42. False

43. 0 ⩾ 8 is false, since neither 0 > 8 nor 0 = 8 is true.

44. True

45. The distance of -3 from 0 is 3, so |-3| = 3.

46. 7

47. The distance of 10 from 0 is 10, so |10| = 10.

48. 11

49. The distance of 0 from 0 is 0, so |0| = 0.

50. 4

51. The distance of -24 from 0 is 24, so |-24| = 24.

52. 325

Chapter 2   (2.1)

53. The distance of $-\frac{2}{3}$ from 0 is $\frac{2}{3}$, so $\left|-\frac{2}{3}\right| = \frac{2}{3}$.

54. $\frac{10}{7}$

55. The distance of $\frac{0}{4}$ from 0 is $\frac{0}{4}$, or 0, so $\left|\frac{0}{4}\right| = 0$.

56. 14.8

57. When $x = 5$, $|x| = |5| = 5$.

58. $\frac{7}{8}$

59. Answers may vary. $-\frac{9}{7}$, 0, $4\frac{1}{2}$, -1.97, -491, 128, $\frac{3}{11}$, $-\frac{1}{7}$, 0.000011, $-26\frac{1}{3}$

60. Answers may vary. 1.26, $9\frac{1}{5}$, $\frac{3}{2}$, 0.17, $\frac{6}{11}$, $-\frac{1}{10,000}$, -0.1, -5.6283, -8.3, $-47\frac{1}{2}$

61. Answers may vary. $-\pi$, $\sqrt{42}$, 8.4262262226...

62. Answers may vary. -8.2, $\frac{11}{3}$, $\sqrt{91}$

63. We first find decimal notation for $\frac{3}{8}$. Since $\frac{3}{8}$ means $3 \div 8$, we divide.

$$\begin{array}{r} 0.375 \\ 8\overline{)3.000} \\ \underline{24} \\ 60 \\ \underline{56} \\ 40 \\ \underline{40} \\ 0 \end{array}$$

Thus $\frac{3}{8} = 0.375$, so $-\frac{3}{8} = -0.375$.

64. -0.125

65. $\frac{5}{3}$ means $5 \div 3$, so we divide.

$$\begin{array}{r} 1.66... \\ 3\overline{)5.00} \\ \underline{3} \\ 20 \\ \underline{18} \\ 20 \\ \underline{18} \\ 2 \end{array}$$

We have $\frac{5}{3} = 1.\overline{6}$.

66. $0.8\overline{3}$

67. $\frac{7}{6}$ means $7 \div 6$, so we divide.

$$\begin{array}{r} 1.166... \\ 6\overline{)7.000} \\ \underline{6} \\ 10 \\ \underline{6} \\ 40 \\ \underline{36} \\ 40 \\ \underline{36} \\ 4 \end{array}$$

We have $\frac{7}{6} = 1.1\overline{6}$.

68. $0.41\overline{6}$

69. $\frac{2}{3}$ means $2 \div 3$, so we divide.

$$\begin{array}{r} 0.666... \\ 3\overline{)2.000} \\ \underline{18} \\ 20 \\ \underline{18} \\ 20 \\ \underline{18} \\ 2 \end{array}$$

We have $\frac{2}{3} = 0.\overline{6}$.

70. 0.25

71. We first find decimal notation for $\frac{1}{2}$. Since $\frac{1}{2}$ means $1 \div 2$, we divide.

$$\begin{array}{r} 0.5 \\ 2\overline{)1.0} \\ \underline{10} \\ 0 \end{array}$$

Thus, $\frac{1}{2} = 0.5$, so $-\frac{1}{2} = -0.5$.

72. 0.625

73. $\frac{1}{10}$ means $1 \div 10$, so we divide.

$$\begin{array}{r} 0.1 \\ 10\overline{)1.0} \\ \underline{10} \\ 0 \end{array}$$

We have $\frac{1}{10} = 0.1$

74. -0.35

75. The number $\frac{10}{3}$ can be named $3\frac{1}{3}$, or $3.3\overline{3}$. The graph is $\frac{1}{3}$ of the way from 3 to 4.

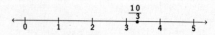

76.

77. The graph of -4.3 is $\frac{3}{10}$ of the way from -4 to -5.

Chapter 2   (2.1)

78.

79. Using a calculator, we find $\sqrt{5} \approx 2.236$. The graph of $\sqrt{5}$ is about $\frac{2}{10}$ of the way from 2 to 3.

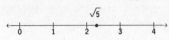

80.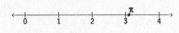

81. $\pi \approx 3.14$, so the graph of π is about $\frac{1}{10}$ of the way from 3 to 4.

82.

83. $x < -6$ has the same meaning as $-6 > x$.

84. $8 > x$

85. $y \geq -10$ has the same meaning as $-10 \leq y$.

86. $t \leq 12$

87. $\frac{21}{5} \cdot \frac{1}{7} = \frac{21 \cdot 1}{5 \cdot 7}$ Multiplying numerators and denominators

$= \frac{3 \cdot 7 \cdot 1}{5 \cdot 7}$ Factoring the numerator

$= \frac{7 \cdot 3 \cdot 1}{7 \cdot 5} = \frac{7}{7} \cdot \frac{3 \cdot 1}{5}$ Factoring the fractional expression

$= \frac{3}{5}$ Removing a factor of 1

88. 1000

89. $5 + ab$ is equivalent to $ab + 5$ by the commutative law of addition.

$ba + 5$ is equivalent to $ab + 5$ by the commutative law of multiplication.

$5 + ba$ is equivalent to $ab + 5$ by both commutative laws.

90. $3(x + 3 + 4y)$

91. List the numbers as they occur on the number line, from left to right: $-17, -12, 5, 13$

92. $-23, -17, 0, 4$

93. In either case, a lower number is better than a higher one.

94. $7, -7$

95. $|x| < 2$

x represents an integer whose distance from 0 is less than 2. Thus, $x = -1$ or $x = 0$ or $x = 1$.

96. $\frac{13}{5}, -\frac{13}{5}$

97. $|x| = -4$

Absolute value cannot be negative. (The distance of a number from 0 cannot be negative.) The equation has no solution.

98. $|-3| < 5$

99. $|-4| = 4$, so $2 < |-4|$.

100. $0 = |0|$

101. $|-5| = 5$ and $|-2| = 2$, so $|-5| > |-2|$.

102. $|4| < |-7|$

103. $|-8| = 8$ and $|8| = 8$, so $|-8| = |8|$.

104. $-\frac{8}{8}, -\frac{4}{8}, \frac{1}{8}, \frac{3}{8}, \frac{5}{8}, \frac{7}{8}$

105. Converting to decimal notation, we can write

$\frac{4}{5}, \frac{4}{3}, \frac{4}{8}, \frac{4}{6}, \frac{4}{9}, \frac{4}{2}, -\frac{4}{3}$ as

$0.8, 1.3\overline{3}, 0.5, 0.6\overline{6}, 0.4\overline{4}, 2, -1.3\overline{3}$, respectively. List the numbers (in fractional form) as they occur on the number line, from left to right:

$-\frac{4}{3}, \frac{4}{9}, \frac{4}{8}, \frac{4}{6}, \frac{4}{5}, \frac{4}{3}, \frac{4}{2}$

106. $-\frac{5}{6}, -\frac{3}{4}, -\frac{2}{3}, \frac{1}{6}, \frac{3}{8}, \frac{1}{2}$

107. Note that $7^1 = 7$, $|-6| = 6$, $|3| = 3$, $1^7 = 1$, and $\frac{14}{4} = \frac{7}{2}$, or $3\frac{1}{2}$. List the numbers as they occur on the number line, from left to right:

$-100, -5, 0, 1^7, |3|, \frac{14}{4}, 4, |-6|, 7^1$

108. 1

109. For $x = 5$, $y = -3$, and $z = 4$,

$2|x| + |y| - |z| = 2|5| + |-3| - |4| = 2 \cdot 5 + 3 - 4 = 10 + 3 - 4 = 9$.

110. 2

111. For $x = 5$, $y = -3$, and $z = 4$,

$\frac{|x| - |y|}{2z} = \frac{|5| - |-3|}{2 \cdot 4} = \frac{5 - 3}{8} = \frac{2}{8} = \frac{1}{4}$.

112. Answers may vary. $-1\frac{3}{4}$

113. Answers may vary. Point B appears to be about $\frac{9}{10}$ of the way from 0 to 1, so we estimate that $\frac{9}{10}$, or 0.9, corresponds to it.

25

Chapter 2 (2.2)

114. a) 1
 b) $\frac{1}{9}$
 c) $\frac{2}{9}$
 d) $\frac{5}{9}$

Exercise Set 2.2

1. Start at -9. Move 2 units to the right.

 -9 + 2 = -7

2. -3

3. Start at -10. Move 6 units to the right.
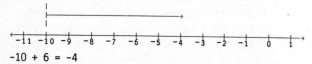
 -10 + 6 = -4

4. 5

5. Start at -8. Move 8 units to the right.
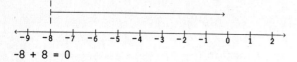
 -8 + 8 = 0

6. 0

7. Start at -3. Move 5 units to the left.
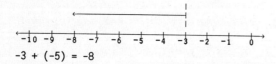
 -3 + (-5) = -8

8. -10

9. -7 + 0   One number is 0. The answer is the other number.   -7 + 0 = -7

10. -13

11. 0 + (-27)   One number is 0. The answer is the other number.   0 + (-27) = -27

12. -35

13. 17 + (-17)   The numbers have the same absolute value. The sum is 0.   17 + (-17) = 0

14. 0

15. -17 + (-25)   Two negatives. Add the absolute values, getting 42. Make the answer negative.   -17 + (-25) = -42

16. -41

17. 18 + (-18)   The numbers have the same absolute value. The sum is 0.   18 + (-18) = 0

18. 0

19. -18 + 18   The numbers have the same absolute value. The sum is 0.   -18 + 18 = 0

20. 0

21. 8 + (-5)   The absolute values are 8 and 5. The difference is 8 - 5, or 3. The positive number has the larger absolute value, so the answer is positive.   8 + (-5) = 3

22. 1

23. -4 + (-5)   Two negatives. Add the absolute values, getting 9. Make the answer negative.   -4 + (-5) = -9

24. -2

25. 13 + (-6)   The absolute values are 13 and 6. The difference is 13 - 6, or 7. The positive number has the larger absolute value, so the answer is positive.   13 + (-6) = 7

26. 11

27. 11 + (-9)   The absolute values are 11 and 9. The difference is 11 - 9, or 2. The positive number has the larger absolute value, so the answer is positive.   11 + (-9) = 2

28. -33

29. -20 + (-6)   Two negatives. Add the absolute values, getting 26. Make the answer negative.   -20 + (-6) = -26

30. 0

31. -15 + (-7)   Two negatives. Add the absolute values, getting 22. Make the answer negative.   -15 + (-7) = -22

32. 18

33. 40 + (-8)   The absolute values are 40 and 8. The difference is 40 - 8, or 32. The positive number has the larger absolute value, so the answer is positive.   40 + (-8) = 32

34. -32

35. -25 + 25   The numbers have the same absolute value. The sum is 0.   -25 + 25 = 0

Chapter 2   (2.2)

36. 0

37. 63 + (-18)   The absolute values are 63 and 18. The difference is 63 - 18, or 45. The positive number has the larger absolute value, so the answer is positive.   63 + (-18) = 45

38. 20

39. -6.5 + 4.7   The absolute values are 6.5 and 4.7. The difference is 6.5 - 4.7, or 1.8. The negative number has the larger absolute value, so the answer is negative.   -6.5 + 4.7 = -1.8

40. -1.7

41. -2.8 + (-5.3)   Two negatives. Add the absolute values, getting 8.1. Make the answer negative.   -2.8 + (-5.3) = -8.1

42. -14.4

43. $-\frac{3}{5} + \frac{2}{5}$ The absolute values are $\frac{3}{5}$ and $\frac{2}{5}$. The difference is $\frac{3}{5} - \frac{2}{5}$, or $\frac{1}{5}$. The negative number has the larger absolute value, so the answer is negative. $-\frac{3}{5} + \frac{2}{5} = -\frac{1}{5}$

44. $-\frac{2}{3}$

45. $-\frac{3}{7} + \left(-\frac{5}{7}\right)$ Two negatives. Add the absolute values, getting $\frac{8}{7}$. Make the answer negative. $-\frac{3}{7} + \left(-\frac{5}{7}\right) = -\frac{8}{7}$

46. $-\frac{10}{9}$

47. $-\frac{5}{8} + \frac{1}{4}$ The absolute values are $\frac{5}{8}$ and $\frac{1}{4}$. The difference is $\frac{5}{8} - \frac{2}{8}$, or $\frac{3}{8}$. The negative number has the larger absolute value, so the answer is negative. $-\frac{5}{8} + \frac{1}{4} = -\frac{3}{8}$

48. $-\frac{1}{6}$

49. $-\frac{3}{7} + \left(-\frac{2}{5}\right)$ Two negatives. Add the absolute values, getting $\frac{15}{35} + \frac{14}{35}$, or $\frac{29}{35}$. Make the answer negative. $-\frac{3}{7} + \left(-\frac{2}{5}\right) = -\frac{29}{35}$

50. $-\frac{23}{24}$

51. $-\frac{3}{5} + \left(-\frac{2}{15}\right)$ Two negatives. Add the absolute values, getting $\frac{9}{15} + \frac{2}{15}$, or $\frac{11}{15}$. Make the answer negative. $-\frac{3}{5} + \left(-\frac{2}{15}\right) = -\frac{11}{15}$

52. $-\frac{11}{18}$

53. -5.7 + (-7.2) + 6.6 = -12.9 + 6.6   Adding the positive numbers
   = -6.3   Adding the results

54. -14.7

55. -8.5 + 7.9 + (-3.7) = -0.6 + (-3.7)   Adding from left to right
   = -4.3

56. -13

57. $-\frac{7}{16} + \frac{7}{8}$ The absolute values are $\frac{7}{16}$ and $\frac{7}{8}$. The difference is $\frac{14}{16} - \frac{7}{16}$, or $\frac{7}{16}$. The positive number has the larger absolute value, so the answer is positive. $-\frac{7}{16} + \frac{7}{8} = \frac{7}{16}$

58. $\frac{1}{84}$

59. 75 + (-14) + (-17) + (-5)
   a) -14 + (-17) + (-5) = -36   Adding the negative numbers
   b) 75 + (-36) = 39   Adding the results

60. -62

61. $-44 + \left(-\frac{3}{8}\right) + 95 + \left(-\frac{5}{8}\right)$
   a) $-44 + \left(-\frac{3}{8}\right) + \left(-\frac{5}{8}\right) = -45$ Adding the negative numbers
   b) -45 + 95 = 50   Adding the results

62. 37.9

63. 98 + (-54) + 113 + (-998) + 44 + (-612) + (-18) + 334
   a) 98 + 113 + 44 + 334 = 589   Adding the positive numbers
   b) -54 + (-998) + (-612) + (-18) = -1682   Adding the negative numbers
   c) 589 + (-1682) = -1093   Adding the results

64. -1021

65. The additive inverse of 24 is -24 because 24 + (-24) = 0.

66. 64

67. The additive inverse of -9 is 9 because -9 + 9 = 0.

68. $-\frac{7}{2}$

27

Chapter 2   (2.2)

69. The additive inverse of -26.9 is 26.9 because -26.9 + 26.9 = 0.

70. -48.2

71. If x = 9, then -x = -(9) = -9. (The additive inverse of 9 is -9.)

72. 26

73. If $x = -\frac{14}{3}$, then $-x = -\left(-\frac{14}{3}\right) = \frac{14}{3}$.

    $\left(\text{The additive inverse of } -\frac{14}{3} \text{ is } \frac{14}{3}.\right)$

74. $-\frac{1}{328}$

75. If x = 0.101, then -x = -(0.101) = -0.101.
    (The additive inverse of 0.101 is -0.101.)

76. 0

77. If x = -65, then -(-x) = -[-(-65)] = -65
    (The opposite of the opposite of -65 is -65.)

78. 29

79. If $x = \frac{5}{3}$, then $-(-x) = -\left(-\frac{5}{3}\right) = \frac{5}{3}$.

    $\left(\text{The opposite of the opposite of } \frac{5}{3} \text{ is } \frac{5}{3}.\right)$

80. -9.1

81. -(-1) = 1

82. 7

83. -(7) = -7

84. -10

85. -(-14) = 14

86. 22.4

87. -0 = 0

88. $\frac{7}{8}$

89. Familiarize. We can represent a gain with a positive number, a loss with a negative number, and an incomplete pass with 0. Thus we have the following gains, losses, and incompletions:

    13, 0, -12, 21, -14

    We let x = the total gain or loss.

    Translate. We write an addition sentence which represents the total gain or loss:

    13 + 0 + (-12) + 21 + (-14) = x

89. (continued)

    Carry out. First add the nonnegative numbers, and do the same for the negative numbers. Then add the results:

    13 + 0 + (-12) + 21 + (-14) = x
    34 + (-26) = x
    8 = x

    Check. Go over the addition to be sure it is correct.

    State. Since 8 is positive, the total gain was 8 yd.

90. $77,320 profit

91. Familiarize. We can represent a drop in pressure with a negative number and a rise in pressure with a positive number. We have the following drops and rises:

    -6, 3, -14, 4

    We let x represent the pressure after the drops and rises occur. Since the original pressure was 1012 mb, x is the sum of this pressure and the total drop or rise in pressure.

    Translate. We write an addition sentence which represents the final pressure.

    x = 1012 + (-6) + 3 + (-14) + 4

    Carry out. We add the positive numbers, and we add the negative numbers. Then we add the results:

    x = 1012 + (-6) + 3 + (-14) + 4
      = 1019 + (-20)
      = 999

    Check. Go over the addition to be sure it is correct.

    State. The pressure was 999 mb.

92. $$253\frac{1}{2}$

93. Familiarize. We let a negative number represent the amount poured out and a positive number represent the amount added. We have the following amounts:

    -16, 27.3

    Let x = the final amount of acid in the beaker. Since the original amount was 500 mL, x is the sum of 500 and the total of the amounts removed and added.

    Translate. We write an addition sentence which represents the final amount:

    x = 500 + (-16) + 27.3

    Carry out. We add from left to right:

    x = 500 + (-16) + 27.3
      = 484 + 27.3
      = 511.3

    Check. Go over the addition to be sure it is correct.

    State. The beaker contained 511.3 mL.

94. -$0.01

Chapter 2   (2.3)

95. $7(3z + y + 2) = 7 \cdot 3z + 7 \cdot y + 7 \cdot 2 = 21z + 7y + 14$

96. $\frac{28}{3}$

97. Familiarize. We let x = the number that is 25% of 14.
    Translate. We write an equation.
    What is 25% of 14?
    x   = 25% · 14
    Carry out. We solve the equation.
    x = 25%·14
    x = 25(0.01)(14)
    x = 3.5
    Check. We check by finding what percent of 14 the number 3.5 is:
    $\frac{3.5}{14} = 0.25$, or 25%
    State. The answer is 3.5.

98. 144

99. When x is positive, the inverse of x, -x, is negative.

100. When x is negative, the inverse of x, -x, is positive.

101. Positive

102. Positive

103. If n is positive, -n is negative. Thus -n + m, the sum of two negatives, is negative.

104. Positive

105. If n = m, then n + (-m) = n + (-n) = 0 because -n is the additive inverse of n.

Exercise Set 2.3

1. $3 - 7 = 3 + (-7) = -4$

2. -5

3. $0 - 7 = 0 + (-7) = -7$

4. -10

5. $-8 - (-2) = -8 + 2 = -6$

6. 2

7. $-10 - (-10) = -10 + 10 = 0$

8. 0

9. $12 - 16 = 12 + (-16) = -4$

10. -5

11. $20 - 27 = 20 + (-27) = -7$

12. 26

13. $-9 - (-3) = -9 + 3 = -6$

14. 2

15. $-40 - (-40) = -40 + 40 = 0$

16. 0

17. $7 - 7 = 7 + (-7) = 0$

18. 0

19. $7 - (-7) = 7 + 7 = 14$

20. 8

21. $8 - (-3) = 8 + 3 = 11$

22. -11

23. $-6 - 8 = -6 + (-8) = -14$

24. 16

25. $-4 - (-9) = -4 + 9 = 5$

26. -16

27. $2 - 9 = 2 + (-9) = -7$

28. -6

29. $-6 - (-5) = -6 + 5 = -1$

30. -1

31. $8 - (-10) = 8 + 10 = 18$

32. 11

33. $0 - 5 = 0 + (-5) = -5$

34. -6

35. $-5 - (-2) = -5 + 2 = -3$

36. -2

37. $-7 - 14 = -7 + (-14) = -21$

38. -25

39. $0 - (-5) = 0 + 5 = 5$

40. 1

Chapter 2   (2.3)

41. −8 − 0 = −8 + 0 = −8
42. −9
43. 7 − (−5) = 7 + 5 = 12
44. 35
45. 2 − 25 = 2 + (−25) = −23
46. −45
47. −42 − 26 = −42 + (−26) = −68
48. −81
49. −71 − 2 = −71 + (−2) = −73
50. −52
51. 24 − (−92) = 24 + 92 = 116
52. 121
53. −50 − (−50) = −50 + 50 = 0
54. 0
55. $\frac{3}{8} - \frac{5}{8} = \frac{3}{8} + \left(-\frac{5}{8}\right) = -\frac{2}{8} = -\frac{1}{4}$
56. $-\frac{2}{3}$
57. $\frac{3}{4} - \frac{2}{3} = \frac{9}{12} - \frac{8}{12} = \frac{9}{12} + \left(-\frac{8}{12}\right) = \frac{1}{12}$
58. $-\frac{1}{8}$
59. $-\frac{3}{4} - \frac{2}{3} = -\frac{9}{12} - \frac{8}{12} = -\frac{9}{12} + \left(-\frac{8}{12}\right) = -\frac{17}{12}$
60. $-\frac{11}{8}$
61. $-\frac{5}{8} - \left(-\frac{3}{4}\right) = -\frac{5}{8} - \left(-\frac{6}{8}\right) = -\frac{5}{8} + \frac{6}{8} = \frac{1}{8}$
62. $-\frac{1}{12}$
63. 6.1 − (−13.8) = 6.1 + 13.8 = 19.9
64. 5
65. −3.2 − 5.8 = −3.2 + (−5.8) = −9
66. −8.6
67. 0.99 − 1 = 0.99 + (−1) = −0.01
68. −0.13
69. −79 − 114 = −79 + (−114) = −193

70. −413
71. 0 − (−500) = 0 + 500 = 500
72. 1500
73. −2.8 − 0 = −2.8 + 0 = −2.8
74. 4.94
75. 7 − 10.53 = 7 + (−10.53) = −3.53
76. 17.3
77. $\frac{1}{6} - \frac{2}{3} = \frac{1}{6} - \frac{4}{6} = \frac{1}{6} + \left(-\frac{4}{6}\right) = -\frac{3}{6} = -\frac{1}{2}$
78. $\frac{1}{8}$
79. $-\frac{4}{7} - \left(-\frac{10}{7}\right) = -\frac{4}{7} + \frac{10}{7} = \frac{6}{7}$
80. 0
81. $-\frac{7}{10} - \frac{10}{15} = -\frac{21}{30} - \frac{20}{30} = -\frac{21}{30} + \left(-\frac{20}{30}\right) = -\frac{41}{30}$
82. 0
83. $\frac{1}{13} - \frac{1}{12} = \frac{12}{156} - \frac{13}{156} = \frac{12}{156} + \left(-\frac{13}{156}\right) = -\frac{1}{156}$
84. $\frac{1}{42}$
85. 18 − (−15) − 3 − (−5) + 2 =
    18 + 15 + (−3) + 5 + 2 = 37
86. −22
87. −31 + (−28) − (−14) − 17 =
    (−31) + (−28) + 14 + (−17) = −62
88. 22
89. −34 − 28 + (−33) − 44 =
    (−34) + (−28) + (−33) + (−44) = −139
90. 5
91. −93 − (−84) − 41 − (−56) =
    (−93) + 84 + (−41) + 56 = 6
92. 4
93. −5 − (−3x) + 3x + 4x − (−12) =
    (−5) + 3x + 3x + 4x + 12 = 10x + 7
94. 7x + 46
95. 13x − (−2x) + 45 − (−21) = 13x + 2x + 45 + 21 =
    15x + 66

Chapter 2 (2.4)

96. 15x + 39

97. Familiarize. Let x = the assets after the purchase of the stereo system.
Translate. We write a subtraction equation.
x = 619.46 - 950
Carry out. We carry out the subtraction.
x = 619.46 - 950
x = 619.46 + (-950)
x = -330.54
Check. We can check by adding the difference to the amount borrowed:
950 + (-330.54) = 619.46
State. The total assets now are -$330.54.

98. $264

99. Familiarize. Represent the debt as a negative number. Let x = the amount of money required to make the total assets y dollars.
Translate. We write a subtraction equation.
y = x - 215.50
Carry out. We solve the equation for x.
y = x - 215.50
y + 215.50 = x
Check. We check by adding the value found for x and the amount of the debt:
y + 215.50 - 215.50 = y + 215.50 + (-215.50) = y
State. You will need y + $215.50.

100. 7° C

101. Familiarize. We draw a picture of the situation. We let x = the number of meters that Lake Assal is lower than the Valdes Peninsula.

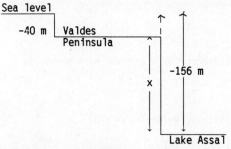

Translate. We subtract the lower altitude from the higher altitude.
x = -40 - (-156)
Carry out. We carry out the subtraction.
x = -40 - (-156)
x = -40 + 156
x = 116
Check. We can check by adding the difference to the lower altitude.
116 + (-156) = -40
The result, -40, is the altitude of the Valdes Peninsula.

101. (continued)
State. Lake Assal is 116 m lower than the Valdes Peninsula.

102. 1767 m

103. Area = ℓw = (36 ft)(12 ft) = 432 ft²

104. $2^5 \cdot 3^3$

105. Use a calculator to do this exercise.
123,907 - 433,789 = -309,882

106. 83,443

107. -(-3 + 2) = -(-1) = 1

108. 0

109. -5|2| + 3|4| = -5·2 + 3·4 = -10 + 12 = 2

110. -9

111. False. 3 - 0 = 3, 0 - 3 = -3, 3 - 0 ≠ 0 - 3

112. False. 0 - 3 ≠ 3

113. True

114. True

115. True by definition of additive inverses.

116. False, m = n. If 3 - 3 = 0, 3 ≠ -3.

Exercise Set 2.4

1. -16
2. -10
3. -42
4. -18
5. -24
6. -45
7. -72
8. -30
9. 16
10. 10
11. 42

Chapter 2 (2.4)

12. 18

13. −120

14. 120

15. −238

16. 195

17. 1200

18. −1677

19. 98

20. −203.7

21. −72

22. −63

23. −12.4

24. −6.6

25. 24

26. 30

27. 21.7

28. 12.8

29. $\frac{2}{3} \cdot \left(-\frac{3}{5}\right) = -\left(\frac{2 \cdot 3}{3 \cdot 5}\right) = -\left(\frac{2}{5} \cdot \frac{3}{3}\right) = -\frac{2}{5}$

30. $-\frac{10}{21}$

31. $-\frac{3}{8} \cdot \left(-\frac{2}{9}\right) = \frac{3 \cdot 2 \cdot 1}{4 \cdot 2 \cdot 3 \cdot 3} = \frac{1}{12}$

32. $\frac{1}{4}$

33. −17.01

34. −38.95

35. $-\frac{5}{9} \cdot \frac{3}{4} = -\frac{5 \cdot 3}{3 \cdot 3 \cdot 4} = -\frac{5}{12}$

36. −6

37. $7 \cdot (-4) \cdot (-3) \cdot 5 = 7 \cdot 12 \cdot 5 = 7 \cdot 60 = 420$

38. 756

39. $-\frac{2}{3} \cdot \frac{1}{2} \cdot \left(-\frac{6}{7}\right) = -\frac{2}{6} \cdot \left(-\frac{6}{7}\right) = \frac{2 \cdot 6}{7 \cdot 6} = \frac{2}{7}$

40. $-\frac{3}{160}$

41. $-3 \cdot (-4) \cdot (-5) = 12 \cdot (-5) = -60$

42. −70

43. $-2 \cdot (-5) \cdot (-3) \cdot (-5) = 10 \cdot 15 = 150$

44. 30

45. $-\frac{2}{45}$

46. $\frac{6}{35}$

47. $-7 \cdot (-21) \cdot 13 = 147 \cdot 13 = 1911$

48. −5712

49. $-4 \cdot (-1.8) \cdot 7 = (7.2) \cdot 7 = 50.4$

50. −52

51. $-\frac{1}{9} \cdot \left(-\frac{2}{3}\right) \cdot \left(\frac{5}{7}\right) = \frac{2}{27} \cdot \frac{5}{7} = \frac{10}{189}$

52. −1

53. $4 \cdot (-4) \cdot (-5) \cdot (-12) = -16 \cdot (60) = -960$

54. 120

55. $0.07 \cdot (-7) \cdot 6 \cdot (-6) = 0.07 \cdot 6 \cdot (-7) \cdot (-6) =$
    $0.42 \cdot (42) = 17.64$

56. −518.4

57. $\left(-\frac{5}{6}\right)\left(\frac{1}{8}\right)\left(-\frac{3}{7}\right)\left(-\frac{1}{7}\right) = \left(-\frac{5}{48}\right)\left(\frac{3}{49}\right) = -\frac{5 \cdot 3}{16 \cdot 3 \cdot 49} =$
    $-\frac{5}{784}$

58. $\frac{4}{7}$

59. 0, The product of 0 and any real number is 0.

60. 0

61. $(-8)(-9)(-10) = 72(-10) = -720$

62. 5040

63. $(-6)(-7)(-8)(-9)(-10) = 42 \cdot 72 \cdot (-10) =$
    $3024 \cdot (-10) = -30,240$

64. 151,200

65. $1 \cdot 10 = 10$
    $0 \cdot 10 = 0$
    $-1 \cdot 10 = -10$
    $-2 \cdot 10 = -20$
    $-3 \cdot 10 = -30$

Chapter 2 (2.5)

66. -10, 0, 10, 20, 30

67. $(-3x)^2 = (-3 \cdot 7)^2$ Substituting
    $= (-21)^2$ Multiplying inside the parentheses
    $= (-21)(-21)$ Evaluating the power
    $= 441$

    $-3x^2 = -3(7)^2$ Substituting
    $= -3 \cdot 49$ Evaluating the power
    $= -147$

68. 36; -18

69. $\frac{3x}{y} = \frac{3 \cdot 4}{12} = \frac{12}{12} = 1$

70. 64%t, or 0.64t

71. $a(3b) = (a \cdot 3)b$ Associative law of multiplication

72. $\frac{1}{6}$

73. $-6[(-5) + (-7)] = -6[-12] = 72$

74. -49

75. $-3[(-8) + (-6)]\left[-\frac{1}{7}\right] = -3[-14]\left[-\frac{1}{7}\right] = -3(2) = -6$

76. -40

77. $-(3^5) \cdot [-(2^3)] = -243[-8] = 1944$

78. -10,368

79. $(-2)^4 = (-2)(-2)(-2)(-2) = 4 \cdot 4 = 16$

80. -16

81. $(-1)^{23} = -1$ (There are 23 factors of -1. The product of an odd number of negative numbers is negative.)

82. -1

83. $-5^3 = -(5 \cdot 5 \cdot 5) = -125$

84. -125

85. $|-8| \cdot (-5) + 2 \cdot |10| = 8 \cdot (-5) + 2 \cdot 10 = -40 + 20 = -20$

86. 77

87. $|(-2)^3 + 4^2| - (2 - 7)^2 = |(-2)^3 + 4^2| - (-5)^2 = |-8 + 16| - 25 = |8| - 25 = 8 - 25 = -17$

88. 276

89. $xy + z = -2(-4) + 5 = 8 + 5 = 13$

90. 15

91. $-6(3x - 5y) + z = -6[3(-2) - 5(-4)] + 5 = -6[-6 + 20] + 5 = -6[14] + 5 = -84 + 5 = -79$

92. -12,600

93. a) m and n have different signs;
    b) either m or n is zero;
    c) m and n have the same sign

94. $a(-b) + ab = a[-b + b]$ Distributive law
    $= a(0)$ Additive inverse
    $= 0$ Multiplication property of zero
    Therefore, $-(ab) = a(-b)$.   Additive inverse

Exercise Set 2.5

1. $36 \div (-6) = -6$ Check: $-6 \cdot (-6) = 36$

2. -4

3. $\frac{26}{-2} = -13$ Check: $-13 \cdot (-2) = 26$

4. -2

5. $\frac{-16}{8} = -2$ Check: $-2 \cdot 8 = -16$

6. 11

7. $\frac{-48}{-12} = 4$ Check: $4(-12) = -48$

8. 7

9. $\frac{-72}{9} = -8$ Check: $-8 \cdot 9 = -72$

10. -2

11. $-100 \div (-50) = 2$ Check: $2(-50) = -100$

12. -25

13. $-108 \div 9 = -12$ Check: $9(-12) = -108$

14. $\frac{64}{7}$

15. $\frac{200}{-25} = -8$ Check: $-8(-25) = 200$

16. $\frac{300}{13}$

17. Undefined

18. 0

Chapter 2 (2.5)

19. $\frac{88}{-9} = -\frac{88}{9}$ Check: $-\frac{88}{9} \cdot (-9) = 88$

20. 29

21. The reciprocal of $\frac{15}{7}$ is $\frac{7}{15}$ because $\frac{15}{7} \cdot \frac{7}{15} = 1$.

22. $\frac{8}{3}$

23. The reciprocal of $-\frac{47}{13}$ is $-\frac{13}{47}$ because $\left(-\frac{47}{13}\right) \cdot \left(-\frac{13}{47}\right) = 1$.

24. $-\frac{12}{31}$

25. The reciprocal of 13 is $\frac{1}{13}$ because $13 \cdot \frac{1}{13} = 1$.

26. $-\frac{1}{10}$

27. The reciprocal of 4.3 is $\frac{1}{4.3}$ (or $\frac{10}{43}$) because $4.3 \cdot \frac{1}{4.3} = 1$.

28. $-\frac{1}{8.5}$, or $-\frac{10}{85}$

29. The reciprocal of $-\frac{1}{7.1}$ is -7.1 because $\left(-\frac{1}{7.1}\right)(-7.1) = 1$.

30. -4.9

31. The reciprocal of $\frac{p}{q}$ is $\frac{q}{p}$ because $\frac{p}{q} \cdot \frac{q}{p} = 1$.

32. $\frac{t}{s}$

33. The reciprocal of $\frac{1}{4y}$ is $4y$ because $\frac{1}{4y} \cdot 4y = 1$.

34. $-8a$

35. The reciprocal of $\frac{2a}{3b}$ is $\frac{3b}{2a}$ because $\frac{2a}{3b} \cdot \frac{3b}{2a} = 1$.

36. $-\frac{3x}{4y}$

37. $3 \cdot \frac{1}{19}$

38. $4\left(-\frac{1}{9}\right)$

39. $6\left(-\frac{1}{13}\right)$

40. $-12\left(\frac{1}{41}\right)$

41. $13.9\left(-\frac{1}{1.5}\right)$

42. $-47.3\left(\frac{1}{21.4}\right)$

43. $x \cdot y$

44. $13 \cdot \frac{1}{x}$

45. $(3x + 4)\frac{1}{5}$

46. $(4y - 8)\left(-\frac{1}{7}\right)$

47. $(5a - b)\left(\frac{1}{5a + b}\right)$

48. $(2x + x^2)\left(\frac{1}{x - 5}\right)$

49. $\frac{3}{4} \div \left(-\frac{2}{3}\right) = \frac{3}{4} \cdot \left(-\frac{3}{2}\right) = -\frac{9}{8}$

50. $-\frac{7}{4}$

51. $-\frac{5}{4} \div \left(-\frac{3}{4}\right) = -\frac{5}{4} \cdot \left(-\frac{4}{3}\right) = \frac{20}{12} = \frac{5 \cdot 4}{3 \cdot 4} = \frac{5}{3}$

52. $\frac{2}{3}$

53. $-\frac{2}{7} \div \left(-\frac{4}{9}\right) = -\frac{2}{7} \cdot \left(-\frac{9}{4}\right) = \frac{18}{28} = \frac{9 \cdot 2}{14 \cdot 2} = \frac{9}{14}$

54. $\frac{24}{25}$

55. $-\frac{3}{8} \div \left(-\frac{8}{3}\right) = -\frac{3}{8} \cdot \left(-\frac{3}{8}\right) = \frac{9}{64}$

56. $\frac{25}{48}$

57. $-6.6 \div 3.3 = -2$ Do the long division. Make the answer negative.

58. 7

59. $\frac{-11}{-13} = \frac{11}{13}$ The opposite of a number divided by the opposite of another number is the quotient of the two numbers.

60. -0.095

61. $\frac{48.6}{-3} = -16.2$ Do the long division. Make the answer negative.

62. -5.5625

63. $\frac{-9}{17 - 17} = \frac{-9}{0}$
Division by zero is undefined.

64. Undefined

Chapter 2   (2.5)

65. To find the additive inverse we change the sign. To find the reciprocal we invert but do not change the sign.

Number	Additive inverse	Reciprocal
$\frac{2}{3}$	$-\frac{2}{3}$	$\frac{3}{2}$
$-\frac{5}{4}$	$\frac{5}{4}$	$-\frac{4}{5}$
0	0	undefined
1	-1	1
-4.5	4.5	$-\frac{1}{4.5}$, or $-\frac{10}{45}$

66. 706.5 cm²

67. $m^3 + 2 = 5^3 + 2 = 125 + 2 = 127$

68. $8(3x + 4y + 8)$

69. $x + 12y + 11x + 14y + 9 =$
    $(1 + 11)x + (12 + 14)y + 9 = 12x + 26y + 9$

70. -60

71. $-(-3 + 2) = -(-1) = 1$

72. $-\frac{1}{6}$

73. $\frac{1}{3} - 4 = \frac{1}{3} - \frac{12}{3} = -\frac{11}{3}$

74. $\frac{69}{25}$

75. $\frac{-3(-9) + 7}{-4} = \frac{27 + 7}{-4} = \frac{34}{-4} = -\frac{34}{4} = -\frac{17 \cdot 2}{2 \cdot 2} = -\frac{17}{2}$

76. -8

77. $\frac{(-3)^4}{-9} = \frac{81}{-9} = -9$

78. 11

79. $\frac{(-3)(-4) - (-6)2}{(-2)5 + (-7)(-6)} = \frac{12 - (-12)}{-10 + 42} = \frac{12 + 12}{-10 + 42} = \frac{24}{32} =$
    $\frac{3 \cdot 8}{4 \cdot 8} = \frac{3}{4}$

80. $-\frac{1}{2}$

81. $\frac{(-2)^3(-5)^2}{5(-2) + (-2)(-6) - 12 + 20} = \frac{-8 \cdot 25}{-10 + 12 - 12 + 20} =$
    $\frac{-200}{10} = -20$

82. $-\frac{95}{77}$

83. $\frac{10}{7} \div (-0.25) = \frac{10}{7} \div \left(-\frac{1}{4}\right)$ Converting from decimal to fractional notation
    $= \frac{10}{7}\left(-\frac{4}{1}\right)$
    $= -\frac{40}{7}$

84. 1

85. $-6[(-5) + (-7)] = -6[-12] = 72$

86. -40

87. $-3[(-8)9 + (-6)(-7) - 8 \cdot 13] =$
    $-3[-72 + 42 - 104] = -3[-134] = 402$

88. About -0.095238095

89. No; $\frac{4}{2} \neq \frac{2}{4}$.

90. No; $16 \div (8 \div 4) = 16 \div 2 = 8$ but $(16 \div 8) \div 4 =$
    $2 \div 4 = \frac{1}{2}$

91. -1 and 1 are their own reciprocals.
    $[-1(-1) = 1$ and $1 \cdot 1 = 1]$

92. There are none. A reciprocal has the same sign as the number. Zero has no reciprocal.

93. You get the original number. The reciprocal of the reciprocal of a number is the original number.

94. Negative

95. When m and n are positive, -m and -n are negative. Then $\frac{-n}{-m}$ is the quotient of two negative numbers and, thus, is positive.

96. Positive

97. n is positive and -m is negative, so $\frac{n}{-m}$ is the quotient of a positive and a negative number and, thus, is negative. Then $-\left(\frac{n}{-m}\right)$ is the opposite, or additive inverse, of a negative number and, thus, is positive.

98. Negative

99. When n is negative, -n is positive so $\frac{-n}{m}$ is the quotient of a positive and a negative number and, thus, is negative.

100. Positive

101. $\frac{-n}{m}$ is negative (see Exercise 99), so $-\left(\frac{-n}{m}\right)$ is the opposite, or additive inverse, of a negative number and, thus, is positive.

Chapter 2   (2.6)

102. Positive

103. When n and m are negative, -n and -m are positive, so $\frac{-n}{-m}$ is the quotient of two positive numbers and, thus, is positive. Then $-\left[\frac{-n}{-m}\right]$ is the opposite, or additive inverse, of a positive number and, thus, is negative.

Exercise Set 2.6

1. 4x + 3z   Parts are separated by plus signs
   The terms are 4x and 3z.

2. 8x, -1.4y

3. 7x + 8y - 9z = 7x + 8y + (-9z)   Separating parts with plus signs
   The terms are 7x, 8y, and -9z.

4. 8a, 10b, -18c

5. $12x - 13.2y + \frac{5}{8}z - 4.5 = 12x + (-13.2y) + \frac{5}{8}z + (-4.5)$
   The terms are 12x, -13.2y, $\frac{5}{8}z$, and -4.5.

6. -7.8a, -3.4y, -8.7z, -12.4

7. 7(4 - 3) = 7·4 - 7·3 = 28 - 21 = 7

8. 30

9. -3(3 - 7) = -3·3 - (-3)·7 = -9 - (-21) = -9 + 21 = 12

10. 3.48

11. 4.1(6.3 - 9.4) = 4.1(6.3) - 4.1(9.4) = 25.83 - 38.54 = -12.71

12. $\frac{8}{9}$

13. 7(x - 2) = 7·x - 7·2 = 7x - 14

14. 5x - 40

15. -7(y - 2) = -7·y - (-7)·2 = -7y - (-14) = -7y + 14

16. -9y + 63

17. -9(-5x - 6y + 8) = -9(-5x) - (-9)6y + (-9)8 = 45x - (-54)y + (-72) = 45x + 54y - 72

18. 14x + 35y - 63

19. -4(x - 3y - 2z) = -4·x - (-4)3y - (-4)2z = -4x - (-12)y - (-8)z = -4x + 12y + 8z

20. 16x - 40y - 64z

21. 3.1(-1.2x + 3.2y - 1.1) = 3.1(-1.2x) + (3.1)3.2y - 3.1(1.1) = -3.72x + 9.92y - 3.41

22. 8.82x + 9.03y + 4.62

23. 8x - 24 = 8·x - 8·3 = 8(x - 3)

24. 10(x - 5)

25. 32 - 4y = 4·8 - 4·y = 4(8 - y)

26. 6(4 - m)

27. 8x + 10y - 22 = 2·4x + 2·5y - 2·11 = 2(4x + 5y - 11)

28. 3(3a + 2b - 5)

29. ax - 7a = a·x - 7·a = a(x - 7)

30. b(y - 9)

31. ax - ay - az = a·x - a·y - a·z = a(x - y - z)

32. c(x + y - z)

33. 18x - 12y + 6 = 6·3x - 6·2y + 6·1 = 6(3x - 2y + 1)

34. -7(2x - 3y - 1), or 7(-2x + 3y + 1)

35. 11x - 3x = (11 - 3)x = 8x

36. -8t

37. 6n - n = (6 - 1)n = 5n

38. 9t

39. y - 17y = (1 - 17)y = -16y

40. -6m + 4

41. -8 + 11a - 5b + 6a - 7b + 7 = 11a + 6a - 5b - 7b - 8 + 7 = (11 + 6)a + (-5 - 7)b + (-8 + 7) = 17a - 12b - 1

42. 3x + y + 2

43. 9x + 2y - 5x = (9 - 5)x + 2y = 4x + 2y

44. 12y - 3z

45. 11x + 2y - 4x - y = (11 - 4)x + (2 - 1)y = 7x + y

46. 11a + 5b

Chapter 2   (2.7)

47. $2.7x + 2.3y - 1.9x - 1.8y = (2.7 - 1.9)x + (2.3 - 1.8)y = 0.8x + 0.5y$

48. $2.6a + 1.4b$

49. $\frac{1}{5}x + \frac{4}{5}y + \frac{2}{5}x - \frac{1}{5}y = \left(\frac{1}{5} + \frac{2}{5}\right)x + \left(\frac{4}{5} - \frac{1}{5}\right)y = \frac{3}{5}x + \frac{3}{5}y$

50. $x + \frac{1}{4}y$

51. Familiarize. Let x = the unknown number. We draw a picture of the situation.

    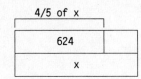

    Translate.
    Four-fifths of what number is 624?
    $\frac{4}{5} \cdot x = 624$

    Carry out. We solve the equation.
    $\frac{4}{5}x = 624$
    $x = \frac{624}{\frac{4}{5}}$
    $x = \frac{624}{1} \cdot \frac{5}{4} = \frac{4 \cdot 156 \cdot 5}{1 \cdot 4}$
    $x = 780$

    Check. We find 4/5 of 780:
    $\frac{4}{5} \cdot 780 = \frac{4 \cdot 5 \cdot 156}{5 \cdot 1} = 624$

    State. The number is 780.

52. 144

53. $2y + 3x$

54. 6.25%

55. $2\pi r + \pi rs = \pi r \cdot 2 + \pi r \cdot s = \pi r(2 + s)$

56. $\frac{1}{2}h(a + b)$

57. $2\pi rh - 2\pi r = 2\pi r \cdot h - 2\pi r \cdot 1 = 2\pi r(h - 1)$

58. $\frac{1}{4}a(h - b)$

59. $8x - 9 - 2(7 - 5x) = 8x - 9 + (-2)(7 - 5x) = 8x - 9 + (-2)7 - (-2)5x = 8x - 9 + (-14) - (-10x) = 8x - 9 - 14 + 10x = (8 + 10)x - 9 - 14 = 18x - 23$

60. $-15y - 81$

61. $\frac{5x - 15}{5} + \frac{2x + 6}{2} = \frac{5(x - 3)}{5} + \frac{2(x + 3)}{2} = \frac{5}{5} \cdot (x - 3) + \frac{2}{2} \cdot (x + 3) = (x - 3) + (x + 3) = x + x - 3 + 3 = 2x$

62. $-2x - 1$

63. $\frac{3a - 2b}{6} + \frac{2b + 4a}{8} = \frac{3a - 2b}{6} \cdot \frac{4}{4} + \frac{2b + 4a}{8} \cdot \frac{3}{3} = \frac{12a - 8b}{24} + \frac{6b + 12a}{24} = \frac{12a - 8b + 6b + 12a}{24} = \frac{24a - 2b}{24} = \frac{2(12a - b)}{2 \cdot 12} = \frac{12a - b}{12}$

64. $\frac{-19x + 14y}{10}$

65. $P + 8\%P = 1 \cdot P + 0.08 \cdot P = (1 + 0.08)P = 1.08P$

66. $1.06P$

67. $8(x - y)$

68. $9y - 6z$

69. $3(a + b) - 7a = 3a + 3b - 7a = 3b - 4a$

70. $\$2.95(x + y)$

71. $\$2500x + \$2500y = \$2500(x + y)$

72. $5420\left(\$41\frac{1}{8} - \$37\frac{3}{4}\right)$ or $5420\left(\$41\frac{1}{8}\right) - 5420\left(\$37\frac{3}{4}\right)$
    Her loss is $18,292.50.

Exercise Set 2.7

1. $-(2x + 7) = -2x - 7$ Changing the sign of each term

2. $-3x - 5$

3. $-(5x - 8) = -5x + 8$ Changing the sign of each term

4. $-6x + 7$

5. $-4a + 3b - 7c$

6. $-5x + 2y + 3z$

7. $-6x + 8y - 5$

8. $-8x - 3y - 9$

9. $-3x + 5y + 6$

10. $-6a + 4b + 7$

Chapter 2   (2.7)

11. 8x + 6y + 43

12. 2a - 9b + 5c

13. 9x - (4x + 3) = 9x - 4x - 3   Removing parentheses by changing the sign of every term
    = 5x - 3   Collecting like terms

14. 5y - 9

15. 2a - (5a - 9) = 2a - 5a + 9 = -3a + 9

16. 8n + 7

17. 2x + 7x - (4x + 6) = 2x + 7x - 4x - 6 = 5x - 6

18. a - 7

19. 2x - 4y - 3(7x - 2y) = 2x - 4y - 21x + 6y =
    -19x + 2y

20. -a - 4b

21.   15x - y - 5(3x - 2y + 5z)
    = 15x - y - 15x + 10y - 25z   Multiplying each term in parentheses by -5
    = 9y - 25z

22. -16a + 27b - 32c

23. (3x + 2y) - 2(5x - 4y) = 3x + 2y - 10x + 8y
    = -7x + 10y

24. -9a - 13b

25.   (12a - 3b + 5c) - 5(-5a + 4b - 6c)
    = 12a - 3b + 5c + 25a - 20b + 30c
    = 37a - 23b + 35c

26. -20x + 29y + 48

27. [9 - 2(5 - 4)] = [9 - 2·1]   Computing 5 - 4
    = [9 - 2]   Computing 2·1
    = 7

28. -14

29. 8[7 - 6(4 - 2)] = 8[7 - 6(2)] = 8[7 - 12] =
    8[-5] = -40

30. -10

31.   [4(9 - 6) + 11] - [14 - (6 + 4)]
    = [4(3) + 11] - [14 - 10]
    = [12 + 11] - [14 - 10]
    = 23 - 4
    = 19

32. 39

33.   [10(x + 3) - 4] + [2(x - 1) + 6]
    = [10x + 30 - 4] + [2x - 2 + 6]
    = [10x + 26] + [2x + 4]
    = 10x + 26 + 2x + 4
    = 12x + 30

34. 13x - 1

35.   [7(x + 5) - 19] - [4(x - 6) + 10]
    = [7x + 35 - 19] - [4x - 24 + 10]
    = [7x + 16] - [4x - 14]
    = 7x + 16 - 4x + 14
    = 3x + 30

36. x + 41

37.   3{[7(x - 2) + 4] - [2(2x - 5) + 6]}
    = 3{[7x - 14 + 4] - [4x - 10 + 6]}
    = 3{[7x - 10] - [4x - 4]}
    = 3{7x - 10 - 4x + 4}
    = 3{3x - 6}
    = 9x - 18

38. -16x + 44

39.   4{[5(x - 3) + 2] - 3[2(x + 5) - 9]}
    = 4{[5x - 15 + 2] - 3[2x + 10 - 9]}
    = 4{[5x - 13] - 3[2x + 1]}
    = 4{5x - 13 - 6x - 3}
    = 4{-x - 16}
    = -4x - 64

40. -12x - 237

41. 8 - 2·3 - 9 = 8 - 6 - 9   Multiplying
    = 2 - 9   Doing all additions and subtractions in order from left to right
    = -7

42. 11

43. (8 - 2·3) - 9 = (8 - 6) - 9   Multiplying inside the parentheses
    = 2 - 9   Subtracting inside the parentheses
    = -7

44. -36

45. $[(-24) \div (-3)] \div \left[-\frac{1}{2}\right] = 8 \div \left[-\frac{1}{2}\right] = 8 \cdot (-2) = -16$

46. 8

47. 16·(-24) + 50 = -384 + 50 = -334

Chapter 2   (2.7)

48. -160

49. $2^4 + 2^3 - 10 = 16 + 8 - 10 = 24 - 10 = 14$

50. 23

51. $5^3 + 26 \cdot 71 - (16 + 25 \cdot 3) =$
    $5^3 + 26 \cdot 71 - (16 + 75) = 5^3 + 26 \cdot 71 - 91 =$
    $125 + 26 \cdot 71 - 91 = 125 + 1846 - 91 =$
    $1971 - 91 = 1880$

52. 305

53. $3000 \cdot (1 + 0.16)^3 = 3000 \cdot (1.16)^3 =$
    $3000(1.560896) = 4682.688$

54. 34,279.2

55. $4 \cdot 5 - 2 \cdot 6 + 4 = 20 - 12 + 4 = 8 + 4 = 12$

56. 8

57. $4 \cdot (6 + 8)/(4 + 3) = 4 \cdot 14/7$ Adding inside the parentheses
    $= 56/7$ Multiplying and dividing in order from left to right
    $= 8$

58. 8

59. $[2 \cdot (5 - 3)]^2 = [2 \cdot 2]^2 = 4^2 = 16$

60. 76

61. $7 + 2(-6) = 7 - 12 = -5$

62. 27

63. $8(-7) + 6(-5) = -56 - 30 = -86$

64. -51

65. $19 - 5(-3) + 3 = 19 + 15 + 3 = 34 + 3 = 37$

66. 33

67. $9 \div (-3) + 16 \div 8 = -3 + 2 = -1$

68. -32

69. $7 + 10 - (-10 \div 2) = 7 + 10 - (-5) =$
    $7 + 10 + 5 = 17 + 5 = 22$

70. -10

71. $(2 - 5)^2 = (-3)^2 = 9$

72. 25

73. $3 - 3^2 = 3 - 9 = -6$

74. -7988

75. $20 + 4^3 \div (-8) = 20 + 64 \div (-8) = 20 - 8 = 12$

76. -3000

77. $-7(3)^4 + 18 = -7 \cdot 81 + 18 = -567 + 18 = -549$

78. 60

79. $8[(6 - 13) - 11] = 8[-7 - 11] = 8[-18] = -144$

80. 10

81. $(8 - 7) - 9 = 1 - 9 = -8$

82. 1

83. $256 \div (-32) \div (-4) = -8 \div (-4)$ Doing the divisions in order from left to right
    $= 2$

84. $-\frac{13}{45}$

85. $\frac{5^2 - 4^3 - 3}{9^2 - 2^2 - 1^5} = \frac{25 - 64 - 3}{81 - 4 - 1} = \frac{-39 - 3}{77 - 1} = \frac{-42}{76} =$
    $-\frac{2 \cdot 21}{2 \cdot 38} = -\frac{21}{38}$

86. $-\frac{23}{18}$

87. $\frac{20(8 - 3) - 4(10 - 3)}{10(2 - 6) - 2(5 + 2)} = \frac{20 \cdot 5 - 4 \cdot 7}{10(-4) - 2 \cdot 7} =$
    $\frac{100 - 28}{-40 - 14} = \frac{72}{-54} = -\frac{18 \cdot 4}{18 \cdot 3} = -\frac{4}{3}$

88. -118

89. $6y + 2x - 3a + c = 6y - (-2x) - 3a - (-c)$
    $= 6y - (-2x + 3a - c)$

90. $x - (y + a + b)$

91. $6m + 3n - 5m + 4b = 6m - (-3n) - 5m - (-4b)$
    $= 6m - (-3n + 5m - 4b)$

92. $(a + b) + [-a + (-b)] = a + b + (-a) + (-b) =$
    $a + (-a) + b + (-b) = 0 + 0 = 0$

93. $z - \{2z - [3z - (4z - 5z) - 6z] - 7z\} - 8z$
    $= z - \{2z - [3z - (-z) - 6z] - 7z\} - 8z$
    $= z - \{2z - [3z + z - 6z] - 7z\} - 8z$
    $= z - \{2z - [-2z] - 7z\} - 8z$
    $= z - \{2z + 2z - 7z\} - 8z$
    $= z - \{-3z\} - 8z$
    $= z + 3z - 8z$
    $= -4z$

94. $-2x - f$

95. $x-\{x-1-[x-2-(x-3-\{x-4-[x-5-(x-6)]\})]\}$
   $= x-\{x-1-[x-2-(x-3-\{x-4-[x-5-x+6]\})]\}$
   $= x - \{x - 1 - [x - 2 - (x - 3 - \{x - 4 - 1\})]\}$
   $= x - \{x - 1 - [x - 2 - (x - 3 - \{x - 5\})]\}$
   $= x - \{x - 1 - [x - 2 - (x - 3 - x + 5)]\}$
   $= x - \{x - 1 - [x - 2 - 2]\}$
   $= x - \{x - 1 - [x - 4]\}$
   $= x - \{x - 1 - x + 4\}$
   $= x - 3$

96. False

97. False; $-n + m = -(n - m) \neq -(n + m)$ for $m > 0$

98. True

99. $-n - m = -(n + m) \neq -(n - m)$ for $m > 0$

100. False

101. False; $-m(n - m) = -mn + m^2 = -(mn - m^2) \neq -(mn + m^2)$ for $m > 0$

102. True

103. True; $-n(-n - m) = n^2 + nm = n(n + m)$

104. Yes; $(-a)(-b) = (-1)(a)(-1)(b) = (-1)^2(ab) = ab$

105. See the answer section in the text.

CHAPTER 3   SOLVING EQUATIONS AND PROBLEMS

Exercise Set 3.1

1. $x + 2 = 6$
   $x + 2 + (-2) = 6 + (-2)$ Adding -2 on both sides
            $x = 4$ Simplifying

   Check: $\dfrac{x + 2 = 6}{\,4 + 2\,\big|\,6\,}$
                6

2. 3

3. $x + 15 = -5$
   $x + 15 + (-15) = -5 + (-15)$ Adding -15 on both sides
            $x = -20$

   Check: $\dfrac{x + 15 = -5}{-20 + 15\,\big|\,-5}$
               -5

4. 34

5. $x + 6 = -8$ Check: $\dfrac{x + 6 = -8}{-14 + 6\,\big|\,-8}$
   $x + 6 + (-6) = -8 + (-6)$ -8
            $x = -14$

6. -21

7. $x + 16 = -2$ Check: $\dfrac{x + 16 = -2}{-18 + 16\,\big|\,-2}$
   $x + 16 + (-16) = -2 + (-16)$ -2
            $x = -18$

8. -31

9. $x - 9 = 6$ Check: $\dfrac{x - 9 = 6}{15 - 9\,\big|\,6}$
   $x - 9 + 9 = 6 + 9$ 6
            $x = 15$

10. 13

11. $x - 7 = -21$ Check: $\dfrac{x - 7 = -21}{-14 - 7\,\big|\,-21}$
    $x - 7 + 7 = -21 + 7$ -21
             $x = -14$

12. -11

13. $5 + t = 7$ Check: $\dfrac{5 + t = 7}{5 + 2\,\big|\,7}$
    $-5 + 5 + t = -5 + 7$ 7
             $t = 2$

14. 4

15. $-7 + y = 13$ Check: $\dfrac{-7 + y = 13}{-7 + 20\,\big|\,13}$
    $7 + (-7) + y = 7 + 13$ 13
             $y = 20$

16. 24

17. $-3 + t = -9$ Check: $\dfrac{-3 + t = -9}{-3 + (-6)\,\big|\,-9}$
    $3 + (-3) + t = 3 + (-9)$ -9
             $t = -6$

18. -15

19. $r + \dfrac{1}{3} = \dfrac{8}{3}$ Check: $\dfrac{r + \tfrac{1}{3} = \tfrac{8}{3}}{\tfrac{7}{3} + \tfrac{1}{3}\,\big|\,\tfrac{8}{3}}$
    $r + \dfrac{1}{3} + \left(-\dfrac{1}{3}\right) = \dfrac{8}{3} + \left(-\dfrac{1}{3}\right)$ $\tfrac{8}{3}$
             $r = \dfrac{7}{3}$

20. $\dfrac{1}{4}$

21. $m + \dfrac{5}{6} = -\dfrac{11}{12}$
    $m + \dfrac{5}{6} + \left(-\dfrac{5}{6}\right) = -\dfrac{11}{12} + \left(-\dfrac{5}{6}\right)$
             $m = -\dfrac{11}{12} + \left(-\dfrac{5}{6}\right)\left(\dfrac{2}{2}\right)$
             $m = -\dfrac{11}{12} + \left(-\dfrac{10}{12}\right)$
             $m = -\dfrac{21}{12} = -\dfrac{3 \cdot 7}{3 \cdot 4}$
             $m = -\dfrac{7}{4}$

    Check: $\dfrac{m + \tfrac{5}{6} = -\tfrac{11}{12}}{-\tfrac{7}{4} + \tfrac{5}{6}\,\big|\,-\tfrac{11}{12}}$
             $-\tfrac{21}{12} + \tfrac{10}{12}$
             $-\tfrac{11}{12}$

22. $-\dfrac{3}{2}$

23. $x - \dfrac{5}{6} = \dfrac{7}{8}$ Check: $\dfrac{x - \tfrac{5}{6} = \tfrac{7}{8}}{\tfrac{41}{24} - \tfrac{5}{6}\,\big|\,\tfrac{7}{8}}$
    $x - \dfrac{5}{6} + \dfrac{5}{6} = \dfrac{7}{8} + \dfrac{5}{6}$ $\tfrac{41}{24} - \tfrac{20}{24}\,\big|\,\tfrac{21}{24}$
           $x = \dfrac{7}{8} \cdot \dfrac{3}{3} + \dfrac{5}{6} \cdot \dfrac{4}{4}$ $\tfrac{21}{24}$
           $x = \dfrac{21}{24} + \dfrac{20}{24}$
           $x = \dfrac{41}{24}$

24. $\dfrac{19}{12}$

41

Chapter 3 (3.1)

25. $-\frac{1}{5} + z = -\frac{1}{4}$

$\frac{1}{5} - \frac{1}{5} + z = \frac{1}{5} - \frac{1}{4}$

$z = \frac{1}{5} \cdot \frac{4}{4} - \frac{1}{4} \cdot \frac{5}{5}$

$z = \frac{4}{20} - \frac{5}{20}$

$z = -\frac{1}{20}$

Check: $\dfrac{-\frac{1}{5} + z = -\frac{1}{4}}{\begin{array}{c|c} -\frac{1}{5} + \left(-\frac{1}{20}\right) & -\frac{1}{4} \\ -\frac{4}{20} + \left(-\frac{1}{20}\right) & -\frac{5}{20} \\ -\frac{5}{20} & \end{array}}$

26. $-\frac{5}{8}$

27. $x + 2.3 = 7.4$
$x + 2.3 + (-2.3) = 7.4 + (-2.3)$
$x = 5.1$

Check: $\dfrac{x + 2.3 = 7.4}{\begin{array}{c|c} 5.1 + 2.3 & 7.4 \\ 7.4 & \end{array}}$

28. 4.7

29. $x - 4.8 = 7.6$ Check: $\dfrac{x - 4.8 = 7.6}{\begin{array}{c|c} 12.4 - 4.8 & 7.6 \\ 7.6 & \end{array}}$
$x - 4.8 + 4.8 = 7.6 + 4.8$
$x = 12.4$

30. 17.8

31. $-9.7 = -4.7 + y$
$4.7 + (-9.7) = 4.7 + (-4.7) + y$
$-5 = y$

Check: $\dfrac{-9.7 = -4.7 + y}{\begin{array}{c|c} -9.7 & -4.7 + (-5) \\ & -9.7 \end{array}}$

32. -10.6

33. $5\frac{1}{6} + x = 7$ Check: $\dfrac{5\frac{1}{6} + x = 7}{\begin{array}{c|c} 5\frac{1}{6} + 1\frac{5}{6} & 7 \\ 7 & \end{array}}$
$-5\frac{1}{6} + 5\frac{1}{6} + x = -5\frac{1}{6} + 7$
$x = -\frac{31}{6} + \frac{42}{6}$
$x = \frac{11}{6}$, or $1\frac{5}{6}$

34. $\frac{7}{12}$

35. $q + \frac{1}{3} = -\frac{1}{7}$

$q + \frac{1}{3} + \left(-\frac{1}{3}\right) = -\frac{1}{7} + \left(-\frac{1}{3}\right)$

$q = -\frac{1}{7} \cdot \frac{3}{3} + \left(-\frac{1}{3}\right)\left(\frac{7}{7}\right)$

$q = -\frac{3}{21} - \frac{7}{21}$

$q = -\frac{10}{21}$

Check: $\dfrac{q + \frac{1}{3} = -\frac{1}{7}}{\begin{array}{c|c} -\frac{10}{21} + \frac{1}{3} & -\frac{1}{7} \\ -\frac{10}{21} + \frac{7}{21} & -\frac{3}{21} \\ -\frac{3}{21} & \end{array}}$

36. $123\frac{1}{8}$

37. Familiarize. Let t = the average daily low temperature in Key West in January.

Translate. We reword the problem.

             Key West
-31° C is temperature less 50° C.
-31   =   t   -   50

Carry out. We solve the equation.
$-31 = t - 50$
$-31 + 50 = t$
$19 = t$

Check. We subtract 50 from the value we found. The result should be -31.
$19 - 50 = -31$

State. The average daily low temperature in Key West in January is 19° C.

38. -11

39. $-\frac{2}{3} \cdot \frac{5}{8} = -\frac{2 \cdot 5}{3 \cdot 2 \cdot 4} = -\frac{5}{12}$

40. $\frac{1}{3}$

41. $-356.788 = -699.034 + t$
$699.034 + (-356.788) = 699.034 + (-699.034) + t$
$342.246 = t$

42. $\frac{13}{20}$

43. $x + \frac{4}{5} = -\frac{2}{3} - \frac{4}{15}$

$x + \frac{4}{5} + \left(-\frac{4}{5}\right) = -\frac{2}{3} - \frac{4}{15} + \left(-\frac{4}{5}\right)$

$x = -\frac{2}{3} \cdot \frac{5}{5} - \frac{4}{15} - \frac{4}{5} \cdot \frac{3}{3}$

$x = -\frac{10}{15} - \frac{4}{15} - \frac{12}{15}$

$x = -\frac{26}{15}$

Chapter 3   (3.2)

44. -4

45. $16 + x - 22 = -16$
    $x - 6 = -16$ Adding on the left side
    $x - 6 + 6 = -16 + 6$
    $x = -10$

46. 0

47. $x + 3 = 3 + x$
    $x + 3 + (-3) = 3 + x + (-3)$
    $x = x$
    $x = x$ is true for all real numbers. Thus the solution is all real numbers.

48. No solution

49. $-\frac{3}{2} + x = -\frac{5}{17} - \frac{3}{2}$
    $\frac{3}{2} - \frac{3}{2} + x = \frac{3}{2} - \frac{5}{17} - \frac{3}{2}$
    $x = \left(\frac{3}{2} - \frac{3}{2}\right) - \frac{5}{17}$
    $x = -\frac{5}{17}$

50. $b + 3$

51. $1 - c = a + x$
    $1 - c + (-a) = a + x + (-a)$
    $1 - c - a = x$

52. $a + 4$

53. $|x| + 6 = 19$
    $|x| = 13$
    x represents a number whose distance from 0 is 13. Thus, $x = -13$ or $x = 13$.

54. No solution

55. $x - 4720 = 1634$
    $x - 4720 + 4720 = 1634 + 4720$
    $x = 6354$
    $x + 4720 = 6354 + 4720$
    $x + 4720 = 11{,}074$

56. We subtract by adding an inverse.

Exercise Set 3.2

1. $6x = 36$ Check: $\begin{array}{c|c} 6x = 36 \\ \hline 6 \cdot 6 & 36 \\ 36 & \end{array}$
   $\frac{1}{6} \cdot 6x = \frac{1}{6} \cdot 36$
   $1 \cdot x = 6$
   $x = 6$

2. 13

3. $5x = 45$ Check: $\begin{array}{c|c} 5x = 45 \\ \hline 5 \cdot 9 & 45 \\ 45 & \end{array}$
   $\frac{1}{5} \cdot 5x = \frac{1}{5} \cdot 45$
   $1 \cdot x = 9$
   $x = 9$

4. 8

5. $84 = 7x$ Check: $\begin{array}{c|c} 84 = 7x \\ \hline 84 & 7 \cdot 12 \\ & 84 \end{array}$
   $\frac{1}{7} \cdot 84 = \frac{1}{7} \cdot 7x$
   $12 = x$

6. 7

7. $-x = 40$ Check: $\begin{array}{c|c} -x = 40 \\ \hline -(-40) & 40 \\ 40 & \end{array}$
   $-1 \cdot x = 40$
   $-1 \cdot (-1 \cdot x) = -1 \cdot 40$
   $1 \cdot x = -40$
   $x = -40$

8. -100

9. $-x = -1$ Check: $\begin{array}{c|c} -x = -1 \\ \hline -(1) & -1 \\ -1 & \end{array}$
   $-1 \cdot x = -1$
   $-1 \cdot (-1 \cdot x) = -1 \cdot (-1)$
   $1 \cdot x = 1$
   $x = 1$

10. 68

11. $7x = -49$ Check: $\begin{array}{c|c} 7x = -49 \\ \hline 7(-7) & -49 \\ -49 & \end{array}$
    $\frac{1}{7} \cdot 7x = \frac{1}{7} \cdot (-49)$
    $x = -\frac{49}{7}$
    $x = -7$

12. -4

13. $-12x = 72$ Check: $\begin{array}{c|c} -12x = 72 \\ \hline -12(-6) & 72 \\ 72 & \end{array}$
    $-\frac{1}{12} \cdot (-12x) = -\frac{1}{12} \cdot 72$
    $x = -\frac{72}{12}$
    $x = -6$

14. -7

15. $-21x = -126$ Check: $\begin{array}{c|c} -21x = -126 \\ \hline -21 \cdot 6 & -126 \\ -126 & \end{array}$
    $-\frac{1}{21} \cdot (-21x) = -\frac{1}{21} \cdot (-126)$
    $x = \frac{126}{21}$
    $x = 6$

Chapter 3   (3.2)

16. 8

17. $\dfrac{t}{7} = -9$ Check: $\dfrac{t}{7} = -9$
   $7 \cdot \left[\dfrac{1}{7}t\right] = 7 \cdot (-9)$ $\dfrac{-63}{7}\,\bigg|\,-9$
   $t = -63$ $-9\,\bigg|$

18. −88

19. $\dfrac{3}{4}x = 27$ Check: $\dfrac{3}{4}x = 27$
   $\dfrac{4}{3} \cdot \dfrac{3}{4}x = \dfrac{4}{3} \cdot 27$ $\dfrac{3}{4} \cdot 36\,\bigg|\,27$
   $x = \dfrac{4 \cdot 3 \cdot 3 \cdot 3}{3 \cdot 1}$ $27\,\bigg|$
   $x = 36$

20. 20

21. $\dfrac{-t}{3} = 7$ Check: $\dfrac{-t}{3} = 7$
   $3 \cdot \dfrac{1}{3} \cdot (-t) = 3 \cdot 7$ $\dfrac{-(-21)}{3}\,\bigg|\,7$
   $-t = 21$ $\dfrac{21}{3}\,\bigg|$
   $-1 \cdot (-1 \cdot t) = -1 \cdot 21$ $7\,\bigg|$
   $1 \cdot t = -21$
   $t = -21$

22. −54

23. $-\dfrac{m}{3} = \dfrac{1}{5}$ Check: $-\dfrac{m}{3} = \dfrac{1}{5}$
   $-\dfrac{1}{3} \cdot m = \dfrac{1}{5}$ $-\dfrac{-\frac{3}{5}}{3}\,\bigg|\,\dfrac{1}{5}$
   $-3 \cdot \left[-\dfrac{1}{3} \cdot m\right] = -3 \cdot \dfrac{1}{5}$ $-\left[-\dfrac{3}{5} \div 3\right]$
   $m = -\dfrac{3}{5}$ $-\left[-\dfrac{3}{5} \cdot \dfrac{1}{3}\right]$
   $-\left[-\dfrac{1}{5}\right]$
   $\dfrac{1}{5}\,\bigg|$

24. $-\dfrac{7}{9}$

25. $-\dfrac{3}{5}r = -\dfrac{9}{10}$ Check: $-\dfrac{3}{5}r = -\dfrac{9}{10}$
   $-\dfrac{5}{3} \cdot \left[-\dfrac{3}{5}r\right] = -\dfrac{5}{3} \cdot \left[-\dfrac{9}{10}\right]$ $-\dfrac{3}{5} \cdot \dfrac{3}{2}\,\bigg|\,-\dfrac{9}{10}$
   $r = \dfrac{5 \cdot 3 \cdot 3}{3 \cdot 5 \cdot 2}$ $-\dfrac{9}{10}\,\bigg|$
   $r = \dfrac{3}{2}$

26. $\dfrac{2}{3}$

27. $-\dfrac{3}{2}r = -\dfrac{27}{4}$ Check: $-\dfrac{3}{2}r = -\dfrac{27}{4}$
   $-\dfrac{2}{3} \cdot \left[-\dfrac{3}{2}r\right] = -\dfrac{2}{3} \cdot \left[-\dfrac{27}{4}\right]$ $-\dfrac{3}{2} \cdot \dfrac{9}{2}\,\bigg|\,-\dfrac{27}{4}$
   $r = \dfrac{2 \cdot 3 \cdot 3 \cdot 3}{3 \cdot 2 \cdot 2}$ $-\dfrac{27}{4}\,\bigg|$
   $r = \dfrac{9}{2}$

28. −1

29. $6.3x = 44.1$ Check: $6.3x = 44.1$
   $\dfrac{1}{6.3} \cdot 6.3x = \dfrac{1}{6.3} \cdot 44.1$ $6.3 \cdot 7\,\bigg|\,44.1$
   $x = \dfrac{44.1}{6.3}$ $44.1\,\bigg|$
   $x = 7$

30. 20

31. $-3.1y = 21.7$
   $-\dfrac{1}{3.1} \cdot (-3.1y) = -\dfrac{1}{3.1} \cdot (21.7)$
   $y = -\dfrac{21.7}{3.1}$
   $y = -7$
   Check: $\dfrac{-3.1y = 21.7}{-3.1(-7)\,\bigg|\,21.7}$
   $21.7\,\bigg|$

32. −2

33. $38.7m = 309.6$
   $\dfrac{1}{38.7} \cdot (38.7m) = \dfrac{1}{38.7} \cdot (309.6)$
   $m = \dfrac{309.6}{38.7}$
   $m = 8$
   Check: $\dfrac{38.7m = 309.6}{38.7 \cdot 8\,\bigg|\,309.6}$
   $309.6\,\bigg|$

34. 8

35. $-\dfrac{2}{3}y = -10.6$
   $-\dfrac{3}{2} \cdot \left[-\dfrac{2}{3}y\right] = -\dfrac{3}{2} \cdot (-10.6)$
   $y = \dfrac{31.8}{2}$
   $y = 15.9$
   Check: $-\dfrac{2}{3}y = -10.6$
   $-\dfrac{2}{3}(15.9)\,\bigg|\,-10.6$
   $-\dfrac{31.8}{3}\,\bigg|$
   $-10.6\,\bigg|$

Chapter 3   (3.3)

36. -9.38

37. <u>Familiarize</u>. Let s = the speed of the first supersonic flight.
<u>Translate</u>. We reword the problem.

37 times what speed, is 24,790 mph?
↓   ↓   ↓   ↓   ↓
37 ·   s   =   24,790

<u>Carry out</u>. We solve the equation.
$$37s = 24,790$$
$$\frac{1}{37} \cdot 37s = \frac{1}{37} \cdot 24,790$$
$$s = \frac{24,790}{37}$$
$$s = 670$$

<u>Check</u>. Find 37 times 670.
$$37 \cdot 670 = 24,790$$

<u>State</u>. The speed of the first supersonic flight was 670 mph.

38. About 2956

39.
$$-0.2344m = 2028.732$$
$$-\frac{1}{0.2344} \cdot (-0.2344m) = -\frac{1}{0.2344} \cdot (2028.732)$$
$$m = -\frac{2028.732}{0.2344}$$
$$m = -8655$$

40. All real numbers

41. For all x, $0 \cdot x = 0$. There is no solution to $0 \cdot x = 9$.

42. -12, 12

43.
$$2|x| = -12$$
$$\frac{1}{2} \cdot 2|x| = \frac{1}{2} \cdot (-12)$$
$$|x| = -6$$

Absolute value cannot be negative. The equation has no solution.

44. 5

45.
$$3x = \frac{b}{a}$$
$$\frac{1}{3} \cdot 3x = \frac{1}{3} \cdot \frac{b}{a}$$
$$x = \frac{b}{3a}$$

46. $\frac{a^2 + 1}{c}$

47.
$$\frac{a}{b}x = 4$$
$$\frac{b}{a} \cdot \frac{a}{b}x = \frac{b}{a} \cdot 4$$
$$x = \frac{4b}{a}$$

48. Yes; if a = b, then $a^2 = b^2$

49. To "undo" the last step, divide 22.5 by 0.3.
$$22.5 \div 0.3 = 75$$
Now divide 75 by 0.3.
$$75 \div 0.3 = 250$$
The answer should be 250 not 22.5.

50. We divide by multiplying by a reciprocal.

Exercise Set 3.3

1.
$$5x + 6 = 31$$
$$5x + 6 + (-6) = 31 + (-6)$$
$$5x = 25$$
$$\frac{1}{5} \cdot 5x = \frac{1}{5} \cdot 25$$
$$x = 5$$

Check: $5x + 6 = 31$
$5 \cdot 5 + 6$ | 31
$25 + 6$
31

2. 8

3.
$$8x + 4 = 68$$
$$8x + 4 + (-4) = 68 + (-4)$$
$$8x = 64$$
$$\frac{1}{8} \cdot 8x = \frac{1}{8} \cdot 64$$
$$x = 8$$

Check: $8x + 4 = 68$
$8 \cdot 8 + 4$ | 68
$64 + 4$
68

4. 9

5.
$$4x - 6 = 34$$
$$4x - 6 + 6 = 34 + 6$$
$$4x = 40$$
$$\frac{1}{4} \cdot 4x = \frac{1}{4} \cdot 40$$
$$x = 10$$

Check: $4x - 6 = 34$
$4 \cdot 10 - 6$ | 34
$40 - 6$
34

6. 3

7.
$$3x - 9 = 33$$
$$3x - 9 + 9 = 33 + 9$$
$$3x = 42$$
$$\frac{1}{3} \cdot 3x = \frac{1}{3} \cdot 42$$
$$x = 14$$

Check: $3x - 9 = 33$
$3 \cdot 14 - 9$ | 33
$42 - 9$
33

8. 11

9.
$$7x + 2 = -54$$
$$7x + 2 + (-2) = -54 + (-2)$$
$$7x = -56$$
$$\frac{1}{7} \cdot 7x = \frac{1}{7} \cdot (-56)$$
$$x = -8$$

Check: $7x + 2 = -54$
$7(-8) + 2$ | -54
$-56 + 2$
-54

Chapter 3   (3.3)

10. -9

11. $\quad 6y + 3 = -45$ Check: $\dfrac{6y + 3 = -45}{6(-8) + 3 \mid -45}$
   $6y + 3 + (-3) = -45 + (-3)$
   $\qquad\qquad 6y = -48$ $-48 + 3$
   $\quad \dfrac{1}{6} \cdot 6y = \dfrac{1}{6} \cdot (-48)$ -45
   $\qquad\qquad\; y = -8$

12. -11

13. $\quad -4x + 7 = 35$ Check: $\dfrac{-4x + 7 = 35}{-4(-7) + 7 \mid 35}$
   $-4x + 7 + (-7) = 35 + (-7)$
   $\qquad\qquad\; -4x = 28$ $28 + 7$
   $\quad -\dfrac{1}{4} \cdot (4x) = -\dfrac{1}{4} \cdot 28$ 35
   $\qquad\qquad\qquad x = -7$

14. -23

15. $\quad -7x - 24 = -129$
   $-7x - 24 + 24 = -129 + 24$
   $\qquad\qquad\; -7x = -105$
   $\quad -\dfrac{1}{7} \cdot (-7x) = -\dfrac{1}{7} \cdot (-105)$
   $\qquad\qquad\qquad x = 15$

   Check: $\dfrac{-7x - 24 = -129}{-7 \cdot 15 - 24 \mid -129}$
   $\qquad\qquad\;\; -105 - 24$
   $\qquad\qquad\qquad\; -129$

16. 19

17. $\quad -4x + 71 = -1$
   $-4x + 71 + (-71) = -1 + (-71)$
   $\qquad\qquad\; -4x = -72$
   $\quad -\dfrac{1}{4} \cdot (-4x) = -\dfrac{1}{4} \cdot (-72)$
   $\qquad\qquad\qquad x = 18$

   Check: $\dfrac{-4x + 71 = -1}{-4 \cdot 18 + 71 \mid -1}$
   $\qquad\qquad -72 + 71$
   $\qquad\qquad\qquad -1$

18. 21

19. $\quad 5x + 7x = 72$ Check: $\dfrac{5x + 7x = 72}{5 \cdot 6 + 7 \cdot 6 \mid 72}$
   $\qquad\;\; 12x = 72$ $30 + 42$
   $\dfrac{1}{12} \cdot 12x = \dfrac{1}{12} \cdot 72$ 72
   $\qquad\quad\;\; x = 6$

20. 5

21. $\quad 8x + 7x = 60$ Check: $\dfrac{8x + 7x = 60}{8 \cdot 4 + 7 \cdot 4 \mid 60}$
   $\qquad\;\; 15x = 60$ $32 + 28$
   $\dfrac{1}{15} \cdot 15x = \dfrac{1}{15} \cdot 60$ 60
   $\qquad\quad\;\; x = 4$

22. 8

23. $\quad 4x + 3x = 42$ Check: $\dfrac{4x + 3x = 42}{4 \cdot 6 + 3 \cdot 6 \mid 42}$
   $\qquad\;\; 7x = 42$ $24 + 18$
   $\dfrac{1}{7} \cdot 7x = \dfrac{1}{7} \cdot 42$ 42
   $\qquad\quad x = 6$

24. 4

25. $\quad 4y - 2y = 10$ Check: $\dfrac{4y - 2y = 10}{4 \cdot 5 - 2 \cdot 5 \mid 10}$
   $\qquad\;\; 2y = 10$ $20 - 10$
   $\dfrac{1}{2} \cdot 2y = \dfrac{1}{2} \cdot 10$ 10
   $\qquad\quad y = 5$

26. 16

27. $\quad -6y - 3y = 27$ Check: $\dfrac{-6y - 3y = 27}{-6(-3) - 3(-3) \mid 27}$
   $\qquad\;\; -9y = 27$ $18 + 9$
   $-\dfrac{1}{9} \cdot (-9y) = -\dfrac{1}{9} \cdot 27$ 27
   $\qquad\quad\;\; y = -3$

28. -4

29. $\quad -7y - 8y = -15$ Check: $\dfrac{-7y - 8y = -15}{-7 \cdot 1 - 8 \cdot 1 \mid -15}$
   $\qquad\;\; -15y = -15$ $-7 - 8$
   $-\dfrac{1}{15} \cdot (-15y) = -\dfrac{1}{15} \cdot (-15)$ -15
   $\qquad\quad\;\; y = 1$

30. 3

31. $\quad 10.2y - 7.3y = -58$
   $\qquad\qquad\; 2.9y = -58$
   $\quad \dfrac{1}{2.9} \cdot 2.9y = \dfrac{1}{2.9} \cdot (-58)$
   $\qquad\qquad\qquad y = -\dfrac{58}{2.9}$
   $\qquad\qquad\qquad y = -20$

   Check: $\dfrac{10.2y - 7.3y = -58}{10.2(-20) - 7.3(-20) \mid -58}$
   $\qquad\qquad\qquad -204 + 146$
   $\qquad\qquad\qquad\qquad -58$

32. -20

Chapter 3   (3.3)

33. $x + \frac{1}{3}x = 8$

$\left(1 + \frac{1}{3}\right)x = 8$

$\frac{4}{3}x = 8$

$\frac{3}{4} \cdot \frac{4}{3}x = \frac{3}{4} \cdot 8$

$x = 6$

Check: $\begin{array}{c|c} x + \frac{1}{3}x = 8 \\ \hline 6 + \frac{1}{3} \cdot 6 & 8 \\ 6 + 2 & \\ 8 & \end{array}$

34. 8

35. $8y - 35 = 3y$

$8y = 3y + 35$

$8y - 3y = 35$

$5y = 35$

$y = \frac{35}{5}$

$y = 7$

Check: $\begin{array}{c|c} 8y - 35 = 3y \\ \hline 8 \cdot 7 - 35 & 3 \cdot 7 \\ 56 - 35 & 21 \\ 21 & \end{array}$

36. −3

37. $4x - 7 = 3x$

$4x = 3x + 7$

$4x - 3x = 7$

$x = 7$

Check: $\begin{array}{c|c} 4x - 7 = 3x \\ \hline 4 \cdot 7 - 7 & 3 \cdot 7 \\ 28 - 7 & 21 \\ 21 & \end{array}$

38. 1

39. $8x - 1 = 23 - 4x$

$8x + 4x = 23 + 1$

$12x = 24$

$x = \frac{24}{12}$

$x = 2$

Check: $\begin{array}{c|c} 8x - 1 = 23 - 4x \\ \hline 8 \cdot 2 - 1 & 23 - 4 \cdot 2 \\ 16 - 1 & 23 - 8 \\ 15 & 15 \end{array}$

40. 5

41. $2x - 1 = 4 + x$

$2x - x = 4 + 1$

$x = 5$

Check: $\begin{array}{c|c} 2x - 1 = 4 + x \\ \hline 2 \cdot 5 - 1 & 4 + 5 \\ 10 - 1 & 9 \\ 9 & \end{array}$

42. 2

43. $6x + 3 = 2x + 11$

$6x - 2x = 11 - 3$

$4x = 8$

$x = \frac{8}{4}$

$x = 2$

Check: $\begin{array}{c|c} 6x + 3 = 2x + 11 \\ \hline 6 \cdot 2 + 3 & 2 \cdot 2 + 11 \\ 12 + 3 & 4 + 11 \\ 15 & 15 \end{array}$

44. 4

45. $5 - 2x = 3x - 7x + 25$

$5 - 2x = -4x + 25$

$4x - 2x = 25 - 5$

$2x = 20$

$x = \frac{20}{2}$

$x = 10$

Check: $\begin{array}{c|c} 5 - 2x = 3x - 7x + 25 \\ \hline 5 - 2 \cdot 10 & 3 \cdot 10 - 7 \cdot 10 + 25 \\ 5 - 20 & 30 - 70 + 25 \\ -15 & -40 + 25 \\ & -15 \end{array}$

46. 10

47. $4 + 3x - 6 = 3x + 2 - x$

$3x - 2 = 2x + 2$ Collecting like terms on each side

$3x - 2x = 2 + 2$

$x = 4$

Check: $\begin{array}{c|c} 4 + 3x - 6 = 3x + 2 - x \\ \hline 4 + 3 \cdot 4 - 6 & 3 \cdot 4 + 2 - 4 \\ 4 + 12 - 6 & 12 + 2 - 4 \\ 16 - 6 & 14 - 4 \\ 10 & 10 \end{array}$

48. 0

49. $4y - 4 + y + 24 = 6y + 20 - 4y$

$5y + 20 = 2y + 20$

$5y - 2y = 20 - 20$

$3y = 0$

$y = 0$

Check: $\begin{array}{c|c} 4y - 4 + y + 24 = 6y + 20 - 4y \\ \hline 4 \cdot 0 - 4 + 0 + 24 & 6 \cdot 0 + 20 - 4 \cdot 0 \\ 0 - 4 + 0 + 24 & 0 + 20 - 0 \\ 20 & 20 \end{array}$

50. 7

Chapter 3   (3.3)

51. $\frac{7}{2}x + \frac{1}{2}x = 3x + \frac{3}{2} + \frac{5}{2}x$, LCM is 2

$2\left(\frac{7}{2}x + \frac{1}{2}x\right) = 2\left(3x + \frac{3}{2} + \frac{5}{2}x\right)$

$2 \cdot \frac{7}{2}x + 2 \cdot \frac{1}{2}x = 2 \cdot 3x + 2 \cdot \frac{3}{2} + 2 \cdot \frac{5}{2}x$

$7x + x = 6x + 3 + 5x$

$8x = 11x + 3$

$8x - 11x = 3$

$-3x = 3$

$x = \frac{3}{-3}$

$x = -1$

Check: $\frac{7}{2}x + \frac{1}{2}x = 3x + \frac{3}{2} + \frac{5}{2}x$

$\frac{7}{2}(-1) + \frac{1}{2}(-1)$	$3(-1) + \frac{3}{2} + \frac{5}{2}(-1)$
$-\frac{7}{2} - \frac{1}{2}$	$-3 + \frac{3}{2} - \frac{5}{2}$
$-\frac{8}{2}$	$-\frac{3}{2} - \frac{5}{2}$
-4	$-\frac{8}{2}$
	-4

52. $\frac{1}{2}$

53. $\frac{2}{3} + \frac{1}{4}t = \frac{1}{3}$, LCM is 12

$12\left(\frac{2}{3} + \frac{1}{4}t\right) = 12 \cdot \frac{1}{3}$

$12 \cdot \frac{2}{3} + 12 \cdot \frac{1}{4}t = 12 \cdot \frac{1}{3}$

$8 + 3t = 4$

$3t = 4 - 8$

$3t = -4$

$t = \frac{-4}{3}$

$t = -\frac{4}{3}$

Check: $\frac{2}{3} + \frac{1}{4}t = \frac{1}{3}$

$\frac{2}{3} + \frac{1}{4}\left(-\frac{4}{3}\right)$	$\frac{1}{3}$
$\frac{2}{3} - \frac{1}{3}$	
$\frac{1}{3}$	

54. $-\frac{2}{3}$

55. $\frac{2}{3} + 3y = 5y - \frac{2}{15}$, LCM is 15

$15\left(\frac{2}{3} + 3y\right) = 15\left(5y - \frac{2}{15}\right)$

$15 \cdot \frac{2}{3} + 15 \cdot 3y = 15 \cdot 5y - 15 \cdot \frac{2}{15}$

$10 + 45y = 75y - 2$

$10 + 2 = 75y - 45y$

$12 = 30y$

$\frac{12}{30} = y$

$\frac{2}{5} = y$

Check: $\frac{2}{3} + 3y = 5y - \frac{2}{15}$

$\frac{2}{3} + 3 \cdot \frac{2}{5}$	$5 \cdot \frac{2}{5} - \frac{2}{15}$
$\frac{2}{3} + \frac{6}{5}$	$2 - \frac{2}{15}$
$\frac{10}{15} + \frac{18}{15}$	$\frac{30}{15} - \frac{2}{15}$
$\frac{28}{15}$	$\frac{28}{15}$

56. -3

57. $\frac{5}{3} + \frac{2}{3}x = \frac{25}{12} + \frac{5}{4}x + \frac{3}{4}$, LCM is 12

$12\left(\frac{5}{3} + \frac{2}{3}x\right) = 12\left(\frac{25}{12} + \frac{5}{4}x + \frac{3}{4}\right)$

$12 \cdot \frac{5}{3} + 12 \cdot \frac{2}{3}x = 12 \cdot \frac{25}{12} + 12 \cdot \frac{5}{4}x + 12 \cdot \frac{3}{4}$

$20 + 8x = 25 + 15x + 9$

$20 + 8x = 15x + 34$

$20 - 34 = 15x - 8x$

$-14 = 7x$

$\frac{-14}{7} = 7x$

$-2 = x$

Check: $\frac{5}{3} + \frac{2}{3}x = \frac{25}{12} + \frac{5}{4}x + \frac{3}{4}$

$\frac{5}{3} + \frac{2}{3}(-2)$	$\frac{25}{12} + \frac{5}{4}(-2) + \frac{3}{4}$
$\frac{5}{3} - \frac{4}{3}$	$\frac{25}{12} - \frac{5}{2} + \frac{3}{4}$
$\frac{1}{3}$	$\frac{25}{12} - \frac{30}{12} + \frac{9}{12}$
	$\frac{4}{12}$
	$\frac{1}{3}$

58. -3

Chapter 3 (3.3)

59.
$$2.1x + 45.2 = 3.2 - 8.4x \quad \text{Greatest number of decimal places is 1}$$
$$10(2.1x + 45.2) = 10(3.2 - 8.4)x \quad \text{Multiplying by 10 to clear decimals}$$
$$10(2.1x) + 10(45.2) = 10(3.2) - 10(8.4x)$$
$$21x + 452 = 32 - 84x$$
$$21x + 84x = 32 - 452$$
$$105x = -420$$
$$x = \frac{-420}{105}$$
$$x = -4$$

Check:
$$\begin{array}{c|c} 2.1x + 45.2 = 3.2 - 8.4x \\ \hline 2.1(-4) + 45.2 & 3.2 - 8.4(-4) \\ -8.4 + 45.2 & 3.2 + 33.6 \\ 36.8 & 36.8 \end{array}$$

60. $\frac{5}{3}$

61.
$$1.03 - 0.62x = 0.71 - 0.22x$$
Greatest number of decimal places is 2
$$100(1.03 - 0.62x) = 100(0.71 - 0.22x)$$
Multiplying by 100 to clear decimals
$$100(1.03) - 100(0.62x) = 100(0.71) - 100(0.22x)$$
$$103 - 62x = 71 - 22x$$
$$32 = 40x$$
$$\frac{32}{40} = x$$
$$\frac{4}{5} = x, \text{ or}$$
$$0.8 = x$$

Check:
$$\begin{array}{c|c} 1.03 - 0.62x = 0.71 - 0.22x \\ \hline 1.03 - 0.62(0.8) & 0.71 - 0.22(0.8) \\ 1.03 - 0.496 & 0.71 - 0.176 \\ 0.534 & 0.534 \end{array}$$

62. 1

63.
$$0.42 - 0.03y = 3.33 - y$$
$$100(0.42 - 0.03y) = 100(3.33 - y)$$
$$100(0.42) - 100(0.03y) = 100(3.33) - 100 \cdot y$$
$$42 - 3y = 333 - 100y$$
$$100y - 3y = 333 - 42$$
$$97y = 291$$
$$y = \frac{291}{97}$$
$$y = 3$$

Check:
$$\begin{array}{c|c} 0.42 - 0.03y = 3.33 - y \\ \hline 0.42 - 0.03(3) & 3.33 - 3 \\ 0.42 - 0.09 & 0.33 \\ 0.33 & \end{array}$$

64. 70

65.
$$\frac{2}{7}x + \frac{1}{2}x = \frac{3}{4}x + 1, \quad \text{LCM is 28}$$
$$28\left[\frac{2}{7}x + \frac{1}{2}x\right] = 28\left[\frac{3}{4}x + 1\right]$$
$$28 \cdot \frac{2}{7}x + 28 \cdot \frac{1}{2}x = 28 \cdot \frac{3}{4}x + 28 \cdot 1$$
$$8x + 14x = 21x + 28$$
$$22x = 21x + 28$$
$$22x - 21x = 28$$
$$x = 28$$

Check:
$$\begin{array}{c|c} \frac{2}{7}x + \frac{1}{2}x = \frac{3}{4}x + 1 \\ \hline \frac{2}{7} \cdot 28 + \frac{1}{2} \cdot 28 & \frac{3}{4} \cdot 28 + 1 \\ 8 + 14 & 21 + 1 \\ 22 & 22 \end{array}$$

66. $\frac{32}{7}$

67.
$$\frac{4}{5}x - \frac{3}{4}x = \frac{3}{10}x - 1, \quad \text{LCM is 20}$$
$$20\left[\frac{4}{5}x - \frac{3}{4}x\right] = 20\left[\frac{3}{10}x - 1\right]$$
$$20 \cdot \frac{4}{5}x - 20 \cdot \frac{3}{4}x = 20 \cdot \frac{3}{10}x - 20 \cdot 1$$
$$16x - 15x = 6x - 20$$
$$x = 6x - 20$$
$$20 = 6x - x$$
$$20 = 5x$$
$$\frac{20}{5} = x$$
$$4 = x$$

Check:
$$\begin{array}{c|c} \frac{4}{5}x - \frac{3}{4}x = \frac{3}{10}x - 1 \\ \hline \frac{4}{5} \cdot 4 - \frac{3}{4} \cdot 4 & \frac{3}{10} \cdot 4 - 1 \\ \frac{16}{5} - 3 & \frac{6}{5} - 1 \\ \frac{16}{5} - \frac{15}{5} & \frac{6}{5} - \frac{5}{5} \\ \frac{1}{5} & \frac{1}{5} \end{array}$$

68. 23

69. $-\frac{2}{3} - \frac{5}{8} = -\frac{2}{3} + \left(-\frac{5}{8}\right) = -\frac{2}{3} \cdot \frac{8}{8} + \left(-\frac{5}{8}\right)\left(\frac{3}{3}\right) = -\frac{16}{24} + \left(-\frac{15}{24}\right) = -\frac{31}{24}$

70. $\frac{5}{6}$

71. If $x = -14$, then $-(-x) = -[-(-14)] = -14$. (The opposite of the opposite of -14 is -14.)

72. $9x - 35$

Chapter 3   (3.4)

73. Since we are using a calculator we will not clear the decimals.
$$0.008 + 9.62x - 42.8 = 0.944x + 0.0083 - x$$
$$9.62x - 42.792 = -0.056x + 0.0083$$
$$9.62x + 0.056x = 0.0083 + 42.792$$
$$9.676x = 42.8003$$
$$x = \frac{42.8003}{9.676}$$
$$x \approx 4.42$$

74. $x = 3, y = -5$

75. $9x + 2 = -1$ $4x - y = \frac{11}{3}$
$9x = -3$
$x = -\frac{3}{9}$ $4\left(-\frac{1}{3}\right) - y = \frac{11}{3}$
$x = -\frac{1}{3}$ $-\frac{4}{3} - y = \frac{11}{3}$
                $-y = \frac{15}{3}$
                $-y = 5$
                $y = -5$

76. $x = 0.13, y = -0.324$

77. $\frac{y - 2}{3} = \frac{2 - y}{5}$, LCM is 15
$$15\left[\frac{y-2}{3}\right] = 15\left[\frac{2-y}{5}\right]$$
$$5(y - 2) = 3(2 - y)$$
$$5y - 10 = 6 - 3y$$
$$5y + 3y = 6 + 10$$
$$8y = 16$$
$$y = 2$$

78. $\frac{a}{1 - 3a}$

79. $0 = y - (-14) - (-3y)$
$0 = y + 14 + 3y$
$0 = 4y + 14$
$-14 = 4y$
$\frac{-14}{4} = y$
$-\frac{7}{2} = y$

80. -2

81. $0.05y - 1.82 = 0.708y - 0.504$
$1000(0.05y - 1.82) = 1000(0.708y - 0.504)$
$50y - 1820 = 708y - 504$
$-1820 + 504 = 708y - 50y$
$-1316 = 658y$
$\frac{-1316}{658} = y$
$-2 = y$

82. 0

83. $-2y + 5y = 6y$
$3y = 6y$
$0 = 3y$
$0 = y$

84. $\frac{52}{45}$

85. $\frac{2x - 5}{6} + \frac{4 - 7x}{8} = \frac{10 + 6x}{3}$, LCM is 24
$$24\left[\frac{2x-5}{6} + \frac{4-7x}{8}\right] = 24\left[\frac{10+6x}{3}\right]$$
$$4(2x - 5) + 3(4 - 7x) = 8(10 + 6x)$$
$$8x - 20 + 12 - 21x = 80 + 48x$$
$$-13x - 8 = 80 + 48x$$
$$-13x - 48x = 80 + 8$$
$$-61x = 88$$
$$x = -\frac{88}{61}$$

86. $4x - 8 = 32, 4x = 40, x = 10$; $4x - 8 = 32,$
$x - 2 = 8, x = 0$

Exercise Set 3.4

1. $3(2y - 3) = 27$ Check: $\frac{3(2y - 3) = 27}{3(2 \cdot 6 - 3) \;|\; 27}$
$6y - 9 = 27$
$6y = 27 + 9$ $3(12 - 3)$
$6y = 36$ $3 \cdot 9$
$y = 6$ 27

2. 5

3. $40 = 5(3x + 2)$ Check: $\frac{40 = 5(3x + 2)}{40 \;|\; 5(3 \cdot 2 + 2)}$
$40 = 15x + 10$
$40 - 10 = 15x$ $5(6 + 2)$
$30 = 15x$ $5 \cdot 8$
$2 = x$ 40

4. 1

5. $2(3 + 4m) - 9 = 45$ Check: $\frac{2(3 + 4m) - 9 = 45}{2(3 + 4 \cdot 6) - 9 \;|\; 45}$
$6 + 8m - 9 = 45$
$8m - 3 = 45$ $2(3 + 24) - 9$
$8m = 45 + 3$ $2 \cdot 27 - 9$
$8m = 48$ $54 - 9$
$m = 6$ 45

6. 9

Chapter 3   (3.4)

7. $5r - (2r + 8) = 16$ Check: $\dfrac{5r - (2r + 8) = 16}{5 \cdot 8 - (2 \cdot 8 + 8) \mid 16}$
    $5r - 2r - 8 = 16$ $40 - (16 + 8)$
    $3r - 8 = 16$ $40 - 24$
    $3r = 16 + 8$ 16
    $3r = 24$
    $r = 8$

8. 8

9. $3g - 3 = 3(7 - g)$ Check: $\dfrac{3g - 3 = 3(7 - g)}{3 \cdot 4 - 3 \mid 3(7 - 4)}$
    $3g - 3 = 21 - 3g$ $12 - 3 \mid 3 \cdot 3$
    $3g + 3g = 21 + 3$ $9 \mid 9$
    $6g = 24$
    $g = 4$

10. 5

11. $6 - 2(3x - 1) = 2$ Check: $\dfrac{6 - 2(3x - 1) = 2}{6 - 2(3 \cdot 1 - 1) \mid 2}$
    $6 - 6x + 2 = 2$ $6 - 2(3 - 1)$
    $8 - 6x = 2$ $6 - 2 \cdot 2$
    $8 - 2 = 6x$ $6 - 4$
    $6 = 6x$ 2
    $1 = x$

12. 2

13. $5(d + 4) = 7(d - 2)$ Check: $\dfrac{5(d + 4) = 7(d - 2)}{5(17 + 4) \mid 7(17 - 2)}$
    $5d + 20 = 7d - 14$ $5 \cdot 21 \mid 7 \cdot 15$
    $20 + 14 = 7d - 5d$ $105 \mid 105$
    $34 = 2d$
    $17 = d$

14. -4

15. $3(x - 2) = 5(x + 2)$ Check: $\dfrac{3(x - 2) = 5(x + 2)}{3(-8 - 2) \mid 5(-8 + 2)}$
    $3x - 6 = 5x + 10$ $3(-10) \mid 5(-6)$
    $3x - 5x = 10 + 6$ $-30 \mid -30$
    $-2x = 16$
    $x = -8$

16. -13

17. $8(2t + 1) = 4(7t + 7)$
    $16t + 8 = 28t + 28$
    $16t - 28t = 28 - 8$
    $-12t = 20$
    $t = -\dfrac{20}{12}$
    $t = -\dfrac{5}{3}$

17. (continued)
    Check: $\dfrac{8(2t + 1) = 4(7t + 7)}{8\left[2\left(-\dfrac{5}{3}\right) + 1\right] \mid 4\left[7\left(-\dfrac{5}{3}\right) + 7\right]}$
    $8\left[-\dfrac{10}{3} + 1\right] \mid 4\left[-\dfrac{35}{3} + 7\right]$
    $8\left[-\dfrac{7}{3}\right] \mid 4\left[-\dfrac{14}{3}\right]$
    $-\dfrac{56}{3} \mid -\dfrac{56}{3}$

18. -8

19. $3(r - 6) + 2 = 4(r + 2) - 21$
    $3r - 18 + 2 = 4r + 8 - 21$
    $3r - 16 = 4r - 13$
    $13 - 16 = 4r - 3r$
    $-3 = r$

    Check: $\dfrac{3(r - 6) + 2 = 4(r + 2) - 21}{3(-3 - 6) + 2 \mid 4(-3 + 2) - 21}$
    $3(-9) + 2 \mid 4(-1) - 21$
    $-27 + 2 \mid -4 - 21$
    $-25 \mid -25$

20. -12

21. $19 - (2x + 3) = 2(x + 3) + x$
    $19 - 2x - 3 = 2x + 6 + x$
    $16 - 2x = 3x + 6$
    $16 - 6 = 3x + 2x$
    $10 = 5x$
    $2 = x$

    Check: $\dfrac{19 - (2x + 3) = 2(x + 3) + x}{19 - (2 \cdot 2 + 3) \mid 2(2 + 3) + 2}$
    $19 - (4 + 3) \mid 2 \cdot 5 + 2$
    $19 - 7 \mid 10 + 2$
    $12 \mid 12$

22. 1

Chapter 3   (3.4)

23. $\frac{1}{4}(8y + 4) - 17 = -\frac{1}{2}(4y - 8)$
    $2y + 1 - 17 = -2y + 4$
    $2y - 16 = -2y + 4$
    $2y + 2y = 4 + 16$
    $4y = 20$
    $y = 5$

    Check: $\frac{1}{4}(8y + 4) - 17 = -\frac{1}{2}(4y - 8)$

    | $\frac{1}{4}(8 \cdot 5 + 4) - 17$ | $-\frac{1}{2}(4 \cdot 5 - 8)$ |
    | $\frac{1}{4}(40 + 4) - 17$ | $-\frac{1}{2}(20 - 8)$ |
    | $\frac{1}{4} \cdot 44 - 17$ | $-\frac{1}{2}(12)$ |
    | $11 - 17$ | -6 |
    | -6 | |

24. 6

25. $2[4 - 2(3 - x)] - 1 = 4[2(4x - 3) + 7] - 25$
    $2[4 - 6 + 2x] - 1 = 4[8x - 6 + 7] - 25$
    $2[-2 + 2x] - 1 = 4[8x + 1] - 25$
    $-4 + 4x - 1 = 32x + 4 - 25$
    $4x - 5 = 32x - 21$
    $-5 + 21 = 32x - 4x$
    $16 = 28x$
    $\frac{16}{28} = x$
    $\frac{4}{7} = x$

    The check is left to the student.

26. $-\frac{27}{19}$

27. $\frac{2}{3}(2x - 1) = 10$
    $3 \cdot \frac{2}{3}(2x - 1) = 3 \cdot 10$ Multiplying by 3 to clear the fraction
    $2(2x - 1) = 30$
    $4x - 2 = 30$
    $4x = 30 + 2$
    $4x = 32$
    $x = 8$

    Check: $\frac{2}{3}(2x - 1) = 10$

    | $\frac{2}{3}(2 \cdot 8 - 1)$ | 10 |
    | $\frac{2}{3}(16 - 1)$ | |
    | $\frac{2}{3} \cdot 15$ | |
    | 10 | |

28. 7

29. $\frac{3}{4}(3 + 2x) + 1 = 13$
    $4\left[\frac{3}{4}(3 + 2x) + 1\right] = 4 \cdot 13$ Clearing the fraction
    $3(3 + 2x) + 4 = 52$
    $9 + 6x + 4 = 52$
    $6x + 13 = 52$
    $6x = 52 - 13$
    $6x = 39$
    $x = \frac{39}{6}$
    $x = \frac{13}{2}$

    Check: $\frac{3}{4}(3 + 2x) + 1 = 13$

    | $\frac{3}{4}\left(3 + 2 \cdot \frac{13}{2}\right) + 1$ | 13 |
    | $\frac{3}{4}(3 + 13) + 1$ | |
    | $\frac{3}{4} \cdot 16 + 1$ | |
    | $12 + 1$ | |
    | 13 | |

30. $-\frac{405}{28}$

31. $\frac{3}{4}\left(3x - \frac{1}{2}\right) - \frac{2}{3} = \frac{1}{3}$
    $\frac{9}{4}x - \frac{3}{8} - \frac{2}{3} = \frac{1}{3}$,   LCM is 24
    $24\left(\frac{9}{4}x - \frac{3}{8} - \frac{2}{3}\right) = 24 \cdot \frac{1}{3}$
    $24 \cdot \frac{9}{4}x - 24 \cdot \frac{3}{8} - 24 \cdot \frac{2}{3} = 8$
    $54x - 9 - 16 = 8$
    $54x - 25 = 8$
    $54x = 8 + 25$
    $54x = 33$
    $x = \frac{33}{54}$
    $x = \frac{11}{18}$

    The check is left to the student.

32. $-\frac{5}{32}$

33. $0.7(3x + 6) = 1.1 - (x + 2)$
    $2.1x + 4.2 = 1.1 - x - 2$
    $10(2.1x + 4.2) = 10(1.1 - x - 2)$ Clearing decimals
    $21x + 42 = 11 - 10x - 20$
    $21x + 42 = -10x - 9$
    $21x + 10x = -9 - 42$
    $31x = -51$
    $x = -\frac{51}{31}$

    The check is left to the student.

52

Chapter 3 (3.5)

34. $\dfrac{39}{14}$

35. $a + (a - 3) = (a + 2) - (a + 1)$
$a + a - 3 = a + 2 - a - 1$
$2a - 3 = 1$
$2a = 1 + 3$
$2a = 4$
$a = 2$

Check: $\dfrac{a + (a - 3) = (a + 2) - (a + 1)}{\begin{array}{c|c} 2 + (2 - 3) & (2 + 2) - (2 + 1) \\ 2 - 1 & 4 - 3 \\ 1 & 1 \end{array}}$

36. -7.4

37. Familiarize. Let b = the number of bottles produced.

Translate. We reword the problem.
59 is 1.3% of what?
59 = 1.3% · b

Carry out. We solve the equation.
$59 = 1.3\% \cdot b$
$59 = 1.3 \times 0.01 \times b$
$59 = 0.013b$
$\dfrac{59}{0.013} = b$
$4538 = b$ Rounding

Check. We find 1.3% of 4538:
$1.3\% \cdot 4538 = 0.013(4538) = 59$ (Rounding)

State. About 4538 bottles were produced.

38. -6.5

39. $7x - 21 - 14y = 7 \cdot x - 7 \cdot 3 - 7 \cdot 2y = 7(x - 3 - 2y)$

40. $-15 < -13$

41. $475(54x + 7856) + 9762 = 402(83x + 975)$
$25{,}650x + 3{,}731{,}600 + 9762 = 33{,}366x + 391{,}950$
$25{,}650x + 3{,}741{,}362 = 33{,}366x + 391{,}950$
$3{,}741{,}362 - 391{,}950 = 33{,}366x - 25{,}650x$
$3{,}349{,}412 = 7716x$
$\dfrac{3{,}349{,}412}{7716} = x$
$\dfrac{837{,}353}{1929} = x$

42. -4

43. $x(x - 4) = 3x(x + 1) - 2(x^2 + x - 5)$
$x^2 - 4x = 3x^2 + 3x - 2x^2 - 2x + 10$
$x^2 - 4x = x^2 + x + 10$
$-x^2 + x^2 - 4x = -x^2 + x^2 + x + 10$
$-4x = x + 10$
$-4x - x = 10$
$-5x = 10$
$x = -2$

44. All real numbers

45. $4(x - a) = 16$
$4x - 4a = 16$
$4x = 16 + 4a$
$x = \dfrac{16 + 4a}{4} = \dfrac{4(4 + a)}{4}$
$x = 4 + a$

46. About -0.000036364

Exercise Set 3.5

1. Let x = the number. Then "twice the number" translates to 2x, and "three less than 2x" translates to 2x - 3.

2. $\dfrac{x}{8} - 5$

3. Let y = the number. Then "the difference of a number and 1" translates to y - 1, and "one-half of y - 1" translates to $\dfrac{1}{2}(y - 1)$.

4. $10n - 2$

5. Let a = the number. Then "the sum of 3 and some number" translates to 3 + a, and "5 times 3 + a" translates to 5(3 + a).

6. $6(x + y)$

7. The longer piece is 2 ft longer than L, or L + 2.

8. $x \div 3$, or $\dfrac{x}{3}$

9. The amount of the reduction is 30%b. Then the sale price is b - 30%b, or b - 0.3b.

10. $p - 20\%p$, or $p - 0.2p$

11. Each even integer is 2 greater than the one preceding it. If we let x = the first of the even integers, then x + 2 = the second and x + 4 = the third. We can express their sum as x + (x + 2) + (x + 4).

12. $x + (x + 1) + (x + 2)$

53

Chapter 3 (3.5)

13. The cost is
    (Initial charge) plus (mileage charge), or
    $34.95 plus Cost per mile times Number of miles driven
    $34.95 + $0.27 · m
    which is $34.95 + $0.27m.

14. $\frac{t}{2} + 2$

15. a) Twice the width is 2w.
    b) The width is one-half the length, or $\frac{1}{2}\ell$.

16. 3(h + 5)

17. Second angle: Three times the first is 3x
    Third angle: 30° more than the first is x + 30

18. Second: 4x; third: x + 4x - 45, or 5x - 45

19. Familiarize. Let n = the number.
    Translate. We reword the problem.
    Six times a number less 18 is 96.
    6 · n - 18 = 96
    Carry out. We solve the equation.
        6n - 18 = 96
        6n = 114   Adding 18
        n = 19   Multiplying by $\frac{1}{6}$
    Check. Six times 19 is 114. Subtracting 18 from 114, we get 96. This checks.
    State. The number is 19.

20. 52

21. Familiarize. Let y = the number.
    Translate. We reword the problem.
    Two times a number plus 16 is $\frac{2}{5}$ of the number.
    2 · y + 16 = $\frac{2}{5}$ · y
    Carry out. We solve the equation.
        2y + 16 = $\frac{2}{5}$y
        5(2y + 16) = 5 · $\frac{2}{5}$y   Clearing the fraction
        10y + 80 = 2y
        80 = -8y   Adding -10y
        -10 = y   Dividing by -8
    Check. We double -10 and get -20. Adding 16, we get -4. Also, $\frac{2}{5}$(-10) = -4. The answer checks.
    State. The number is -10.

22. -68

23. Familiarize. Let x = the number.
    Translate. We reword the problem.
    A number plus two-fifths of the number is 56.
    x + $\frac{2}{5}$ · x = 56
    Carry out. We solve the equation.
        x + $\frac{2}{5}$x = 56
        $\frac{7}{5}$x = 56   Collecting like terms
        x = $\frac{5}{7}$ · 56 Multiplying by $\frac{5}{7}$
        x = 40
    Check. $\frac{2}{5}$ · 40 = 16, and 40 + 16 = 56. The answer checks.
    State. The number is 40.

24. 36

25. Familiarize. First draw a picture.

    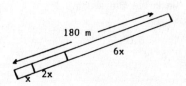

    We use x for the first length, 2x for the second length, and 3·2x, or 6x, for the third length.

    Translate. The lengths of the three pieces add up to 180 m. This gives us the equation.

    Length of 1st piece plus Length of 2nd piece plus Length of 3rd piece is 180
    x + 2x + 6x = 180

    Carry out. We solve the equation.
        x + 2x + 6x = 180
        9x = 180
        x = 20
    Check. If the first piece is 20 m long, then the second is 2·20 m, or 40 m and the third is 6·20 m, or 120 m. The lengths of these pieces add up to 180 m (20 + 40 + 120 = 180). This checks.
    State. The first piece measures 20 m. The second measures 40 m, and the third measures 120 m.

26. 30 m, 90 m, 360 m

27. <u>Familiarize</u>. The page numbers are consecutive integers. (See Example 10.) If we let p = the smaller number, then p + 1 = the larger number.

<u>Translate</u>. We reword the problem.

<u>First integer</u> + <u>Second integer</u> = 73
    x   +   (x + 1)   = 73

<u>Carry out</u>. We solve the equation.

   x + (x + 1) = 73
      2x + 1 = 73   Collecting like terms
        2x = 72   Adding -1
         x = 36   Dividing by 2

<u>Check</u>. If x = 36, then x + 1 = 37. These are consecutive integers, and 36 + 37 = 73. The answer checks.

<u>State</u>. The page numbers are 36 and 37.

28. 40, 41

29. <u>Familiarize</u>. We let x = the smaller even integer. Then x + 2 = the next even integer.

<u>Translate</u>.

<u>Smaller even integer</u> plus <u>next even integer</u> is 114.
     x   +   (x + 2)   = 114

<u>Carry out</u>. We solve the equation.

   x + (x + 2) = 114
      2x + 2 = 114
        2x = 112
         x = 56

<u>Check</u>. If the smaller even integer is 56, then the larger is 56 + 2, or 58. They are consecutive even integers. Their sum, 56 + 58, is 114. This checks.

<u>State</u>. The integers are 56 and 58.

30. 52, 54

31. <u>Familiarize</u>. Let x = the first integer. Then x + 1 = the second integer, and x + 2 = the third.

<u>Translate</u>.

First integer plus Second integer plus Third integer is 108.
   x   +   (x + 1)   +   (x + 2)   = 108

<u>Carry out</u>. We solve the equation.

   x + (x + 1) + (x + 2) = 108
        3x + 3 = 108
          3x = 105
           x = 35

<u>Check</u>. If the first integer is 35, then the second is 35 + 1, or 36, and the third is 35 + 2, or 37. They are consecutive integers. Their sum, 35 + 36 + 37, is 108. This checks.

<u>State</u>. The integers are 35, 36, and 37.

32. 41, 42, 43

33. <u>Familiarize</u>. Let x = the first odd integer. Then x + 2 = the second odd integer, and x + 4 = the third.

<u>Translate</u>.

First integer plus Second integer plus Third integer is 189.
   x   +   (x + 2)   +   (x + 4)   = 189

<u>Carry out</u>. We solve the equation.

   x + (x + 2) + (x + 4) = 189
        3x + 6 = 189
          3x = 183
           x = 61

<u>Check</u>. If the first odd integer is 61, then the second odd integer is 61 + 2, or 63, and the third is 61 + 4, or 65. They are consecutive odd integers. Their sum, 61 + 63 + 65, is 189. This checks.

<u>State</u>. The integers are 61, 63, and 65.

34. 274, 276, 278

35. <u>Familiarize</u>. We draw a picture. Let w = the width of the rectangle. Then w + 60 = the length.

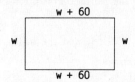

The perimeter of a rectangle is the sum of the lengths of the sides. The area is the product of the length and the width.

<u>Translate</u>. We use the definition of perimeter to write an equation that will allow us to find the width and length.

Width + Width + Length + Length = Perimeter
  w + w + (w + 60) + (w + 60) = 520

To find the area we will compute the product of the length and width, or (w + 60)w.

<u>Carry out</u>. We solve the equation.

  w + w + (w + 60) + (w + 60) = 520
          4w + 120 = 520
             4w = 400
              w = 100

If w = 100, then w + 60 = 100 + 60 = 160, and the area is 160(100) = 16,000.

<u>Check</u>. The length is 60 ft more than the width. The perimeter is 100 + 100 + 160 + 160 = 520 ft. This checks. To check the area we recheck the computation. This also checks.

<u>State</u>. The width is 100 ft, the length is 160 ft, and the area is 16,000 ft².

36. ℓ = 265 ft, w = 165 ft; 43,725 ft²

Chapter 3   (3.5)

37. <u>Familiarize</u>. We draw a picture. Let ℓ = the length of the paper. Then ℓ - 6.3 = the width.

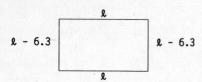

The perimeter is the sum of the lengths of the sides.

<u>Translate</u>. We use the definition of perimeter to write an equation.

Width + Width + Length + Length is 99.
(ℓ - 6.3) + (ℓ - 6.3) + ℓ + ℓ = 99

<u>Carry out</u>. We solve the equation.
(ℓ - 6.3) + (ℓ - 6.3) + ℓ + ℓ = 99
4ℓ - 12.6 = 99
4ℓ = 111.6
ℓ = 27.9

Then ℓ - 6.3 = 21.6.

<u>Check</u>. The width, 21.6 cm, is 6.3 cm less than the length, 27.9 cm. The perimeter is 21.6 cm + 21.6 cm + 27.9 cm + 27.9 cm, or 99 cm. This checks.

<u>State</u>. The length is 27.9 cm, and the width is 21.6 cm.

38. ℓ = 365 mi, w = 275 mi

39. <u>Familiarize</u>. We draw a picture. We let x = the measure of the first angle. Then 4x = the measure of the second angle, and (x + 4x) - 45, or 5x - 45 = the measure of the third angle.

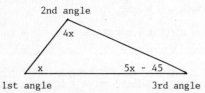

Recall that the measures of the angles of any triangle add up to 180°.

<u>Translate</u>.
Measure of + Measure of + Measure of = 180.
1st angle   2nd angle   3rd angle
  x   +   4x   +   (5x - 45)   = 180

<u>Carry out</u>. We solve the equation.
x + 4x + (5x - 45) = 180
10x - 45 = 180
10x = 225
x = 22.5

Possible answers for the angle measures are as follows:
1st angle: x = 22.5°
2nd angle: 4x = 4(22.5) = 90°
3rd angle: 5x - 45 = 5(22.5) - 45 = 112.5 - 45
                    = 67.5°

39. (continued)

<u>Check</u>. Consider 22.5°, 90°, and 67.5°. The second is four times the first, and the third is 45° less than five times the first. The sum is 180°. These numbers check.

<u>State</u>. The measure of the first angle is 22.5°.

40. 25.625°

41. <u>Familiarize</u>. Let x = the original price. Then, 40%x = the reduction. The sale price is found by subtracting the amount of reduction from the original price.

<u>Translate</u>.
Original price, minus reduction is $9.60.
       x   -   40%x   = 9.60

<u>Carry out</u>. We solve the equation.
x - 40%x = 9.60
1·x - 0.40x = 9.60
(1 - 0.40)x = 9.60
0.6x = 9.60
x = $\frac{9.60}{0.6}$
x = 16

<u>Check</u>. 40% of $16 is $6.40. Subtracting this from $16 we get $9.60. This checks.

<u>State</u>. The original price was $16.

42. $14

43. <u>Familiarize</u>. Let x = the original investment. Interest earned in 1 year is found by taking 12% of the original investment. Then 12%x = the interest. The amount in the account at the end of the year is the sum of the original investment and the interest earned.

<u>Translate</u>.
Original investment, plus interest earned, is $4928.
       x   +   12%x   = 4928

<u>Carry out</u>. We solve the equation.
x + 12%x = 4928
1·x + 0.12x = 4928
1.12x = 4928
x = $\frac{4928}{1.12}$
x = 4400

<u>Check</u>. 12% of $4400 is $528. Adding this to $4400 we get $4928. This checks.

<u>State</u>. The original investment was $4400.

44. $6540

56

Chapter 3 (3.5)

45. Familiarize. The total cost is the daily charge plus the mileage charge. The mileage charge is the cost per mile times the number of miles driven. Let m = the number of miles that can be driven for $80.

Translate.

Daily rate	plus	Cost per mile	times	Number of miles driven	is	Amount
34.95	+	0.10	·	m	=	80

Carry out. We solve the equation.

$$34.95 + 0.10m = 80$$
$$100(34.95 + 0.10m) = 100(80) \quad \text{Clearing the decimals}$$
$$3495 + 10m = 8000$$
$$10m = 4505$$
$$m = 450.5$$

Check. The mileage cost is found by multiplying 450.5 by $0.10 obtaining $45.05. Then we add $45.05 to $34.95, the daily rate, and get $80.

State. The businessperson can drive 450.5 mi on the car-rental allotment.

46. 460.5 mi

47. Familiarize. We draw a picture. Let x = the measure of the first angle. Then 3x = the measure of the second angle, and x + 40 = the measure of the third angle.

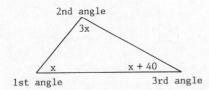

Recall that the sum of the measures of the angles of a triangle is 180°.

Translate.

Measure of 1st angle + Measure of 2nd angle + Measure of 3rd angle = 180.

$$x + 3x + (x + 40) = 180$$

Carry out.

$$x + 3x + (x + 40) = 180$$
$$5x + 40 = 180$$
$$5x = 140$$
$$x = 28$$

Possible answers for the angle measures are as follows:

1st angle: x = 28°
2nd angle: 3x = 3(28) = 84°
3rd angle: x + 40 = 28 + 40 = 68°

Check. Consider 28°, 84°, and 68°. The second angle is three times the first, and the third is 40° more than the first. The sum, 28° + 84° + 68°, is 180°. These numbers check.

State. The measures of the angles are 28°, 84°, and 68°.

48. 5°, 160°, 15°

49. Familiarize. We will use the equation R = -0.028t + 20.8 where R is in seconds and t is the number of years since 1920. We want to find t when R = 18.0 sec.

Translate.

Record	is	18.0 sec.
-0.028t + 20.8	=	18.0

Carry out.

$$-0.028t + 20.8 = 18.0$$
$$1000(-0.028t + 20.8) = 1000(18.0) \quad \text{Clearing the decimals}$$
$$-28t + 20,800 = 18,000$$
$$-28t = -2800$$
$$t = 100$$

Check. Substitute 100 for t in the given equation:

$$R = -0.028(100) + 20.8 = -2.8 + 20.8 = 18.0$$

This checks.

State. The record will be 18.0 sec 100 years after 1920, or in 2020.

50. 160

51. Familiarize. Let s = one score. Then four score = 4s and four score and seven = 4s + 7.

Translate. We reword.

1776	plus	four score and seven	is	1863.
1776	+	(4s + 7)	=	1863

Carry out. We solve the equation.

$$1776 + (4s + 7) = 1863$$
$$4s + 1783 = 1863$$
$$4s = 80$$
$$s = 20$$

Check. If a score is 20 years, then four score and seven represents 87 years. Adding 87 to 1776 we get 1863. This checks.

State. A score is 20.

52. 16, 4

Chapter 3   (3.5)

53. <u>Familiarize</u>. The cost of the rental is the daily charge plus the mileage charge. Let c = the cost per mile that will make the total cost equal to the budget amount. Then c is the highest price per mile the person can afford.

<u>Translate</u>.

Daily rate	plus	Cost per mile	times	Number of miles driven	is	Budget amount.
18.90	+	c	·	190	=	55

<u>Carry out</u>.

$$18.90 + 190c = 55$$
$$10(18.90 + 190c) = 10(55) \quad \text{Clearing the decimal}$$
$$189 + 1900c = 550$$
$$1900c = 361$$
$$c = 0.19$$

<u>Check</u>. The mileage cost is found by multiplying 190 by $0.19 obtaining $36.10. Adding $36.10 to $18.90, the daily rate, we get $55.

<u>State</u>. The cost per mile cannot exceed $0.19 to stay within a $55 budget.

54. 19

55. <u>Familiarize</u>. We let x = the length of the original rectangle. Then $\frac{3}{4}$x = the width. We draw a picture of the enlarged rectangle. Each dimension is increased by 2 cm, so x + 2 = the length of the enlarged rectangle and $\frac{3}{4}$x + 2 = the width.

$\frac{3}{4}$x + 2   [x + 2 / x + 2] $\frac{3}{4}$x + 2

<u>Translate</u>. We use the perimeter of the enlarged rectangle to write an equation.

Width + Width + Length + Length is Perimeter.
$\left(\frac{3}{4}x + 2\right) + \left(\frac{3}{4}x + 2\right) + (x + 2) + (x + 2) = 50$

<u>Carry out</u>.

$$\left(\frac{3}{4}x + 2\right) + \left(\frac{3}{4}x + 2\right) + (x + 2) + (x + 2) = 50$$
$$\frac{7}{2}x + 8 = 50$$
$$2\left(\frac{7}{2}x + 8\right) = 2 \cdot 50$$
$$7x + 16 = 100$$
$$7x = 84$$
$$x = 12$$

Then $\frac{3}{4}x = \frac{3}{4}(12) = 9$.

55. (continued)

<u>Check</u>. If the dimensions of the original rectangle are 12 cm and 9 cm, then the dimensions of the enlarged rectangle are 14 cm and 11 cm. The perimeter of the enlarged rectangle is 11 + 11 + 14 + 14 = 50 cm. Also, 9 is $\frac{3}{4}$ of 12. These values check.

<u>State</u>. The length is 12 cm, and the width is 9 cm.

56. 120

57. <u>Familiarize</u>. Let x = the number of additional games the Falcons will have to play. Then $\frac{x}{2}$ = the number of those games they will win, 15 + $\frac{x}{2}$ = the total number of games won, and 20 + x = the total number of games played.

<u>Translate</u>.

The number of games won	is 60% of	the number of games played.
$\left(15 + \frac{x}{2}\right)$	= 60% ·	(20 + x)

<u>Carry out</u>.

$$15 + \frac{x}{2} = 60\%(20 + x)$$
$$15 + \frac{x}{2} = 0.6(20 + x)$$
$$15 + 0.5x = 12 + 0.6x \quad \text{Expressing } \frac{1}{2} \text{ as } 0.5$$
$$10(15 + 0.5x) = 10(12 + 0.6x) \quad \text{Clearing decimals}$$
$$150 + 5x = 120 + 6x$$
$$30 = x$$

<u>Check</u>. If 30 more games are played of which $15\left(\frac{30}{2} = 15\right)$ are won, then the total games played will be 20 + 30, or 50, and the total games won will be 15 + 15, or 30. 30 is 60% of 50. The numbers check.

<u>State</u>. The Falcons must play 30 more games.

58. $9.17, not $9.10

59. <u>Familiarize</u>. The interest paid is $2150 - $2000, or $150. Let p = the percent being charged.

<u>Translate</u>. We reword.

$150	is	what percent	of	$2000?
150	=	p %	·	2000

<u>Carry out</u>.

$$150 = p\% \cdot 2000$$
$$150 = p \times 0.01 \times 2000$$
$$150 = p(20)$$
$$7.5 = p$$

Chapter 3   (3.6)

59. (continued)

   Check. 7.5% of $2000 is 0.075(2000) = 150.
   State. The interest rate is 7.5%.

60. 76

61. Familiarize. Let x = number of half dollars.
   Then 2x = number of quarters,
         4x = number of dimes (2·2x = 4x), and
         12x = number of nickels (3·4x = 12x).
   The value of x half dollars is 0.50(x).
   The value of 2x quarters is 0.25(2x).
   The value of 4x dimes is 0.10(4x).
   The value of 12x nickels is 0.05(12x).
   Translate. The total value is $10.
   $0.50(x) + 0.25(2x) + 0.10(4x) + 0.05(12x) = 10$
   Carry out.
   $0.50(x) + 0.25(2x) + 0.10(4x) + 0.05(12x) = 10$
   $0.5x + 0.5x + 0.4x + 0.6x = 10$
   $2x = 10$
   $x = 5$
   Possible answers for the number of each coin:
   Half dollars = x = 5
   Quarters = 2x = 2·5 = 10
   Dimes = 4x = 4·5 = 20
   Nickels = 12x = 12·5 = 60
   Check. The value of
   5 half dollars = $2.50
   10 quarters   = 2.50
   20 dimes   = 2.00
   60 nickels   = 3.00
   The total value is $10. The numbers check.
   State. The storekeeper got 5 half dollars, 10 quarters, 20 dimes, and 60 nickels.

62. About 0.65 in.

Exercise Set 3.6

1. a) $A = \ell w = 17 \cdot 4 = 68$ ft^2

   b) $A = \ell w$
       $\frac{1}{\ell} \cdot A = \frac{1}{\ell} \cdot \ell w$ Multiplying by $\frac{1}{\ell}$
       $\frac{A}{\ell} = w$

   c) $w = \frac{A}{\ell} = \frac{48}{6} = 8$ cm

2. a) $720

   b) $t = \frac{I}{Pr}$

   c) 4 years

3. a) $A = \frac{1}{2}bh = \frac{1}{2} \cdot 51 \cdot 62 = 1581$ m^2

   b) $A = \frac{1}{2}bh$
      $2A = bh$ Multiplying by 2
      $\frac{2A}{b} = h$ Multiplying by $\frac{1}{b}$

   c) $h = \frac{2A}{b} = \frac{2 \cdot 504}{12} = 84$ yd

4. a) 138 cm

   b) $\ell = \frac{P - 2w}{2}$, or $\frac{P}{2} - w$

   c) 111 mi

5. a) $d = rt = 55 \cdot 3 = 165$ mi

   b) $d = rt$
      $\frac{d}{r} = t$ Multiplying by $\frac{1}{r}$

   c) $t = \frac{273}{65} = 4.2$ hr

6. a) 351 mi

   b) $\frac{d}{t} = r$

   c) 55 mph

7. $A = bh$
   $\frac{A}{h} = b$ Dividing by h

8. $h = \frac{A}{b}$

9. $I = Prt$
   $\frac{I}{rt} = P$ Dividing by rt

10. $r = \frac{I}{Pt}$

11. $F = ma$
    $\frac{F}{m} = a$

12. $m = \frac{F}{a}$

13. $P = 2\ell + 2w$
    $P - 2\ell = 2w$ Subtracting 2ℓ
    $\frac{P - 2\ell}{2} = w$ Dividing by 2

14. $b = \frac{2A}{h}$

15. $A = \pi r^2$
    $\frac{A}{\pi} = r^2$

16. $\pi = \frac{A}{r^2}$

59

17. $E = mc^2$

    $\dfrac{E}{c^2} = m$ Dividing by c^2

18. $c^2 = \dfrac{E}{m}$

19. $A = \dfrac{a + b + c}{3}$

    $3A = a + b + c$ Multiplying by 3

    $3A - a - c = b$ Subtracting a and c

20. $c = 3A - a - b$

21. $v = \dfrac{3k}{t}$

    $tv = 3k$ Multiplying by t

    $t = \dfrac{3k}{v}$ Dividing by v

22. $P = \dfrac{ab}{c}$

    $Pc = ab$

    $c = \dfrac{ab}{P}$

23. $A = \tfrac{1}{2}ah + \tfrac{1}{2}bh$

    $2A = 2\left[\tfrac{1}{2}ah + \tfrac{1}{2}bh\right]$ Clearing the fractions

    $2A = ah + bh$

    $2A - ah = bh$ Subtracting ah

    $\dfrac{2A - ah}{h} = b$ Dividing by h

24. $h = \dfrac{2A}{a + b}$

25. $H = \dfrac{D^2 N}{2.5}$

    $\dfrac{2.5}{N} \cdot H = \dfrac{2.5}{N} \cdot \dfrac{D^2 N}{2.5}$ Multiplying by $\dfrac{2.5}{N}$

    $\dfrac{2.5H}{N} = D^2$

26. $N = \dfrac{2.5H}{D^2}$

27. $A = \dfrac{\pi r^2 S}{360}$

    $\dfrac{360}{\pi r^2} \cdot A = \dfrac{360}{\pi r^2} \cdot \dfrac{\pi r^2 S}{360}$

    $\dfrac{360A}{\pi r^2} = S$

28. $r^2 = \dfrac{360A}{\pi S}$

29. $R = -0.0075t + 3.85$

    $R - 3.85 = -0.0075t$

    $\dfrac{R - 3.85}{-0.0075} = t$, or

    $\dfrac{38.5 - R}{0.0075} = t$ Multiplying by $\dfrac{-1}{-1}$ on the left

30. $C = \dfrac{5F - 160}{9}$, or $\tfrac{5}{9}(F - 32)$

31. $A = pq + pw + 7$

    $A - 7 = pq + pw$ Subtracting 7

    $A - 7 = p(q + w)$ Factoring out p

    $\dfrac{A - 7}{q + w} = p$ Multiplying by $\dfrac{1}{q + w}$

32. $w = \dfrac{Q}{a - b}$

33. $y = a - ab$

    $y = a(1 - b)$ Factoring out a

    $\dfrac{y}{1 - b} = a$ Multiplying by $\dfrac{1}{1 - b}$

34. $a = \dfrac{x - b}{1 - 2b}$

35. <u>Familiarize</u>. Let y = the percent.

    <u>Translate</u>.

    What percent of 7500 is 2500?

    y   % · 7500 = 2500

    <u>Carry out</u>.

    $y\% \cdot 7500 = 2500$

    $y \times 0.01 \times 7500 = 2500$

    $y(75) = 2500$

    $y = 33\tfrac{1}{3}$

    <u>Check</u>. Find $33\tfrac{1}{3}\%$ of 7500:

    $33\tfrac{1}{3}\% \cdot 7500 = \tfrac{1}{3} \cdot 7500 = 2500$

    <u>State</u>. The answer is $33\tfrac{1}{3}\%$.

36. -90

37. $-45.8 - (-32.6) = -45.8 + 32.6 = -13.2$

38. $-21a + 12b$

39. $ax + b = c$

    $b = c - ax$

40. $a = \dfrac{c - b}{x}$

41. $ax + b = 0$

    $ax = -b$

    $x = -\dfrac{b}{a}$

42. $R = \dfrac{1}{A}$

Chapter 3   (3.6)

43. $\dfrac{s}{t} = \dfrac{t}{v}$

    $t \cdot \dfrac{s}{t} = t \cdot \dfrac{t}{v}$

    $s = \dfrac{t^2}{v}$

44. $\dfrac{a}{c} = \dfrac{b}{d}$

45. $g = 40n + 20k$

    $g - 40n = 20k$

    $\dfrac{g - 40n}{20} = k$, or

    $\dfrac{g}{20} - 2n = k$

46. $f = 8h - 4r$

47. $d = \dfrac{1}{e + f}$

    $d(e + f) = 1$ Multiplying by $e + f$

    $e + f = \dfrac{1}{d}$ Dividing by d

    $f = \dfrac{1}{d} - e$ Subtracting e

48. $y = \dfrac{xz^2}{t}$

49. $m = ax^2 + bx + c$

    $m - ax^2 - c = bx$

    $\dfrac{m - ax^2 - c}{x} = b$

50. No; $3^2 = (-3)^2$, but $3 \neq -3$.

51. $A = 180°(n - 2)$
    $A = 180°(6 - 2)$ Substituting 6 for n
    $A = 180° \cdot 4$
    $A = 720°$

52. 10

53. $A = 180°(n - 2)$
          $A = 180°n - 360°$
    $A + 360° = 180°n$
    $\dfrac{A + 360°}{180°} = n$, or
    $\dfrac{A}{180°} + 2 = n$

54. $180° = \dfrac{A}{n - 2}$

55. Not necessarily; $6 = 2 \cdot 2 + 2 \cdot 1$, but $2 \cdot 6$, or 12, can be expressed as $2 \cdot 5 + 2 \cdot 1$.

56. A quadruples

57. $A = \dfrac{1}{2}bh$

    When b increases by 4 units we have $\dfrac{1}{2}(b + 4)h = \dfrac{1}{2}bh + 2h = A + 2h$, so A increases by $2h$ units.

58. An increase in a

CHAPTER 4   POLYNOMIALS

Exercise Set 4.1

1. $3^{-2} = \frac{1}{3^2} = \frac{1}{9}$

2. $\frac{1}{2^3} = \frac{1}{8}$

3. $10^{-4} = \frac{1}{10^4} = \frac{1}{10,000}$

4. $\frac{1}{5^6} = \frac{1}{15,625}$

5. $7^{-3} = \frac{1}{7^3} = \frac{1}{343}$

6. $\frac{1}{5^2} = \frac{1}{25}$

7. $a^{-3} = \frac{1}{a^3}$

8. $\frac{1}{x^2}$

9. $\frac{1}{y^{-4}} = y^4$

10. t^7

11. $\frac{1}{z^{-n}} = z^n$

12. h^m

13. $2^{-1} = \frac{1}{2^1} = \frac{1}{2}$

14. $\frac{3}{2}$

15. $\left(\frac{1}{4}\right)^{-2} = \frac{1}{\left(\frac{1}{4}\right)^2} = \frac{1}{\frac{1}{16}} = 1 \cdot \frac{16}{1} = 16$

16. $\frac{1}{\left(\frac{4}{5}\right)^2} = \frac{25}{16}$

17. $\frac{1}{4^3} = 4^{-3}$

18. 5^{-2}

19. $\frac{1}{x^3} = x^{-3}$

20. y^{-2}

21. $\frac{1}{a^4} = a^{-4}$

22. t^{-5}

23. $\frac{1}{p^n} = p^{-n}$

24. m^{-n}

25. $\frac{1}{5} = \frac{1}{5^1} = 5^{-1}$

26. 8^{-1}

27. $\frac{1}{t} = \frac{1}{t^1} = t^{-1}$

28. m^{-1}

29. When $x = -12$, $x^0 = (-12)^0 = 1$. (Any nonzero number to the 0 power is 1.)

30. 1

31. When $x = -4$, $5x^0 = 5(-4)^0 = 5 \cdot 1 = 5$

32. 7

33. When $n \neq 0$, $n^0 = 1$. (Any nonzero number to the 0 power is 1.)

34. 1

35. $10^0 = 1$

36. 1

37. $5^1 - 5^0 = 5 - 1 = 4$

38. -7

39. $2^4 \cdot 2^3 = 2^{4+3} = 2^7$

40. 3^7

41. $8^5 \cdot 8^9 = 8^{5+9} = 8^{14}$

42. n^{23}

43. $x^4 \cdot x^3 = x^{4+3} = x^7$

44. y^{16}

45. $9^{17} \cdot 9^{21} = 9^{17+21} = 9^{38}$

46. t^{16}

47. $(3y)^4(3y)^8 = (3y)^{4+8} = (3y)^{12}$

48. $(2t)^{25}$

49. $(7y)^1(7y)^{16} = (7y)^{1+16} = (7y)^{17}$

50. $8x$

51. $3^{-5} \cdot 3^8 = 3^{-5+8} = 3^3$

52. 5

53. $x^{-2} \cdot x = x^{-2+1} = x^{-1}$, or $\frac{1}{x}$

54. 1

Chapter 4   (4.1)

55. $x^4 \cdot x^3 = x^{4+3} = x^7$

56. x^{13}

57. $x^{-7} \cdot x^{-6} = x^{-7+(-6)} = x^{-13}$, or $\frac{1}{x^{13}}$

58. y^{-13}, or $\frac{1}{y^{13}}$

59. $t^8 \cdot t^{-8} = t^{8+(-8)} = t^0 = 1$

60. 1

61. $\frac{7^5}{7^2} = 7^{5-2} = 7^3$

62. 4^4

63. $\frac{8^{12}}{8^6} = 8^{12-6} = 8^6$

64. 9^{12}

65. $\frac{y^9}{y^5} = y^{9-5} = y^4$

66. x

67. $\frac{16^2}{16^8} = 16^{2-8} = 16^{-6}$, or $\frac{1}{16^6}$

68. 5^{-6}, or $\frac{1}{5^6}$

69. $\frac{m^6}{m^{12}} = m^{6-12} = m^{-6}$, or $\frac{1}{m^6}$

70. p^{-1}, or $\frac{1}{p}$

71. $\frac{(8x)^6}{(8x)^{10}} = (8x)^{6-10} = (8x)^{-4}$, or $\frac{1}{(8x)^4}$

72. $(9t)^{-7}$, or $\frac{1}{(9t)^7}$

73. $\frac{18^9}{18^9} = 18^{9-9} = 18^0 = 1$

74. 1

75. $\frac{x}{x^{-1}} = x^{1-(-1)} = x^2$

76. x^5

77. $\frac{x^7}{x^{-2}} = x^{7-(-2)} = x^9$

78. t^{11}

79. $\frac{z^{-6}}{z^{-2}} = z^{-6-(-2)} = z^{-4}$, or $\frac{1}{z^4}$

80. y^{-4}, or $\frac{1}{y^4}$

81. $\frac{x^{-5}}{x^{-8}} = x^{-5-(-8)} = x^3$

82. y^5

83. $\frac{m^{-9}}{m^{-9}} = m^{-9-(-9)} = m^0 = 1$

84. 1

85. $8^2 = 64$

$8^{-2} = \frac{1}{8^2} = \frac{1}{64}$

$\left(\frac{1}{8}\right)^2 = \frac{1}{64}$

$\left(\frac{1}{8}\right)^{-2} = \frac{1}{\left(\frac{1}{8}\right)^2} = \frac{1}{\frac{1}{64}} = 1 \cdot \frac{64}{1} = 64$

$-8^2 = -(8)(8) = -64$

$(-8)^2 = (-8)(-8) = 64$

86. $25, \frac{1}{25}, \frac{1}{25}, 25, -25, 25$

87. $3s + 3t + 24 = 3s + 3t + 3 \cdot 8 = 3(s + t + 8)$

88. $-7(x + 2)$

89. $9x + 2y - 4x - 2y = (9 - 4)x + (2 - 2)y = 5x$

90. 37.5%

91. $A = P(1 + r)^t$
We substitute 2000 for P, 12% for r, and 2 for t.
$A = 2000(1 + 12\%)^2$
$A = 2000(1 + 0.12)^2$
$A = 2000(1.12)^2$
$A = 2000(1.2544)$
$A = 2508.80$
At the end of two years, $2508.80 is in the account.

92. $3041.75

93. $A = P(1 + r)^t$
We substitute 10,400 for P, 16.5% for r, and 5 for t.
$A = 10,400(1 + 16.5\%)^5$
$A = 10,400(1 + 0.165)^5$
$A = 10,400(1.165)^5$
$A \approx 22,318.40$ Using a calculator to do the computations
There is $22,318.40 in the account at the end of 5 years.

94. About $63,677.44

95. No; $(5y)^0 = 1$, but $5y^0 = 5 \cdot 1 = 5$.

96. y^{5x}

97. $a^{5k} \div a^{3k} = a^{5k-3k} = a^{2k}$

Chapter 4   (4.2)

98. a^4t

99. $\dfrac{\left(\frac{1}{2}\right)^4}{\left(\frac{1}{2}\right)^5} = \left(\frac{1}{2}\right)^{4-5} = \left(\frac{1}{2}\right)^{-1} = \dfrac{1}{\frac{1}{2}} = 1 \cdot \dfrac{2}{1} = 2$

100. 1

101. No; $(2 + 3)^2 = 5^2 = 25$, but $2^2 + 3^2 = 4 + 9 = 13$.

102. No. See Exercise 101.

103. No; $(2^2)^3 = 4^3 = 64$, but $2^{2^3} = 2^8 = 256$.

104. 25

105. Since the bases are the same, the expression with the larger exponent is larger. Thus, $3^5 > 3^4$.

106. $<$

107. Since the exponents are the same, the expression with the larger base is larger. Thus, $4^3 < 5^3$.

108. $<$

Exercise Set 4.2

1. $(2^3)^2 = 2^{3\cdot 2} = 2^6$

2. 3^{12}

3. $(5^2)^{-3} = 5^{2(-3)} = 5^{-6}$, or $\dfrac{1}{5^6}$

4. 9^{-12}, or $\dfrac{1}{9^{12}}$

5. $(x^{-3})^{-4} = x^{(-3)(-4)} = x^{12}$

6. a^{30}

7. $(4x^3)^2 = 4^2(x^3)^2$ Raising each factor to the second power
   $= 16x^6$

8. $4x^6$

9. $2(m^5)^3 = 2\cdot m^{5\cdot 3} = 2m^{15}$

10. $8m^{15}$

11. $(x^4y^5)^{-3} = (x^4)^{-3}(y^5)^{-3} = x^{4(-3)}y^{5(-3)} = x^{-12}y^{-15}$, or $\dfrac{1}{x^{12}y^{15}}$

12. $t^{-20}x^{-12}$, or $\dfrac{1}{t^{20}x^{12}}$

13. $(x^{-6}y^{-2})^{-4} = (x^{-6})^{-4}(y^{-2})^{-4} = x^{(-6)(-4)}y^{(-2)(-4)} = x^{24}y^8$

14. $x^{10}y^{35}$

15. $(3x^3y^{-8}z^{-3})^2 = 3^2(x^3)^2(y^{-8})^2(z^{-3})^2 = 9x^6y^{-16}z^{-6}$, or $\dfrac{9x^6}{y^{16}z^6}$

16. $8a^6y^{-12}z^{-15}$, or $\dfrac{8a^6}{y^{12}z^{15}}$

17. $\left(\dfrac{a^2}{b^3}\right)^4 = \dfrac{(a^2)^4}{(b^3)^4} = \dfrac{a^8}{b^{12}}$, or a^8b^{-12}

18. $\dfrac{x^{15}}{y^{20}}$, or $x^{15}y^{-20}$

19. $\left(\dfrac{y^3}{2}\right)^2 = \dfrac{(y^3)^2}{2^2} = \dfrac{y^6}{4}$

20. $\dfrac{a^{15}}{27}$

21. $\left(\dfrac{y^2}{2}\right)^{-3} = \dfrac{(y^2)^{-3}}{2^{-3}} = \dfrac{y^{-6}}{2^{-3}} = \dfrac{\frac{1}{y^6}}{\frac{1}{2^3}} = \dfrac{1}{y^6} \cdot \dfrac{2^3}{1} = \dfrac{8}{y^6}$, or $8y^{-6}$

22. $\dfrac{9}{a^8}$, or $9a^{-8}$

23. $\left(\dfrac{3}{a^2}\right)^3 = \dfrac{3^3}{(a^2)^3} = \dfrac{27}{a^6}$, or $27a^{-6}$

24. $\dfrac{49}{x^{14}}$, or $49x^{-14}$

25. $\left(\dfrac{x^2y}{z}\right)^3 = \dfrac{(x^2)^3y^3}{z^3} = \dfrac{x^6y^3}{z^3}$, or $x^6y^3z^{-3}$

26. $\dfrac{m^3}{n^{12}p^3}$

27. $\left(\dfrac{a^2b}{cd^3}\right)^{-2} = \dfrac{(a^2)^{-2}b^{-2}}{c^{-2}(d^3)^{-2}} = \dfrac{a^{-4}b^{-2}}{c^{-2}d^{-6}} = \dfrac{\frac{1}{a^4} \cdot \frac{1}{b^2}}{\frac{1}{c^2} \cdot \frac{1}{d^6}} = \dfrac{\frac{1}{a^4b^2}}{\frac{1}{c^2d^6}} = \dfrac{1}{a^4b^2} \cdot \dfrac{c^2d^6}{1} = \dfrac{c^2d^6}{a^4b^2}$

28. $\dfrac{27b^{12}}{8a^6}$, or $\dfrac{27a^{-6}b^{12}}{8}$

29. 7.8,000,000,000.
    ↑_____| 10 places

    Large number, so the exponent is positive.
    $78,000,000,000 = 7.8 \times 10^{10}$

30. 3.7×10^{12}

31. 9.07,000,000,000,000,000.
    ↑_____| 17 places

    Large number, so the exponent is positive.
    $907,000,000,000,000,000 = 9.07 \times 10^{17}$

32. 1.68×10^{14}

33. 0.000003.74
    └─────↑ 6 places

    Small number, so the exponent is negative.
    0.00000374 = 3.74 × 10⁻⁶

34. 2.75 × 10⁻¹⁰

35. 0.00000001.8
    └──────↑ 8 places

    Small number, so the exponent is negative.
    0.000000018 = 1.8 × 10⁻⁸

36. 2 × 10⁻¹¹

37. 1.0,000,000.
    ↑─────┘ 7 places

    Large number, so the exponent is positive.
    10,000,000 = 1 × 10⁷, or 10⁷

38. 10¹¹

39. 0.000000001.
    └──────↑ 9 places

    Small number, so the exponent is negative.
    0.000000001 = 1 × 10⁻⁹, or 10⁻⁹

40. 10⁻⁷

41. 7.84 × 10⁸

    Positive exponent, so the answer is a large number.
    7.84000000.
    └─────↑ 8 places

    7.84 × 10⁸ = 784,000,000

42. 13,500,000

43. 8.764 × 10⁻¹⁰

    Negative exponent, so the answer is a small number.
    0.0000000008.764
    ↑──────────┘ 10 places

    8.764 × 10⁻¹⁰ = 0.0000000008764

44. 0.009043

45. 10⁸ = 1 × 10⁸

    Positive exponent, so the answer is a large number.
    1.00000000.
    └─────↑ 8 places

    10⁸ = 100,000,000

46. 10,000

47. 10⁻⁴ = 1 × 10⁻⁴

    Negative exponent, so the answer is a small number.
    0.0001.
    ↑───┘ 4 places

    10⁻⁴ = 0.0001

48. 0.0000001

49. (3 × 10⁴)(2 × 10⁵)

    a) Multiply 3 and 2: 3 × 2 = 6
    b) Multiply 10⁴ and 10⁵: 10⁴·10⁵ = 10⁴⁺⁵ = 10⁹
    c) The answer is 6 × 10⁹.

50. 6.46 × 10⁵

51. (5.2 × 10⁵)(6.5 × 10⁻²)

    a) Multiply 5.2 and 6.5: 5.2 × 6.5 = 33.8
    b) Multiply 10⁵ and 10⁻²: 10⁵·10⁻² = 10⁵⁺⁽⁻²⁾ = 10³
    c) The answer at this stage is 33.8 × 10³ but this is not scientific notation since 33.8 is not a number between 1 and 10. We convert 33.8 to scientific notation and simplify.
    33.8 × 10³ = (3.38 × 10) × 10³ = 3.38 × 10⁴

52. 6.106 × 10⁻¹¹

53. (9.9 × 10⁻⁶)(8.23 × 10⁻⁸)

    a) Multiply 9.9 and 8.23: 81.477
    b) Multiply 10⁻⁶ and 10⁻⁸: 10⁻⁶ × 10⁻⁸ = 10⁻⁶⁺⁽⁻⁸⁾ = 10⁻¹⁴
    c) The answer at this stage is 81.477 × 10⁻¹⁴. We convert 81.477 to scientific notation and simplify.
    81.477 × 10⁻¹⁴ = (8.1477 × 10) × 10⁻¹⁴ = 8.1477 × 10⁻¹³

54. 1.123 × 10⁻⁵

55. $\dfrac{8.5 \times 10^8}{3.4 \times 10^{-5}}$

    a) Divide 8.5 by 3.4: 8.5 ÷ 3.4 = 2.5
    b) Divide 10⁸ by 10⁻⁵: 10⁸ ÷ 10⁻⁵ = 10⁸⁻⁽⁻⁵⁾ = 10¹³
    c) The answer is 2.5 × 10¹³.

56. 2.24 × 10⁻⁷

Chapter 4   (4.2)

57. $(3.0 \times 10^6) \div (6.0 \times 10^9)$
   a) Divide 3.0 by 6.0: $3.0 \div 6.0 = 0.5$
   b) Divide 10^6 by 10^9: $10^6 \div 10^9 = 10^{6-9} = 10^{-3}$
   c) The answer at this stage is 0.5×10^{-3}. We convert 0.5 to scientific notation and simplify.
   $0.5 \times 10^{-3} = (5 \times 10^{-1}) \times 10^{-3} = 5 \times 10^{-4}$

58. 9.375×10^2

59. $\dfrac{7.5 \times 10^{-9}}{2.5 \times 10^{12}}$
   a) Divide 7.5 by 2.5: $7.5 \div 2.5 = 3$
   b) Divide 10^{-9} by 10^{12}: $10^{-9} \div 10^{12} = 10^{-9-12} = 10^{-21}$
   c) The answer is 3×10^{-21}.

60. 5×10^{-24}

61. <u>Familiarize.</u> We express 2864 and 243 million in scientific notation

   2.864. = 2.864×10^3
   3 places

   2.43,000,000. = 2.43×10^8
   8 places

   Let p = the part of the population that are members of the Professional Bowlers Association.

   <u>Translate.</u> We reword the problem.

   What is   number of members   divided by   population of the U.S.?
   p   = (2.864×10^3) ÷ (2.43×10^8)

   <u>Carry out.</u> We do the computation.
   $p = (2.864 \times 10^3) \div (2.43 \times 10^8)$
   $p = (2.864 \div 2.43) \times (10^3 \div 10^8)$
   $p \approx 1.179 \times 10^{3-8}$
   $p \approx 1.179 \times 10^{-5}$

   <u>Check.</u> We review our computation. Also, the answer seems reasonable since it is smaller than either of the original numbers.

   <u>State.</u> Approximately 1.179×10^{-5} of the population are members of the Professional Bowlers Association.

62. Approximately 3.3×10^{-2}

63. <u>Familiarize.</u> There are 365 days in one year. Express 6.5 million and 365 in scientific notation.

   6.5 million = $6.5 \times 1,000,000. = 6.5 \times 10^6$
   6 places

   3.65. = 3.65×10^2
   2 places

   Let p = the amount of popcorn Americans eat in one year.

   <u>Translate.</u> We reword the problem.

   What is   daily consumption   times   number of days in a year?
   p   = (6.5×10^6) × (3.65×10^2)

   <u>Carry out.</u> We do the computation.
   $p = (6.5 \times 10^6) \times (3.65 \times 10^2)$
   $p = (6.5 \times 3.65) \times (10^6 \times 10^2)$
   $p = 23.725 \times 10^8$
   $p = (2.3725 \times 10) \times 10^8$
   $p = 2.3725 \times 10^9$

   <u>Check.</u> We review the computation. Also, the answer seems reasonable since it is larger than 6.5 million.

   <u>State.</u> Americans eat 2.3725×10^9 gal of popcorn each year.

64. 1.095×10^9 gal

65. <u>Familiarize.</u> There are 60 seconds in one minute and 60 minutes in one hour, so there are 60(60), or 3600 seconds in one hour. There are 24 hours in one day and 365 days in one year, so there are 3600(24)(365), or 31,536,000 seconds in one year.

   We express 3600, 31,536,000 and 4,200,000 in scientific notation:

   3.600. = 3.6×10^3
   3 places

   3.1,536,000. = 3.1536×10^7
   7 places

   4.200,000. = 4.2×10^6
   6 places

   Let h = the discharge in one hour and y = the discharge in one year.

   <u>Translate.</u> We reword and write two equations. To find the discharge in one hour:

   What is   number of seconds in one hour   times   discharge per second?
   h   = (3.6×10^3) × (4.2×10^6)

   To find the discharge in one year:

   What is   number of seconds in one year   times   discharge per second?
   y   = (3.1536×10^7) × (4.2×10^6)

Chapter 4   (4.3)

65. (continued)

   Carry out. We do the computations.
   $h = (3.6 \times 10^3) \times (4.2 \times 10^6)$
   $h = (3.6 \times 4.2) \times (10^3 \times 10^6)$
   $h = 15.12 \times 10^9 = (1.512 \times 10) \times 10^9$
   $h = 1.512 \times 10^{10}$

   $y = (3.1536 \times 10^7) \times (4.2 \times 10^6)$
   $y = (3.1536 \times 4.2) \times (10^7 \times 10^6)$
   $y = 13.24512 \times 10^{13} = (1.324512 \times 10) \times 10^{13}$
   $y = 1.324512 \times 10^{14}$

   Check. We can review the computations. Also, the answers seem reasonable since they are both larger than the numbers we started with.

   State. In one hour 1.512×10^{10} cu ft of water is discharged. In one year 1.324512×10^{14} cu ft of water is discharged.

66. Approximately 6.7×10^{-2}

67. $\dfrac{(5.2 \times 10^6)(6.1 \times 10^{-11})}{1.28 \times 10^{-3}}$

   a) Compute, using a calculator: $\dfrac{5.2 \times 6.1}{1.28} = 24.78125$

   b) Compute: $\dfrac{10^6 \times 10^{-11}}{10^{-3}} = 10^{6+(-11)-(-3)} = 10^{-2}$

   c) $24.78125 \times 10^{-2} = (2.478125 \times 10) \times 10^{-2} = 2.478125 \times 10^{-1}$

68. 1.5234375×10^7

69. $\{2.1 \times 10^6 [(2.5 \times 10^{-3}) \div (5.0 \times 10^{-5})]\} \div (3.0 \times 10^{17})$
   $= \{2.1 \times 10^6 [0.5 \times 10^2]\} \div (3.0 \times 10^{17})$ Dividing inside the brackets first
   $= \{1.05 \times 10^8\} \div (3.0 \times 10^{17})$ Multiplying inside the braces
   $= 0.35 \times 10^{-9}$ Dividing
   $= (3.5 \times 10^{-1}) \times 10^{-9}$ Writing 0.35 in scientific notation
   $= 3.5 \times 10^{-10}$ Simplifying

70. a) 1.6×10^2
    b) 2.5×10^{-11}

71. $4^3 \cdot 8 \cdot 16 = (2^2)^3 \cdot 2^3 \cdot 2^4 = 2^6 \cdot 2^3 \cdot 2^4 = 2^{13}$

72. 4^{13}

73. $\dfrac{(5^{12})^2}{5^{25}} = \dfrac{5^{24}}{5^{25}} = 5^{24-25} = 5^{-1}$, or $\dfrac{1}{5}$

74. 1

75. $\dfrac{(3^5)^4}{3^5 \cdot 3^4} = \dfrac{3^{5 \cdot 4}}{3^{5+4}} = \dfrac{3^{20}}{3^9} = 3^{20-9} = 3^{11}$

76. 1

77. $\dfrac{a^{22}}{(a^2)^{11}} = \dfrac{a^{22}}{a^{22}} = a^{22-22} = a^0 = 1$

78. 7

79. $\left(\dfrac{1}{2}\right)^{-2} = \dfrac{1^{-2}}{2^{-2}} = \dfrac{\frac{1}{1^2}}{\frac{1}{2^2}} = \dfrac{\frac{1}{1}}{\frac{1}{4}} = \dfrac{1}{1} \cdot \dfrac{4}{1} = 4$

80. $(0.4)^{-1}$, or 2.5

81. $\left(\dfrac{1}{a}\right)^{-n} = \dfrac{1^{-n}}{a^{-n}} = \dfrac{\frac{1}{1^n}}{\frac{1}{a^n}} = \dfrac{1}{1} \cdot \dfrac{a^n}{1} = a^n$

82. 81

83. False; let $x = 2$, $y = 3$, $m = 4$, and $n = 2$:
    $2^4 \cdot 3^2 = 16 \cdot 9 = 144$, but
    $(2 \cdot 3)^{4 \cdot 2} = 6^8 = 1,679,616$

84. True

85. False; let $x = 3$, $y = 4$, and $m = 2$:
    $3^2 \cdot 4^2 = 9 \cdot 16 = 144$, but
    $(3 \cdot 4)^{2 \cdot 2} = 12^4 = 20,736$

86. False

87. True

88. False

Exercise Set 4.3

1. $-5x + 2 = -5 \cdot 4 + 2 = -20 + 2 = -18$

2. -11

3. $2x^2 - 5x + 7 = 2 \cdot 4^2 - 5 \cdot 4 + 7 = 2 \cdot 16 - 20 + 7 = 32 - 20 + 7 = 19$

4. 59

5. $x^3 - 5x^2 + x = 4^3 - 5 \cdot 4^2 + 4 = 64 - 5 \cdot 16 + 4 = 64 - 80 + 4 = -12$

6. 51

7. $3x + 5 = 3(-1) + 5 = -3 + 5 = 2$

8. 8

9. $x^2 - 2x + 1 = (-1)^2 - 2(-1) + 1 = 1 + 2 + 1 = 4$

10. -10

Chapter 4   (4.3)

11. $-3x^3 + 7x^2 - 3x - 2 =$
    $-3(-1)^3 + 7(-1)^2 - 3(-1) - 2 =$
    $-3(-1) + 7 \cdot 1 + 3 - 2 = 3 + 7 + 3 - 2 = 11$

12. -4

13. $0.4a^2 - 40a + 1039 = 0.4(18)^2 - 40(18) + 1039 =$
    $0.4(324) - 720 + 1039 = 129.6 - 720 + 1039 =$
    448.6
    There are approximately 449 accidents daily involving an 18-year-old driver.

14. 399

15. Evaluate the polynomial for $x = 17$:
    $4x = 4 \cdot 17 = 68$
    The perimeter is 68 ft.

16. 114 m

17. Evaluate the polynomial for $t = 8$:
    $16t^2 = 16(8)^2 = 16 \cdot 64 = 1024$
    The cliff is 1024 ft high.

18. 144 ft

19. Evaluate the polynomial for $x = 75$:
    $280x - 0.4x^2 = 280 \cdot 75 - 0.4(75)^2 =$
    $21{,}000 - 0.4(5625) = 21{,}000 - 2250 = 18{,}750$
    The total revenue is $18,750.

20. $24,000

21. Evaluate the polynomial for $x = 75$:
    $5000 + 0.6(75)^2 = 5000 + 0.6(5625) =$
    $5000 + 3375 = 8375$
    The total cost is $8375.

22. $11,000

23. $2 - 3x + x^2 = 2 + (-3x) + x^2$
    The terms are 2, $-3x$, and x^2.

24. $2x^2$, $3x$, -4

25. $5x^3 + 6x^2 - 3x^2$
    Like terms: $6x^2$ and $-3x^2$ Same exponent and variable

26. $3x^2$ and $-2x^2$

27. $2x^4 + 5x - 7x - 3x^4$
    Like terms: $2x^4$ and $-3x^4$ Same exponent
    Like terms: $5x$ and $-7x$ and variable

28. $-3t$ and $-2t$, t^3 and $-5t^3$

29. $2x - 5x = (2 - 5)x = -3x$

30. $10x^2$

31. $x - 9x = 1x - 9x = (1 - 9)x = -8x$

32. $-4x$

33. $5x^3 + 6x^3 + 4 = (5 + 6)x^3 + 4 = 11x^3 + 4$

34. $4x^4 + 5$

35. $5x^3 + 6x - 4x^3 - 7x = (5 - 4)x^3 + (6 - 7)x =$
    $1x^3 + (-1)x = x^3 - x$

36. $4a^4$

37. $6b^5 + 3b^2 - 2b^5 - 3b^2 = (6 - 2)b^5 + (3 - 3)b^2 =$
    $4b^5 + 0b^2 = 4b^5$

38. $6x^2 - 3x$

39. $\frac{1}{4}x^5 - 5 + \frac{1}{2}x^5 - 2x - 37 =$
    $\left(\frac{1}{4} + \frac{1}{2}\right)x^5 - 2x + (-5 - 37) = \frac{3}{4}x^5 - 2x - 42$

40. $\frac{1}{6}x^3 + 2x - 12$

41. $6x^2 + 2x^4 - 2x^2 - x^4 - 4x^2 =$
    $6x^2 + 2x^4 - 2x^2 - 1x^4 - 4x^2 =$
    $(6 - 2 - 4)x^2 + (2 - 1)x^4 = 0x^2 + 1x^4 = 0 + x^4 = x^4$

42. $-x^3$

43. $\frac{1}{4}x^3 - x^2 - \frac{1}{6}x^2 + \frac{3}{8}x^3 + \frac{5}{16}x^3 =$
    $\frac{1}{4}x^3 - 1x^2 - \frac{1}{6}x^2 + \frac{3}{8}x^3 + \frac{5}{16}x^3 =$
    $\left(\frac{1}{4} + \frac{3}{8} + \frac{5}{16}\right)x^3 + \left(-1 - \frac{1}{6}\right)x^2 =$
    $\left(\frac{4}{16} + \frac{6}{16} + \frac{5}{16}\right)x^3 + \left(-\frac{6}{6} - \frac{1}{6}\right)x^2 = \frac{15}{16}x^3 - \frac{7}{6}x^2$

44. 0

45. $x^5 + x + 6x^3 + 1 + 2x^2 = x^5 + 6x^3 + 2x^2 + x + 1$

46. $-5x^6 - 2x^3 + 2x^2 + 3x + 3$

47. $5x^3 + 15x^9 + x - x^2 + 7x^8 =$
    $15x^9 + 7x^8 + 5x^3 - x^2 + x$

48. $x^5 - 5x^4 + 6x^3 + 9x - 5$

49. $8y^3 - 7y^2 + 9y^6 - 5y^8 + y^7 =$
    $-5y^8 + y^7 + 9y^6 + 8y^3 - 7y^2$

50. $p^8 - 7p^4 + p^2 + p - 4$

51. $3x^4 - 5x^6 - 2x^4 + 6x^6 = x^4 + x^6 = x^6 + x^4$

Chapter 4   (4.3)

52. $x^4 - 2x^3 + 1$

53. $-2x + 4x^3 - 7x + 9x^3 + 8 = -9x + 13x^3 + 8 = 13x^3 - 9x + 8$

54. $x^2 - 4x + 1$

55. $3x + 3x + 3x - x^2 - 4x^2 = 9x - 5x^2 = -5x^2 + 9x$

56. $-4x^3 - 6x$

57. $-x + \frac{3}{4} + 15x^4 - x - \frac{1}{2} - 3x^4 = -2x + \frac{1}{4} + 12x^4 = 12x^4 - 2x + \frac{1}{4}$

58. $4x^3 + x - \frac{1}{2}$

59. $2x - 4 = 2x^1 - 4x^0$
    The degree of $2x$ is 1.
    The degree of -4 is 0.
    The degree of the polynomial is 1, the largest exponent.

60. 0, 1; 1

61. $3x^2 - 5x + 2 = 3x^2 - 5x^1 + 2x^0$
    The degree of $3x^2$ is 2.
    The degree of $-5x$ is 1.
    The degree of 2 is 0.
    The degree of the polynomial is 2, the largest exponent.

62. 3, 2, 0; 3

63. $-7x^3 + 6x^2 + 3x + 7 = -7x^3 + 6x^2 + 3x^1 + 7x^0$
    The degree of $-7x^3$ is 3.
    The degree of $6x^2$ is 2.
    The degree of $3x$ is 1.
    The degree of 7 is 0.
    The degree of the polynomial is 3, the largest exponent.

64. 4, 2, 1, 0; 4

65. $x^2 - 3x + x^6 - 9x^4 = x^2 - 3x^1 + x^6 - 9x^4$
    The degree of x^2 is 2.
    The degree of $-3x$ is 1.
    The degree of x^6 is 6.
    The degree of $-9x^4$ is 4.
    The degree of the polynomial is 6, the largest exponent.

66. 1, 2, 0, 3; 3

67. $-3x + 6$
    The coefficient of $-3x$, the first term, is -3.
    The coefficient of 6, the second term, is 6.

68. 2, -4

69. $5x^2 + 3x + 3$
    The coefficient of $5x^2$, the first term, is 5.
    The coefficient of $3x$, the second term, is 3.
    The coefficient of 3, the third term, is 3.

70. 3, -5, 2

71. $-7x^3 + 6x^2 + 3x + 7$
    The coefficient of $-7x^3$, the first term, is -7.
    The coefficient of $6x^2$, the second term, is 6.
    The coefficient of $3x$, the third term, is 3.
    The coefficient of 7, the fourth term, is 7.

72. 5, 1, -1, 2

73. $-5x^4 + 6x^3 - 3x^2 + 8x - 2$
    The coefficient of $-5x^4$, the first term, is -5.
    The coefficient of $6x^3$, the second term, is 6.
    The coefficient of $-3x^2$, the third term, is -3.
    The coefficient of $8x$, the fourth term, is 8.
    The coefficient of -2, the fifth term, is -2.

74. 7, -4, -4, 5

75. See the answer section in the text.

76.

Term	Coefficient	Degree of Term	Degree of Polynomial
$8x^5$	8	5	5
$-\frac{1}{2}x^4$	$-\frac{1}{2}$	4	
$-46x^3$	-46	3	
$3x^2$	3	2	
$6x$	6	1	
-2.4	-2.4	0	

77. In the polynomial $x^3 - 27$, there are no x^2 or x terms. The x^2 term (or second-degree term) and the x term (or first-degree term) are missing.

78. x^4, x^3, x^2, x^0

79. In the polynomial $x^4 - x$, there are no x^3, x^2, or x^0 terms. The x^3 term (or third-degree term), the x^2 term (or second-degree term), and the x^0 term (or zero-degree term) are missing.

80. x^3, x^2

Chapter 4   (4.3)

81. No terms are missing in the polynomial $2x^3 - 5x^2 + x - 3$.

82. x^2, x, x^0

83. The polynomial $x^2 - 10x + 25$ is a <u>trinomial</u> because it has just three terms.

84. Monomial

85. The polynomial $x^3 - 7x^2 + 2x - 4$ is <u>none of these</u> because it has more than three terms.

86. Binomial

87. The polynomial $4x^2 - 25$ is a <u>binomial</u> because it has just two terms.

88. None of these

89. The polynomial $40x$ is a <u>monomial</u> because it has just one term.

90. Trinomial

91. <u>Familiarize</u>. Let a = the number of apples the campers had to begin with. Then the first camper ate $\frac{1}{3}a$ apples and $a - \frac{1}{3}a$, $\frac{2}{3}a$, apples were left. The second camper ate $\frac{1}{3}\left(\frac{2}{3}a\right)$, or $\frac{2}{9}a$, apples, and $\frac{2}{3}a - \frac{2}{9}a$, or $\frac{4}{9}a$, apples were left. The third camper ate $\frac{1}{3}\left(\frac{4}{9}a\right)$, or $\frac{4}{27}a$, apples, and $\frac{4}{9}a - \frac{4}{27}a$, or $\frac{8}{27}a$, apples were left.

<u>Translate</u>. We write an equation for the number of apples left after the third camper eats.

Number of apples left is 8.

$\frac{8}{27}a = 8$

<u>Carry out</u>. We solve the equation.

$\frac{8}{27}a = 8$

$a = \frac{27}{8} \cdot 8$

$a = 27$

<u>Check</u>. If the campers begin with 27 apples, then the first camper eats $\frac{1}{3} \cdot 27$, or 9, and $27 - 9$, or 18, are left. The second camper then eats $\frac{1}{3} \cdot 18$, or 6, apples and $18 - 6$, or 12, are left. Finally, the third camper eats $\frac{1}{3} \cdot 12$, or 4, apples and $12 - 4$, or 8, are left. The answer checks.

<u>State</u>. The campers had 27 apples to begin with.

92. 12.5¢

93. $3x^2 + 2x - 2 + 3x^0 = 3x^2 + 2x - 2 + 3\cdot 1 = 3x^2 + 2x - 2 + 3 = 3x^2 + 2x + 1$

94. $4x^8 + x^2 + 5x^9 + 5x$

95. $(3x^2)^3 + 4x^2 \cdot 4x^4 - x^4(2x)^2 + [(2x)^2]^3 - 100x^2(x^2)^2$
    $= 27x^6 + 4x^2 \cdot 4x^4 - x^4 \cdot 4x^2 + (4x^2)^3 - 100x^2 \cdot x^4$
    $= 27x^6 + 16x^6 - 4x^6 + 64x^6 - 100x^6$
    $= 3x^6$

96. 99, -99; 50, -50; 99, -99

97. Using a calculator evaluate $0.4a^2 - 40a + 1039$ for a = 10, 20, 30, 40, 50, 60, and 70 and list the values in a table.

Age	Average number of accidents per day
a	$0.4a^2 - 40a + 1039$
10	679
20	399
30	199
40	79
50	39
60	79
70	199

The numbers in the chart increase both below and above age 50. We would assume the number of accidents is the smallest near age 50. Now we evaluate for 49 and 51.

49	39.4
50	39
51	39.4

Again the numbers increase below and above 50. We conclude that the smallest number of daily accidents occurs at age 50.

98. Answers may vary. Use any ax^5-term, where a is an integer, and 3 other terms with different degrees each less than degree 5, and integer coefficients. Three answers are
$-6x^5 + 14x^4 - x^2 + 11$, $x^5 - 8x^3 + 3x + 1$, and $23x^5 + 2x^4 - x^2 + 5x$.

99. Answers may vary. Use any ay^4-term, where a is a rational number, and 2 other terms with different degrees, each less than degree 4, and rational coefficients. Three answers are
$2y^4 - y + 5$, $-8y^4 + 5y^3 - 2y^2$, and $29y^4 - 4y^2 - 11$.

100. 10

Chapter 4   (4.4)

101. Answers may vary. The terms must have the same variable and be of degree 4. One answer is $9y^4$, $-\frac{3}{2}y^4$, and $4.2y^4$.

102. $x^3 - 2x^2 - 6x + 3$

Exercise Set 4.4

1. $(3x + 2) + (-4x + 3) = (3 - 4)x + (2 + 3) = -x + 5$

2. $-x + 3$

3. $(-6x + 2) + (x^2 + x - 3) =$
   $x^2 + (-6 + 1)x + (2 - 3) = x^2 - 5x - 1$

4. $x^2 + 3x - 5$

5. $(x^2 - 9) + (x^2 + 9) = (1 + 1)x^2 + (-9 + 9) = 2x^2$

6. $3x^3 - 4x^2$

7. $(3x^2 - 5x + 10) + (2x^2 + 8x - 40) =$
   $(3 + 2)x^2 + (-5 + 8)x + (10 - 40) = 5x^2 + 3x - 30$

8. $-x^2 - 2$

9. $(6x^4 + 3x^3 - 1) + (4x^2 - 3x + 3) =$
   $6x^4 + 3x^3 + 4x^2 - 3x + (-1 + 3) =$
   $6x^4 + 3x^3 + 4x^2 - 3x + 2$

10. $3x^3 - 9x^2 + 2x + 3$

11. $(1.2x^3 + 4.5x^2 - 3.8x) + (-3.4x^3 - 4.7x^2 + 23) =$
    $(1.2 - 3.4)x^3 + (4.5 - 4.7)x^2 - 3.8x + 23 =$
    $-2.2x^3 - 0.2x^2 - 3.8x + 23$

12. $2.8x^4 - 0.6x^2 + 1.8x - 3.2$

13. $(3x^5 + 6x^2 - 1) + (7x^2 + 6x - 2) =$
    $3x^5 + (6 + 7)x^2 + 6x + (-1 - 2) =$
    $3x^5 + 13x^2 + 6x - 3$

14. $7x^3 + 6x - 6$

15. $(-4x^4 + 6x^2 - 3x - 5) + (6x^3 + 5x + 9) =$
    $-4x^4 + 6x^3 + 6x^2 + (-3 + 5)x + (-5 + 9) =$
    $-4x^4 + 6x^3 + 6x^2 + 2x + 4$

16. $5x^4 - x^3 + 6x^2 - x - 4$

17. $(1 + 4x + 6x^2 + 7x^3) + (5 - 4x + 6x^2 - 7x^3) =$
    $(1 + 5) + (4 - 4)x + (6 + 6)x^2 + (7 - 7)x^3 =$
    $6 + 0x + 12x^2 + 0x^3 = 6 + 12x^2$

18. $3x^4 - 4x^3 + x^2 + x + 4$

19. $(5x^4 - 6x^3 - 7x^2 + x - 1) + (4x^3 - 6x + 1) =$
    $5x^4 + (-6 + 4)x^3 - 7x^2 + (1 - 6)x + (-1 + 1) =$
    $5x^4 - 2x^3 - 7x^2 - 5x + 0 = 5x^4 - 2x^3 - 7x^2 - 5x$

20. $8x^5 - 4x^4 - 3x^3 - x + 5$

21. $(9x^8 - 7x^4 + 2x^2 + 5) + (8x^7 + 4x^4 - 2x) =$
    $9x^8 + 8x^7 + (-7 + 4)x^4 + 2x^2 - 2x + 5 =$
    $9x^8 + 8x^7 - 3x^4 + 2x^2 - 2x + 5$

22. $4x^5 + 9x^2 + 1$

23. $\left[\frac{1}{4}x^4 + \frac{2}{3}x^3 + \frac{5}{8}x^2 + 7\right] + \left[-\frac{3}{4}x^4 + \frac{3}{8}x^2 - 7\right] =$
    $\left(\frac{1}{4} - \frac{3}{4}\right)x^4 + \frac{2}{3}x^3 + \left(\frac{5}{8} + \frac{3}{8}\right)x^2 + (7 - 7) =$
    $-\frac{2}{4}x^4 + \frac{2}{3}x^3 + \frac{8}{8}x^2 + 0 =$
    $-\frac{1}{2}x^4 + \frac{2}{3}x^3 + x^2$

24. $\frac{2}{15}x^9 - \frac{2}{5}x^5 + \frac{1}{4}x^4 + \frac{1}{4}x^2 + \frac{15}{2}$

25. $(0.02x^5 - 0.2x^3 + x + 0.08) + (-0.01x^5 + x^4 - 0.8x - 0.02) =$
    $(0.02 - 0.01)x^5 + x^4 - 0.2x^3 + (1 - 0.8)x + (0.08 - 0.02) =$
    $0.01x^5 + x^4 - 0.2x^3 + 0.2x + 0.06$

26. $0.10x^6 + 0.02x^3 + 0.22x + 0.55$

27. $-3x^4 + 6x^2 + 2x - 1$
    $ -3x^2 + 2x + 1$
    $\overline{-3x^4 + 3x^2 + 4x + 0}$
    $-3x^4 + 3x^2 + 4x$

28. $-4x^3 + 4x^2 + 6x$

29. Rewrite the problem so that the coefficients of like terms have a common denominator.
    $-\frac{2}{4}x^4 - \frac{3}{4}x^3 \phantom{+ \frac{2}{4}x^2} + \frac{24}{4}x$
    $\phantom{-\frac{2}{4}x^4} \frac{2}{4}x^3 + \frac{2}{4}x^2 + \frac{1}{4}x$
    $\frac{3}{4}x^4 \phantom{+ \frac{2}{4}x^3} + \frac{1}{2}x^2 + \frac{2}{4}x + \frac{1}{4}$
    $\overline{\frac{1}{4}x^4 - \frac{1}{4}x^3 + \frac{3}{2}x^2 + \frac{27}{4}x + \frac{1}{4}}$

30. $\frac{1}{4}x^4 + \frac{1}{4}x^3 - \frac{1}{2}x^2 + 3x + \frac{1}{2}$

31. $1 + 5x - 6x^2 + 6x^3 - 3x^4$
    $ -5x - 3x^3 + 5x^5$
    $-8 + 7x^2 + 4x^4$
    $1 + 3x - 2x^5$
    $\overline{-6 + 3x + x^2 + 3x^3 + x^4 + 3x^5}$

32. $3 + 2x + 3x^2 + 2x^3 - 2x^4 - 2x^5$

Chapter 4   (4.4)

33. Rewrite the problem so the coefficients of like terms have the same number of decimal places.

$0.15x^4 + 0.10x^3 - 0.90x^2$
$ -0.01x^3 + 0.01x^2 + x$
$1.25x^4 + 0.11x^2 + 0.01$
$ 0.27x^3 + 0.99$
$-0.35x^4 + 15.00x^2 - 0.03$
$\overline{1.05x^4 + 0.36x^3 + 14.22x^2 + x + 0.97}$

34. $1.3x^4 + 0.35x^3 + 9.53x^2 + 2x + 0.96$

35. a)

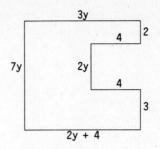

The area of a rectangle is the product of the length and the width. The sum of the areas is found as follows:

Area of A + Area of B + Area of C + Area of D
$= 3x \cdot x + x \cdot x + x \cdot x + 4 \cdot x$
$= 3x^2 + x^2 + x^2 + 4x$
$= 5x^2 + 4x$

A polynomial for the sum of the areas is $5x^2 + 4x$.

b) For $x = 3$: $5x^2 + 4x = 5 \cdot 3^2 + 4 \cdot 3 =$
$5 \cdot 9 + 4 \cdot 3 = 45 + 12 = 57$

When $x = 3$, the sum of the areas is 57 square units.

For $x = 8$: $5x^2 + 4x = 5 \cdot 8^2 + 4 \cdot 8 =$
$5 \cdot 64 + 4 \cdot 8 = 320 + 32 = 352$

When $x = 8$, the sum of the areas is 352 square units.

36. a) $r^2\pi + 13\pi$

b) 38π, 140.69π

37.

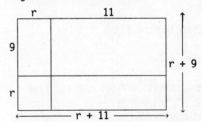

The perimeter is the sum of the lengths of the sides. The sum of the lengths is found as follows:

$3y + 7y + (2y + 4) + 3 + 4 + 2y + 4 + 2 =$
$(3 + 7 + 2 + 2)y + (4 + 3 + 4 + 4 + 2) =$
$14y + 17$

A polynomial for the perimeter of the figure is $14y + 17$.

38. $11\frac{1}{2}a + 10$

39.

```
     r        11
  +-----+------------+
  |     |            |
9 |  A  |     B      | 9
  |     |            |
  +-----+------------+
r |  C  |     D      | r
  +-----+------------+
     r        11
```

The area of the figure can be found by finding the sum of the areas of the four rectangles A, B, C, and D. The area of a rectangle is the product of the length and the width.

Area of A + Area of B + Area of C + Area of D
$= 9 \cdot r + 11 \cdot 9 + r \cdot r + 11 \cdot r$
$= 9r + 99 + r^2 + 11r$
$= r^2 + 20r + 99$

An algebraic expression for the area of the figure is $r^2 + 20r + 99$.

```
     r        11
  +-----+------------+  ↑
  |     |            |
9 |     |            |
  |     |            |  r + 9
  +-----+------------+
r |     |            |
  +-----+------------+  ↓
  ←———— r + 11 ————→
```

The length and width of the figure can be expressed as $r + 11$ and $r + 9$, respectively. The area of this figure (a rectangle) is the product of the length and width. An algebraic expression for the area is $(r + 11) \cdot (r + 9)$.

Both algebraic expressions $r^2 + 20r + 99$ and $(r + 11) \cdot (r + 9)$ represent the same area.

40. $20 + 5(m - 4) + 4(m - 5) + (m - 4)(m - 5) = m^2$

Chapter 4   (4.5)

41.

```
    ←——— x + 3 ———→
         x        3
  ┌──────────┬─────┐
  │          │     │
x │    A     │  B  │ x
  │          │     │    ↑
  │          │     │  x + 3
  ├──────────┼─────┤    ↓
3 │    C     │  D  │ 3
  └──────────┴─────┘
         x        3
```

The length and width of the figure can each be expressed as $x + 3$. The area can be expressed as $(x + 3) \cdot (x + 3)$, or $(x + 3)^2$. Another way to express the area is to find an expression for the sum of the areas of the four rectangles A, B, C, and D. The area of each rectangle is the product of its length and width.

$$\begin{array}{c}\text{Area} \\ \text{of A}\end{array} + \begin{array}{c}\text{Area} \\ \text{of B}\end{array} + \begin{array}{c}\text{Area} \\ \text{of C}\end{array} + \begin{array}{c}\text{Area} \\ \text{of D}\end{array}$$

$$x \cdot x + 3 \cdot x + 3 \cdot x + 3 \cdot 3$$
$$= x^2 + 6x + 9$$

The algebraic expressions $(x + 3)^2$ and $x^2 + 6x + 9$ represent the same area.

$$(x + 3)^2 = x^2 + 6x + 9$$

42. -3.2

43. $9x + 54y - 63 = 9x + 9 \cdot 6y - 9 \cdot 7 = 9(x + 6y - 7)$

44. $-4x + 7y - 3$

45. When $x = 6$, $-(-x) = -(-6) = 6$.

46. $48.544x^6 - 0.795x^5 + 890x$

Exercise Set 4.5

1. Two equivalent expressions for the additive inverse of $-5x$ are

    a) $-(-5x)$ and

    b) $5x$.   (Changing the sign)

2. $-(x^2 - 3x)$, $-x^2 + 3x$

3. Two equivalent expressions for the additive inverse of $-x^2 + 10x - 2$ are

    a) $-(-x^2 + 10x - 2)$ and

    b) $x^2 - 10x + 2$.   (Changing the sign of every term)

4. $-(-4x^3 - x^2 - x)$, $4x^3 + x^2 + x$

5. Two equivalent expressions for the additive inverse of $12x^4 - 3x^3 + 3$ are

    a) $-(12x^4 - 3x^3 + 3)$ and

    b) $-12x^4 + 3x^3 - 3$.   (Changing the sign of every term)

6. $-(4x^3 - 6x^2 - 8x + 1)$, $-4x^3 + 6x^2 + 8x - 1$

7. We change the sign of every term inside parentheses.

    $-(3x - 7) = -3x + 7$

8. $2x - 4$

9. We change the sign of every term inside parentheses.

    $-(4x^2 - 3x + 2) = -4x^2 + 3x - 2$

10. $6a^3 - 2a^2 + 9a - 1$

11. We change the sign of every term inside parentheses.

    $-\left[-4x^4 - 6x^2 + \frac{3}{4}x - 8\right] = 4x^4 + 6x^2 - \frac{3}{4}x + 8$

12. $5x^4 - 4x^3 + x^2 - 0.9$

13. $(3x + 2) - (-4x + 3) = 3x + 2 + 4x - 3$
    Changing the sign of every term inside parentheses
    $= 7x - 1$

14. $13x - 1$

15. $(-6x + 2) - (x^2 + x - 3) = -6x + 2 - x^2 - x + 3$
    $= -x^2 - 7x + 5$

16. $x^2 - 13x + 13$

17. $(x^2 - 9) - (x^2 + 9) = x^2 - 9 - x^2 - 9 = -18$

18. $-x^3 + 6x^2$

19. $(3x^2 - 5x + 10) - (2x^2 + 8x - 40)$
    $= 3x^2 - 5x + 10 - 2x^2 - 8x + 40$
    $= x^2 - 13x + 50$

20. $-5x^2 + 2x$

21. $(6x^4 + 3x^3 - 1) - (4x^2 - 3x + 3)$
    $= 6x^4 + 3x^3 - 1 - 4x^2 + 3x - 3$
    $= 6x^4 + 3x^3 - 4x^2 + 3x - 4$

22. $-3x^3 + x^2 + 2x - 3$

23. $(1.2x^3 + 4.5x^2 - 3.8x) - (-3.4x^3 - 4.7x^2 + 23)$
    $= 1.2x^3 + 4.5x^2 - 3.8x + 3.4x^3 + 4.7x^2 - 23$
    $= 4.6x^3 + 9.2x^2 - 3.8x - 23$

24. $-1.8x^4 - 0.6x^2 - 1.8x + 4.6$

25. $(5x^2 + 6) - (3x^2 - 8) = 5x^2 + 6 - 3x^2 + 8$
    $= 2x^2 + 14$

26. $7x^3 - 9x^2 - 2x + 10$

Chapter 4   (4.5)

27. $(6x^5 - 3x^4 + x + 1) - (8x^5 + 3x^4 - 1)$
$= 6x^5 - 3x^4 + x + 1 - 8x^5 - 3x^4 + 1$
$= -2x^5 - 6x^4 + x + 2$

28. $-x^2 - 2x + 4$

29. $(6x^2 + 2x) - (-3x^2 - 7x + 8)$
$= 6x^2 + 2x + 3x^2 + 7x - 8$
$= 9x^2 + 9x - 8$

30. $7x^3 + 3x^2 + 2x - 1$

31. $\frac{5}{8}x^3 - \frac{1}{4}x - \frac{1}{3} - \left[-\frac{1}{8}x^3 + \frac{1}{4}x - \frac{1}{3}\right]$
$= \frac{5}{8}x^3 - \frac{1}{4}x - \frac{1}{3} + \frac{1}{8}x^3 - \frac{1}{4}x + \frac{1}{3}$
$= \frac{6}{8}x^3 - \frac{2}{4}x$
$= \frac{3}{4}x^3 - \frac{1}{2}x$

32. $\frac{3}{5}x^3 - 0.11$

33. $(0.08x^3 - 0.02x^2 + 0.01x) - (0.02x^3 + 0.03x^2 - 1)$
$= 0.08x^3 - 0.02x^2 + 0.01x - 0.02x^3 - 0.03x^2 + 1$
$= 0.06x^3 - 0.05x^2 + 0.01x + 1$

34. $0.1x^4 - 0.9$

35. $x^2 + 5x + 6$
$x^2 + 2x$

$x^2 + 5x + 6$
$\underline{-x^2 - 2x}$ Changing signs
$ 3x + 6$ Adding

36. $-x^2 + 1$

37. $x^4 - 3x^2 + x + 1$
$\underline{x^4 - 4x^3}$

$x^4 - 3x^2 + x + 1$
$\underline{-x^4 + 4x^3}$ Changing signs
$4x^3 - 3x^2 + x + 1$ Adding

38. $-3x^2 - 14x + 4$

39. $5x^4 + 6x^3 - 9x^2$
$\underline{-6x^4 - 6x^3 + 8x + 9}$

$5x^4 + 6x^3 - 9x^2$
$\underline{6x^4 + 6x^3 - 8x - 9}$ Changing signs
$11x^4 + 12x^3 - 9x^2 - 8x - 9$ Adding

40. $5x^4 - 6x^3 - x^2 + 5x + 15$

41. $3x^4 + 6x^2 + 8x - 1$
$\underline{4x^5 - 6x^4 - 8x - 7}$

$3x^4 + 6x^2 + 8x - 1$
$\underline{-4x^5 + 6x^4 + 8x + 7}$ Changing signs
$-4x^5 + 9x^4 + 6x^2 + 16x + 6$ Adding

42. $-4x^5 - 6x^3 + 8x^2 - 5x - 2$

43. $x^5 - 1$
$\underline{x^5 - x^4 + x^3 - x^2 + x - 1}$

$x^5 - 1$
$\underline{-x^5 + x^4 - x^3 + x^2 - x + 1}$ Changing signs
$x^4 - x^3 + x^2 - x$ Adding

44. $2x^4 - 2x^3 + 2x^2$

45.

Recall that the area of a circle is the product of π and the square of the radius, r^2. $A = \pi r^2$

$\dfrac{\text{Area of circle}}{\text{with radius } r} - \dfrac{\text{Area of circle}}{\text{with radius } 3} = \dfrac{\text{Shaded}}{\text{area}}$

$\pi \cdot r^2 \quad - \quad \pi \cdot 3^2 \quad = \text{Shaded area}$

A polynomial for the shaded area is $\pi r^2 - 9\pi$.

46. $m^2 - 28$

47.

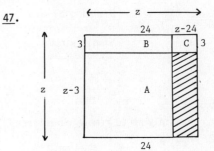

We label the sides of A, B, and C with additional information. The area of the square is $z \cdot z$, or z^2. The area of the shaded section is z^2 minus the areas of sections A, B, and C.

$\begin{array}{l}\text{Area of}\\\text{shaded}\\\text{section}\end{array} = \begin{array}{l}\text{Area}\\\text{of}\\\text{square}\end{array} - \begin{array}{l}\text{Area}\\\text{of}\\\text{A}\end{array} - \begin{array}{l}\text{Area}\\\text{of}\\\text{B}\end{array} - \begin{array}{l}\text{Area}\\\text{of}\\\text{C}\end{array}$

$\begin{array}{l}\text{Area of}\\\text{shaded}\\\text{section}\end{array} = z \cdot z - 24(z - 3) - 3 \cdot 24 - 3(z - 24)$

$= z^2 - 24z + 72 - 72 - 3z + 72$
$= z^2 - 27z + 72$

A polynomial for the shaded area is $z^2 - 27z + 72$.

Chapter 4   (4.6)

48. $(\pi - 2)x^2$

49.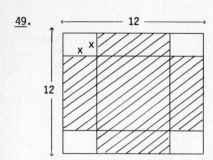

The area of the square is 12·12, or 144.

The area of each corner that is not shaded is x·x, or x^2.

The sum of the areas of all four corners is $4x^2$.

Area of shaded section = Area of square − Area of four corners

Area of shaded section = 144 − $4x^2$

A polynomial for the shaded area is $144 - 4x^2$.

50. $(y - 2)^2 = y^2 - 2(y - 2) - 2(y - 2) - 2\cdot 2$

51. $1.5x - 2.7x = 23 - 5.6x$
    $10(1.5x - 2.7x) = 10(23 - 5.6x)$ Clearing decimals
    $15x - 27x = 230 - 56x$
    $-12x = 230 - 56x$ Collecting like terms
    $44x = 230$ Adding 56x
    $x = \frac{230}{44}$ Dividing by 44
    $x = \frac{115}{22}$ Simplifying

52. 1

53. $8(x - 2) = 16$
    $8x - 16 = 16$ Multiplying to remove parentheses
    $8x = 32$ Adding 16
    $x = 4$ Dividing by 8

54. $-\frac{76}{3}$

55. $(y + 4) + (y - 5) - (y + 8)$
    $= y + 4 + y - 5 - y - 8$
    $= y - 9$

56. $12y^2 - 23y + 21$

57. $(4a^2 - 3a) + (7a^2 - 9a - 13) - (6a - 9)$
    $= 4a^2 - 3a + 7a^2 - 9a - 13 - 6a + 9$
    $= 11a^2 - 18a - 4$

58. $5x^2 - 9x - 1$

59. $(-8y^2 - 4) - (3y + 6) - (2y^2 - y)$
    $= -8y^2 - 4 - 3y - 6 - 2y^2 + y$
    $= -10y^2 - 2y - 10$

60. $4x^3 - 5x^2 + 6$

61. $(-y^4 - 7y^3 + y^2) + (-2y^4 + 5y - 2) - (-6y^3 + y^2)$
    $= -y^4 - 7y^3 + y^2 - 2y^4 + 5y - 2 + 6y^3 - y^2$
    $= -3y^4 - y^3 + 5y - 2$

62. $2 + x + 2x^2 + 4x^3$

63. $(345.099x^3 - 6.178x) - (-224.508x^3 + 8.99x)$
    $= 345.099x^3 - 6.178x + 224.508x^3 - 8.99x$
    $= 569.607x^3 - 15.168x$

64. No, $-3(-x)^2 = -3x^2$

65. a) $R - C = 280x - 0.4x^2 - (5000 + 0.6x^2)$
    $= 280x - 0.4x^2 - 5000 - 0.6x^2$
    $= -x^2 + 280x - 5000$
    A polynomial for total profit is $-x^2 + 280x - 5000$.

    b) Evaluate the polynomial for $x = 75$:
    $-x^2 + 280x - 5000 = -(75)^2 + 280(75) - 5000$
    $= -5625 + 21,000 - 5000$
    $= 10,375$
    The total profit is $10,375.

    c) Evaluate the polynomial for $x = 100$:
    $-x^2 + 280x - 5000 = -(100)^2 + 280(100) - 5000$
    $= -10,000 + 28,000 - 5000$
    $= 13,000$
    The total profit from the production and sale of 100 stereos is $13,000.

Exercise 4.6

1. $(6x^2)(7) = (6\cdot 7)x^2 = 42x^2$

2. $-10x^2$

3. $(-x^3)(-x) = (-1x^3)(-1x) = (-1)(-1)(x^3\cdot x) = x^4$

4. $-x^6$

5. $(-x^5)(x^3) = (-1x^5)(1x^3) = (-1)(1)(x^5\cdot x^3) = -x^8$

6. x^8

7. $(3x^4)(2x^2) = (3\cdot 2)(x^4\cdot x^2) = 6x^6$

8. $20x^8$

Chapter 4   (4.6)

9. $(7t^5)(4t^3) = (7\cdot 4)(t^5\cdot t^3) = 28t^8$

10. $30a^4$

11. $(-0.1x^6)(0.2x^4) = (-0.1)(0.2)(x^6\cdot x^4) = -0.02x^{10}$

12. $-0.12x^9$

13. $\left[-\frac{1}{5}x^3\right]\left[-\frac{1}{3}x\right] = \left[-\frac{1}{5}\right]\left[-\frac{1}{3}\right](x^3\cdot x) = \frac{1}{15}x^4$

14. $-\frac{1}{20}x^{12}$

15. $(-4x^2)(0) = 0$ Any number multiplied by 0 is 0.

16. $4m^5$

17. $(3x^2)(-4x^3)(2x^6) = (3)(-4)(2)(x^2\cdot x^3\cdot x^6) = -24x^{11}$

18. $60y^{12}$

19. $3x(-x + 5) = 3x(-x) + 3x(5)$
    $= -3x^2 + 15x$

20. $8x^2 - 12x$

21. $4x(x + 1) = 4x(x) + 4x(1)$
    $= 4x^2 + 4x$

22. $3x^2 + 6x$

23. $-3x(x - 1) = -3x(x) - 3x(-1)$
    $= -3x^2 + 3x$

24. $5x^2 + 5x$

25. $x^2(x^3 + 1) = x^2(x^3) + x^2(1)$
    $= x^5 + x^2$

26. $-2x^5 + 2x^3$

27. $3x(2x^2 - 6x + 1) = 3x(2x^2) + 3x(-6x) + 3x(1)$
    $= 6x^3 - 18x^2 + 3x$

28. $-8x^4 + 24x^3 + 20x^2 - 4x$

29. $4x^2(3x + 6) = 4x^2(3x) + 4x^2(6)$
    $= 12x^3 + 24x^2$

30. $-10x^3 + 5x^2$

31. $-6x^2(x^2 + x) = -6x^2(x^2) - 6x^2(x)$
    $= -6x^4 - 6x^3$

32. $-4x^4 + 4x^3$

33. $3y^2(6y^4 + 8y^3) = 3y^2(6y^4) + 3y^2(8y^3)$
    $= 18y^6 + 24y^5$

34. $4y^7 - 24y^6$

35. $3x^4(14x^{50} + 20x^{11} + 6x^{57} + 60x^{15})$
    $= 3x^4(14x^{50}) + 3x^4(20x^{11}) + 3x^4(6x^{57}) + 3x^4(60x^{15})$
    $= 42x^{54} + 60x^{15} + 18x^{61} + 180x^{19}$

36. $20x^{38} - 50x^{25} + 25x^{14}$

37. $-4a^7(20a^{19} + 6a^{15} - 5a^{12} + 14a)$
    $= -4a^7(20a^{19}) - 4a^7(6a^{15}) - 4a^7(-5a^{12}) - 4a^7(14a)$
    $= -80a^{26} - 24a^{22} + 20a^{19} - 56a^8$

38. $-66y^{108} + 42y^{58} - 66y^{49} + 360y^{12} - 54y^8$

39. $(x + 6)(x + 3) = (x + 6)x + (x + 6)3$
    $= x\cdot x + 6\cdot x + x\cdot 3 + 6\cdot 3$
    $= x^2 + 6x + 3x + 18$
    $= x^2 + 9x + 18$

40. $x^2 + 7x + 10$

41. $(x + 5)(x - 2) = (x + 5)x + (x + 5)(-2)$
    $= x\cdot x + 5\cdot x + x(-2) + 5(-2)$
    $= x^2 + 5x - 2x - 10$
    $= x^2 + 3x - 10$

42. $x^2 + 4x - 12$

43. $(x - 4)(x - 3) = (x - 4)x + (x - 4)(-3)$
    $= x\cdot x - 4\cdot x + x(-3) - 4(-3)$
    $= x^2 - 4x - 3x + 12$
    $= x^2 - 7x + 12$

44. $x^2 - 10x + 21$

45. $(x + 3)(x - 3) = (x + 3)x + (x + 3)(-3)$
    $= x\cdot x + 3\cdot x + x(-3) + 3(-3)$
    $= x^2 + 3x - 3x - 9$
    $= x^2 - 9$

46. $x^2 - 36$

47. $(5 - x)(5 - 2x) = (5 - x)5 + (5 - x)(-2x)$
    $= 5\cdot 5 - x\cdot 5 + 5(-2x) - x(-2x)$
    $= 25 - 5x - 10x + 2x^2$
    $= 25 - 15x + 2x^2$

48. $18 + 12x + 2x^2$

49. $(2x + 5)(2x + 5) = (2x + 5)2x + (2x + 5)5$
    $= 2x\cdot 2x + 5\cdot 2x + 2x\cdot 5 + 5\cdot 5$
    $= 4x^2 + 10x + 10x + 25$
    $= 4x^2 + 20x + 25$

50. $9x^2 - 24x + 16$

Chapter 4   (4.6)

51. $(3y - 4)(3y + 4) = (3y - 4)3y + (3y - 4)4$
$= 3y \cdot 3y - 4 \cdot 3y + 3y \cdot 4 - 4 \cdot 4$
$= 9y^2 - 12y + 12y - 16$
$= 9y^2 - 16$

52. $4y^2 - 1$

53. $\left(x - \frac{5}{2}\right)\left(x + \frac{2}{5}\right) = \left(x - \frac{5}{2}\right)x + \left(x - \frac{5}{2}\right)\frac{2}{5}$
$= x \cdot x - \frac{5}{2} \cdot x + x \cdot \frac{2}{5} - \frac{5}{2} \cdot \frac{2}{5}$
$= x^2 - \frac{5}{2}x + \frac{2}{5}x - 1$
$= x^2 - \frac{25}{10}x + \frac{4}{10}x - 1$
$= x^2 - \frac{21}{10}x - 1$

54. $x^2 + \frac{17}{6}x + 2$

55. $(x^2 + x + 1)(x - 1)$
$= (x^2 + x + 1)x + (x^2 + x + 1)(-1)$
$= x^2 \cdot x + x \cdot x + 1 \cdot x + x^2(-1) + x(-1) + 1(-1)$
$= x^3 + x^2 + x - x^2 - x - 1$
$= x^3 - 1$

56. $x^3 + x^2 + 4$

57. $(2x^2 + 6x + 1)(2x + 1)$
$= (2x^2 + 6x + 1)2x + (2x^2 + 6x + 1)1$
$= 2x^2 \cdot 2x + 6x \cdot 2x + 1 \cdot 2x + 2x^2 \cdot 1 + 6x \cdot 1 + 1 \cdot 1$
$= 4x^3 + 12x^2 + 2x + 2x^2 + 6x + 1$
$= 4x^3 + 14x^2 + 8x + 1$

58. $12x^3 - 10x^2 - x + 1$

59. $(3y^2 - 6y + 2)(y^2 - 3)$
$= (3y^2 - 6y + 2)y^2 + (3y^2 - 6y + 2)(-3)$
$= 3y^2 \cdot y^2 - 6y \cdot y^2 + 2 \cdot y^2 + 3y^2(-3) - 6y(-3) + 2(-3)$
$= 3y^4 - 6y^3 + 2y^2 - 9y^2 + 18y - 6$
$= 3y^4 - 6y^3 - 7y^2 + 18y - 6$

60. $3y^4 + 18y^3 - 18y - 3$

61. $(x^3 + x^2 - x)(x^3 + x^2)$
$= (x^3 + x^2 - x)x^3 + (x^3 + x^2 - x)x^2$
$= x^3 \cdot x^3 + x^2 \cdot x^3 - x \cdot x^3 + x^3 \cdot x^2 + x^2 \cdot x^2 - x \cdot x^2$
$= x^6 + x^5 - x^4 + x^5 + x^4 - x^3$
$= x^6 + 2x^5 - x^3$

62. $x^6 - 2x^5 + 2x^4 - x^3$

63. $(-5x^3 - 7x^2 + 1)(2x^2 - x)$
$= (-5x^3 - 7x^2 + 1)2x^2 + (-5x^3 - 7x^2 + 1)(-x)$
$= -5x^3 \cdot 2x^2 - 7x^2 \cdot 2x^2 + 1 \cdot 2x^2 - 5x^3(-x) - 7x^2(-x) + 1(-x)$
$= -10x^5 - 14x^4 + 2x^2 + 5x^4 + 7x^3 - x$
$= -10x^5 - 9x^4 + 7x^3 + 2x^2 - x$

64. $-20x^5 + 25x^4 - 4x^3 - 5x^2 - 2$

65. $1 + x + x^2$ Line up like terms
$-1 - x + x^2$ in columns
$x^2 + x^3 + x^4$ Multiplying the top row by x^2
$-x - x^2 - x^3$ Multiplying by $-x$
$\underline{-1 - x - x^2}$ Multiplying by -1
$-1 - 2x - x^2 + x^4$

66. $1 - 2x + 3x^2 - 2x^3 + x^4$

67. $2x^2 + 3x - 4$
$2x^2 + x - 2$
$-4x^2 - 6x + 8$ Multiplying by -2
$2x^3 + 3x^2 - 4x$ Multiplying by x
$\underline{4x^4 + 6x^3 - 8x^2}$ Multiplying by $2x^2$
$4x^4 + 8x^3 - 9x^2 - 10x + 8$

68. $4x^4 - 12x^3 - 5x^2 + 17x + 6$

69. $2t^2 - t - 4$
$3t^2 + 2t - 1$
$-2t^2 + t + 4$ Multiplying by -1
$4t^3 - 2t^2 - 8t$ Multiplying by $2t$
$\underline{6t^4 - 3t^3 - 12t^2}$ Multiplying by $3t^2$
$6t^4 + t^3 - 16t^2 - 7t + 4$

70. $6a^4 - 19a^3 + 31a^2 - 26a + 8$

71. $2x^2 + x - 2$
$-2x^2 + 4x - 5$
$-10x^2 - 5x + 10$ Multiplying by -5
$8x^3 + 4x^2 - 8x$ Multiplying by $4x$
$\underline{-4x^4 - 2x^3 + 4x^2}$ Multiplying by $-2x^2$
$-4x^4 + 6x^3 - 2x^2 - 13x + 10$

72. $-6x^4 + 4x^3 + 36x^2 - 20x + 2$

73. $x - x^3 + x^5$
$\underline{-1 + x^2 + x^4}$ Rewriting in ascending order
$x^5 - x^7 + x^9$ Multiplying by x^4
$x^3 - x^5 + x^7$ Multiplying by x^2
$\underline{-x + x^3 - x^5}$ Multiplying by -1
$-x + 2x^3 - x^5 + x^9$

74. $3x^3 + 3x^7 + 3x^{11}$

78

Chapter 4   (4.6)

75. $x^3 + x^2 + x + 1$
           $x - 1$
       ―――――――――――――――
       $-x^3 - x^2 - x - 1$
       $x^4 + x^3 + x^2 + x$
       ―――――――――――――――
       $x^4 - 1$

76. $x^4 - 3x^3 + 3x^2 - 4x + 4$

77. $x^3 + x^2 - x - 3$
                    $x - 3$
       ―――――――――――――――
       $-3x^3 - 3x^2 + 3x + 9$
       $x^4 + x^3 - x^2 - 3x$
       ―――――――――――――――
       $x^4 - 2x^3 - 4x^2 + 9$

78. $x^4 + 3x^3 - 5x^2 + 16$

79. $-\frac{1}{4} - \frac{1}{2} = -\frac{1}{4} - \frac{1}{2} \cdot \frac{2}{2} = -\frac{1}{4} - \frac{2}{4} = -\frac{3}{4}$

80. $4(4x - 6y + 9)$

81. $(a + b)^2 = (a + b)(a + b)$
             $= (a + b)a + (a + b)b$
             $= a^2 + ab + ab + b^2$
             $= a^2 + 2ab + b^2$

82. $a^2 - 2ab + b^2$

83. $(2x + 3)^2 = (2x + 3)(2x + 3)$
              $= (2x + 3)2x + (2x + 3)3$
              $= 4x^2 + 6x + 6x + 9$
              $= 4x^2 + 12x + 9$

84. $25y^2 + 60y + 36$

85.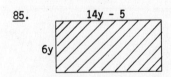

The shaded area is the product of the length and width of the rectangle:
$6y(14y - 5) = 6y \cdot 14y + 6y(-5)$
             $= 84y^2 - 30y$

86. $78t^2 + 40t$

87.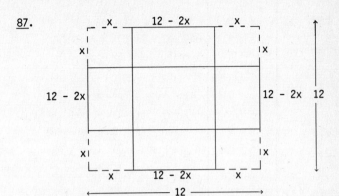

The dimensions of the box are $12 - 2x$ by $12 - 2x$ by x. The volume is the product of the dimensions (volume = length × width × height):

Volume $= (12 - 2x)(12 - 2x)x$
       $= (144 - 48x + 4x^2)x$
       $= 144x - 48x^2 + 4x^3$

The outside surface area is the sum of the area of the bottom and the areas of the four sides. The dimensions of the bottom are $12 - 2x$ by $12 - 2x$, and the dimensions of each side are x by $12 - 2x$.

Surface area = Area of bottom + 4·Area of each side
             $= (12 - 2x)(12 - 2x) + 4 \cdot x(12 - 2x)$
             $= 144 - 48x + 4x^2 + 48x - 8x^2$
             $= 144 - 4x^2$

88. $x^3 - 5x^2 + 8x - 4$

89. Let b = the length of the base. Then b + 4 = the height. Let A represent the area.

Area $= \frac{1}{2} \times$ base $\times$ height

$A = \frac{1}{2} \cdot b \cdot (b + 4)$

$A = \frac{1}{2}b(b + 4)$

$A = \frac{1}{2}b^2 + 2b$

90. 8 ft by 16 ft

91. a) $(x + 3)(x + 6) + (x + 3)(x + 6)$
       $= (x + 3)x + (x + 3)6 + (x + 3)x + (x + 3)6$
       $= x^2 + 3x + 6x + 18 + x^2 + 3x + 6x + 18$
       $= 2x^2 + 18x + 36$

    b) $(x + 4)(x + 5) - (x + 4)(x + 5)$
       $= (x + 4)x + (x + 4)5 - [(x + 4)x + (x + 4)5]$
       $= x^2 + 4x + 5x + 20 - (x^2 + 4x + 5x + 20)$
       $= x^2 + 9x + 20 - x^2 - 9x - 20$
       $= 0$

92. a) $2x^2 - 18x + 28$

    b) 0

Chapter 4   (4.7)

93. a) $(x + 5)(x - 3) + (x + 5)(x - 3)$
    $= (x+5)x + (x+5)(-3) + (x+5)x + (x+5)(-3)$
    $= x^2 + 5x - 3x - 15 + x^2 + 5x - 3x - 15$
    $= 2x^2 + 4x - 30$

    b) $(x + 9)(x - 4) - (x + 9)(x - 4)$
    $= (x+9)x + (x+9)(-4) - [(x+9)x + (x+9)(-4)]$
    $= x^2 + 9x - 4x - 36 - (x^2 + 9x - 4x - 36)$
    $= x^2 + 5x - 36 - x^2 - 5x + 36$
    $= 0$

94. a) $2x^2 - 112$

    b) $-6x$

Exercise Set 4.7

1. $(x + 1)(x^2 + 3)$
    $\ \ \ \ \ \,F\ \ \ \ \ \ \ \ O\ \ \ \ \ \ I\ \ \ \ \ L$
    $= x \cdot x^2 + x \cdot 3 + 1 \cdot x^2 + 1 \cdot 3$
    $= x^3 + 3x + x^2 + 3$

2. $x^3 - x^2 - 3x + 3$

3. $(x^3 + 2)(x + 1)$
    $\ \ \ \ \ \ F\ \ \ \ \ \ \ O\ \ \ \ \ I\ \ \ \ L$
    $= x^3 \cdot x + x^3 \cdot 1 + 2 \cdot x + 2 \cdot 1$
    $= x^4 + x^3 + 2x + 2$

4. $x^5 + 12x^4 + 2x + 24$

5. $(y + 2)(y - 3)$
    $\ \ \ F\ \ \ \ \ \ \ O\ \ \ \ \ \ \ I\ \ \ \ \ \ L$
    $= y \cdot y + y \cdot (-3) + 2 \cdot y + 2 \cdot (-3)$
    $= y^2 - 3y + 2y - 6$
    $= y^2 - y - 6$

6. $a^2 + 4a + 4$

7. $(3x + 2)(3x + 3)$
    $\ \ \ \ F\ \ \ \ \ \ \ O\ \ \ \ \ \ I\ \ \ \ L$
    $= 3x \cdot 3x + 3x \cdot 3 + 2 \cdot 3x + 2 \cdot 3$
    $= 9x^2 + 9x + 6x + 6$
    $= 9x^2 + 15x + 6$

8. $8x^2 + 10x + 2$

9. $(5x - 6)(x + 2)$
    $\ \ \ \ F\ \ \ \ \ \ \ O\ \ \ \ \ \ \ I\ \ \ \ \ \ L$
    $= 5x \cdot x + 5x \cdot 2 + (-6) \cdot x + (-6) \cdot 2$
    $= 5x^2 + 10x - 6x - 12$
    $= 5x^2 + 4x - 12$

10. $x^2 - 64$

11. $(3t - 1)(3t + 1)$
    $\ \ \ F\ \ \ \ \ \ O\ \ \ \ \ \ \ \ I\ \ \ \ \ \ \ L$
    $= 3t \cdot 3t + 3t \cdot 1 + (-1) \cdot 3t + (-1) \cdot 1$
    $= 9t^2 + 3t - 3t - 1$
    $= 9t^2 - 1$

12. $4m^2 + 12m + 9$

13. $(4x - 2)(x - 1)$
    $\ \ \ \ F\ \ \ \ \ \ \ O\ \ \ \ \ \ \ I\ \ \ \ \ \ \ L$
    $= 4x \cdot x + 4x \cdot (-1) + (-2) \cdot x + (-2) \cdot (-1)$
    $= 4x^2 - 4x - 2x + 2$
    $= 4x^2 - 6x + 2$

14. $6x^2 - x - 1$

15. $\left(p - \frac{1}{4}\right)\left(p + \frac{1}{4}\right)$
    $\ \ \ F\ \ \ \ \ \ \ O\ \ \ \ \ \ \ \ \ I\ \ \ \ \ \ \ \ L$
    $= p \cdot p + p \cdot \frac{1}{4} + \left(-\frac{1}{4}\right) \cdot p + \left(-\frac{1}{4}\right) \cdot \frac{1}{4}$
    $= p^2 + \frac{1}{4}p - \frac{1}{4}p - \frac{1}{16}$
    $= p^2 - \frac{1}{16}$

16. $q^2 + \frac{3}{2}q + \frac{9}{16}$

17. $(x - 0.1)(x + 0.1)$
    $\ \ \ F\ \ \ \ \ \ \ O\ \ \ \ \ \ \ \ I\ \ \ \ \ \ \ \ \ L$
    $= x \cdot x + x \cdot (0.1) + (-0.1) \cdot x + (-0.1)(0.1)$
    $= x^2 + 0.1x - 0.1x - 0.01$
    $= x^2 - 0.01$

18. $3x^3 + 3x^2 + x + 1$

19. $(2x^2 + 6)(x + 1)$
    $\ \ \ \ F\ \ \ \ O\ \ \ \ I\ \ \ \ L$
    $= 2x^3 + 2x^2 + 6x + 6$

20. $4x^3 - 2x^2 + 6x - 3$

21. $(-2x + 1)(x + 6)$
    $\ \ \ \ \ F\ \ \ \ \ \ O\ \ \ \ I\ \ \ L$
    $= -2x^2 - 12x + x + 6$
    $= -2x^2 - 11x + 6$

22. $6x^2 - 4x - 16$

23. $(a + 7)(a + 7)$
    $\ \ \ F\ \ \ \ O\ \ \ I\ \ \ L$
    $= a^2 + 7a + 7a + 49$
    $= a^2 + 14a + 49$

24. $4y^2 + 20y + 25$

Chapter 4   (4.7)

25. $(1 + 2x)(1 - 3x)$
$$\ \ \text{F}\ \ \ \ \text{O}\ \ \ \ \text{I}\ \ \ \ \text{L}$$
$= 1 - 3x + 2x - 6x^2$
$= 1 - x - 6x^2$

26. $-3x^2 - 5x - 2$

27. $(x^2 + 3)(x^3 - 1)$
$$\ \ \text{F}\ \ \ \ \text{O}\ \ \ \ \text{I}\ \ \ \ \text{L}$$
$= x^5 - x^2 + 3x^3 - 3$

28. $2x^5 + x^4 - 6x - 3$

29. $(x^2 - 2)(x - 1)$
$$\ \ \text{F}\ \ \ \ \text{O}\ \ \ \ \text{I}\ \ \ \ \text{L}$$
$= x^3 - x^2 - 2x + 2$

30. $x^4 - 3x^3 + 2x - 6$

31. $(3x^2 - 2)(x^4 - 2)$
$$\ \ \text{F}\ \ \ \ \text{O}\ \ \ \ \text{I}\ \ \ \ \text{L}$$
$= 3x^6 - 6x^2 - 2x^4 + 4$

32. $x^{20} - 9$

33. $(3x^5 + 2)(2x^2 + 6)$
$$\ \ \text{F}\ \ \ \ \text{O}\ \ \ \ \text{I}\ \ \ \ \text{L}$$
$= 6x^7 + 18x^5 + 4x^2 + 12$

34. $1 + 3x^2 - 2x - 6x^3$

35. $(8x^3 + 1)(x^3 + 8)$
$= 8x^6 + 64x^3 + x^3 + 8$
$= 8x^6 + 65x^3 + 8$

36. $20 - 8x^2 - 10x + 4x^3$

37. $(4x^2 + 3)(x - 3)$
$$\ \ \text{F}\ \ \ \ \text{O}\ \ \ \ \text{I}\ \ \ \ \text{L}$$
$= 4x^3 - 12x^2 + 3x - 9$

38. $14x^2 - 53x + 14$

39. $(4y^4 + y^2)(y^2 + y)$
$$\ \ \text{F}\ \ \ \ \text{O}\ \ \ \ \text{I}\ \ \ \ \text{L}$$
$= 4y^6 + 4y^5 + y^4 + y^3$

40. $10y^{12} + 16y^9 + 6y^6$

41. $4y(y + 5)(2y + 8)$
$= 4y(2y^2 + 8y + 10y + 40)$
$= 4y(2y^2 + 18y + 40)$
$= 8y^3 + 72y^2 + 160y$

42. $80x^3 + 24x^2 - 216x$

43. $[(x + 1) - x^2][(x - 2) + 2x^2]$
$$\ \ \ \ \ \ \ \ \text{F}\ \ \ \ \ \ \ \ \ \ \ \ \ \text{O}\ \ \ \ \ \ \ \ \ \ \ \ \ \text{I}\ \ \ \ \ \ \ \ \ \ \ \ \ \text{L}$$
$= (x+1)(x-2) + (x+1)(2x^2) + (-x^2)(x-2) + (-x^2)(2x^2)$
$= x^2 - 2x + x - 2 + 2x^3 + 2x^2 - x^3 + 2x^2 - 2x^4$
$= -2x^4 + x^3 + 5x^2 - x - 2$

44. $16x^4 - 1$

45. $(x + 2)(x - 5) = (x + 1)(x - 3)$
$x^2 - 5x + 2x - 10 = x^2 - 3x + x - 3$
$x^2 - 3x - 10 = x^2 - 2x - 3$
$-3x - 10 = -2x - 3 \quad \text{Adding } -x^2$
$-3x + 2x = 10 - 3 \quad \text{Adding } 2x \text{ and } 10$
$-x = 7$
$x = -7$

46. 0

47. $(x + 1)(x + 2) = (x + 3)(x + 4)$
$x^2 + 2x + x + 2 = x^2 + 4x + 3x + 12$
$x^2 + 3x + 2 = x^2 + 7x + 12$
$3x + 2 = 7x + 12 \quad \text{Adding } -x^2$
$2 - 12 = 7x - 3x \quad \text{Adding } -3x \text{ and } -12$
$-10 = 4x$
$-\frac{10}{4} = x$
$-\frac{5}{2} = x$

48. $\frac{13}{4}$

49. If w = the width, then $w + 1$ = the length, and $(w + 1) + 1$, or $w + 2$ = the height.
Volume = length × width × height
$\phantom{\text{Volume}}= (w + 1) \cdot w \cdot (w + 2)$
$\phantom{\text{Volume}}= (w^2 + w)(w + 2)$
$\phantom{\text{Volume}}= w^3 + 2w^2 + w^2 + 2w$
$\phantom{\text{Volume}}= w^3 + 3w^2 + 2w$

50. $\ell(\ell - 1)(\ell + 1) = \ell^3 - \ell$

51. If h = the height, then $h - 1$ = the length and $(h - 1) - 1$, or $h - 2$ = the width.
Volume = length × width × height
$\phantom{\text{Volume}}= (h - 1) \cdot (h - 2) \cdot h$
$\phantom{\text{Volume}}= (h^2 - 2h - h + 2)h$
$\phantom{\text{Volume}}= (h^2 - 3h + 2)h$
$\phantom{\text{Volume}}= h^3 - 3h^2 + 2h$

52. $Q(Q - 14) - 5(Q - 14)$, or $(Q - 5)(Q - 14)$

Chapter 4   (4.8)

53.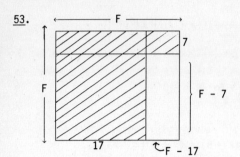

The area of the entire figure is F·F, or F^2.

The area of the section not shaded is
$(F - 7)(F - 17)$

$$\underline{\text{Area of shaded region}} = \underline{\text{Area of entire figure}} - \underline{\text{Area of section not shaded}}$$

Area of shaded region $= F^2 - (F - 7)(F - 17)$

$= F^2 - (F^2 - 24F + 119)$

$= F^2 - F^2 + 24F - 119$

$= 24F - 119$

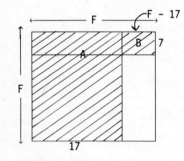

Area of shaded region $=$ Area of A $+$ Area of B

Area of shaded region $= 17F + 7(F - 17)$

$= 17F + 7F - 119$

$= 24F - 119$

54. $y(y - 1)$, or $(y - 1)(y - 1) + 1 \cdot (y - 1)$

Exercise Set 4.8

1. $(x + 4)(x - 4) = x^2 - 4^2$ Product of sum and difference of two terms
   $= x^2 - 16$

2. $x^2 - 1$

3. $(2x + 1)(2x - 1) = (2x)^2 - 1^2$ Product of sum and difference of two terms
   $= 4x^2 - 1$

4. $x^4 - 1$

5. $(5m - 2)(5m + 2) = (5m)^2 - 2^2$ Product of sum and difference of two terms
   $= 25m^2 - 4$

6. $9x^8 - 4$

7. $(2x^2 + 3)(2x^2 - 3) = (2x^2)^2 - 3^2$ Product of sum and difference of two terms
   $= 4x^4 - 9$

8. $36x^{10} - 25$

9. $(3x^4 - 4)(3x^4 + 4) = (3x^4)^2 - 4^2$
   $= 9x^8 - 16$

10. $t^4 - 0.04$

11. $(x^6 - x^2)(x^6 + x^2) = (x^6)^2 - (x^2)^2$
    $= x^{12} - x^4$

12. $4x^6 - 0.09$

13. $(x^4 + 3x)(x^4 - 3x) = (x^4)^2 - (3x)^2$
    $= x^8 - 9x^2$

14. $\frac{9}{16} - 4x^6$

15. $(x^{12} - 3)(x^{12} + 3) = (x^{12})^2 - 3^2$
    $= x^{24} - 9$

16. $144 - 9x^4$

17. $(2y^8 + 3)(2y^8 - 3) = (2y^8)^2 - 3^2$
    $= 4y^{16} - 9$

18. $m^2 - \frac{4}{9}$

19. $(x + 2)^2 = x^2 + 2 \cdot x \cdot 2 + 2^2$ Square of a binomial sum
    $= x^2 + 4x + 4$

20. $4x^2 - 4x + 1$

21. $(3x^2 + 1) = (3x^2)^2 + 2 \cdot 3x^2 \cdot 1 + 1^2$ Square of a binomial sum
    $= 9x^4 + 6x^2 + 1$

22. $9x^2 + \frac{9}{2}x + \frac{9}{16}$

23. $\left(a - \frac{1}{2}\right)^2 = a^2 - 2 \cdot a \cdot \frac{1}{2} + \left(\frac{1}{2}\right)^2$ Square of a binomial difference
    $= a^2 - a + \frac{1}{4}$

24. $4a^2 - \frac{4}{5}a + \frac{1}{25}$

25. $(3 + x)^2 = 3^2 + 2 \cdot 3 \cdot x + x^2$
    $= 9 + 6x + x^2$

Chapter 4   (4.8)

26. $x^6 - 2x^3 + 1$

27. $(x^2 + 1)^2 = (x^2)^2 + 2 \cdot x^2 \cdot 1 + 1^2$
    $= x^4 + 2x^2 + 1$

28. $64x^2 - 16x^3 + x^4$

29. $(2 - 3x^4)^2 = 2^2 - 2 \cdot 2 \cdot 3x^4 + (3x^4)^2$
    $= 4 - 12x^4 + 9x^8$

30. $36x^6 - 24x^3 + 4$

31. $(5 + 6t^2)^2 = 5^2 + 2 \cdot 5 \cdot 6t^2 + (6t^2)^2$
    $= 25 + 60t^2 + 36t^4$

32. $9p^4 - 6p^3 + p^2$

33. $(3 - 2x^3)^2 = 3^2 - 2 \cdot 3 \cdot 2x^3 + (2x^3)^2$
    $= 9 - 12x^3 + 4x^6$

34. $x^2 - 8x^4 + 16x^6$

35. $4x(x^2 + 6x - 3)$ Product of a monomial and a trinomial
    $= 4x \cdot x^2 + 4x \cdot 6x + 4x(-3)$
    $= 4x^3 + 24x^2 - 12x$

36. $-8x^6 + 48x^3 + 72x$

37. $\left(2x^2 - \frac{1}{2}\right)\left(2x^2 - \frac{1}{2}\right)$ Square of a binomial difference
    $= (2x^2)^2 - 2 \cdot 2x^2 \cdot \frac{1}{2} + \left(\frac{1}{2}\right)^2$
    $= 4x^4 - 2x^2 + \frac{1}{4}$

38. $x^4 - 2x^2 + 1$

39. $(-1 + 3p)(1 + 3p)$
    $= (3p - 1)(3p + 1)$ Product of the sum and difference of two terms
    $= (3p)^2 - 1^2$
    $= 9p^2 - 1$, or $-1 + 9p^2$

40. $-9q^2 + 4$

41. $3t^2(5t^3 - t^2 + t)$ Product of a monomial and a trinomial
    $= 3t^2 \cdot 5t^3 + 3t^2(-t^2) + 3t^2 \cdot t$
    $= 15t^5 - 3t^4 + 3t^3$

42. $-6x^5 - 48x^3 + 54x^2$

43. $(6x^4 + 4)^2$ Square of a binomial sum
    $= (6x^4)^2 + 2 \cdot 6x^4 \cdot 4 + 4^2$
    $= 36x^8 + 48x^4 + 16$

44. $64a^2 + 80a + 25$

45. $(3x + 2)(4x^2 + 5)$ Product of two binomials; use FOIL
    $= 3x \cdot 4x^2 + 3x \cdot 5 + 2 \cdot 4x^2 + 2 \cdot 5$
    $= 12x^3 + 15x + 8x^2 + 10$

46. $6x^4 - 3x^2 - 63$

47. $(8 - 6x^4)^2$ Square of binomial difference
    $= 8^2 - 2 \cdot 8 \cdot 6x^4 + (6x^4)^2$
    $= 64 - 96x^4 + 36x^8$

48. $\frac{3}{25}x^4 + 4x^2 - 63$

49.
$$\begin{array}{r} t^2 + t + 1 \\ \underline{t - 1} \\ -t^2 - t - 1 \\ \underline{t^3 + t^2 + t} \\ t^3 - 1 \end{array}$$

50. $y^3 + 125$

51. $3^2 + 4^2 = 9 + 16 = 25$
    $(3 + 4)^2 = 7^2 = 49$

52. 85; 169

53. $9^2 - 5^2 = 81 - 25 = 56$
    $(9 - 5)^2 = 4^2 = 16$

54. 105; 49

55. <u>Familiarize.</u> Let t = the number of watts used by the television set. Then 10t = the number of watts used by the lamps, and 40t = the number of watts used by the air conditioner.
    <u>Translate.</u>

Lamp watts	+	Air conditioner watts	+	Television watts	=	Total watts
10t	+	40t	+	t	=	2550

<u>Carry out.</u> We solve the equation.
$$10t + 40t + t = 2550$$
$$51t = 2550$$
$$t = 50$$

The possible solution is:

   Television, t: 50 watts

   Lamps, 10t: 10·50, or 500 watts

   Air conditioner, 40t: 40·50, or 2000 watts

<u>Check.</u> The number of watts used by the lamps, 500, is 10 times 50, the number used by the television. The number of watts used by the air conditioner, 2000, is 40 times 50, the number used by the television. Also, 50 + 500 + 2000 = 2550, the total wattage used.

<u>State.</u> The television uses 50 watts, the lamps use 500 watts, and the air conditioner uses 2000 watts.

56. $\dfrac{28}{27}$

57. $(67.58x + 3.225)^2$
    $= (67.58x)^2 + 2(67.58x)(3.225) + (3.225)^2$
    $= 4567.0564x^2 + 435.891x + 10.400625$

58. $400 - 4 = 396$

59. 93×107
    $= (100 - 7)(100 + 7)$
    $= 100^2 - 7^2$
    $= 10,000 - 49$
    $= 9951$

60. $16x^4 - 1$

61. $[(a + 5) + 1][(a + 5) - 1]$
    $= (a + 5)^2 - 1^2$
    $= (a^2 + 2 \cdot a \cdot 5 + 5^2) - 1^2$
    $= a^2 + 10a + 25 - 1$
    $= a^2 + 10a + 24$

62. $5a^2 + 12a - 9$

63. $[(x + 3) + 2]^2$
    $= (x + 3)^2 + 2 \cdot (x + 3) \cdot 2 + 2^2$
    $= x^2 + 6x + 9 + 4x + 12 + 4$
    $= x^2 + 10x + 25$

64. $625t^8 - 450t^4 + 81$

65. $x^2 = (x + 10)^2$
    $x^2 = x^2 + 20x + 100$
    $0 = 20x + 100$ Adding $-x^2$
    $-100 = 20x$
    $-5 = x$

66. 6

67. $(x + 4)^2 = (x + 8)(x - 8)$
    $x^2 + 8x + 16 = x^2 - 64$
    $8x + 16 = -64$ Adding $-x^2$
    $8x = -80$
    $x = -10$

68. $\dfrac{1}{17}$

69. a)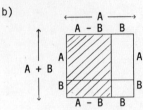

    The area of the entire rectangle is $A(A + B)$, or $A^2 + AB$.

69. (continued)

    b)

    The sum of the areas of the two unshaded rectangles is $A \cdot B + B \cdot B$, or $AB + B^2$.

    c) Area in part a) - area in part b)
    $= A^2 + AB - (AB + B^2)$
    $= A^2 + AB - AB - B^2$
    $= A^2 - B^2$

    d) The area of the shaded region is $(A + B)(A - B) = A^2 - B^2$. This is the same as the polynomial found in part c).

70. 10, 11, 12

71. $(10x + 5)^2 = (10x)^2 + 2 \cdot 10x \cdot 5 + 5^2$
    $= 100x^2 + 100x + 25$
    $= 100(x^2 + x) + 25$

    To square any two-digit number ending in 5 mentally, add the first digit to its square, multiply by 100, and add 25.

Exercise Set 4.9

1. We replace x by 3 and y by -2.
   $x^2 - y^2 + xy = 3^2 - (-2)^2 + 3(-2) = 9 - 4 - 6 = -1$

2. 19

3. We replace x by 2, y by -3, and z by -1.
   $xyz^2 + z = 2(-3)(-1)^2 + (-1) = -6 - 1 = -7$

4. -1

5. Evaluate the polynomial for $P = 10,000$ and $i = 0.08$.
   $A = P(1 + i)^2 = 10,000(1 + 0.08)^2$
   $= 10,000(1.08)^2$
   $= 10,000(1.1664)$
   $= 11,664$

   At 8% interest for 2 years, $10,000 will grow to $11,664.

6. $11,449

Chapter 4   (4.9)

7. Evaluate the polynomial for P = 10,000 and i = 0.08.
$$A = P(1 + i)^3 = 10,000(1 + 0.08)^3$$
$$= 10,000(1.08)^3$$
$$= 10,000(1.259712)$$
$$= 12,597.12$$
At 8% interest for 3 years, $10,000 will grow to $12,597.12.

8. $12,250.43

9. Replace h by 6.5, r by 2.5, and π by 3.14.
$$2\pi rh + 2\pi r^2 \approx 2(3.14)(2.5)(6.5) + 2(3.14)(2.5)^2$$
$$= 2(3.14)(2.5)(6.5) + 2(3.14)(6.25)$$
$$= 102.05 + 39.25$$
$$= 141.3$$
The surface area is 141.3 in².

10. 56.52 in²

11. $x^3y - 2xy + 3x^2 - 5$

Term	Coefficient	Degree	
x^3y	1	4	(Think: $x^3y = x^3y^1$)
$-2xy$	-2	2	(Think: $-2xy = -2x^1y^1$)
$3x^2$	3	2	
-5	-5	0	(Think: $-5 = -5x^0$)

The degree of a polynomial is the degree of the term of highest degree.
The term of highest degree is x^3y. Its degree is 4. The degree of the polynomial is 4.

12. Coefficients: 5, -1, 15, 1
    Degrees: 3, 2, 1, 0; 3

13. $17x^2y^3 - 3x^3yz - 7$

Term	Coefficient	Degree	
$17x^2y^3$	17	5	
$-3x^3yz$	-3	5	(Think: $-3x^3yz = 3x^3y^1z^1$)
-7	-7	0	(Think: $-7 = -7x^0$)

The terms of highest degree are $17x^2y^3$ and $-3x^3yz$. Each has degree 5. The degree of the polynomial is 5.

14. Coefficients: 6, -1, 8, -1
    Degrees: 0, 2, 4, 5; 5

15. $a + b - 2a - 3b = (1 - 2)a + (1 - 3)b = -a - 2b$

16. $y - 7$

17. $3x^2y - 2xy^2 + x^2$
There are no like terms, so none of the terms can be collected.

18. $m^3 + 2m^2n - 3m^2 + 3mn^2$

19. $2u^2v - 3uv^2 + 6u^2v - 2uv^2$
$= (2 + 6)u^2v + (-3 - 2)uv^2$
$= 8u^2v - 5uv^2$

20. $-2x^2 - 4xy - 2y^2$

21. $6au + 3av + 14au + 7av$
$= (6 + 14)au + (3 + 7)av$
$= 20au + 10av$

22. $3x^2y + 3z^2y + 3xy^2$

23. $(2x^2 - xy + y^2) + (-x^2 - 3xy + 2y^2)$
$= (2 - 1)x^2 + (-1 - 3)xy + (1 + 2)y^2$
$= x^2 - 4xy + 3y^2$

24. $6 - z$

25. $(r - 2s + 3) + (2r + s) + (s + 4)$
$= (1 + 2)r + (-2 + 1 + 1)s + (3 + 4)$
$= 3r + 0s + 7$
$= 3r + 7$

26. $-3b^3a^2 - b^2a^3 + 5ba + 3$

27. $(2x^2 - 3xy + y^2) + (-4x^2 - 6xy - y^2) + (x^2 + xy - y^2)$
$= (2 - 4 + 1)x^2 + (-3 - 6 + 1)xy + (1 - 1 - 1)y^2$
$= -x^2 - 8xy - y^2$

28. $3x^3 - x^2y + xy^2 - 3y^3$

29. $(xy - ab) - (xy - 3ab)$
$= xy - ab - xy + 3ab$
$= (1 - 1)xy + (-1 + 3)ab$
$= 0xy + 2ab$
$= 2ab$

30. $x^4y^2 + y + 2x$

31. $(-2a + 7b - c) - (-3b + 4c - 8d)$
$= -2a + 7b - c + 3b - 4c + 8d$
$= -2a + (7 + 3)b + (-1 - 4)c + 8d$
$= -2a + 10b - 5c + 8d$

32. $-5b$

33. $\overset{FOIL}{(3z - u)(2z + 3u)} = 6z^2 + 9zu - 2uz - 3u^2$
$= 6z^2 + 7zu - 3u^2$

34. $a^3 + a^2b - ab^2 - b^3$

35. $\overset{FOIL}{(a^2b - 2)(a^2b - 5)} = a^4b^2 - 5a^2b - 2a^2b + 10$
$= a^4b^2 - 7a^2b + 10$

Chapter 4   (4.9)

36. $x^2y^2 + 3xy - 28$

37. $(a + a^2 - 1)(a^2 + 1 - y)$
    $= (a^2 + a - 1)(a^2 - y + 1)$

    $ a^2 + a - 1$
    $ a^2 - y + 1$
    $ \overline{}$
    $ a^2 + a - 1$
    $ -a^2y - ay + y$
    $\underline{a^4 + a^3 - a^2}$
    $a^4 + a^3 - a^2y - ay + y + a - 1$

38. $rvx + rs + tvx^2 + stx$

39. $(a^3 + bc)(a^3 - bc) = (a^3)^2 - (bc)^2$
    $ [(A + B)(A - B) = A^2 - B^2]$
    $ = a^6 - b^2c^2$

40. $m^4 + m^2n^2 + n^4$

41. $ y^4x + y^2 + 1$
    $ y^2 + 1$
    $ \overline{}$
    $ y^4x + y^2 + 1$
    $\underline{y^6x + y^4 + y^2}$
    $y^6x + y^4 + y^4x + 2y^2 + 1$

42. $a^3 - b^3$

43. $(3xy - 1)(4xy + 2)$
    $ F O I L$
    $= 12x^2y^2 + 6xy - 4xy - 2$
    $= 12x^2y^2 + 2xy - 2$

44. $m^6n^2 + 2m^3n - 48$

45. $(3 - c^2d^2)(4 + c^2d^2)$
    $ F O I L$
    $= 12 + 3c^2d^2 - 4c^2d^2 - c^4d^4$
    $= 12 - c^2d^2 - c^4d^4$

46. $30x^2 - 28xy + 6y^2$

47. $(m^2 - n^2)(m + n)$
    $ F O I L$
    $= m^3 + m^2n - mn^2 - n^3$

48. $0.4p^2q^2 - 0.02pq - 0.02$

49. $(xy + x^5y^5)(x^4y^4 - xy)$
    $ F O I L$
    $= x^5y^5 - x^2y^2 + x^9y^9 - x^6y^6$
    $= x^9y^9 - x^6y^6 + x^5y^5 - x^2y^2$

50. $x^2 + xy^3 - 2y^6$

51. $(x + h)^2$
    $= x^2 + 2xh + h^2 [(A + B)^2 = A^2 + 2AB + B^2]$

52. $9a^2 + 12ab + 4b^2$

53. $(r^3t^2 - 4)^2$
    $= (r^3t^2)^2 - 2\cdot r^3t^2\cdot 4 + 4^2$
    $ [(A - B)^2 = A^2 - 2AB + B^2]$
    $= r^6t^4 - 8r^3t^2 + 16$

54. $9a^4b^2 - 6a^2b^3 + b^4$

55. $(p^4 + m^2n^2)^2$
    $= (p^4)^2 + 2\cdot p^4\cdot m^2n^2 + (m^2n^2)^2$
    $ [(A + B)^2 = A^2 + 2AB + B^2]$
    $= p^8 + 2p^4m^2n^2 + m^4n^4$

56. $a^2b^2 + 2abcd + c^2d^2$

57. $\left(2a^3 - \frac{1}{2}b^3\right)^2$
    $= (2a^3)^2 - 2\cdot 2a^3 \cdot \frac{1}{2}b^3 + \left(\frac{1}{2}b^3\right)^2$
    $ [(A - B)^2 = A^2 - 2AB + B^2]$
    $= 4a^6 - 2a^3b^3 + \frac{1}{4}b^6$

58. $-5x^3 - 30x^2y - 45xy^2$

59. $3a(a - 2b)^2 = 3a(a^2 - 4ab + 4b^2)$
    $ = 3a^3 - 12a^2b + 12ab^2$

60. $a^4 + 2a^2b + 4a^2 + b^2 + 4b + 4$

61. $(2a - b)(2a + b) = (2a)^2 - b^2 = 4a^2 - b^2$

62. $x^2 - y^2$

63. $(c^2 - d)(c^2 + d) = (c^2)^2 - d^2$
    $ = c^4 - d^2$

64. $p^6 - 25q^2$

65. $(ab + cd^2)(ab - cd^2) = (ab)^2 - (cd^2)^2$
    $ = a^2b^2 - c^2d^4$

66. $x^2y^2 - p^2q^2$

67. $(x + y - 3)(x + y + 3)$
    $= [(x + y) - 3][(x + y) + 3]$
    $= (x + y)^2 - 3^2$
    $= x^2 + 2xy + y^2 - 9$

68. $p^2 + 2pq + q^2 - 16$

69. $[x + y + z][x - (y + z)]$
    $= [x + (y + z)][x - (y + z)]$
    $= x^2 - (y + z)^2$
    $= x^2 - (y^2 + 2yz + z^2)$
    $= x^2 - y^2 - 2yz - z^2$

Chapter 4   (4.10)

70. $a^2 - b^2 - 2bc - c^2$

71. $(a + b + c)(a - b - c)$
 $= [a + (b + c)][a - (b + c)]$
 $= a^2 - (b + c)^2$
 $= a^2 - (b^2 + 2bc + c^2)$
 $= a^2 - b^2 - 2bc - c^2$

72. $9x^2 + 12x + 4 - 25y^2$

73. It is helpful to add additional labels to the figure.

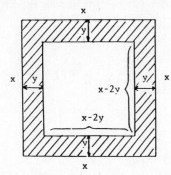

The area of the large square is $x \cdot x$, or x^2. The area of the small square is $(x - 2y)(x - 2y)$, or $(x - 2y)^2$.

Area of shaded region = Area of large square − Area of small square

Area of shaded region $= x^2 - (x - 2y)^2$
$= x^2 - (x^2 - 4xy + 4y^2)$
$= x^2 - x^2 + 4xy - 4y^2$
$= 4xy - 4y^2$

74. $2\pi ab - \pi b^2$

75. It is helpful to add additional labels to the figure.

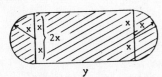

The two semicircles make a circle with radius x. The area of that circle is πx^2. The area of the rectangle is $2x \cdot y$. The sum of the two regions, $\pi x^2 + 2xy$, is the area of the shaded region.

76. $a^2 - 4b^2$

77. Evaluate the polynomial for $G = 162$, $P = 77 + 40$, or 117, and $L = 53 - 40$, or 13:
 $G - P - L + 1 = 162 - 117 - 13 + 1 = 33$

78. 2.4747 liters

79. $(A + B)^3 = (A + B)(A + B)^2$
 $(A + B)^3 = (A + B)(A^2 + 2AB + B^2)$
 $(A + B)^3 = (A + B)A^2 + (A + B)2AB + (A + B)B^2$
 $(A + B)^3 = A^3 + A^2B + 2A^2B + 2AB^2 + AB^2 + B^3$
 $(A + B)^3 = A^3 + 3A^2B + 3AB^2 + B^3$

Exercise Set 4.10

1. $\dfrac{24x^4 - 4x^3 + x^2 - 16}{8}$

 $= \dfrac{24x^4}{8} - \dfrac{4x^3}{8} + \dfrac{x^2}{8} - \dfrac{16}{8}$

 $= 3x^4 - \dfrac{1}{2}x^3 + \dfrac{1}{8}x^2 - 2$

 Check: We multiply.
 $3x^4 - \dfrac{1}{2}x^3 + \dfrac{1}{8}x^2 - 2$
 $\phantom{3x^4 - \dfrac{1}{2}x^3 + \dfrac{1}{8}x^2 - } 8$
 $\overline{24x^4 - 4x^3 + x^2 - 16}$

2. $2a^4 - \dfrac{1}{2}a^2 + \dfrac{1}{6}a - 1$

3. $\dfrac{u - 2u^2 - u^5}{u}$

 $= \dfrac{u}{u} - \dfrac{2u^2}{u} - \dfrac{u^5}{u}$

 $= 1 - 2u - u^4$

 Check: We multiply.
 $1 - 2u - u^4$
 u
 $\overline{u - 2u^2 - u^5}$

4. $50x^4 - 7x^3 + x$

5. $(15t^3 + 24t^2 - 6t) \div 3t$

 $= \dfrac{15t^3 + 24t^2 - 6t}{3t}$

 $= \dfrac{15t^3}{3t} + \dfrac{24t^2}{3t} - \dfrac{6t}{3t}$

 $= 5t^2 + 8t - 2$

 Check: We multiply.
 $5t^2 + 8t - 2$
 $3t$
 $\overline{15t^3 + 24t^2 - 6t}$

6. $5t^2 + 3t - 6$

7. $(20x^6 - 20x^4 - 5x^2) \div (-5x^2)$

 $= \dfrac{20x^6 - 20x^4 - 5x^2}{-5x^2}$

 $= \dfrac{20x^6}{-5x^2} - \dfrac{20x^4}{-5x^2} - \dfrac{5x^2}{-5x^2}$

 $= -4x^4 + 4x^2 + 1$

 Check: We multiply.
 $-4x^4 + 4x^2 + 1$
 $-5x^2$
 $\overline{20x^6 - 20x^4 - 5x^2}$

8. $-3x^4 - 4x^3 + 1$

Chapter 4   (4.10)

9. $(24x^5 - 40x^4 + 6x^3) \div (4x^3)$

   $= \dfrac{24x^5 - 40x^4 + 6x^3}{4x^3}$

   $= \dfrac{24x^5}{4x^3} - \dfrac{40x^4}{4x^3} + \dfrac{6x^3}{4x^3}$

   $= 6x^2 - 10x + \dfrac{3}{2}$

   Check: We multiply.

   $6x^2 - 10x + \dfrac{3}{2}$
   $4x^3$
   $\overline{}$
   $24x^5 - 40x^4 + 6x^3$

10. $2x^3 - 3x^2 - \dfrac{1}{3}$

11. $\dfrac{8x^2 - 3x + 1}{2}$

    $= \dfrac{8x^2}{2} - \dfrac{3x}{2} + \dfrac{1}{2}$

    $= 4x^2 - \dfrac{3}{2}x + \dfrac{1}{2}$

    Check: We multiply.

    $4x^2 - \dfrac{3}{2}x + \dfrac{1}{2}$
    2
    $\overline{}$
    $8x^2 - 3x + 1$

12. $2x^2 + x - \dfrac{2}{3}$

13. $\dfrac{2x^3 + 6x^2 + 4x}{2x}$

    $= \dfrac{2x^3}{2x} + \dfrac{6x^2}{2x} + \dfrac{4x}{2x}$

    $= x^2 + 3x + 2$

    Check: We multiply.

    $x^2 + 3x + 2$
    $2x$
    $\overline{}$
    $2x^3 + 6x^2 + 4x$

14. $2x^2 - 3x + 5$

15. $\dfrac{9r^2s^2 + 3r^2s - 6rs^2}{-3rs}$

    $= \dfrac{9r^2s^2}{-3rs} + \dfrac{3r^2s}{-3rs} - \dfrac{6rs^2}{-3rs}$

    $= -3rs - r + 2s$

    Check: We multiply.

    $-3rs - r + 2s$
    $-3rs$
    $\overline{}$
    $9r^2s^2 + 3r^2s - 6rs^2$

16. $1 - 2x^2y + 3x^4y^5$

17. $\phantom{x+2\overline{)x^2+}}x + 2$
    $x + 2\overline{\smash{)}x^2 + 4x + 4}$
    $\phantom{x+2\overline{)}}x^2 + 2x$
    $\phantom{x+2\overline{)xx}}\overline{}$
    $\phantom{x+2\overline{)xxx}}2x + 4\ \ \leftarrow (x^2+4x)-(x^2+2x)$
    $\phantom{x+2\overline{)xxx}}2x + 4$
    $\phantom{x+2\overline{)xxxx}}\overline{}$
    $\phantom{x+2\overline{)xxxxx}}0\ \ \leftarrow (2x+4)-(2x+4)$

    The answer is $x + 2$.

18. $x - 3$

19. $\phantom{x-5\overline{)x^2-}}x - 5$
    $x - 5\overline{\smash{)}x^2 - 10x - 25}$
    $\phantom{x-5\overline{)}}x^2 - 5x$
    $\phantom{x-5\overline{)xx}}\overline{}$
    $\phantom{x-5\overline{)xxx}}-5x - 25\ \ \leftarrow (x^2-10x)-(x^2-5x)$
    $\phantom{x-5\overline{)xxx}}-5x + 25$
    $\phantom{x-5\overline{)xxxx}}\overline{}$
    $\phantom{x-5\overline{)xxxxx}}-50\ \ \leftarrow (-5x-25)-(-5x+25)$

    The answer is $x - 5 + \dfrac{-50}{x - 5}$, or $x - 5 - \dfrac{50}{x - 5}$.

20. $x + 4 - \dfrac{32}{x + 4}$

21. $\phantom{x+6\overline{)x^2+}}x - 2$
    $x + 6\overline{\smash{)}x^2 + 4x - 14}$
    $\phantom{x+6\overline{)}}x^2 + 6x$
    $\phantom{x+6\overline{)xx}}\overline{}$
    $\phantom{x+6\overline{)xxx}}-2x - 14\ \ \leftarrow (x^2+4x)-(x^2+6x)$
    $\phantom{x+6\overline{)xxx}}-2x - 12$
    $\phantom{x+6\overline{)xxxx}}\overline{}$
    $\phantom{x+6\overline{)xxxxxx}}-2\ \ \leftarrow (-2x-14)-(-2x-12)$

    The answer is $x - 2 + \dfrac{-2}{x + 6}$, or $x - 2 - \dfrac{2}{x + 6}$.

22. $x + 7 + \dfrac{5}{x - 2}$

23. $\phantom{x+3\overline{)x^2+}}x - 3$
    $x + 3\overline{\smash{)}x^2 + 0x - 9}\ \ \leftarrow$ Filling in the missing term
    $\phantom{x+3\overline{)}}x^2 + 3x$
    $\phantom{x+3\overline{)xx}}\overline{}$
    $\phantom{x+3\overline{)xxx}}-3x - 9\ \ \leftarrow x^2 - (x^2 + 3x)$
    $\phantom{x+3\overline{)xxx}}-3x - 9$
    $\phantom{x+3\overline{)xxxx}}\overline{}$
    $\phantom{x+3\overline{)xxxxx}}0\ \ \leftarrow (-3x-9)-(-3x-9)$

    The answer is $x - 3$.

24. $x - 5$

25. $\phantom{x+1\overline{)x^5+}}x^4 - x^3 + x^2 - x + 1$
    $x + 1\overline{\smash{)}x^5 + 0x^4 + 0x^3 + 0x^2 + 0x + 1}\ \ \leftarrow$ Filling in missing terms
    $\phantom{x+1\overline{)}}x^5 + x^4$
    $\phantom{x+1\overline{)xx}}\overline{}$
    $\phantom{x+1\overline{)xxx}}-x^4\ \ \ \ \leftarrow x^5 - (x^5 + x^4)$
    $\phantom{x+1\overline{)xxx}}-x^4 - x^3$
    $\phantom{x+1\overline{)xxxx}}\overline{}$
    $\phantom{x+1\overline{)xxxxxx}}x^3\ \ \ \ \leftarrow -x^4 - (-x^4 - x^3)$
    $\phantom{x+1\overline{)xxxxxx}}x^3 + x^2$
    $\phantom{x+1\overline{)xxxxxxx}}\overline{}$
    $\phantom{x+1\overline{)xxxxxxxx}}-x^2\ \ \ \ \leftarrow x^3 - (x^3 + x^2)$
    $\phantom{x+1\overline{)xxxxxxxx}}-x^2 - x$
    $\phantom{x+1\overline{)xxxxxxxxx}}\overline{}$
    $\phantom{x+1\overline{)xxxxxxxxxx}}x + 1\ \ \leftarrow -x^2 - (-x^2 - x)$
    $\phantom{x+1\overline{)xxxxxxxxxx}}x + 1$
    $\phantom{x+1\overline{)xxxxxxxxxxx}}\overline{}$
    $\phantom{x+1\overline{)xxxxxxxxxxxx}}0\ \ \leftarrow (x+1)-(x+1)$

    The answer is $x^4 - x^3 + x^2 - x + 1$.

26. $x^4 + x^3 + x^2 + x + 1$

27. $\phantom{4x+3\overline{)8x^3-}}2x^2 - 7x + 4$
    $4x + 3\overline{\smash{)}8x^3 - 22x^2 - 5x + 12}$
    $\phantom{4x+3\overline{)}}8x^3 + 6x^2$
    $\phantom{4x+3\overline{)xx}}\overline{}$
    $\phantom{4x+3\overline{)xxx}}-28x^2 - 5x\ \ \leftarrow (8x^3-22x^2)-(8x^3+6x^2)$
    $\phantom{4x+3\overline{)xxx}}-28x^2 - 21x$
    $\phantom{4x+3\overline{)xxxx}}\overline{}$
    $\phantom{4x+3\overline{)xxxxxxxx}}16x + 12\ \ \leftarrow (-28x^2 - 5x) - (-28x^2 - 21x)$
    $\phantom{4x+3\overline{)xxxxxxxx}}16x + 12$
    $\phantom{4x+3\overline{)xxxxxxxxx}}\overline{}$
    $\phantom{4x+3\overline{)xxxxxxxxxx}}0\ \ \leftarrow (16x+12)-(16x+12)$

    The answer is $2x^2 - 7x + 4$.

28. $x^2 - 3x + 1$

29. $x^3 - 7 \overline{\smash{\big)}\, \begin{array}{c} x^3 - 6 \\ x^6 - 13x^3 + 42 \end{array}}$

$\underline{x^6 - 7x^3}$
$-6x^3 + 42 \leftarrow (x^6 - 13x^3) - (x^6 - 7x^3)$
$\underline{-6x^3 + 42}$
$0 \leftarrow (-6x^3 + 42) - (-6x^3 + 42)$

The answer is $x^3 - 6$.

30. $x^3 + 8$

31. $x - 2 \overline{\smash{\big)}\, \begin{array}{c} x^3 + 2x^2 + 4x + 8 \\ x^4 + 0x^3 + 0x^2 + 0x - 16 \end{array}}$

$\underline{x^4 - 2x^3}$
$2x^3 \leftarrow x^4 - (x^4 - 2x^3)$
$\underline{2x^3 - 4x^2}$
$4x^2 \leftarrow 2x^3 - (2x^3 - 4x^2)$
$\underline{4x^2 - 8x}$
$8x - 16 \leftarrow 4x^2 - (4x^2 - 8x)$
$\underline{8x - 16}$
$0 \leftarrow (8x-16)-(8x-16)$

The answer is $x^3 + 2x^2 + 4x + 8$.

32. $x^3 + 3x^2 + 9x + 27$

33. $t - 1 \overline{\smash{\big)}\, \begin{array}{c} t^2 + 1 \\ t^3 - t^2 + t - 1 \end{array}}$

$\underline{t^3 - t^2} \leftarrow (t^3 - t^2) - (t^3 - t^2)$
$0 + t - 1$
$\underline{t - 1} \leftarrow (t - 1) - (t - 1)$
0

The answer is $t^2 + 1$.

34. $t^2 - 2t + 3 - \dfrac{4}{t + 1}$

35. <u>Familiarize</u>. Let w = the width. Then $w + 15$ = the length. We draw a picture.

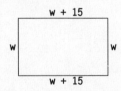

We will use the fact that the perimeter is 640 ft to find w (the width). Then we can find $w + 15$ (the length) and multiply the length and the width to find the area.

<u>Translate</u>.

Width + Width + Length + Length = Perimeter
$w + w + (w + 15) + (w + 15) = 640$

35. (continued)

<u>Carry out</u>.
$w + w + (w + 15) + (w + 15) = 640$
$4w + 30 = 640$
$4w = 610$
$w = 152.5$

If the width is 152.5, then the length is $152.5 + 15$, or 167.5. The area is $(167.5)(152.5)$, or 25,543.75 ft².

<u>Check</u>. The length, 167.5 ft, is 15 ft greater than the width, 152.5 ft. The perimeter is $152.5 + 152.5 + 167.5 + 167.5$, or 640 ft. We should also recheck the computation we used to find the area. The answer checks.

<u>State</u>. The area is 25,543.75 ft².

36. 24 in.

37. $p^3 + p^2 - p - 1 = p^2(p + 1) - (p + 1)$
$ = (p + 1)(p^2 - 1)$
$ = (p + 1)(p - 1)(p + 1)$
$ = (p - 1)(p + 1)^2$

38. $p^3(3p - 1)^2$

39. $x^2 + 4 \overline{\smash{\big)}\, \begin{array}{c} x^2 + 5 \\ x^4 + 9x^2 + 20 \end{array}}$

$\underline{x^4 + 4x^2}$
$5x^2 + 20$
$\underline{5x^2 + 20}$
0

The answer is $x^2 + 5$.

40. $y^3 - ay^2 + a^2y - a^3 + \dfrac{a^2 + a^4}{y + a}$

41. $5a^2 - 7a - 2 \overline{\smash{\big)}\, \begin{array}{c} a + 3 \\ 5a^3 + 8a^2 - 23a - 1 \end{array}}$

$\underline{5a^3 - 7a^2 - 2a}$
$15a^2 - 21a - 1$
$\underline{15a^2 - 21a - 6}$
5

The answer is $a + 3 + \dfrac{5}{5a^2 - 7a - 2}$.

42. $5y + 2 + \dfrac{-10y + 11}{3y^2 - 5y - 2}$

43. We rewrite the dividend in descending order.

$3x^3 - 2x - 1 \overline{\smash{\big)}\, \begin{array}{c} 2x^2 + x - 3 \\ 6x^5 + 3x^4 - 13x^3 - 4x^2 + 5x + 3 \end{array}}$

$\underline{6x^5 - 4x^3 - 2x^2}$
$3x^4 - 9x^3 - 2x^2 + 5x$
$\underline{3x^4 - 2x^2 - x}$
$-9x^3 + 6x + 3$
$\underline{-9x^3 + 6x + 3}$
0

The answer is $2x^2 + x - 3$.

44. $5x^5 + 5x^4 - 8x^2 - 8x + 2$

45.
$$\require{enclose}
\begin{array}{r}
a^5 + a^4b + a^3b^2 + a^2b^3 + ab^4 + b^5 \\
a-b \enclose{longdiv}{a^6 + 0a^5b + 0a^4b^2 + 0a^3b^3 + 0a^2b^4 + 0ab^5 - b^6}
\end{array}$$

$\quad\quad\quad\underline{a^6 - a^5b}$
$\quad\quad\quad\quad a^5b$
$\quad\quad\quad\quad\underline{a^5b - a^4b^2}$
$\quad\quad\quad\quad\quad\quad a^4b^2$
$\quad\quad\quad\quad\quad\quad\underline{a^4b^2 - a^3b^3}$
$\quad\quad\quad\quad\quad\quad\quad\quad a^3b^3$
$\quad\quad\quad\quad\quad\quad\quad\quad\underline{a^3b^3 - a^2b^4}$
$\quad\quad\quad\quad\quad\quad\quad\quad\quad\quad a^2b^4$
$\quad\quad\quad\quad\quad\quad\quad\quad\quad\quad\underline{a^2b^4 - ab^5}$
$\quad\quad\quad\quad\quad\quad\quad\quad\quad\quad\quad\quad ab^5 - b^6$
$\quad\quad\quad\quad\quad\quad\quad\quad\quad\quad\quad\quad\underline{ab^5 - b^6}$
$\quad\quad\quad\quad\quad\quad\quad\quad\quad\quad\quad\quad\quad\quad 0$

The answer is $a^5 + a^4b + a^3b^2 + a^2b^3 + ab^4 + b^5$.

46. $x^4 - x^3y + x^2y^2 - xy^3 + y^4$

47.
$$\begin{array}{r}
3a^{2h} + 2a^h - 5 \\
2a^h + 3 \enclose{longdiv}{6a^{3h} + 13a^{2h} - 4a^h - 15}
\end{array}$$

$\quad\quad\quad\underline{6a^{3h} + 9a^{2h}}$
$\quad\quad\quad\quad\quad 4a^{2h} - 4a^h$
$\quad\quad\quad\quad\quad\underline{4a^{2h} + 6a^h}$
$\quad\quad\quad\quad\quad\quad\quad -10a^h - 15$
$\quad\quad\quad\quad\quad\quad\quad\underline{-10a^h - 15}$
$\quad\quad\quad\quad\quad\quad\quad\quad\quad\quad 0$

The answer is $3a^{2h} + 2a^h - 5$.

48. -5

49.
$$\begin{array}{r}
2x + (3c + 2) \\
x - 1 \enclose{longdiv}{2x^2 + 3cx - 8}
\end{array}$$

$\quad\quad\underline{2x^2 - 2x}$
$\quad\quad\quad (3c + 2)x - 8$
$\quad\quad\quad\underline{(3c + 2)x - (3c + 2)}$
$\quad\quad\quad\quad\quad\quad -8 + (3c + 2)$

We set the remainder equal to 0:
$\quad -8 + 3c + 2 = 0$
$\quad\quad\quad 3c - 6 = 0$
$\quad\quad\quad\quad\quad 3c = 6$
$\quad\quad\quad\quad\quad\, c = 2$

Thus, c must be 2.

50. 1

CHAPTER 5   POLYNOMIALS AND FACTORING

Exercise Set 5.1

1. Answers may vary. $6x^3 = (6x)(x^2) = (3x^2)(2x) = (2x^2)(3x)$

2. Answers may vary. $(3x^2)(3x^2)$, $(9x)(x^3)$, $(3x)(3x^3)$

3. Answers may vary. $-9x^5 = (-3x^2)(3x^3) = (-x)(9x^4) = (3x^2)(-3x^3)$

4. Answers may vary. $(-4x)(3x^5)$, $(-6x^2)(2x^4)$, $(12x^3)(-x^3)$

5. Answers may vary. $24x^4 = (6x)(4x^3) = (-3x^2)(-8x^2) = (2x^3)(12x)$

6. Answers may vary. $(3x)(5x^4)$, $(x^3)(15x^2)$, $(3x^2)(5x^3)$

7. $x^2 - 4x = x \cdot x - x \cdot 4$
   $= x(x - 4)$

8. $x(x + 8)$

9. $2x^2 + 6x = 2x \cdot x + 2x \cdot 3$
   $= 2x(x + 3)$

10. $3x(x - 1)$

11. $x^3 + 6x^2 = x^2 \cdot x + x^2 \cdot 6$
    $= x^2(x + 6)$

12. $x^2(4x^2 + 1)$

13. $8x^4 - 24x^2 = 8x^2 \cdot x^2 - 8x^2 \cdot 3$
    $= 8x^2(x^2 - 3)$

14. $5x^3(x^2 + 2)$

15. $2x^2 + 2x - 8 = 2 \cdot x^2 + 2 \cdot x - 2 \cdot 4$
    $= 2(x^2 + x - 4)$

16. $3(2x^2 + x - 5)$

17. $17x^5y^3 + 34x^3y^2 + 51xy$
    $= 17xy \cdot x^4y^2 + 17xy \cdot 2x^2y + 17xy \cdot 3$
    $= 17xy(x^4y^2 + 2x^2y + 3)$

18. $16xy^2(x^5y^2 - 2x^4y - 3)$

19. $6x^4 - 10x^3 + 3x^2 = x^2 \cdot 6x^2 - x^2 \cdot 10x + x^2 \cdot 3$
    $= x^2(6x^2 - 10x + 3)$

20. $x(5x^4 + 10x - 8)$

21. $x^5y^5 + x^4y^3 + x^3y^3 - x^2y^2$
    $= x^2y^2 \cdot x^3y^3 + x^2y^2 \cdot x^2y + x^2y^2 \cdot xy + x^2y^2(-1)$
    $= x^2y^2(x^3y^3 + x^2y + xy - 1)$

22. $x^3y^3(x^6y^3 - x^4y^2 + xy + 1)$

23. $2x^7 - 2x^6 - 64x^5 + 4x^3$
    $= 2x^3 \cdot x^4 - 2x^3 \cdot x^3 - 2x^3 \cdot 32x^2 + 2x^3 \cdot 2$
    $= 2x^3(x^4 - x^3 - 32x^2 + 2)$

24. $5(2x^3 + 5x^2 + 3x - 4)$

25. $1.6x^4 - 2.4x^3 + 3.2x^2 + 6.4x$
    $= 0.8x(2x^3) - 0.8x(3x^2) + 0.8x(4x) + 0.8x(8)$
    $= 0.8x(2x^3 - 3x^2 + 4x + 8)$

26. $0.5x^2(5x^4 - x^2 + 10x + 20)$

27. $\frac{5}{3}x^6 + \frac{4}{3}x^5 + \frac{1}{3}x^4 + \frac{1}{3}x^3$
    $= \frac{1}{3}x^3(5x^3) + \frac{1}{3}x^3(4x^2) + \frac{1}{3}x^3(x) + \frac{1}{3}x^3(1)$
    $= \frac{1}{3}x^3(5x^3 + 4x^2 + x + 1)$

28. $\frac{1}{7}x(5x^6 + 3x^4 - 6x^2 - 1)$

29. $x^2 - 4$
    a) The first expression is a square: x^2
       The second expression is a square: $4 = 2^2$
    b) The terms have different signs.
       $x^2 - 4$ is a difference of squares.

30. Yes

31. $x^2 + 36$
    The terms do not have different signs.
    $x^2 + 36$ is not a difference of squares.

32. No

33. $x^2 - 35$
    The second expression, 35, is not a square.
    $x^2 - 35$ is not a difference of squares.

34. No

35. $16x^2 - 25$
    a) The first expression is a square:
       $16x^2 = (4x)^2$
       The second expression is a square: $25 = 5^2$
    b) The terms have different signs.
       $16x^2 - 25$ is a difference of squares.

36. Yes

Chapter 5   (5.1)

37. $49x^2 - 2$
The second expression, 2, is not a square.
$49x^2 - 2$ is not a difference of squares.

38. Yes

39. $y^2 - 4 = y^2 - 2^2 = (y + 2)(y - 2)$

40. $(x + 6)(x - 6)$

41. $p^2 - 9 = p^2 - 3^2 = (p + 3)(p - 3)$

42. $(q + 1)(q - 1)$

43. $-49 + t^2 = t^2 - 49 = t^2 - 7^2 = (t + 7)(t - 7)$

44. $(m + 8)(m - 8)$

45. $a^2 - b^2 = (a + b)(a - b)$

46. $(p + q)(p - q)$

47. $25t^2 - m^2 = (5t)^2 - m^2 = (5t + m)(5t - m)$

48. $(w + 7z)(w - 7z)$

49. $100 - k^2 = 10^2 - k^2 = (10 + k)(10 - k)$

50. $(9 + w)(9 - w)$

51. $16a^2 - 9 = (4a)^2 - 3^2 = (4a + 3)(4a - 3)$

52. $(5x + 2)(5x - 2)$

53. $4x^2 - 25y^2 = (2x)^2 - (5y)^2 = (2x + 5y)(2x - 5y)$

54. $(3a + 4b)(3a - 4b)$

55. $8x^2 - 98 = 2(4x^2 - 49) = 2[(2x)^2 - 7^2] = 2(2x + 7)(2x - 7)$

56. $6(2x + 3)(2x - 3)$

57. $36x - 49x^3 = x(36 - 49x^2) = x[6^2 - (7x)^2] = x(6 + 7x)(6 - 7x)$

58. $x(4 + 9x)(4 - 9x)$

59. $16x^2 - 25x^4 = x^2(16 - 25x^2) = x^2[4^2 - (5x)^2] = x^2(4 + 5x)(4 - 5x)$

60. $x^2(x^7 + 3)(x^7 - 3)$

61. $49a^4 - 81 = (7a^2)^2 - 9^2 = (7a^2 + 9)(7a^2 - 9)$

62. $(5a^2 + 3)(5a^2 - 3)$

63. $a^{12} - 4a^2 = a^2(a^{10} - 4) = a^2[(a^5)^2 - 2^2] = a^2(a^5 + 2)(a^5 - 2)$

64. $(11a^4 + 10)(11a^4 - 10)$

65. $81y^6 - 25q^6 = (9y^3)^2 - (5q^3)^2 = (9y^3 + 5q^3)(9y^3 - 5q^3)$

66. $(10y^3 + 7m^3)(10y^3 - 7m^3)$

67. $x^4 - 1 = (x^2)^2 - 1^2$
$= (x^2 + 1)(x^2 - 1)$
$= (x^2 + 1)(x + 1)(x - 1)$ Factoring further; $x^2 - 1$ is a difference of squares

68. $(x^2 + 4)(x + 2)(x - 2)$

69. $4x^4 - 64 = 4(x^4 - 16) = 4[(x^2)^2 - 4^2]$
$= 4(x^2 + 4)(x^2 - 4)$
$= 4(x^2 + 4)(x + 2)(x - 2)$ Factoring further; $x^2 - 4$ is a difference of squares

70. $5(x^2 + 4)(x + 2)(x - 2)$

71. $1 - y^8 = 1^2 - (y^4)^2$
$= (1 + y^4)(1 - y^4)$
$= (1 + y^4)(1 + y^2)(1 - y^2)$ Factoring $1 - y^4$
$= (1 + y^4)(1 + y^2)(1 + y)(1 - y)$
Factoring $1 - y^2$

72. $(x^4 + 1)(x^2 + 1)(x + 1)(x - 1)$

73. $x^{12} - 16 = (x^6)^2 - 4^2$
$= (x^6 + 4)(x^6 - 4)$
$= (x^6 + 4)(x^3 + 2)(x^3 - 2)$ Factoring $x^6 - 4$

74. $(x^4 + 9)(x^2 + 3)(x^2 - 3)$

75. $\frac{1}{16} - y^2 = \left(\frac{1}{4}\right)^2 - y^2$
$= \left(\frac{1}{4} + y\right)\left(\frac{1}{4} - y\right)$

76. $\left(\frac{1}{5} + x\right)\left(\frac{1}{5} - x\right)$

77. $25 - \frac{1}{49}x^2 = 5^2 - \left(\frac{1}{7}x\right)^2$
$= \left(5 + \frac{1}{7}x\right)\left(5 - \frac{1}{7}x\right)$

78. $\left(2 + \frac{1}{3}y\right)\left(2 - \frac{1}{3}y\right)$

79. $16m^4 - t^4 = (4m^2)^2 - (t^2)^2$
$= (4m^2 + t^2)(4m^2 - t^2)$
$= (4m^2 + t^2)(2m + t)(2m - t)$ Factoring $4m^2 - t^2$

Chapter 5   (5.2)

80. $(1 + a^2b^2)(1 + ab)(1 - ab)$

81. <u>Familiarize</u>. Let n = the number of people arrested the year before.
    <u>Translate</u>. We reword the problem.

    Number arrested last year   less 1.2% of that number   is 29,090.
         n   -   1.2% ·   n   = 29,090

    <u>Carry out</u>. We solve the equation.
    $n - 1.2\% \cdot n = 29{,}090$
    $1 \cdot n - 0.012n = 29{,}090$
    $0.988n = 29{,}090$
    $n \approx 29{,}443$ Rounding

    <u>Check</u>. 1.2% of 29,443 is 0.012(29,443) ≈ 353 and 29,443 - 353 = 29,090. The answer checks.
    <u>State</u>. Approximately 29,443 people were arrested last year.

82. 100°, 25°, 55°

83. $2 - 5(x + 5) = 3(x + 2) - 1$
    $2 - 5x - 25 = 3x + 6 - 1$ Removing parentheses
    $-5x - 23 = 3x + 5$ Collecting like terms
    $-5x - 3x = 5 + 23$ Adding -3x and 23
    $-8x = 28$ Collecting like terms
    $x = -\frac{28}{8}$ Dividing by -8
    $x = -\frac{7}{2}$ Simplifying

84. $p = 2A - q$

85. $4x^4 - 4x^2 = 4x^2(x^2 - 1) = 4x^2(x + 1)(x - 1)$

86. $3x^3(x + 2)(x - 2)$

87. $3x^2 - \frac{1}{3} = 3x^2 - 3 \cdot \frac{1}{9} = 3\left(x^2 - \frac{1}{9}\right)$
    $= 3\left(x + \frac{1}{3}\right)\left(x - \frac{1}{3}\right)$

88. $2x\left(3x + \frac{2}{5}\right)\left(3x - \frac{2}{5}\right)$

89. $x^2 - 2.25 = x^2 - (1.5)^2 = (x + 1.5)(x - 1.5)$

90. $p(0.7 + p)(0.7 - p)$

91. $3.24x^2 - 0.81 = 0.81(4x^2 - 1)$
    $= 0.81(2x + 1)(2x - 1)$

92. $(0.8x + 1.1)(0.8x - 1.1)$

93. $1.28x^2 - 2 = 2(0.64x^2 - 1) = 2[(0.8x)^2 - 1^2] =$
    $2(0.8x + 1)(0.8x - 1)$

94. $x(x + 6)$

95. $(y - 5)^2 - 36q^2 = (y - 5)^2 - (6q)^2$
    $= (y - 5 + 6q)(y - 5 - 6q)$

96. $(3a + 4 + 7b)(3a + 4 - 7b)$

97. $(2y - 7)^2 - 1 = (2y - 7 + 1)(2y - 7 - 1) =$
    $(2y - 6)(2y - 8) = 2(y - 3)(2)(y - 4) =$
    $4(y - 3)(y - 4)$

98. $(y^4 + 16)(y^2 + 4)(y + 2)(y - 2)$

99. $x^{16} - 1 = (x^8 + 1)(x^8 - 1) =$
    $(x^8 + 1)(x^4 + 1)(x^4 - 1) =$
    $(x^8 + 1)(x^4 + 1)(x^2 + 1)(x^2 - 1) =$
    $(x^8 + 1)(x^4 + 1)(x^2 + 1)(x + 1)(x - 1)$

100. $\left(x + \frac{1}{x}\right)\left(x - \frac{1}{x}\right)$

101. $a^{2n} - 49b^{2n} = (a^n)^2 - (7b^n)^2 =$
     $(a^n + 7b^n)(a^n - 7b^n)$

102. $(9 + b^{2k})(3 + b^k)(3 - b^k)$

103. $3x^3 + 9x = 3x(x^2 + 3)$
     Since both terms contain a factor of x, the polynomial is not irreducible and hence is not prime.

104. Irreducible

105. $4x^2 + 16y^2 = 4(x^2 + 4y^2)$
     The polynomial cannot be factored except for removing a common constant factor, 4. Therefore, the polynomial is irreducible. Since the coefficient of the leading term is not 1, the irreducible polynomial is not prime.

106. Prime

107. $16x^3 - 9x = x(16x^2 - 9) = x(4x + 3)(4x - 3)$
     Since the polynomial can be factored as shown, it is not irreducible and hence not prime.

108. Irreducible

Exercise Set 5.2

1. $x^2 + 8x + 15$
   Since the constant term and coefficient of the middle term are both positive, we look for a factorization of 15 in which both factors are positive. Their sum must be 8.

Pairs of factors	Sums of factors
1, 15	16
3, 5	8

   The numbers we want are 3 and 5.
   $x^2 + 8x + 15 = (x + 3)(x + 5)$

Chapter 5   (5.2)

2. $(x + 2)(x + 3)$

3. $x^2 + 7x + 12$

   Since the constant term is positive and the coefficient of the middle term is positive, we look for a factorization of 12 in which both factors are positive. Their sum must be 7.

Pairs of factors	Sums of factors
1, 12	13
2, 6	8
3, 4	7

   The numbers we want are 3 and 4.
   $x^2 + 7x + 12 = (x + 3)(x + 4)$.

4. $(x + 1)(x + 8)$

5. $x^2 - 6x + 9$

   Since the constant term is positive and the coefficient of the middle term is negative, we look for a factorization of 9 in which both factors are negative. Their sum must be -6.

Pairs of factors	Sums of factors
-1, -9	-10
-3, -3	-6

   The numbers we want are -3 and -3.
   $x^2 - 6x + 9 = (x - 3)(x - 3)$, or $(x - 3)^2$.

6. $(y + 4)(y + 7)$

7. $x^2 + 9x + 14$

   Since the constant term is positive and the coefficient of the middle term is positive, we look for a factorization of 14 in which both factors are positive. Their sum must be 9.

Pairs of factors	Sums of factors
1, 14	15
2, 7	9

   The numbers we want are 2 and 7.
   $x^2 + 9x + 14 = (x + 2)(x + 7)$.

8. $(a + 5)(a + 6)$

9. $b^2 + 5b + 4$

   Since the constant term is positive and the coefficient of the middle term is positive, we look for a factorization of 4 in which both factors are positive. Their sum must be 5.

Pairs of factors	Sums of factors
1, 4	5
2, 2	4

   The numbers we want are 1 and 4.
   $b^2 + 5b + 4 = (b + 1)(b + 4)$.

10. $\left(x - \frac{1}{5}\right)^2$

11. $x^2 + \frac{2}{3}x + \frac{1}{9}$

    Since the constant term is positive and the coefficient of the middle term is positive, we look for a factorization of $\frac{1}{9}$ in which both factors are positive. Their sum must be $\frac{2}{3}$.

Pairs of factors	Sums of factors
1, $\frac{1}{9}$	$\frac{10}{9}$
$\frac{1}{3}, \frac{1}{3}$	$\frac{2}{3}$

    The numbers we want are $\frac{1}{3}$ and $\frac{1}{3}$.
    $x^2 + \frac{2}{3}x + \frac{1}{9} = \left(x + \frac{1}{3}\right)\left(x + \frac{1}{3}\right)$, or $\left(x + \frac{1}{3}\right)^2$.

12. $(z - 1)(z - 7)$

13. $d^2 - 7d + 10$

    Since the constant term is positive and the coefficient of the middle term is negative, we look for a factorization of 10 in which both factors are negative. Their sum must be -7.

Pairs of factors	Sums of factors
-1, -10	-11
-2, -5	-7

    The numbers we want are -2 and -5.
    $d^2 - 7d + 10 = (d - 2)(d - 5)$.

14. $(x - 3)(x - 5)$

15. $y^2 - 11y + 10$

    Since the constant term is positive and the coefficient of the middle term is negative, we look for a factorization of 10 in which both factors are negative. Their sum must be -11.

Pairs of factors	Sums of factors
-1, -10	-11
-2, -5	-7

    The numbers we want are -1 and -10.
    $y^2 - 11y + 10 = (y - 1)(y - 10)$.

16. $(x + 3)(x - 5)$

Chapter 5   (5.2)

17. $x^2 + x - 42$

Since the constant term is negative, we look for a factorization of -42 in which one factor is positive and one factor is negative. Their sum must be 1, the coefficient of the middle term.

Pairs of factors	Sums of factors
-1, 42	41
1, -42	-41
-2, 21	19
2, -21	-19
-3, 14	11
3, -14	-11
-6, 7	1
6, -7	-1

The numbers we want are -6 and 7.
$x^2 + x - 42 = (x - 6)(x + 7)$.

18. $(x - 3)(x + 5)$

19. $x^2 - 7x - 18$

Since the constant term is negative, we look for a factorization of -18 in which one factor is positive and one factor is negative. Their sum must be -7, the coefficient of the middle term.

Pairs of factors	Sums of factors
-1, 18	17
1, -18	-17
-2, 9	7
2, -9	-7
-3, 6	3
3, -6	-3

The numbers we want are 2 and -9.
$x^2 - 7x - 18 = (x + 2)(x - 9)$.

20. $(x + 4)(x - 7)$

21. $x^3 - 6x^2 - 16x = x(x^2 - 6x - 16)$

After factoring out the common factor, x, we consider $x^2 - 6x - 16$. Since the constant term is negative, we look for a factorization of -16 in which one factor is positive and one factor is negative. Their sum must be -6, the coefficient of the middle term.

Pairs of factors	Sums of factors
-1, 16	15
1, -16	-15
-2, 8	6
2, -8	-6
-4, 4	0

The numbers we want are 2 and -8.
Then $x^2 - 6x - 16 = (x + 2)(x - 8)$, so
$x^3 - 6x^2 - 16x = x(x + 2)(x - 8)$.

22. $x(x + 6)(x - 7)$

23. $y^2 - 4y - 45$

Since the constant term is negative, we look for a factorization of -45 in which one factor is positive and one factor is negative. Their sum must be -4, the coefficient of the middle term.

Pairs of factors	Sums of factors
-1, 45	44
1, -45	-44
-3, 15	12
3, -15	-12
-5, 9	4
5, -9	-4

The numbers we want are 5 and -9.
$y^2 - 4y - 45 = (y + 5)(y - 9)$.

24. $(x + 5)(x - 12)$

25. $-2x - 99 + x^2 = x^2 - 2x - 99$

Since the constant term is negative, we look for a factorization of -99 in which one factor is positive and one factor is negative. Their sum must be -2, the coefficient of the middle term.

Pairs of factors	Sums of factors
-1, 99	98
1, -99	-98
-3, 33	30
3, -33	-30
-9, 11	2
9, -11	-2

The numbers we want are 9 and -11.
$-2x - 99 + x^2 = (x + 9)(x - 11)$.

26. $(x - 6)(x + 12)$

27. $c^4 + c^2 - 56$

Consider this trinomial as $(c^2)^2 + c^2 - 56$. We look for numbers p and q such that $c^4 + c^2 - 56 = (c^2 + p)(c^2 + q)$. Since the constant term is negative, we look for a factorization of -56 in which one factor is positive and one factor is negative. Their sum must be 1.

Pairs of factors	Sums of factors
-1, 56	55
1, -56	-55
-2, 28	26
2, -28	-26
-4, 14	12
4, -14	-12
-7, 8	1
7, -8	-1

The numbers we want are -7 and 8.
$c^4 + c^2 - 56 = (c^2 - 7)(c^2 + 8)$.

28. $(b^2 - 3)(b^2 + 8)$

Chapter 5   (5.2)

29. $a^4 + 2a^2 - 35$

Consider this trinomial as $(a^2)^2 + 2a^2 - 35$. We look for numbers p and q such that $a^4 + 2a^2 - 35 = (a^2 + p)(a^2 + q)$. Since the constant term is negative, we look for a factorization of -35 in which one factor is positive and one factor is negative. Their sum must be 2.

Pairs of factors	Sums of factors
-1, 35	34
1, -35	-34
-5, 7	2
5, -7	-2

The numbers we want are -5 and 7.
$a^4 + 2a^2 - 35 = (a^2 - 5)(a^2 + 7)$.

30. $(2 + x^2)(1 + x)(1 - x)$

31. $x^2 + x + 1$

Since the constant term and the coefficient of the middle term are both positive, we look for a factorization of 1 in which both factors are positive. The sum must be 1. The only possible pair of factors is 1 and 1, but their sum is not 1. Thus, this polynomial is not factorable into polynomials with integer coefficients.

32. Not factorable

33. $7 - 2p + p^2 = p^2 - 2p + 7$

Since the constant term is positive and the coefficient of the middle term is negative, we look for a factorization of 7 in which both factors are negative. Their sum must be -2. The only possible pair of factors is -1 and -7, but their sum is not -2. Thus, this polynomial is not factorable into polynomials with integer coefficients.

34. Not factorable

35. $x^2 + 20x + 100$

We look for two factors, both positive, whose product is 100 and whose sum is 20.
They are 10 and 10. $10 \cdot 10 = 100$ and $10 + 10 = 20$.
$x^2 + 20x + 100 = (x + 10)(x + 10)$, or $(x + 10)^2$.

36. $(x + 9)(x + 11)$

37. $x^2 - 21x - 100$

We look for two factors, one positive and one negative, whose product is -100 and whose sum is -21.
They are 4 and -25. $4 \cdot (-25) = -100$ and $4 + (-25) = -21$.
$x^2 - 21x - 100 = (x + 4)(x - 25)$

38. $(x - 8)(x - 12)$

39. $x^2 - 21x - 72$

We look for two factors, one positive and one negative, whose product is -72 and whose sum is -21. They are 3 and -24.
$x^2 - 21x - 72 = (x + 3)(x - 24)$

40. $4(x + 5)^2$

41. $x^2 - 25x + 144$

We look for two factors, both negative, whose product is 144 and whose sum is -25. They are -9 and -16.
$x^2 - 25x + 144 = (x - 9)(x - 16)$

42. $(y - 9)(y - 12)$

43. $a^2 + a - 132$

We look for two factors, one positive and one negative, whose product is -132 and whose sum is 1. They are -11 and 12.
$a^2 + a - 132 = (a - 11)(a + 12)$

44. $(a - 6)(a + 15)$

45. $120 - 23x + x^2 = x^2 - 23x + 120$

We look for two factors, both negative, whose product is 120 and whose sum is -23. They are -8 and -15.
$x^2 - 23x + 120 = (x - 8)(x - 15)$

46. $(d + 6)(d + 16)$

47. First write the polynomial in descending order and factor out -1.
$108 - 3x - x^2 = -x^2 - 3x + 108 = -1(x^2 + 3x - 108)$
Now we factor the polynomial $x^2 + 3x - 108$. We look for two factors, one positive and one negative, whose product is -108 and whose sum is 3. They are -9 and 12.
$x^2 + 3x - 108 = (x - 9)(x + 12)$
The final answer must include the -1 which was factored out above.
$-x^2 - 3x + 108 = -1(x - 9)(x + 12)$.
Using the distributive law to find $-1(x - 9)$, we see that $-1(x - 9)(x + 12)$ can also be expressed as $(-x + 9)(x + 12)$, or $(9 - x)(12 + x)$.

48. $-1(y - 16)(y + 7)$, or $(16 - y)(7 + y)$

49. $y^2 - 0.2y - 0.08$

We look for two factors, one positive and one negative, whose product is -0.08 and whose sum is -0.2. They are -0.4 and 0.2.
$y^2 - 0.2y - 0.08 = (y + 0.4)(y - 0.2)$

50. $(t - 0.5)(t + 0.2)$

Chapter 5   (5.2)

51. $p^2 + 3pq - 10q^2 = p^2 + 3qp - 10q^2$

    Think of $3q$ as a "coefficient" of p. Then we look for factors of $-10q^2$ whose sum is $3q$. They are $5q$ and $-2q$.

    $p^2 + 3pq - 10q^2 = (p + 5q)(p - 2q)$.

52. $(a - 3b)(a + b)$

53. $m^2 + 5mn + 5n^2 = m^2 + 5nm + 5n^2$

    We look for factors of $5n^2$ whose sum is $5n$. The only reasonable possibilities are shown below.

Pairs of factors	Sums of factors
$5n, n$	$6n$
$-5n, -n$	$-6n$

    There are no factors whose sum is $5n$. Thus, the polynomial is not factorable into polynomials with integer coefficients.

54. $(x - 8y)(x - 3y)$

55. $s^2 - 2st - 15t^2 = s^2 - 2ts - 15t^2$

    We look for factors of $-15t^2$ whose sum is $-2t$. They are $-5t$ and $3t$.

    $s^2 - 2st - 15t^2 = (s - 5t)(s + 3t)$

56. $(b + 10c)(b - 2c)$

57. $8x(2x^2 - 6x + 1) = 8x \cdot 2x^2 - 8x \cdot 6x + 8x \cdot 1 = 16x^3 - 48x^2 + 8x$

58. $28w^2 - 53w - 66$

59. $(7w + 6)^2 = (7w)^2 + 2 \cdot 7w \cdot 6 + 6^2 = 49w^2 + 84w + 36$

60. $27x^{12}$

61. $y^2 + my + 50$

    We look for pairs of factors whose product is 50. The sum of each pair is represented by m.

Pairs of factors whose product is 50	Sums of factors
$1, 50$	51
$-1, -50$	-51
$2, 25$	27
$-2, -25$	-27
$5, 10$	15
$-5, -10$	-15

    The polynomial $y^2 + my + 50$ can be factored if m is $51, -51, 27, -27, 15,$ or -15.

62. $49, -49, 23, -23, 5, -5$

63. $x^2 - \frac{1}{2}x - \frac{3}{16}$

    We look for two factors, one positive and one negative, whose product is $-\frac{3}{16}$ and whose sum is $-\frac{1}{2}$.

    They are $-\frac{3}{4}$ and $\frac{1}{4}$.

    $-\frac{3}{4} \cdot \frac{1}{4} = -\frac{3}{16}$ and $-\frac{3}{4} + \frac{1}{4} = -\frac{2}{4} = -\frac{1}{2}$.

    $x^2 - \frac{1}{2}x - \frac{3}{16} = \left(x - \frac{3}{4}\right)\left(x + \frac{1}{4}\right)$

64. $\left(x - \frac{1}{2}\right)\left(x + \frac{1}{4}\right)$

65. $x^2 + \frac{30}{7}x - \frac{25}{7}$

    We look for two factors, one positive and one negative, whose product is $-\frac{25}{7}$ and whose sum is $\frac{30}{7}$.

    They are 5 and $-\frac{5}{7}$.

    $5 \cdot \left(-\frac{5}{7}\right) = -\frac{25}{7}$ and $5 + \left(-\frac{5}{7}\right) = \frac{35}{7} + \left(-\frac{5}{7}\right) = \frac{30}{7}$.

    $x^2 + \frac{30}{7}x - \frac{25}{7} = (x + 5)\left(x - \frac{5}{7}\right)$

66. $\frac{1}{3}x(x + 3)(x - 2)$

67. $b^{2n} + 7b^n + 10$

    Consider this trinomial as $(b^n)^2 + 7b^n + 10$. We look for numbers p and q such that $b^{2n} + 7b^n + 10 = (b^n + p)(b^n + q)$. We find two factors, both positive, whose product is 10 and whose sum is 7. They are 5 and 2.

    $b^{2n} + 7b^n + 10 = (b^n + 5)(b^n + 2)$

68. $(a^m - 7)(a^m - 4)$

69. We first label the drawing with additional information.

    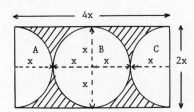

    $4x$ represents the length of the rectangle and $2x$ the width. The area of the rectangle is $4x \cdot 2x$, or $8x^2$.

    The area of semicircle A is $\frac{1}{2}\pi x^2$.

    The area of circle B is πx^2.

    The area of semicircle C is $\frac{1}{2}\pi x^2$.

Chapter 5   (5.3)

69. (continued)

Area of shaded region = Area of rectangle − Area of A − Area of B − Area of C

Area of shaded region = $8x^2 - \frac{1}{2}\pi x^2 - \pi x^2 - \frac{1}{2}\pi x^2$

= $8x^2 - 2\pi x^2$

= $2x^2(4 - \pi)$

The shaded area can be represented by $2x^2(4 - \pi)$.

70. $x^2(\pi - 1)$

Exercise Set 5.3

---

1. $2x^2 - 7x - 4$

   (1) Look for a common factor. There is none (other than 1 or −1).

   (2) Factor the first term, $2x^2$. The factor pair is $2x$, x. We have this possibility:

   $(2x\ \ )(x\ \ )$

   (3) Factor the last term, −4. The factor pairs are −4, 1 and 4, −1 and 2, −2.

   (4) Look for factors in steps (2) and (3) such that the sum of their products is the middle term, −7x. We try some possibilities:

   $(2x - 4)(x + 1) = 2x^2 - 2x - 4$
   $(2x + 4)(x - 1) = 2x^2 + 2x - 4$
   $(2x + 2)(x - 2) = 2x^2 - 2x - 4$
   $(2x + 1)(x - 4) = 2x^2 - 7x - 4$

   The factorization is $(2x + 1)(x - 4)$.

2. $(3x - 4)(x + 1)$

3. $5x^2 + x - 18$

   (1) There is no common factor (other than 1 or −1).

   (2) Factor $5x^2$. The factor pair is $5x$, x. We have this possibility:

   $(5x\ \ )(x\ \ )$

   (3) Factor −18. The factor pairs are −18, 1 and 18, −1 and −9, 2 and 9, −2 and −6, 3 and 6, −3.

   (4) Look for factors in steps (2) and (3) such that the sum of their products is x. We try some possibilities:

   $(5x - 18)(x + 1) = 5x^2 - 13x - 18$
   $(5x + 18)(x - 1) = 5x^2 + 13x - 18$
   $(5x - 9)(x + 2) = 5x^2 + x - 18$

   The factorization is $(5x - 9)(x + 2)$.

4. $(3x + 5)(x - 3)$

5. $6x^2 + 23x + 7$

   (1) There is no common factor (other than 1 or −1).

   (2) Factor $6x^2$. The factor pairs are $6x$, x and $3x$, $2x$. We have these possibilities:

   $(6x\ \ )(x\ \ )$ and $(3x\ \ )(2x\ \ )$

   (3) Factor 7. The factor pairs are 7, 1 and −7, −1.

   (4) Look for factors in steps (2) and (3) such that the sum of their products is 23x. Since all signs are positive, we need consider only plus signs. We try some possibilities:

   $(6x + 7)(x + 1) = 6x^2 + 13x + 7$
   $(3x + 7)(2x + 1) = 6x^2 + 17x + 7$
   $(6x + 1)(x + 7) = 6x^2 + 43x + 7$
   $(3x + 1)(2x + 7) = 6x^2 + 23x + 7$

   The factorization is $(3x + 1)(2x + 7)$.

6. $(2x + 3)(3x + 2)$

7. $3x^2 + 4x + 1$

   (1) There is no common factor (other than 1 or −1).

   (2) Factor $3x^2$. The factor pair is $3x$, x. We have this possibility:

   $(3x\ \ )(x\ \ )$

   (3) Factor 1. The factor pairs are 1, 1 and −1, −1.

   (4) Look for factors in steps (2) and (3) such that the sum of their products is 4x. Since all signs are positive, we need consider only plus signs. There is only one such possibility:

   $(3x + 1)(x + 1) = 3x^2 + 4x + 1$

   The factorization is $(3x + 1)(x + 1)$.

8. $(7x + 1)(x + 2)$

9. $4x^2 + 4x - 15$

   (1) There is no common factor (other than 1 or −1).

   (2) Factor $4x^2$. The factor pairs are $4x$, x and $2x$, $2x$. We have these possibilities:

   $(4x\ \ )(x\ \ )$ and $(2x\ \ )(2x\ \ )$

   (3) Factor −15. The factor pairs are 15, −1 and −15, 1 and 5, −3 and −5, 3.

   (4) We try some possibilities:

   $(4x + 15)(x - 1) = 4x^2 + 11x - 15$
   $(2x + 15)(2x - 1) = 4x^2 + 28x - 15$
   $(4x - 15)(x + 1) = 4x^2 - 11x - 15$
   $(2x - 15)(2x + 1) = 4x^2 - 28x - 15$
   $(4x + 5)(x - 3) = 4x^2 - 7x - 15$
   $(2x + 5)(2x - 3) = 4x^2 + 4x - 15$

   The factorization is $(2x + 5)(2x - 3)$.

10. $(3x - 2)(3x + 4)$

11. $2x^2 - x - 1$
    (1) There is no common factor (other than 1 or -1).
    (2) Factor $2x^2$. The factor pair is $2x, x$. We have this possibility:
        $(2x\quad)(x\quad)$
    (3) Factor -1. The factor pair is $-1, 1$.
    (4) We try some possibilities:
        $(2x - 1)(x + 1) = 2x^2 + x - 1$
        $(2x + 1)(x - 1) = 2x^2 - x - 1$
    The factorization is $(2x + 1)(x - 1)$.

12. $(3x - 5)(5x + 2)$

13. $9x^2 + 18x - 16$
    (1) There is no common factor (other than 1 or -1).
    (2) Factor $9x^2$. The factor pairs are $9x, x$ and $3x, 3x$. We have these possibilities:
        $(9x\quad)(x\quad)$ and $(3x\quad)(3x\quad)$
    (3) Factor -16. The factor pairs are $16, -1$ and $-16, 1$ and $8, -2$ and $-8, 2$ and $4, -4$.
    (4) We try some possibilities:
        $(9x + 16)(x - 1) = 9x^2 + 7x - 16$
        $(3x + 16)(3x - 1) = 9x^2 + 45x - 16$
        $(9x - 16)(x + 1) = 9x^2 - 7x - 16$
        $(3x - 16)(3x + 1) = 9x^2 - 45x - 16$
        $(9x + 8)(x - 2) = 9x^2 - 10x - 16$
        $(3x + 8)(3x - 2) = 9x^2 + 18x - 16$
    The factorization is $(3x + 8)(3x - 2)$.

14. $(2x + 1)(x + 2)$

15. $3x^2 - 5x - 2$
    (1) There is no common factor (other than 1 or -1).
    (2) Factor $3x^2$. The factor pair is $3x, x$. We have this possibility:
        $(3x\quad)(x\quad)$
    (3) Factor -2. The factor pairs are $2, -1$ and $-2, 1$.
    (4) We try some possibilities:
        $(3x + 2)(x - 1) = 3x^2 - x - 2$
        $(3x - 2)(x + 1) = 3x^2 + x - 2$
        $(3x - 1)(x + 2) = 3x^2 + 5x - 2$
        $(3x + 1)(x - 2) = 3x^2 - 5x - 2$
    The factorization is $(3x + 1)(x - 2)$.

16. $(6x - 5)(3x + 2)$

17. $12x^2 + 31x + 20$
    (1) There is no common factor (other than 1 or -1).
    (2) Factor $12x^2$. The factor pairs are $12x, x$ and $6x, 2x$ and $4x, 3x$. We have these possibilities:
        $(12x\quad)(x\quad)$ and $(6x\quad)(2x\quad)$ and $(4x\quad)(3x\quad)$
    (3) Factor 20. Since all signs are positive, we need consider only positive pairs of factors. Those factor pairs are $20, 1$ and $10, 2$ and $5, 4$.
    (4) We can immediately reject all possibilities in which either factor has a common factor, such as $(12x + 20)$ or $(6x + 4)$, because we determined at the outset that there are no common factors. We try some of the remaining possibilities:
        $(12x + 1)(x + 20) = 12x^2 + 241x + 20$
        $(12x + 5)(x + 4) = 12x^2 + 53x + 20$
        $(6x + 1)(2x + 20) = 12x^2 + 122x + 20$
        $(4x + 5)(3x + 4) = 12x^2 + 31x + 20$
    The factorization is $(4x + 5)(3x + 4)$.

18. $(3x + 5)(5x - 2)$

19. $14x^2 + 19x - 3$
    (1) There is no common factor (other than 1 or -1).
    (2) Factor $14x^2$. The factor pairs are $14x, x$ and $7x, 2x$. We have these possibilities:
        $(14x\quad)(x\quad)$ and $(7x\quad)(2x\quad)$
    (3) Factor -3. The factor pairs are $-1, 3$ and $3, -1$.
    (4) We try some possibilities:
        $(14x - 1)(x + 3) = 14x^2 + 41x - 3$
        $(7x - 1)(2x + 3) = 7x^2 + 19x - 3$
    The factorization is $(7x - 1)(2x + 3)$.

20. $(7x + 4)(5x + 2)$

21. $9x^2 + 18x + 8$
    (1) There is no common factor (other than 1 or -1).
    (2) Factor $9x^2$. The factor pairs are $9x, x$ and $3x, 3x$. We have these possibilities:
        $(9x\quad)(x\quad)$ and $(3x\quad)(3x\quad)$
    (3) Factor 8. Since all signs are positive, we need consider only positive pairs of factors. Those factor pairs are $8, 1$ and $4, 2$.
    (4) We try some possibilities:
        $(9x + 8)(x + 1) = 9x^2 + 17x + 8$
        $(3x + 8)(3x + 1) = 9x^2 + 27x + 8$
        $(9x + 4)(x + 2) = 9x^2 + 22x + 8$
        $(3x + 4)(3x + 2) = 9x^2 + 18x + 8$
    The factorization is $(3x + 4)(3x + 2)$.

22. $(2 - 3x)(3 - 2x)$

Chapter 5   (5.3)

23. $49 - 42x + 9x^2 = 9x^2 - 42x + 49$
    (1) There is no common factor (other than 1 or -1).
    (2) Factor $9x^2$. The factor pairs are $9x, x$ and $3x, 3x$. We have these possibilities:
    $(9x\quad)(x\quad)$ and $(3x\quad)(3x\quad)$
    (3) Factor 49. Since 49 is positive and the middle term is negative, we need consider only negative pairs of factors. Those factor pairs are $-49, -1$ and $-7, -7$.
    (4) We try some possibilities:
    $(9x - 49)(x - 1) = 9x^2 - 58x + 49$
    $(3x - 49)(3x - 1) = 9x^2 - 150x + 49$
    $(9x - 7)(x - 7) = 9x^2 - 70x + 49$
    $(3x - 7)(3x - 7) = 9x^2 - 42x + 49$
    The factorization is $(3x - 7)(3x - 7)$, or $(3x - 7)^2$. This can also be expressed as follows:
    $(3x - 7)^2 = (-1)^2(3x - 7)^2 = [-1 \cdot (3x - 7)]^2 = (-3x + 7)^2$, or $(7 - 3x)^2$

24. $(5x + 4)^2$

25. $24x^2 + 47x - 2$
    (1) There is no common factor (other than 1 or -1).
    (2) Factor $24x^2$. The factor pairs are $24x, x$ and $12x, 2x$ and $6x, 4x$ and $3x, 8x$. We have these possibilities:
    $(24x\quad)(x\quad)$ and $(12x\quad)(2x\quad)$ and $(6x\quad)(4x\quad)$ and $(3x\quad)(8x\quad)$
    (3) Factor -2. The factor pairs are $2, -1$ and $-2, 1$.
    (4) We can immediately reject all possibilities in which either factor has a common factor, such as $(24x + 2)$ or $(12x - 2)$, because we determined at the outset that there are no common factors. We try some of the remaining possibilities:
    $(24x - 1)(x + 2) = 24x^2 + 47x - 2$
    The factorization is $(24x - 1)(x + 2)$.

26. $(8a + 3)(2a + 9)$

27. $35x^2 - 57x - 44$
    (1) There is no common factor (other than 1 or -1).
    (2) Factor $35x^2$. The factor pairs are $35x, x$ and $7x, 5x$. We have these possibilities:
    $(35x\quad)(x\quad)$ and $(7x\quad)(5x\quad)$
    (3) Factor -44. The factor pairs are $1, -44$ and $-1, 44$ and $2, -22$ and $-2, 22$ and $4, -11$, and $-4, 11$.

27. (continued)
    (4) We try some possibilities:
    $(35x + 1)(x - 44) = 35x^2 - 1539x - 44$
    $(7x + 1)(5x - 44) = 35x^2 - 303x - 44$
    $(35x + 2)(x - 22) = 35x^2 - 768x - 44$
    $(7x + 2)(5x - 22) = 35x^2 - 144x - 44$
    $(35x + 4)(x - 11) = 35x^2 - 381x - 44$
    $(7x + 4)(5x - 11) = 35x^2 - 57x - 44$
    The factorization is $(7x + 4)(5x - 11)$.

28. $(3a - 1)(3a + 5)$

29. $20 + 6x - 2x^2$
    (1) We factor out the common factor, 2:
    $2(10 + 3x - x^2)$
    Then we factor the trinomial $10 + 3x - x^2$.
    (2) Factor 10. The factor pairs are $10, 1$ and $5, 2$. We have these possibilities:
    $(10\quad)(1\quad)$ and $(5\quad)(2\quad)$
    Note that the second term of each factor is an x-term.
    (3) Factor $-x^2$. The factor pair is $x, -x$.
    (4) We try some possibilities:
    $(10 + x)(1 - x) = 10 - 9x - x^2$
    $(5 + x)(2 - x) = 10 - 3x - x^2$
    $(5 - x)(2 + x) = 10 + 3x - x^2$
    The factorization of $10 + 3x - x^2$ is $(5 - x)(2 + x)$. We must include the common factor in order to get a factorization of the original trinomial.
    $20 + 6x - 2x^2 = 2(5 - x)(2 + x)$

30. $(5 + 2x)(3 - x)$

31. $12x^2 + 28x - 24$
    (1) We factor out the common factor, 4:
    $4(3x^2 + 7x - 6)$
    Then we factor the trinomial $3x^2 + 7x - 6$.
    (2) Factor $3x^2$. The factor pair is $3x, x$. We have this possibility:
    $(3x\quad)(x\quad)$
    (3) Factor -6. The factor pairs are $6, -1$ and $-6, 1$ and $3, -2$ and $-3, 2$.
    (4) We can immediately reject all possibilities in which either factor has a common factor, such as $(3x + 6)$ or $(3x - 3)$, because we factored out the largest common factor at the outset. We try some of the remaining possibilities:
    $(3x - 1)(x + 6) = 3x^2 + 17x - 6$
    $(3x - 2)(x + 3) = 3x^2 + 7x - 6$
    The factorization of $3x^2 + 7x - 6$ is $(3x - 2)(x + 3)$. We must include the common factor in order to get a factorization of the original trinomial.
    $12x^2 + 28x - 24 = 4(3x - 2)(x + 3)$

Chapter 5   (5.3)

32. $3(2x + 1)(x + 5)$

33. $30x^2 - 24x - 54$
    (1) Factor out the common factor, 6:
    $6(5x^2 - 4x - 9)$
    Then we factor the trinomial $5x^2 - 4x - 9$.
    (2) Factor $5x^2$. The factor pair is $5x, x$. We have this possibility:
    $(5x\ \ \ )(x\ \ \ )$
    (3) Factor -9. The factor pairs are 9, -1 and -9, 1 and 3, -3.
    (4) We try some possibilities:
    $(5x + 9)(x - 1) = 5x^2 + 4x - 9$
    $(5x - 9)(x + 1) = 5x^2 - 4x - 9$
    The factorization of $5x^2 - 4x - 9$ is $(5x - 9)(x + 1)$. We must include the common factor in order to get a factorization of the original trinomial.
    $30x^2 - 24x - 54 = 6(5x - 9)(x + 1)$

34. $5(4x - 1)(x - 1)$

35. $4x + 6x^2 - 10 = 6x^2 + 4x - 10$
    (1) Factor out the common factor, 2:
    $2(3x^2 + 2x - 5)$
    Then we factor the trinomial $3x^2 + 2x - 5$.
    (2) Factor $3x^2$. The factor pair is $3x, x$. We have this possibility:
    $(3x\ \ \ )(x\ \ \ )$
    (3) Factor -5. The factor pairs are 5, -1 and -5, 1.
    (4) We try some possibilities:
    $(3x + 5)(x - 1) = 3x^2 + 2x - 5$
    Then $3x^2 + 2x - 5 = (3x + 5)(x - 1)$, so
    $6x^2 + 4x - 10 = 2(3x + 5)(x - 1)$.

36. $3(2x - 3)(3x + 1)$

37. $3x^2 - 4x + 1$
    (1) There is no common factor (other than 1 or -1).
    (2) Factor $3x^2$. The factor pair is $3x, x$. We have this possibility:
    $(3x\ \ \ )(x\ \ \ )$
    (3) Factor 1. Since 1 is positive and the middle term is negative, we need consider only negative factor pairs. The only such pair is $-1, -1$.
    (4) There is only one possibility:
    $(3x - 1)(x - 1) = 3x^2 - 4x + 1$
    The factorization is $(3x - 1)(x - 1)$.

38. $(2x - 3)(3x - 2)$

39. $12x^2 - 28x - 24$
    (1) Factor out the common factor, 4:
    $4(3x^2 - 7x - 6)$
    Then we factor the trinomial $3x^2 - 7x - 6$.
    (2) Factor $3x^2$. The factor pair is $3x, x$. We have this possibility:
    $(3x\ \ \ )(x\ \ \ )$
    (3) Factor -6. The factor pairs are 6, -1 and -6, 1 and 3, -2 and -3, 2.
    (4) We can immediately reject all possibilities in which either factor has a common factor, such as $(3x - 6)$ or $(3x + 3)$, because we factored out the largest common factor at the outset. We try some of the remaining possibilities:
    $(3x - 1)(x + 6) = 3x^2 + 17x - 6$
    $(3x - 2)(x + 3) = 3x^2 + 7x - 6$
    $(3x + 2)(x - 3) = 3x^2 - 7x - 6$
    Then $3x^2 - 7x - 6 = (3x + 2)(x - 3)$, so
    $12x^2 - 28x - 24 = 4(3x + 2)(x - 3)$.

40. $3(2x - 1)(x - 5)$

41. $-1 + 2x^2 - x = 2x^2 - x - 1$
    (1) There is no common factor (other than 1 or -1).
    (2) Factor $2x^2$. The factor pair is $2x, x$. We have this possibility:
    $(2x\ \ \ )(x\ \ \ )$
    (3) Factor -1. The factor pair is 1, -1.
    (4) We try some possibilities:
    $(2x + 1)(x - 1) = 2x^2 - x - 1$
    The factorization is $(2x + 1)(x - 1)$.

42. $(5x - 3)(3x - 2)$

43. $9x^2 - 18x - 16$
    (1) There is no common factor (other than 1 or -1).
    (2) Factor $9x^2$. The factor pairs are $9x, x$ and $3x, 3x$. We have these possibilities:
    $(9x\ \ \ )(x\ \ \ )$ and $(3x\ \ \ )(3x\ \ \ )$
    (3) Factor -16. The factor pairs are 16, -1 and -16, 1 and 8, -2 and -8, 2 and 4, -4.
    (4) We try some possibilities:
    $(9x + 16)(x - 1) = 9x^2 + 7x - 16$
    $(3x + 16)(3x - 1) = 9x^2 + 45x - 16$
    $(9x + 8)(x - 2) = 9x^2 - 10x - 16$
    $(3x + 8)(3x - 2) = 9x^2 + 18x - 16$
    $(3x - 8)(3x + 2) = 9x^2 - 18x - 16$
    The factorization is $(3x - 8)(3x + 2)$.

44. $7(2x + 1)(x + 2)$

Chapter 5   (5.3)

45. $15x^2 - 25x - 10$

    (1) Factor out the common factor, 5:

    $5(3x^2 - 5x - 2)$

    Then we factor the trinomial $3x^2 - 5x - 2$. This was done in Exercise 15. We know that $3x^2 - 5x - 2 = (3x + 1)(x - 2)$, so $15x^2 - 25x - 10 = 5(3x + 1)(x - 2)$.

46. $(6x + 5)(3x - 2)$

47. $12x^3 + 31x^2 + 20x$

    (1) We factor out the common factor, x:

    $x(12x^2 + 31x + 20)$

    Then we factor the trinomial $12x^2 + 31x + 20$. This was done in Exercise 17. We know that $12x^2 + 31x + 20 = (3x + 4)(4x + 5)$, so $12x^3 + 31x^2 + 20x = x(3x + 4)(4x + 5)$.

48. $x(5x - 2)(3x + 5)$

49. $14x^4 + 19x^3 - 3x^2$

    (1) Factor out the common factor, x^2:

    $x^2(14x^2 + 19x - 3)$

    Then we factor the trinomial $14x^2 + 19x - 3$. This was done in Exercise 19. We know that $14x^2 + 19x - 3 = (7x - 1)(2x + 3)$, so $14x^4 + 19x^3 - 3x^2 = x^2(7x - 1)(2x + 3)$.

50. $2x^2(5x + 2)(7x + 4)$

51. $168x^3 - 45x^2 + 3x$

    (1) Factor out the common factor, 3x:

    $3x(56x^2 - 15x + 1)$

    Then we factor the trinomial $56x^2 - 15x + 1$.

    (2) Factor $56x^2$. The factor pairs are 56x, x and 28x, 2x and 14x, 4x and 7x, 8x. We have these possibilities:

    (56x   )(x   ) and (28x   )(x   ) and
    (14x   )(4x   ) and (7x   )(8x   )

    (3) Factor 1. Since 1 is positive and the middle term is negative we need consider only the negative factor pair -1, -1.

    (4) We try some possibilities:

    $(56x - 1)(x - 1) = 56x^2 - 57x + 1$
    $(28x - 1)(2x - 1) = 56x^2 - 30x + 1$
    $(14x - 1)(4x - 1) = 56x^2 - 18x + 1$
    $(7x - 1)(8x - 1) = 56x^2 - 15x + 1$

    Then $56x^2 - 15x + 1 = (7x - 1)(8x - 1)$, so $168x^3 - 45x^2 + 3x = 3x(7x - 1)(8x - 1)$.

52. $24x^3(3x + 2)(2x + 1)$

53. $15x^4 - 19x^2 + 6 = 15(x^2)^2 - 19x^2 + 6$

    (1) There is no common factor (other than 1 or -1).

    (2) Factor $15x^4$. The factor pairs are $15x^2$, x^2 and $5x^2$, $3x^2$. We have these possibilities:

    $(15x^2\ \ )(x^2\ \ )$ and $(5x^2\ \ )(3x^2\ \ )$

53. (continued)

    (3) Factor 6. Since 6 is positive and the middle term is negative, we need consider only negative factor pairs. Those pairs are -6, -1 and -3, -2.

    (4) We can immediately reject all possibilities in which either factor has a common factor, such as $(15x^2 - 6)$ or $(3x^2 - 3)$, because we determined at the outset that there is no common factor. We try some of the remaining possibilities:

    $(15x^2 - 1)(x^2 - 6) = 15x^4 - 91x^2 + 6$
    $(15x^2 - 2)(x^2 - 3) = 15x^4 - 47x^2 + 6$
    $(5x^2 - 6)(3x^2 - 1) = 15x^4 - 23x^2 + 6$
    $(5x^2 - 3)(3x^2 - 2) = 15x^4 - 19x^2 + 6$

    The factorization is $(5x^2 - 3)(3x^2 - 2)$.

54. $(3x^2 + 2)(3x^2 + 4)$

55. $25t^2 + 80t + 64$

    (1) There is no common factor (other than 1 or -1).

    (2) Factor $25t^2$. The factor pairs are 25t, t and 5t, 5t. We have these possibilities:

    (25t   )(t   ) and (5t   )(5t   )

    (3) Factor 64. Since all signs are positive, we need consider only positive pairs of factors. Those factor pairs are 64, 1 and 32, 2 and 16, 4 and 8, 8.

    (4) We try some possibilities:

    $(25t + 64)(t + 1) = 25t^2 + 89t + 64$
    $(5t + 32)(5t + 2) = 25t^2 + 170t + 64$
    $(25t + 16)(t + 4) = 25t^2 + 116t + 64$
    $(5t + 8)(5t + 8) = 25t^2 + 80t + 64$

    The factorization is $(5t + 8)(5t + 8)$, or $(5t + 8)^2$.

56. $(3x - 7)^2$

57. $6x^3 + 4x^2 - 10x$

    (1) Factor out the common factor, 2x:

    $2x(3x^2 + 2x - 5)$

    Then we factor the trinomial $3x^2 + 2x - 5$. We did this in Exercise 35 (after we factored 2 out of the original trinomial). We know that $3x^2 + 2x - 5 = (3x + 5)(x - 1)$, so $6x^3 + 4x^2 - 10x = 2x(3x + 5)(x - 1)$.

58. $3x(3x + 1)(2x - 3)$

59. $25x^2 + 89x + 64$

    We follow the same procedure as in Exercise 55. The factorization is the first possibility we tried in step (4): $(25x + 64)(x + 1)$

60. Not factorable

102

Chapter 5 (5.3)

61. $x^2 + 3x - 7$
    (1) There is no common factor (other than 1 or -1).
    (2) Factor x^2. The factor pair is x, x. We have this possibility:
        $(x\quad)(x\quad)$
    (3) Factor -7. The factor pairs are $7, -1$ and $-7, 1$.
    (4) We try the possibilities:
        $(x + 7)(x - 1) = x^2 + 6x - 7$
        $(x - 7)(x + 1) = x^2 - 6x - 7$
    Neither possibility works. Thus, $x^2 + 3x - 7$ is not factorable. (Note that we could have also used the method of Section 5.2 since the leading coefficient of this trinomial is 1.)

62. Not factorable

63. $12m^2 + mn - 20n^2$
    (1) There is no common factor (other than 1 or -1).
    (2) Factor $12m^2$. The factor pairs are $12m, m$ and $6m, 2m$ and $3m, 4m$. We have these possibilities:
        $(12m\quad)(m\quad)$ and $(6m\quad)(2m\quad)$ and $(3m\quad)(4m\quad)$
    (3) Factor $-20n^2$. The factor pairs are $20n, -n$ and $-20n, n$ and $10n, -2n$ and $-10n, 2n$ and $5n, -4n$ and $-5n, 4n$.
    (4) We can immediately reject all possibilities in which either factor has a common factor, such as $(12m + 20n)$ or $(4m - 2n)$, because we determined at the outset that there is no common factor. We try some of the remaining possibilities:
        $(12m - n)(m + 20n) = 12m^2 + 239mn - 20n^2$
        $(12m + 5n)(m - 4n) = 12m^2 - 43mn - 20n^2$
        $(3m - 20n)(4m + n) = 12m^2 - 77mn - 20n^2$
        $(3m - 4n)(4m + 5n) = 12m^2 - mn - 20n^2$
        $(3m + 4n)(4m - 5n) = 12m^2 + mn - 20n^2$
    The factorization is $(3m + 4n)(4m - 5n)$.

64. $(4a + 3b)(3a + 2b)$

65. $6a^2 - ab - 15b^2$
    (1) There is no common factor (other than 1 or -1).
    (2) Factor $6a^2$. The factor pairs are $6a, a$ and $3a, 2a$. We have these possibilities:
        $(6a\quad)(a\quad)$ and $(3a\quad)(2a\quad)$
    (3) Factor $-15b^2$. The factor pairs are $15b, -b$ and $-15b, b$ and $5b, -3b$ and $-5b, 3b$.
    (4) We can immediately reject all possibilities in which either factor has a common factor, such as $(6a + 15b)$ or $(3a - 3b)$, because we determined at the outset that there is no common factor. We try some of the remaining possibilities:

65. (continued)
    $(6a - b)(a + 15b) = 6a^2 + 89ab - 15b^2$
    $(3a - b)(2a + 15b) = 6a^2 + 43ab - 15b^2$
    $(6a + 5b)(a - 3b) = 6a^2 - 13ab - 15b^2$
    $(3a + 5b)(2a - 3b) = 6a^2 + ab - 15b^2$
    $(3a - 5b)(2a + 3b) = 6a^2 - ab - 15b^2$
    The factorization is $(3a - 5b)(2a + 3b)$.

66. $(3p + 2q)(p - 6q)$

67. $8a^2 - 6ab - 9b^2$
    (1) There is no common factor (other than 1 or -1).
    (2) Factor $8a^2$. The factor pairs are $8a, a$ and $4a, 2a$. We have these possibilities:
        $(8a\quad)(a\quad)$ and $(4a\quad)(2a\quad)$
    (3) Factor $-9b^2$. The factor pairs are $9b, -b$ and $-9b, b$ and $3b, -3b$.
    (4) We try some possibilities:
        $(8a + 9b)(a - b) = 8a^2 + ab - 9b^2$
        $(4a + 9b)(2a - b) = 8a^2 + 14ab - 9b^2$
        $(8a + 3b)(a - 3b) = 8a^2 - 21ab - 9b^2$
        $(4a + 3b)(2a - 3b) = 8a^2 - 6ab - 9b^2$
    The factorization is $(4a + 3b)(2a - 3b)$.

68. $(2m - 3n)(m + 2n)$

69. $9a^2 + 18ab + 8b^2$
    (1) There is no common factor (other than 1 or -1).
    (2) Factor $9a^2$. The factor pairs are $9a, a$ and $3a, 3a$. We have these possibilities:
        $(9a\quad)(a\quad)$ and $(3a\quad)(3a\quad)$
    (3) Factor $8b^2$. Since all signs are positive, we need consider only pairs of factors with positive coefficients. Those factor pairs are $8b, b$ and $4b, 2b$.
    (4) We try some possibilities:
        $(9a + 8b)(a + b) = 9a^2 + 17ab + 8b^2$
        $(3a + 8b)(3a + b) = 9a^2 + 27ab + 8b^2$
        $(9a + 4b)(a + 2b) = 9a^2 + 22ab + 8b^2$
        $(3a + 4b)(3a + 2b) = 9a^2 + 18ab + 8b^2$
    The factorization is $(3a + 4b)(3a + 2b)$.

70. $2(5s - 3t)(s + t)$

Chapter 5   (5.3)

71. $35p^2 + 34pq + 8q^2$

   (1) There is no common factor (other than 1 or −1).

   (2) Factor $35p^2$. The factor pairs are $35p$, p and $7p$, $5p$. We have these possibilities:

   $(35p\ \ )(p\ \ )$ and $(7p\ \ )(5p\ \ )$

   (3) Factor $8q^2$. Since all signs are positive, we need consider only pairs of factors with positive coefficients. Those factor pairs are $8q$, q and $4q$, $2q$.

   (4) We try some possibilities:

   $(35p + 8q)(p + q) = 35p^2 + 43pq + 8q^2$
   $(7p + 8q)(5p + q) = 35p^2 + 47pq + 8q^2$
   $(35p + 4q)(p + 2q) = 35p^2 + 74pq + 8q^2$
   $(7p + 4q)(5p + 2q) = 35p^2 + 34pq + 8q^2$

   The factorization is $(7p + 4q)(5p + 2q)$.

72. $(5x - 7y)(2x + 3y)$

73. $18x^2 - 6xy - 24y^2$

   (1) Factor out the common factor, 6:

   $6(3x^2 - xy - 4y^2)$

   Then we factor the trinomial $3x^2 - xy - 4y^2$.

   (2) Factor $3x^2$. The factor pair is $3x$, x. We have this possibility:

   $(3x\ \ )(x\ \ )$

   (3) Factor $-4y^2$. The factor pairs are $4y$, $-y$ and $-4y$, y and $2y$, $-2y$.

   (4) We try some possibilities:

   $(3x + 4y)(x - y) = 3x^2 + xy - 4y^2$
   $(3x - 4y)(x + y) = 3x^2 - xy - 4y^2$

   Then $3x^2 - xy - 4y^2 = (3x - 4y)(x + y)$, so $18x^2 - 6xy - 24y^2 = 6(3x - 4y)(x + y)$.

74. $5(3a - 4b)(a + b)$

75. Familiarize. We will use the formula $C = 2\pi r$, where C is circumference and r is radius, to find the radius in kilometers. Then we will multiply that number by 0.62 to find the radius in miles.

   Translate.

   Circumference = $2 \cdot \pi \cdot$ radius
   $40,000 \approx 2(3.14)r$

   Carry out. First we solve the equation.

   $40,000 \approx 2(3.14)r$
   $40,000 \approx 6.28r$
   $6369 \approx r$

   Then we multiply to find the radius in miles:

   $6369(0.62) \approx 3949$

   Check. If $r = 6369$, then $2\pi r = 2(3.14)(6369) \approx 40,000$. We should also recheck the multiplication we did to find the radius in miles. Both values check.

   State. The radius of the earth is about 6369 km or 3949 mi. (These values may differ slightly if a different approximation is used for π.)

76. 40°

77. $3(2y + 3) = 21$
    $6y + 9 = 21$
    $6y = 12$
    $y = 2$

78. y^{-4}

79. $20x^{2n} + 16x^n + 3 = 20(x^n)^2 + 16x^n + 3$

   (1) There is no common factor (other than 1 or −1).

   (2) Factor $20x^{2n}$. The factor pairs are $20x^n$, x^n and $10x^n$, $2x^n$ and $5x^n$, $4x^n$. We have these possibilities:

   $(20x^n\ \ )(x^n\ \ )$ and $(10x^n\ \ )(2x^n\ \ )$ and $(5x^n\ \ )(4x^n\ \ )$

   (3) Factor 3. Since all signs are positive, we need consider only the positive factor pair 3, 1.

   (4) We try some possibilities:

   $(20x^n + 3)(x^n + 1) = 20x^{2n} + 23x^n + 3$
   $(10x^n + 3)(2x^n + 1) = 20x^{2n} + 16x^n + 3$

   The factorization is $(10x^n + 3)(2x^n + 1)$.

80. $-(3x^m - 4)(5x^m - 2)$

81. $3x^{6a} - 2x^{3a} - 1 = 3(x^{3a})^2 - 2x^{3a} - 1$

   (1) There is no common factor (other than 1 or −1).

   (2) Factor $3x^{6a}$. The factor pair is $3x^{3a}$, x^{3a}. We have this possibility:

   $(3x^{3a}\ \ )(x^{3a}\ \ )$

   (3) Factor −1. The factor pair is −1, 1.

   (4) We try these possibilities:

   $(3x^{3a} - 1)(x^{3a} + 1) = 3x^{6a} + 2x^{3a} - 1$
   $(3x^{3a} + 1)(x^{3a} - 1) = 3x^{6a} - 2x^{3a} - 1$

   The factorization is $(3x^{3a} + 1)(x^{3a} - 1)$.

82. $x(x^n - 1)^2$

83. $3(a + 1)^{n+1}(a + 3)^2 - 5(a + 1)^n(a + 3)^3$
    $= (a + 1)^n(a + 3)^2[3(a + 1) - 5(a + 3)]$ Removing the common factors
    $= (a + 1)^n(a + 3)^2[3a + 3 - 5a - 15]$ Simplifying
    $= (a + 1)^n(a + 3)^2(-2a - 12)$ inside the brackets
    $= (a + 1)^n(a + 3)^2(-2)(a + 6)$ Removing the common factor
    $= -2(a + 1)^n(a + 3)^2(a + 6)$ Rearranging

Chapter 5 (5.4)

Exercise Set 5.4

1. $x^2 - 14x + 49$

   a) We know that x^2 and 49 are squares.

   b) There is no minus sign before either x^2 or 49.

   c) If we multiply the square roots, x and 7, and double the product, we get $2 \cdot x \cdot 7 = 14x$. This is the additive inverse of the remaining term, $-14x$.

   Thus, $x^2 - 14x + 49$ is a trinomial square.

2. Yes

3. $x^2 + 16x - 64$

   Both x^2 and 64 are squares, but there is a minus sign before 64. Thus, $x^2 + 16x - 64$ is not a trinomial square.

4. No

5. $x^2 - 3x + 9$

   a) Both x^2 and 9 are squares.

   b) There is no minus sign before either x^2 or 9.

   c) If we multiply the square roots, x and 3, and double the product, we get $2 \cdot x \cdot 3 = 6x$. This is neither the remaining term nor its additive inverse.

   Thus, $x^2 - 3x + 9$ is not a trinomial square.

6. No

7. $8x^2 + 40x + 25$

   Only one term, 25, is a square. Thus, $8x^2 + 40x + 25$ is not a trinomial square.

8. No

9. $36x^2 - 24x + 16$

   a) Both $36x^2$ and 16 are squares.

   b) There is no minus sign before either $36x^2$ or 16.

   c) If we multiply the square roots, 6x and 4, and double the product, we get $2 \cdot 6x \cdot 4 = 48x$. This is neither the remaining term nor its additive inverse.

   Thus, $36x^2 - 24x + 16$ is not a trinomial square.

10. $(x - 7)^2$

11. $x^2 - 16x + 64 = x^2 - 2 \cdot x \cdot 8 + 8^2 = (x - 8)^2$
    $A^2 - 2\ A\ B + B^2 = (A - B)^2$

12. $(x + 8)^2$

13. $x^2 + 14x + 49 = x^2 + 2 \cdot x \cdot 7 + 7^2 = (x + 7)^2$
    $A^2 + 2\ A\ B + B^2 = (A + B)^2$

14. $(x - 1)^2$

15. $x^2 + 2x + 1 = x^2 + 2 \cdot x \cdot 1 + 1^2 = (x + 1)^2$

16. $(x + 2)^2$

17. $4 + x^2 - 4x = x^2 - 4x + 4$ Changing the order
    $x^2 - 4x + 4 = x^2 - 2 \cdot x \cdot 2 + 2^2 = (x - 2)^2$

18. $(y - 3)^2$

19. $y^4 + 6y^2 + 9 = (y^2)^2 + 2 \cdot y^2 \cdot 3 + 3^2 = (y^2 + 3)^2$

20. $(3 + p)^2(3 - p)^2$

21. $k^4 - 24k^2 + 144 = (k^2)^2 - 2 \cdot k^2 \cdot 12 + 12^2$
    $= (k^2 - 12)^2$

22. $(w^2 + 11)^2$

23. $49 - 56y + 16y^2 = 16y^2 - 56y + 49$
    $= (4y)^2 - 2 \cdot 4y \cdot 7 + 7^2$
    $= (4y - 7)^2$

24. $3(4m + 5)^2$

25. $2x^2 - 4x + 2 = 2(x^2 - 2x + 1)$
    $= 2(x^2 - 2 \cdot x \cdot 1 + 1^2)$
    $= 2(x - 1)^2$

26. $2(x - 10)^2$

27. $x^3 - 18x^2 + 81x = x(x^2 - 18x + 81)$
    $= x(x^2 - 2 \cdot x \cdot 9 + 9^2)$
    $= x(x - 9)^2$

28. $x(x + 12)^2$

29. $20x^2 + 100x + 125 = 5(4x^2 + 20x + 25)$
    $= 5[(2x)^2 + 2 \cdot 2x \cdot 5 + 5^2]$
    $= 5(2x + 5)^2$

30. $3(2x + 3)^2$

31. $49 - 42x + 9x^2 = 7^2 - 2 \cdot 7 \cdot 3x + (3x)^2 = (7 - 3x)^2$

32. $(8 - 7x)^2$, or $(7x - 8)^2$

33. $5y^4 + 10y^2 + 5 = 5(y^4 + 2y^2 + 1)$
    $= 5[(y^2)^2 + 2 \cdot y^2 \cdot 1 + 1^2]$
    $= 5(y^2 + 1)^2$

34. $(a^2 + 7)^2$

35. $y^6 + 26y^3 + 169 = (y^3)^2 + 2 \cdot y^3 \cdot 13 + 13^2$
    $= (y^3 + 13)^2$

36. $(y^3 - 8)^2$

Chapter 5   (5.4)

37. $16x^{10} - 8x^5 + 1 = (4x^5)^2 - 2\cdot 4x^5\cdot 1 + 1^2$
    $= (4x^5 - 1)^2$

38. $(3x^5 + 2)^2$

39. $1 + 4x^4 + 4x^2 = 4x^4 + 4x^2 + 1$
    $= (2x^2)^2 + 2\cdot 2x^2\cdot 1 + 1^2$
    $= (2x^2 + 1)^2$

40. $(1 - a^3)^2$, or $(a^3 - 1)^2$

41. $\frac{1}{81}x^6 + \frac{8}{27}x^3 + \frac{16}{9} = \left(\frac{1}{9}x^3\right)^2 + 2\cdot\frac{1}{9}x^3\cdot\frac{4}{3} + \left(\frac{4}{3}\right)^2$
    $= \left(\frac{1}{9}x^3 + \frac{4}{3}\right)^2$

42. $\left(\frac{1}{3}a + \frac{1}{2}\right)^2$

43. $4p^2 + 12pq + 9q^2 = (2p)^2 + 2\cdot 2p\cdot 3q + (3q)^2$
    $= (2p + 3q)^2$

44. $(5m + 2n)^2$

45. $a^2 - 14ab + 49b^2 = a^2 - 2\cdot a\cdot 7b + (7b)^2$
    $= (a - 7b)^2$

46. $(x - 3y)^2$

47. $64m^2 + 16mn + n^2 = (8m)^2 + 2\cdot 8m\cdot n + n^2$
    $= (8m + n)^2$

48. $(9p - q)^2$

49. $16s^2 - 40st + 25t^2 = (4s)^2 - 2\cdot 4s\cdot 5t + (5t)^2$
    $= (4s - 5t)^2$

50. $4(3a + 4b)^2$

51. Familiarize. Let a = the amount of oxygen that can be dissolved in 100 L of water at 20° C.
    Translate.

Oxygen dissolved at 0° C	is	1.6	times	oxygen dissolved at 20° C
5	=	1.6	·	a

    Carry out. We solve the equation.
    $5 = 1.6a$
    $\frac{5}{1.6} = a$
    $3.125 = a$
    Check. We find 1.6 times 3.125: $1.6(3.125) = 5$
    The answer checks.
    State. 3.125 L of oxygen can be dissolved in 100 L of water at 20° C.

52. $\ell = 67$ m, $w = 42$ m

53. $(x + 6)(x - 4) = x^2 - 4x + 6x - 24$ Using FOIL
    $= x^2 + 2x - 24$ Simplifying

54. $-12x^{13}$

55. $49x^2 - 216$
    There is no common factor. Also, $49x^2$ is a square, but 216 is not so this expression is not the difference of squares. It is not factorable.

56. $x(27x^2 - 13)$

57. $x^2 + 22x + 121 = x^2 + 2\cdot x\cdot 11 + 11^2$
    $= (x + 11)^2$

58. Not factorable

59. $x^2 - 5x + 25$
    There is no common factor. We look for factors of the form $(x \quad)(x \quad)$. Since the constant term is positive and the middle term is negative, we look for a factorization of 25 in which both factors are negative. Their sum must be -5.

Pairs of factors	Sums of factors
$-25, -1$	-26
$-5, -5$	-10

    There are no factors whose sum is -5. Thus, $x^2 - 5x + 25$ is not factorable.

60. $2x(3x + 1)^2$

61. $63x - 28 = 7(9x - 4)$

62. $2(81x^2 - 41)$

63. $x^4 - 9 = (x^2 + 3)(x^2 - 3)$

64. $0.1(9x + 8)(9x - 8)$

65. $x^8 - 2^8 = (x^4 + 2^4)(x^4 - 2^4) =$
    $(x^4 + 2^4)(x^2 + 2^2)(x^2 - 2^2) =$
    $(x^4 + 2^4)(x^2 + 2^2)(x + 2)(x - 2)$, or
    $(x^4 + 16)(x^4 + 4)(x + 2)(x - 2)$

66. $(9 + x^2)(3 + x)(3 - x)$

67. $(y + 3)^2 + 2(y + 3) + 1$
    $= (y + 3)^2 + 2\cdot(y + 3)\cdot 1 + 1^2$
    $= [(y + 3) + 1]^2$
    $= (y + 4)^2$

68. $(a + 3)^2$

69. $4(a + 5)^2 + 20(a + 5) + 25$
    $= [2(a + 5)]^2 + 2\cdot[2(a + 5)]\cdot 5 + 5^2$
    $= [2(a + 5) + 5]^2$
    $= (2a + 10 + 5)^2$
    $= (2a + 15)^2$

70. $(7x + 4)^2$

Chapter 5   (5.5)

71. $(x + 7)^2 - 4x - 24 = (x^2 + 14x + 49) - 4x - 24$
    $= x^2 + 10x + 25$
    $= x^2 + 2 \cdot x \cdot 5 + 5^2$
    $= (x + 5)^2$

72. $(a + 1)^2$

73. See the answer section in the text.

74. No. $(x + 3)^2(x - 3)^2 = [(x + 3)(x - 3)]^2 = (x^2 - 9)^2 = x^4 - 18x^2 + 81 \neq x^4 + 18x^2 + 81$

75. $9x^{18} + 48x^9 + 64 = (3x^9)^2 + 2 \cdot 3x^9 \cdot 8 + 8^2$
    $= (3x^9 + 8)^2$

76. $(x^n + 5)^2$

77. $a^{2n} - 18a^n + 81 = (a^n)^2 - 2 \cdot a^n \cdot 9 + 9^2$
    $= (a^n - 9)^2$

78. $(3b^n + 2)^2$

79. If $cy^2 + 6y + 1$ is the square of a binomial, then $2 \cdot a \cdot 1 = 6$ where $a^2 = c$. Then $a = 3$, so $c = a^2 = 3^2 = 9$. (The polynomial is $9y^2 + 6y + 1$.)

80. 16

81. See the answer section in the text.

82. 0, 2

Exercise Set 5.5

1. $7(y + 3) + 4(y + 3)$
   $= (7 + 4)(y + 3) = 11(y + 3)$

2. $5(b - 5)$

3. $x^2(x + 3) + 2(x + 3)$
   $= (x^2 + 2)(x + 3)$

4. $(3z^2 + 1)(2z + 1)$

5. $y^2(y + 8) - (y + 8) = y^2(y + 8) - 1(y + 8)$
   $= (y^2 - 1)(y + 8)$
   $= (y + 1)(y - 1)(y + 8)$
   Factoring the difference of squares

6. $(x + 2)(x - 2)(x - 7)$

7. $x^3 + 3x^2 + 2x + 6 = x^3 + 3x^2 + 2x + 6$
   $= x^2(x + 3) + 2(x + 3)$
   Factoring each binomial
   $= (x^2 + 2)(x + 3)$
   Factoring out the common factor

8. $(3z^2 + 1)(2z + 1)$

9. $2x^3 + 6x^2 + x + 3 = 2x^3 + 6x^2 + x + 3$
   $= 2x^2(x + 3) + 1(x + 3)$
   Factoring each binomial
   $= (2x^2 + 1)(x + 3)$

10. $(x^2 + 1)(3x + 2)$

11. $8x^3 - 12x^2 + 6x - 9 = 4x^2(2x - 3) + 3(2x - 3)$
    $= (4x^2 + 3)(2x - 3)$

12. $(5x^2 + 2)(2x - 5)$

13. $12x^3 - 16x^2 + 3x - 4 = 4x^2(3x - 4) + 1(3x - 4)$
    Factoring 1 out of the second binomial
    $= (4x^2 + 1)(3x - 4)$

14. $(3x^2 + 5)(6x - 7)$

15. $x^3 + 8x^2 - 3x - 24 = x^2(x + 8) - 3(x + 8)$
    $= (x^2 - 3)(x + 8)$

16. $(2x^2 - 5)(x + 6)$

17. $w^3 - 7w^2 - 4w + 28 = w^2(w - 7) - 4(w - 7)$
    $= (w^2 - 4)(w - 7)$
    $= (w + 2)(w - 2)(w - 7)$
    Factoring the difference of squares

18. $(y + 1)(y - 1)(y + 8)$

19. $x^3 - x^2 - 2x + 5 = x^2(x - 1) - 1(2x - 5)$
    This polynomial is not factorable using factoring by grouping.

20. Not factorable by grouping

21. $2x^3 - 8x^2 - 9x + 36 = 2x^2(x - 4) - 9(x - 4)$
    $= (2x^2 - 9)(x - 4)$

22. $(4g^2 - 5)(5g - 1)$

23. $y^2 + 4y + y + 4 = y(y + 4) + 1(y + 4)$
    $= (y + 1)(y + 4)$

24. $(x + 2)(x + 5)$

107

Chapter 5   (5.5)

25. $x^2 - 4x - x + 4 = x(x - 4) - 1(x - 4)$
    $= (x - 1)(x - 4)$

26. $(a - 2)(a + 5)$

27. $6x^2 + 4x + 9x + 6 = 2x(3x + 2) + 3(3x + 2)$
    $= (2x + 3)(3x + 2)$

28. $(x + 1)(3x - 2)$

29. $3x^2 - 4x - 12x + 16 = x(3x - 4) - 4(3x - 4)$
    $= (x - 4)(3x - 4)$

30. $(6 - 5y)(4 - 3y)$

31. $35x^2 - 40x + 21x - 24 = 5x(7x - 8) + 3(7x - 8)$
    $= (5x + 3)(7x - 8)$

32. $(2x - 7)(4x - 3)$

33. $4x^2 + 6x - 6x - 9 = 2x(2x + 3) - 3(2x + 3)$
    $= (2x - 3)(2x + 3)$

34. $(2x^2 - 5)(x^2 - 3)$

35. $2x^4 + 6x^2 + 5x^2 + 15 = 2x^2(x^2 + 3) + 5(x^2 + 3)$
    $= (2x^2 + 5)(x^2 + 3)$

36. $(2x^2 - 3)^2$

37. $2x^2 - 7x - 4$
    a) First look for a common factor. There is none (other than 1).
    b) Multiply the leading coefficient and the constant, 2 and -4: $2(-4) = -8$.
    c) Try to factor -8 so that the sum of the factors is -7.

Pairs of factors	Sums of factors
-1, 8	7
1, -8	-7
-2, 4	2
2, -4	-2

    d) Split the middle term: $-7x = 1x - 8x$
    e) Factor by grouping:
    $2x^2 - 7x - 4 = 2x^2 + x - 8x - 4$
    $= x(2x + 1) - 4(2x + 1)$
    $= (x - 4)(2x + 1)$

38. $(3x - 4)(x + 1)$

39. $5x^2 + x - 18$
    a) First look for a common factor. There is none (other than 1).
    b) Multiply the leading coefficient and the constant, 5 and -18: $5(-18) = -90$.
    c) Try to factor -90 so that the sum of the factors is 1.

Pairs of factors	Sums of factors
-1, 90	89
1, -90	-89
-2, 45	43
2, -45	-43
-3, 30	27
3, -30	-27
-5, 18	13
5, -18	-13
-6, 15	9
6, -15	-9
-9, 10	1
9, -10	-1

    d) Split the middle term: $x = -9x + 10x$
    e) Factor by grouping:
    $5x^2 + x - 18 = 5x^2 - 9x + 10x - 18$
    $= x(5x - 9) + 2(5x - 9)$
    $= (x + 2)(5x - 9)$

40. $(3x + 5)(x - 3)$

41. $6x^2 + 23x + 7$
    a) First look for a common factor. There is none (other than 1).
    b) Multiply the leading coefficient and the constant, 6 and 7: $6 \cdot 7 = 42$.
    c) Try to factor 42 so that the sum of the factors is 23. We only need to consider positive factors.

Pairs of factors	Sums of factors
1, 42	43
2, 21	23
3, 14	17
6, 7	13

    d) Split the middle term: $23x = 2x + 21x$
    e) Factor by grouping:
    $6x^2 + 23x + 7 = 6x^2 + 2x + 21x + 7$
    $= 2x(3x + 1) + 7(3x + 1)$
    $= (2x + 7)(3x + 1)$

42. $(3x + 2)(2x + 3)$

Chapter 5   (5.5)

43. $3x^2 + 4x + 1$

   a) First look for a common factor. There is none (other than 1).

   b) Multiply the leading coefficient and the constant, 3 and 1: $3 \cdot 1 = 3$.

   c) Try to factor 3 so that the sum of the factors is 4. The numbers we want are 1 and 3: $1 \cdot 3 = 3$ and $1 + 3 = 4$.

   d) Split the middle term: $4x = 1x + 3x$

   e) Factor by grouping:
   $$3x^2 + 4x + 1 = 3x^2 + x + 3x + 1$$
   $$= x(3x + 1) + 1(3x + 1)$$
   $$= (x + 1)(3x + 1)$$

44. $(7x + 1)(x + 2)$

45. $4x^2 + 4x - 15$

   a) First look for a common factor. There is none (other than 1).

   b) Multiply the leading coefficient and the constant, 4 and -15: $4(-15) = -60$.

   c) Try to factor -60 so that the sum of the factors is 4.

Pairs of factors	Sums of factors
-1, 60	59
1, -60	-59
-2, 30	28
2, -30	-28
-3, 20	17
3, -20	-17
-4, 15	11
4, -15	-11
-5, 12	7
5, -12	-7
-6, 10	4
6, -10	-4

   d) Split the middle term: $4x = -6x + 10x$

   e) Factor by grouping:
   $$4x^2 + 4x - 15 = 4x^2 - 6x + 10x - 15$$
   $$= 2x(2x - 3) + 5(2x - 3)$$
   $$= (2x + 5)(2x - 3)$$

46. $(3x - 2)(3x + 4)$

47. $2x^2 - x - 1$

   a) First look for a common factor. There is none (other than 1).

   b) Multiply the leading coefficient and the constant, 2 and -1: $2(-1) = -2$.

   c) Try to factor -2 so that the sum of the factors is -1. The numbers we want are -2 and 1: $-2 \cdot 1 = -2$ and $-2 + 1 = -1$.

47. (continued)

   d) Split the middle term: $-x = -2x + 1x$

   e) Factor by grouping:
   $$2x^2 - x - 1 = 2x^2 - 2x + x - 1$$
   $$= 2x(x - 1) + 1(x - 1)$$
   $$= (2x + 1)(x - 1)$$

48. $(5x + 2)(3x - 5)$

49. $9x^2 + 18x - 16$

   a) First look for a common factor. There is none (other than 1).

   b) Multiply the leading coefficient and the constant, 9 and -16: $9(-16) = -144$.

   c) Try to factor -144 so that the sum of the factors is 18.

Pairs of factors	Sums of factors
-1, 144	143
1, -144	-143
-2, 72	70
2, -72	-70
-3, 48	45
3, -48	-45
-4, 36	32
4, -36	-32
-6, 24	18
6, -24	-18
-8, 18	10
8, -18	-10
-9, 16	7
9, -16	-7
-12, 12	0

   d) Split the middle term: $18x = -6x + 24x$

   e) Factor by grouping:
   $$9x^2 + 18x - 16 = 9x^2 - 6x + 24x - 16$$
   $$= 3x(3x - 2) + 8(3x - 2)$$
   $$= (3x + 8)(3x - 2)$$

50. $(2x + 1)(x + 2)$

Chapter 5   (5.5)

51. $3x^2 - 5x - 2$
    a) First look for a common factor. There is none (other than 1).
    b) Multiply the leading coefficient and the constant, 3 and -2: $3(-2) = -6$.
    c) Try to factor -6 so that the sum of the factors is -5. The numbers we want are 1 and -6: $1(-6) = -6$ and $1 + (-6) = -5$.
    d) Split the middle term: $-5x = 1x - 6x$
    e) Factor by grouping:
    $3x^2 - 5x - 2 = 3x^2 + x - 6x - 2$
    $= x(3x + 1) - 2(3x + 1)$
    $= (x - 2)(3x + 1)$

52. $(6x - 5)(3x + 2)$

53. $12x^2 + 31x + 20$
    a) First look for a common factor. There is none (other than 1).
    b) Multiply the leading coefficient and the constant, 12 and 20: $12 \cdot 20 = 240$.
    c) Try to factor 240 so that the sum of the factors is 31. We only need to consider positive factors.

Pairs of factors	Sums of factors
1, 240	241
2, 120	122
3, 80	83
4, 60	64
5, 48	53
6, 40	46
8, 30	38
10, 24	34
12, 20	32
15, 16	31

    d) Split the middle term: $31x = 15x + 16x$
    e) Factor by grouping:
    $12x^2 + 31x + 20 = 12x^2 + 15x + 16x + 20$
    $= 3x(4x + 5) + 4(4x + 5)$
    $= (3x + 4)(4x + 5)$

54. $(5x - 2)(3x + 5)$

55. $14x^2 + 19x - 3$
    a) First look for a common factor. There is none (other than 1).
    b) Multiply the leading coefficient and the constant, 14 and -3: $14(-3) = -42$.
    c) Try to factor -42 so that the sum of the factors is 19.

Pairs of factors	Sums of factors
-1, 42	41
1, -42	-41
-2, 21	19
2, -21	-19
-3, 14	11
3, -14	-11
-6, 7	1
6, -7	-1

    d) Split the middle term: $19x = -2x + 21x$
    e) Factor by grouping:
    $14x^2 + 19x - 3 = 14x^2 - 2x + 21x - 3$
    $= 2x(7x - 1) + 3(7x - 1)$
    $= (2x + 3)(7x - 1)$

56. $(5x + 2)(7x + 4)$

57. $9x^2 + 18x + 8$
    a) First look for a common factor. There is none (other than 1).
    b) Multiply the leading coefficient and the constant, 9 and 8: $9 \cdot 8 = 72$.
    c) Try to factor 72 so that the sum of the factors is 18. We only need to consider positive factors.

Pairs of factors	Sums of factors
1, 72	73
2, 36	38
3, 24	27
4, 18	22
6, 12	18
8, 9	17

    d) Split the middle term: $18x = 6x + 12x$.
    e) Factor by grouping:
    $9x^2 + 18x + 8 = 9x^2 + 6x + 12x + 8$
    $= 3x(3x + 2) + 4(3x + 2)$
    $= (3x + 4)(3x + 2)$

58. $(2x - 3)(3x - 2)$

Chapter 5   (5.5)

59. $49 - 42x + 9x^2 = 9x^2 - 42x + 49$

   a) First look for a common factor. There is none (other than 1).

   b) Multiply the leading coefficient and the constant, 9 and 49: $9 \cdot 49 = 441$

   c) Try to factor 441 so that the sum of the factors is -42. We only need to consider negative factors.

Pairs of factors	Sums of factors
-1, -441	-442
-3, -147	-150
-7, -63	-70
-9, -49	-58
-21, -21	-42

   d) Split the middle term: $-42x = -21x - 21x$

   e) Factor by grouping:
   $9x^2 - 42x + 49 = 9x^2 - 21x - 21x + 49$
   $= 3x(3x - 7) - 7(3x - 7)$
   $= (3x - 7)(3x - 7),$ or
   $(3x - 7)^2$

   You may have observed at the outset that $9x^2 - 42x + 49$ is a trinomial square and, hence, factored it using the method of Section 5.4.

60. $(5x + 4)^2$

61. $24x^2 + 47x - 2$

   a) First look for a common factor. There is none (other than 1).

   b) Multiply the leading coefficient and the constant, 24 and -2: $24(-2) = -48$.

   c) Try to factor -48 so that the sum of the factors is 47. The numbers we want are 48 and -1: $48(-1) = -48$ and $48 + (-1) = 47$.

   d) Split the middle term: $47x = 48x - 1x$

   e) Factor by grouping:
   $24x^2 + 47x - 2 = 24x^2 + 48x - x - 2$
   $= 24x(x + 2) - 1(x + 2)$
   $= (24x - 1)(x + 2)$

62. $(8a + 3)(2a + 9)$

63. $35x^2 - 57x - 44$

   a) First look for a common factor. There is none (other than 1).

   b) Multiply the leading coefficient and the constant, 35 and -44: $35(-44) = -1540$

   c) Try to factor -1540 so that the sum of the factors is -57.

Pairs of factors	Sums of factors
7, -220	-213
10, -154	-144
11, -140	-129
14, -110	-96
20, -77	-57

   d) Split the middle term: $-57x = 20x - 77x$

   e) Factor by grouping:
   $35x^2 - 57x - 44 = 35x^2 + 20x - 77x - 44$
   $= 5x(7x + 4) - 11(7x + 4)$
   $= (5x - 11)(7x + 4)$

64. $(3a - 1)(3a + 5)$

65. $20 + 6x - 2x^2$

   a) We first factor out the common factor, 2.
   $20 + 6x - 2x^2 = 2(10 + 3x - x^2)$

   b) Now we factor the trinomial $10 + 3x - x^2$. Multiply the leading coefficient and the constant, -1 and 10: $-1(10) = -10$.

   c) Try to factor -10 so that the sum of the factors is 3. The numbers we want are 5 and -2: $5(-2) = -10$ and $5 + (-2) = 3$.

   d) Split the middle term: $3x = 5x - 2x$

   e) Factor by grouping:
   $10 + 3x - x^2 = 10 + 5x - 2x - x^2$
   $= 5(2 + x) - x(2 + x)$
   $= (5 - x)(2 + x)$

   We must include the common factor to get a factorization of the original trinomial.
   $20 + 6x - 2x^2 = 2(5 - x)(2 + x)$

66. $(3 - x)(5 + 2x)$

67. $(y + 5)(y + 7) = y^2 + 7y + 5y + 35$ Using FOIL
    $= y^2 + 12y + 35$

68. $y^2 + 14y + 49$

69. $(y + 7)(y - 7) = y^2 - 7^2 = y^2 - 49$
    $[(A + B)(A - B) = A^2 - B^2]$

70. $y^2 - 14y + 49$

71. $4x^5 + 6x^3 + 6x^2 + 9 = 2x^3(2x^2 + 3) + 3(2x^2 + 3)$
    $= (2x^3 + 3)(2x^2 + 3)$

Chapter 5   (5.6)

72. $x^2(2x^2 + 3)(2x + 3)$

73. $x^6 + x^4 + x^2 + 1 = x^4(x^2 + 1) + 1(x^2 + 1)$
$= (x^4 + 1)(x^2 + 1)$

74. $(x^7 + 1)(x^6 + 1)$

75. $c^6 - c^4 - c^2 + 1 = c^4(c^2 - 1) - 1(c^2 - 1)$
$= (c^4 - 1)(c^2 - 1)$
$= (c^2 + 1)(c^2 - 1)(c^2 - 1)$
  Factoring $c^4 - 1$
$= (c^2+1)(c+1)(c-1)(c+1)(c-1)$
  Factoring both factors of $c^2 - 1$
$= (c^2 + 1)(c + 1)^2(c - 1)^2$

76. $(3x + 7)(3x - 7)^2$

77. $x^2(x + 1)^2 - (x^2 + 1)^2$
$= x^2(x^2 + 2x + 1) - (x^4 + 2x^2 + 1)$
$= x^4 + 2x^3 + x^2 - x^4 - 2x^2 - 1$
$= 2x^3 + x^2 - 2x^2 - 1$
$= (2x^3 - 2x^2) + (x^2 - 1)$
$= 2x^2(x - 1) + (x + 1)(x - 1)$
$= [2x^2 + (x + 1)](x - 1)$
$= (2x^2 + x + 1)(x - 1)$

78. $(a + 4)(a - 2)$

79. $y^2 + 6y + 9 - x^2 - 8x - 16$
$= (y^2 + 6y + 9) - (x^2 + 8x + 16)$
$= (y + 3)^2 - (x + 4)^2$
$= [(y + 3) + (x + 4)][(y + 3) - (x + 4)]$
$= (y + 3 + x + 4)(y + 3 - x - 4)$
$= (y + x + 7)(y - x - 1)$

Exercise Set 5.6

1. $2x^2 - 128 = 2(x^2 - 64)$ 2 is a common factor
$= 2(x^2 - 8^2)$ Difference of squares
$= 2(x + 8)(x - 8)$

2. $3(t + 3)(t - 3)$

3. $a^2 + 25 - 10a = a^2 - 10a + 25$
$= a^2 - 2 \cdot a \cdot 5 + 5^2$ Trinomial square
$= (a - 5)^2$

4. $(y + 7)^2$

5. $2x^2 - 11x + 12$
There is no common factor (other than 1). This polynomial has three terms, but it is not a trinomial square. Multiply the leading coefficient and the constant, 2 and 12: $2 \cdot 12 = 24$. Try to factor 24 so that the sum of the factors is -11. The numbers we want are -3 and -8: $-3(-8) = 24$ and $-3 + (-8) = -11$. Split the middle term and factor by grouping.
$2x^2 - 11x + 12 = 2x^2 - 3x - 8x + 12$
$= x(2x - 3) - 4(2x - 3)$
$= (x - 4)(2x - 3)$

6. $(2y - 5)(4y + 1)$

7. $x^3 + 24x^2 + 144x = x(x^2 + 24x + 144)$
    x is a common factor
$= x(x^2 + 2 \cdot x \cdot 12 + 12^2)$
    Trinomial square
$= x(x + 12)^2$

8. $x(x - 9)^2$

9. $x^3 + 3x^2 - 4x - 12$
$= x^2(x + 3) - 4(x + 3)$ Factoring by
$= (x^2 - 4)(x + 3)$ grouping
$= (x + 2)(x - 2)(x + 3)$ Factoring the difference of squares

10. $(x + 5)(x - 5)^2$

11. $24x^2 - 54 = 6(4x^2 - 9)$ 6 is a common factor
$= 6[(2x)^2 - 3^2]$ Difference of squares
$= 6(2x + 3)(2x - 3)$

12. $2(2x + 7)(2x - 7)$

13. $20x^3 - 4x^2 - 72x = 4x(5x^2 - x - 18)$ $4x$ is a common factor
$= 4x(5x + 9)(x - 2)$ Factoring the trinomial using trial and error

14. $3x(x + 3)(3x - 5)$

15. $x^2 + 4$ is a sum of squares. It cannot be factored.

16. Not factorable

17. $x^4 + 7x^2 - 3x^3 - 21x = x(x^3 + 7x - 3x^2 - 21)$
$= x[x(x^2 + 7) - 3(x^2 + 7)]$
$= x[(x - 3)(x^2 + 7)]$
$= x(x - 3)(x^2 + 7)$

18. $m(m^2 + 8)(m + 8)$

Chapter 5   (5.6)

19. $x^5 - 14x^4 + 49x^3 = x^3(x^2 - 14x + 49)$
    $\qquad$ x^3 is a common factor
    $\qquad = x^3(x^2 - 2 \cdot x \cdot 7 + 7^2)$
    $\qquad$ Trinomial square
    $\qquad = x^3(x - 7)^2$

20. $2x^4(x + 2)^2$

21. $20 - 6x - 2x^2 = -2(-10 + 3x + x^2)$ -2 is a common factor
    $\qquad = -2(x^2 + 3x - 10)$ Writing in descending order
    $\qquad = -2(x + 5)(x - 2)$ Using trial and error

22. $-3(2x - 5)(x + 3)$, or $3(5 - 2x)(3 + x)$

23. $x^2 + 3x + 1$
    There is no common factor (other than 1). This is not a trinomial square, because $2 \cdot x \cdot 1 \neq 3x$. We try factoring by trial and error. We look for two factors whose product is 1 and whose sum is 3. There are none. The polynomial cannot be factored.

24. Not factorable

25. $4x^4 - 64 = 4(x^4 - 16)$ 4 is a common factor
    $\qquad = 4[(x^2)^2 - 4^2]$ Difference of squares
    $\qquad = 4(x^2 + 4)(x^2 - 4)$ Difference of squares
    $\qquad = 4(x^2 + 4)(x + 2)(x - 2)$

26. $5x(x^2 + 4)(x + 2)(x - 2)$

27. $1 - y^8$ Difference of squares
    $= (1 + y^4)(1 - y^4)$ Difference of squares
    $= (1 + y^4)(1 + y^2)(1 - y^2)$ Difference of squares
    $= (1 + y^4)(1 + y^2)(1 + y)(1 - y)$

28. $(t^4 + 1)(t^2 + 1)(t + 1)(t - 1)$

29. $x^5 - 4x^4 + 3x^3 = x^3(x^2 - 4x + 3)$ x^3 is a common factor
    $\qquad = x^3(x - 3)(x - 1)$ Factoring the trinomial using trial and error

30. $x^4(x^2 - 2x + 7)$

31. $36a^2 - 15a + \frac{25}{16} = (6a)^2 - 2 \cdot 6a \cdot \frac{5}{4} + \left(\frac{5}{4}\right)^2$
    $\qquad = \left(6a - \frac{5}{4}\right)^2$

32. $\left(\frac{1}{9}x^3 - \frac{4}{3}\right)^2$

33. $12n^2 + 24n^3 = 12n^2(1 + 2n)$

34. $a(x^2 + y^2)$

35. $9x^2y^2 - 36xy = 9xy(xy - 4)$

36. $xy(x - y)$

37. $2\pi rh + 2\pi r^2 = 2\pi r(h + r)$

38. $5p^2q^2(2p^2q^2 + 7pq + 2)$

39. $(a + b)(x - 3) + (a + b)(x + 4)$
    $= (a + b)[(x - 3) + (x + 4)]$ $(a + b)$ is a common factor
    $= (a + b)(2x + 1)$

40. $(a^3 + b)(5c - 1)$

41. $(x - 1)(x + 1) - y(x + 1) = (x - 1 - y)(x + 1)$
    $\qquad (x + 1)$ is a common factor

42. $(x + y)(x + 1)$

43. $n^2 + 2n + np + 2p = n(n + 2) + p(n + 2)$ Factoring
    $\qquad = (n + p)(n + 2)$ by grouping

44. $(a + y)(a - 3)$

45. $2x^2 - 4x + xz - 2z$
    $= 2x(x - 2) + z(x - z)$ Factoring
    $= (2x + z)(x - 2)$ by grouping

46. $(3y + p)(2y - 1)$

47. $x^2 + y^2 - 2xy = x^2 - 2xy + y^2$ Trinomial square
    $\qquad = (x - y)^2$

48. $(a - 2b)^2$, or $(2b - a)^2$

49. $9c^2 + 6cd + d^2 = (3c)^2 + 2 \cdot 3c \cdot d + d^2$ Trinomial square
    $\qquad = (3c + d)^2$

50. $(4x + 3y)^2$

51. $49m^4 - 112m^2n + 64n^2$
    $= (7m^2)^2 - 2 \cdot 7m^2 \cdot 8n + (8n)^2$ Trinomial square
    $= (7m^2 - 8n)^2$

52. $(2xy + 3z)^2$

53. $y^4 + 10y^2z^2 + 25z^4$
    $= (y^2)^2 + 2 \cdot y^2 \cdot 5z^2 + (5z^2)^2$ Trinomial square
    $= (y^2 + 5z^2)^2$

54. $(0.1x^2 - 0.5y^2)^2$

55. $\frac{1}{9}a^2 + \frac{1}{3}ab + \frac{1}{9}b^2$
    $= \frac{1}{9}(a^2 + 3ab + b^2)$ $\frac{1}{9}$ is a common factor

56. $p(2p + q)^2$

Chapter 5   (5.6)

57. $a^2 - ab - 2b^2 = (a - 2b)(a + b)$ Using trial and error

58. $(3b + a)(b - 6a)$

59. $2mn - 360n^2 + m^2 = m^2 + 2mn - 360n^2$ Rewriting
    $= (m + 20n)(m - 18n)$ Using trial and error

60. $(xy + 5)(xy + 3)$

61. $m^2n^2 - 4mn - 32 = (mn - 8)(mn + 4)$ Using trial and error

62. $(pq + 6)(pq + 1)$

63. $a^5b^2 + 3a^4b - 10a^3$
    $= a^3(a^2b^2 + 3ab - 10)$ a^3 is a common factor
    $= a^3(ab + 5)(ab - 2)$ Factoring the trinomial

64. $n^4(mn + 8)(mn - 4)$

65. $a^5 + 4a^4b - 5a^3b^2$
    $= a^3(a^2 + 4ab - 5b^2)$ a^3 is a common factor
    $= a^3(a + 5b)(a - b)$ Factoring the trinomial

66. $2t^2(s^3 + 3t)(s^3 + 2t)$

67. $x^6 + x^3y - 2y^2 = (x^3 + 2y)(x^3 - y)$ Using trial and error

68. $(a^2 + 2bc)(a^2 - bc)$

69. $x^2 - y^2 = (x + y)(x - y)$ Difference of squares

70. $(pq + r)(pq - r)$

71. $7p^4 - 7q^4$
    $= 7(p^4 - q^4)$ 7 is a common factor
    $= 7(p^2 + q^2)(p^2 - q^2)$ Factoring a difference of squares
    $= 7(p^2 + q^2)(p + q)(p - q)$ Factoring a difference of squares

72. $(a^2b^2 + 4)(ab + 2)(ab - 2)$

73. $81a^4 - b^4 = (9a^2)^2 - (b^2)^2$ Difference of squares
    $= (9a^2 + b^2)(9a^2 - b^2)$ Difference of squares
    $= (9a^2 + b^2)(3a + b)(3a - b)$

74. $(1 + 4x^6y^6)(1 + 2x^3y^3)(1 - 2x^3y^3)$

75. $\dfrac{7}{5} \div \left(-\dfrac{11}{10}\right) = \dfrac{7}{5} \cdot \left(-\dfrac{10}{11}\right)$ Multiplying by the reciprocal of the divisor
    $= -\dfrac{7 \cdot 10}{5 \cdot 11}$
    $= -\dfrac{7 \cdot 5 \cdot 2}{5 \cdot 11} = -\dfrac{7 \cdot 2}{11} \cdot \dfrac{5}{5}$
    $= -\dfrac{14}{11}$

76. Area: 205 cm², perimeter: 57.8 cm

77. $A = aX + bX - 7$
    $A + 7 = aX + bX$
    $A + 7 = X(a + b)$
    $\dfrac{A + 7}{a + b} = X$

78. 32

79. $a^4 - 2a^2 + 1 = (a^2)^2 - 2 \cdot a^2 \cdot 1 + 1^2$
    $= (a^2 - 1)^2$
    $= [(a + 1)(a - 1)]^2$
    $= (a + 1)^2(a - 1)^2$, or
    $(a + 1)(a + 1)(a - 1)(a - 1)$

80. Not factorable

81. $12.25x^2 - 7x + 1 = (3.5x)^2 - 2 \cdot (3.5x) \cdot 1 + 1^2$
    $= (3.5x - 1)^2$

82. $\dfrac{1}{5}(x - 4)(x - 1)$

83. $5x^2 + 13x + 7.2$
    Multiply the leading coefficient and the constant, 5 and 7.2: $5(7.2) = 36$. Try to factor 36 so that the sum of the factors is 13. The numbers we want are 9 and 4. Split the middle term and factor by grouping:
    $5x^2 + 13x + 7.2 = 5x^2 + 9x + 4x + 7.2$
    $= 5x(x + 1.8) + 4(x + 1.8)$
    $= (5x + 4)(x + 1.8)$, or
    $= 5(x + 0.8)(x + 1.8)$ Factoring 5 out of $(5x + 4)$

84. $(x + 3)(x + 1)(x - 1)$

85. $18 + y^3 - 9y - 2y^2$
    $= y^3 - 2y^2 - 9y + 18$
    $= y^2(y - 2) - 9(y - 2)$
    $= (y^2 - 9)(y - 2)$
    $= (y + 3)(y - 3)(y - 2)$

86. $-(x^2 + 2)(x + 3)(x - 3)$

87. $a^3 + 4a^2 + a + 4 = a^2(a + 4) + 1(a + 4)$
    $= (a^2 + 1)(a + 4)$

88. $(x + 2)(x - 2)(x + 1)$

89. $x^4 - 7x^2 - 18 = (x^2 - 9)(x^2 + 2)$
    $= (x + 3)(x - 3)(x^2 + 2)$

90. $3(x + 2)(x - 2)(x + 1)(x - 1)$

91. $x^3 - x^2 - 4x + 4 = x^2(x - 1) - 4(x - 1)$
    $= (x^2 - 4)(x - 1)$
    $= (x + 2)(x - 2)(x - 1)$

Chapter 5   (5.7)

92. $(y + 1)(y - 7)(y + 3)$

93. $y^2(y - 1) - 2y(y - 1) + (y - 1)$
    $= (y - 1)(y^2 - 2y + 1)$
    $= (y - 1)(y - 1)^2$
    $= (y - 1)^3$

94. $(2x + 3y - 2)(3x - y - 3)$

95. $(y + 4)^2 + 2x(y + 4) + x^2$
    $= (y + 4)^2 + 2 \cdot (y + 4) \cdot x + x^2$ Trinomial square
    $= [(y + 4) + x]^2$

96. $(2a + b + 4)(a - b + 5)$

97. $x^{2k} - 2^{2k} = x^{2 \cdot 4} - 2^{2 \cdot 4}$ Substituting 4 for k
    $= x^8 - 2^8$
    $= x^8 - 256$
    $= (x^4 + 16)(x^4 - 16)$
    $= (x^4 + 16)(x^2 + 4)(x^2 - 4)$
    $= (x^4 + 16)(x^2 + 4)(x + 2)(x - 2)$

98. $(a^2 + 9)(a + 3)(a - 3)$

Exercise Set 5.7

1. $(x + 8)(x + 6) = 0$
   $x + 8 = 0$ or $x + 6 = 0$
   $x = -8$ or $x = -6$

   Check:
   For -8
$(x + 8)(x + 6) = 0$	
$(-8 + 8)(-8 + 6)$	0
$0 \cdot (-2)$	
0	

   For -6
$(x + 8)(x + 6) = 0$	
$(-6 + 8)(-6 + 6)$	0
$2 \cdot 0$	
0	

   The solutions are -8 and -6.

2. -3, -2

3. $(x - 3)(x + 5) = 0$
   $x - 3 = 0$ or $x + 5 = 0$
   $x = 3$ or $x = -5$
   Check:
   For 3
$(x - 3)(x + 5) = 0$	
$(3 - 3)(3 + 5)$	0
$0 \cdot 8$	
0	

   For -5
$(x - 3)(x + 5) = 0$	
$(-5 - 3)(-5 + 5)$	0
$-8 \cdot 0$	
0	

   The solutions are 3 and -5.

4. -9, 3

5. $(x + 12)(x - 11) = 0$
   $x + 12 = 0$ or $x - 11 = 0$
   $x = -12$ or $x = 11$
   The solutions are -12 and 11.

6. 13, -53

7. $x(x + 5) = 0$
   $x = 0$ or $x + 5 = 0$
   $x = 0$ or $x = -5$
   The solutions are 0 and -5.

8. 0, -7

9. $y(y - 13) = 0$
   $y = 0$ or $y - 13 = 0$
   $y = 0$ or $y = 13$
   The solutions are 0 and 13.

10. 0, 4

11. $0 = y(y + 10)$
    $y = 0$ or $y + 10 = 0$
    $y = 0$ or $y = -10$
    The solutions are 0 and -10.

12. 0, 21

13. $(2x + 5)(x + 4) = 0$
    $2x + 5 = 0$ or $x + 4 = 0$
    $2x = -5$ or $x = -4$
    $x = -\frac{5}{2}$ or $x = -4$
    The solutions are $-\frac{5}{2}$ and -4.

14. $-\frac{9}{2}$, -8

15. $(3x - 1)(x + 2) = 0$
    $3x - 1 = 0$ or $x + 2 = 0$
    $3x = 1$ or $x = -2$
    $x = \frac{1}{3}$ or $x = -2$
    The solutions are $\frac{1}{3}$ and -2.

16. 3, -3

17. $(5x + 1)(4x - 12) = 0$
    $5x + 1 = 0$ or $4x - 12 = 0$
    $5x = -1$ or $4x = 12$
    $x = -\frac{1}{5}$ or $x = 3$
    The solutions are $-\frac{1}{5}$ and 3.

115

Chapter 5   (5.7)

18. $-\frac{9}{4}, \frac{1}{2}$

19. $(7x - 28)(28x - 7) = 0$
    $7x - 28 = 0$ or $28x - 7 = 0$
    $7x = 28$ or $28x = 7$
    $x = 4$ or $x = \frac{7}{28} = \frac{1}{4}$
    The solutions are 4 and $\frac{1}{4}$.

20. $\frac{11}{12}, \frac{5}{8}$

21. $2x(3x - 2) = 0$
    $2x = 0$ or $3x - 2 = 0$
    $x = 0$ or $3x = 2$
    $x = 0$ or $x = \frac{2}{3}$
    The solutions are 0 and $\frac{2}{3}$.

22. $0, \frac{9}{8}$

23. $\frac{1}{2}x\left(\frac{2}{3}x - 12\right) = 0$
    $\frac{1}{2}x = 0$ or $\frac{2}{3}x - 12 = 0$
    $x = 0$ or $\frac{2}{3}x = 12$
    $x = 0$ or $x = \frac{3}{2} \cdot 12 = 18$
    The solutions are 0 and 18.

24. 0, 8

25. $\left(\frac{1}{3} - 3x\right)\left(\frac{1}{5} - 2x\right) = 0$
    $\frac{1}{3} - 3x = 0$ or $\frac{1}{5} - 2x = 0$
    $\frac{1}{3} = 3x$ or $\frac{1}{5} = 2x$
    $\frac{1}{9} = x$ or $\frac{1}{10} = x$
    The solutions are $\frac{1}{9}$ and $\frac{1}{10}$.

26. $-\frac{1}{10}, \frac{1}{27}$

27. $\left(\frac{1}{3}y - \frac{2}{3}\right)\left(\frac{1}{4}y - \frac{3}{2}\right) = 0$
    $\frac{1}{3}y - \frac{2}{3} = 0$ or $\frac{1}{4}y - \frac{3}{2} = 0$
    $\frac{1}{3}y = \frac{2}{3}$ or $\frac{1}{4}y = \frac{3}{2}$
    $y = 2$ or $y = 6$
    The solutions are 2 and 6.

28. $\frac{1}{21}, \frac{18}{11}$

29. $(0.3x - 0.1)(0.05x - 1) = 0$
    $0.3x - 0.1 = 0$ or $0.05x - 1 = 0$
    $0.3x = 0.1$ or $0.05x = 1$
    $x = \frac{0.1}{0.3}$ or $x = \frac{1}{0.05}$
    $x = \frac{1}{3}$ or $x = 20$
    The solutions are $\frac{1}{3}$ and 20.

30. 3, 50

31. $9x(3x - 2)(2x - 1) = 0$
    $9x = 0$ or $3x - 2 = 0$ or $2x - 1 = 0$
    $x = 0$ or $3x = 2$ or $2x = 1$
    $x = 0$ or $x = \frac{2}{3}$ or $x = \frac{1}{2}$
    The solutions are 0, $\frac{2}{3}$, and $\frac{1}{2}$.

32. $5, -55, \frac{1}{5}$

33. $x^2 + 6x + 5 = 0$
    $(x + 5)(x + 1) = 0$ Factoring
    $x + 5 = 0$ or $x + 1 = 0$ Using the principle of zero product
    $x = -5$ or $x = -1$
    The solutions are -5 and -1.

34. $-6, -1$

35. $x^2 + 7x - 18 = 0$
    $(x + 9)(x - 2) = 0$ Factoring
    $x + 9 = 0$ or $x - 2 = 0$ Using the principal of zero products
    $x = -9$ or $x = 2$
    The solutions are -9 and 2.

36. $-7, 3$

37. $x^2 - 8x + 15 = 0$
    $(x - 5)(x - 3) = 0$
    $x - 5 = 0$ or $x - 3 = 0$
    $x = 5$ or $x = 3$
    The solutions are 5 and 3.

38. 7, 2

39. $x^2 - 8x = 0$
    $x(x - 8) = 0$
    $x = 0$ or $x - 8 = 0$
    $x = 0$ or $x = 8$
    The solutions are 0 and 8.

40. 0, 3

Chapter 5 (5.7)

41. $x^2 + 19x = 0$
    $x(x + 19) = 0$
    $x = 0$ or $x + 19 = 0$
    $x = 0$ or $x = -19$
    The solutions are 0 and -19.

42. 0, -12

43. $x^2 = 16$
    $x^2 - 16 = 0$ Adding -16
    $(x - 4)(x + 4) = 0$
    $x - 4 = 0$ or $x + 4 = 0$
    $x = 4$ or $x = -4$
    The solutions are 4 and -4.

44. -10, 10

45. $9x^2 - 4 = 0$
    $(3x - 2)(3x + 2) = 0$
    $3x - 2 = 0$ or $3x + 2 = 0$
    $3x = 2$ or $3x = -2$
    $x = \frac{2}{3}$ or $x = -\frac{2}{3}$
    The solutions are $\frac{2}{3}$ and $-\frac{2}{3}$.

46. $-\frac{3}{2}, \frac{3}{2}$

47. $0 = 6x + x^2 + 9$
    $0 = x^2 + 6x + 9$ Writing in descending order
    $0 = (x + 3)(x + 3)$
    $x + 3 = 0$ or $x + 3 = 0$
    $x = -3$ or $x = -3$
    There is only one solution, -3.

48. -5

49. $x^2 + 16 = 8x$
    $x^2 - 8x + 16 = 0$ Adding -8x
    $(x - 4)(x - 4) = 0$
    $x - 4 = 0$ or $x - 4 = 0$
    $x = 4$ or $x = 4$
    There is only one solution, 4.

50. 1

51. $5x^2 = 6x$
    $5x^2 - 6x = 0$
    $x(5x - 6) = 0$
    $x = 0$ or $5x - 6 = 0$
    $x = 0$ or $5x = 6$
    $x = 0$ or $x = \frac{6}{5}$
    The solutions are 0 and $\frac{6}{5}$.

52. $0, \frac{8}{7}$

53. $6x^2 - 4x = 10$
    $6x^2 - 4x - 10 = 0$
    $2(3x^2 - 2x - 5) = 0$
    $2(3x - 5)(x + 1) = 0$
    $3x - 5 = 0$ or $x + 1 = 0$
    $3x = 5$ or $x = -1$
    $x = \frac{5}{3}$ or $x = -1$
    The solutions are $\frac{5}{3}$ and -1.

54. $-\frac{5}{3}, 4$

55. $12y^2 - 5y = 2$
    $12y^2 - 5y - 2 = 0$
    $(4y + 1)(3y - 2) = 0$
    $4y + 1 = 0$ or $3y - 2 = 0$
    $4y = -1$ or $3y = 2$
    $y = -\frac{1}{4}$ or $y = \frac{2}{3}$
    The solutions are $-\frac{1}{4}$ and $\frac{2}{3}$.

56. -5, -1

57. $x(x - 5) = 14$
    $x^2 - 5x = 14$ Multiplying on the left side
    $x^2 - 5x - 14 = 0$ Adding -14
    $(x - 7)(x + 2) = 0$
    $x - 7 = 0$ or $x + 2 = 0$
    $x = 7$ or $x = -2$
    The solutions are 7 and -2.

58. $\frac{2}{3}, -1$

Chapter 5   (5.7)

59. $64m^2 = 81$
$64m^2 - 81 = 0$
$(8m - 9)(8m + 9) = 0$
$8m - 9 = 0$ or $8m + 9 = 0$
$8m = 9$ or $8m = -9$
$m = \frac{9}{8}$ or $m = -\frac{9}{8}$
The solutions are $\frac{9}{8}$ and $-\frac{9}{8}$.

60. $-\frac{7}{10}, \frac{7}{10}$

61. $3x^2 + 8x = 9 + 2x$
$3x^2 + 8x - 2x - 9 = 0$ Adding $-2x$ and -9
$3x^2 + 6x - 9 = 0$ Collecting like terms
$3(x^2 + 2x - 3) = 0$
$3(x + 3)(x - 1) = 0$
$x + 3 = 0$ or $x - 1 = 0$
$x = -3$ or $x = 1$
The solutions are -3 and 1.

62. $9, -2$

63. $10x^2 - 23x + 12 = 0$
$(5x - 4)(2x - 3) = 0$
$5x - 4 = 0$ or $2x - 3 = 0$
$5x = 4$ or $2x = 3$
$x = \frac{4}{5}$ or $x = \frac{3}{2}$
The solutions are $\frac{4}{5}$ and $\frac{3}{2}$.

64. $\frac{5}{3}, -\frac{1}{4}$

65. $(a + b)^2$

66. $a^2 + b^2$

67. $(-9)(16) = -144$
The two numbers have different signs, so their product is negative.

68. -4.5

69. $b(b + 9) = 4(5 + 2b)$
$b^2 + 9b = 20 + 8b$
$b^2 + 9b - 8b - 20 = 0$
$b^2 + b - 20 = 0$
$(b + 5)(b - 4) = 0$
$b + 5 = 0$ or $b - 4 = 0$
$b = -5$ or $b = 4$
The solutions are -5 and 4.

70. 4

71. $(t - 3)^2 = 36$
$t^2 - 6t + 9 = 36$
$t^2 - 6t - 27 = 0$
$(t - 9)(t + 3) = 0$
$t - 9 = 0$ or $t + 3 = 0$
$t = 9$ or $t = -3$
The solutions are 9 and -3.

72. $5, 3$

73. $x^2 - \frac{1}{64} = 0$
$\left(x - \frac{1}{8}\right)\left(x + \frac{1}{8}\right) = 0$
$x - \frac{1}{8} = 0$ or $x + \frac{1}{8} = 0$
$x = \frac{1}{8}$ or $x = -\frac{1}{8}$
The solutions are $\frac{1}{8}$ and $-\frac{1}{8}$.

74. $-\frac{5}{6}, \frac{5}{6}$

75. $\frac{5}{16}x^2 = 5$
$\frac{5}{16}x^2 - 5 = 0$
$5\left(\frac{1}{16}x^2 - 1\right) = 0$
$5\left(\frac{1}{4}x - 1\right)\left(\frac{1}{4}x + 1\right) = 0$
$\frac{1}{4}x - 1 = 0$ or $\frac{1}{4}x + 1 = 0$
$\frac{1}{4}x = 1$ or $\frac{1}{4}x = -1$
$x = 4$ or $x = -4$
The solutions are 4 and -4.

76. $-\frac{5}{9}, \frac{5}{9}$

77. a) $x = 1$ or $x = 3$
$x - 1 = 0$ or $x - 3 = 0$
$(x - 1)(x - 3) = 0$ Principle of zero products
$x^2 - 4x + 3 = 0$ Multiplying

b) $x = 3$ or $x = -1$
$x - 3 = 0$ or $x + 1 = 0$
$(x - 3)(x + 1) = 0$
$x^2 - 2x - 3 = 0$

c) $x = 2$ or $x = 2$
$x - 2 = 0$ or $x - 2 = 0$
$(x - 2)(x - 2) = 0$
$x^2 - 4x + 4 = 0$

Chapter 5   (5.8)

77. (continued)

  d) $x = 3$ or $x = 4$
       $x - 3 = 0$ or $x - 4 = 0$
       $(x - 3)(x - 4) = 0$
       $x^2 - 7x + 12 = 0$

  e) $(x - 3)(x + 4) = 0$
      $x^2 + x - 12 = 0$

  f) $(x + 3)(x - 4) = 0$
      $x^2 - x - 12 = 0$

  g) $(x + 3)(x + 4) = 0$
      $x^2 + 7x + 12 = 0$

  h) $\left(x - \frac{1}{2}\right)\left(x - \frac{1}{2}\right) = 0$
      $x^2 - x + \frac{1}{4} = 0$, or
      $4x^2 - 4x + 1 = 0$ Multiplying by 4

  i) $(x - 5)(x + 5) = 0$
      $x^2 - 25 = 0$

  j) $(x - 0)(x - 0.1)\left(x - \frac{1}{4}\right) = 0$
      $x\left(x - \frac{1}{10}\right)\left(x - \frac{1}{4}\right) = 0$
      $x\left(x^2 - \frac{7}{20}x + \frac{1}{40}\right) = 0$
      $x^3 - \frac{7}{20}x^2 + \frac{1}{40}x = 0$, or
      $40x^3 - 14x^2 + x = 0$ Multiplying by 40

78. The principle of zero products cannot be used unless there is 0 on one side of the equation.
    $x^2 + x - 12 = 8$, $x^2 + x - 20 = 0$,
    $(x + 5)(x - 4) = 0$, $x = -5$ or $x = 4$

79. $(0.00005x + 0.1)(0.0097x + 0.5) = 0$

    $0.00005x + 0.1 = 0$ or $0.0097x + 0.5 = 0$
         $0.00005x = -0.1$ or $0.0097x = -0.5$
            $x = -\frac{0.1}{0.00005}$ or $x = -\frac{0.5}{0.0097}$
            $x = -2000$ or $x \approx -51.546392$

    The solutions are -2000 and -51.546392.

80. a) i, b) k, c) g, d) h, e) j, f) ℓ

Exercise Set 5.8

1. Familiarize. Let x = the number (or numbers).
   Translate. We reword the problem.

   Four times the square of a number minus the number is 3.
   $4x^2$ $-$ x $= 3$

   Carry out. We solve the equation.
   $4x^2 - x = 3$
   $4x^2 - x - 3 = 0$
   $(4x + 3)(x - 1) = 0$
   $4x + 3 = 0$ or $x - 1 = 0$
      $4x = -3$ or $x = 1$
       $x = -\frac{3}{4}$ or $x = 1$

   Check. For $-\frac{3}{4}$: Four times the square of $-\frac{3}{4}$ is $4\left(-\frac{3}{4}\right)^2 = 4\left(\frac{9}{16}\right) = \frac{9}{4}$. If we subtract $-\frac{3}{4}$ from $\frac{9}{4}$ we get $\frac{9}{4} - \left(-\frac{3}{4}\right) = \frac{9}{4} + \frac{3}{4} = \frac{12}{4} = 3$.

   For 1: Four times the square of 1 is $4(1)^2 = 4$. If we subtract 1 from 4 we get $4 - 1 = 3$. Both numbers check.

   State. There are two such numbers, $-\frac{3}{4}$ and 1.

2. 5, -5

3. Familiarize. Let x = the number (or numbers).
   Translate. We reword the problem.

   The square of a number plus 8 is six times the number.
   x^2 $+$ 8 $=$ $6x$

   Carry out. We solve the equation.
   $x^2 + 8 = 6x$
   $x^2 - 6x + 8 = 0$
   $(x - 4)(x - 2) = 0$
   $x - 4 = 0$ or $x - 2 = 0$
      $x = 4$ or $x = 2$

   Check. The square of 4 is 16, and six times the number 4 is 24. Since $16 + 8 = 24$, the number 4 checks. The square of 2 is 4, and six times the number 2 is 12. Since $4 + 8 = 12$, the number 2 checks.

   State. There are two such numbers, 4 and 2.

4. 3, 5

Chapter 5   (5.8)

5. Familiarize. The page numbers on facing pages are consecutive integers. Let x = the smaller integer. Then x + 1 = the larger integer.
Translate. We reword the problem.

Smaller integer, times larger integer, is 182.
   x   ·   (x + 1)   = 182

Carry out.
$$x(x + 1) = 182$$
$$x^2 + x = 182$$
$$x^2 + x - 182 = 0$$
$$(x + 14)(x - 13) = 0$$

x + 14 = 0   or   x - 13 = 0
   x = -14 or   x = 13

Check. The solutions of the equation are -14 and 13. Since a page number cannot be negative, -14 cannot be a solution of the original problem. We only need to check 13. When x = 13, then x + 1 = 14, and 13·14 = 182. This checks.
State. The page numbers are 13 and 14.

6. 11 and 12

7. Familiarize. Let x = the smaller even integer. Then x + 2 = the larger even integer.
Translate. We reword the problem.

Smaller even integer, times larger even integer, is 168.
   x   ·   (x + 2)   = 168

Carry out.
$$x(x + 2) = 168$$
$$x^2 + 2x = 168$$
$$x^2 + 2x - 168 = 0$$
$$(x + 14)(x - 12) = 0$$

x + 14 = 0   or   x - 12 = 0
   x = -14 or   x = 12

Check. The solutions of the equation are -14 and 12. When x is -14, then x + 2 is -12 and -14(-12) = 168. The numbers -14 and -12 are consecutive even integers which are solutions to the problem. When x is 12, then x + 2 is 14 and 12·14 = 168. The numbers 12 and 14 are also consecutive even integers which are solutions to the problem.
State. We have two solutions each of which consists of a pair of numbers: -14 and -12, and 12 and 14.

8. 14 and 16, -16 and -14

9. Familiarize. Let x = the smaller odd integer. Then x + 2 = the larger odd integer.
Translate. We reword the problem.

Smaller odd integer, times larger odd integer, is 255.
   x   ·   x + 2   = 255

Carry out.
$$x(x + 2) = 255$$
$$x^2 + 2x = 255$$
$$x^2 + 2x - 255 = 0$$
$$(x - 15)(x + 17) = 0$$

x - 15 = 0   or   x + 17 = 0
   x = 15   or   x = -17

Check. The solutions of the equation are 15 and -17. When x is 15, then x + 2 is 17 and 15·17 = 255. The numbers 15 and 17 are consecutive odd integers which are solutions to the problem. When x is -17, then x + 2 is -15 and -17(-15) = 255. The numbers -17 and -15 are also consecutive odd integers which are solutions to the problem.
State. We have two solutions each of which consists of a pair of numbers: 15 and 17, and -17 and -15.

10. 11 and 13, -13 and -11

11. Familiarize. Using the labels shown on the drawing in the text, we let w = the width of the rectangle and w + 4 = the length. Recall that the area of a rectangle is length times width.
Translate. We reword the problem.

Length times width is area.
(w + 4) ·   w   = 96

Carry out.
$$(w + 4) \cdot w = 96$$
$$w^2 + 4w = 96$$
$$w^2 + 4w - 96 = 0$$
$$(w + 12)(w - 8) = 0$$

w + 12 = 0   or   w - 8 = 0
   w = -12 or   w = 8

Check. The solutions of the equation are -12 and 8. The width of a rectangle cannot have a negative measure, so -12 cannot be a solution. Suppose the width is 8 m. The length is 4 m greater than the width, so the length is 12 m and the area is 12·8, or 96 m². The numbers check in the original problem.
State. The length is 12 m, and the width is 8 m.

12. Length: 12 cm, width: 7 cm

13. Familiarize. First draw a picture. Let x = the length of a side of the square.

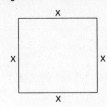

The area of the square is x·x, or x².
The perimeter of the square is x + x + x + x, or 4x.

Translate.

Area of bookcase, is 5 more than perimeter of bookcase.
$\quad x^2 \quad\quad = \quad 5 \quad + \quad\quad 4x$

Carry out.
$$x^2 = 5 + 4x$$
$$x^2 - 4x - 5 = 0$$
$$(x - 5)(x + 1) = 0$$
$$x - 5 = 0 \text{ or } x + 1 = 0$$
$$x = 5 \text{ or } x = -1$$

Check. The solutions of the equation are 5 and -1. The length of a side cannot be negative, so we only check 5. The area is 5·5, or 25. The perimeter is 5 + 5 + 5 + 5, or 20. The area, 25, is 5 more than the perimeter, 20. This checks.

State. The length of a side is 5.

14. 3 or 1

15. Familiarize. Using the labels shown on the drawing in the text, we let h = the height and h + 10 = the base. Recall that the formula for the area of a triangle is $\frac{1}{2}$ · (base)·(height).

Translate.
Area is $\frac{1}{2}$ times the base, times the height.
$\quad 28 \quad = \quad \frac{1}{2} \quad \cdot \quad (h + 10) \quad \cdot \quad h$

Carry out.
$$28 = \tfrac{1}{2}h(h + 10)$$
$$56 = h(h + 10)$$
$$56 = h^2 + 10h$$
$$0 = h^2 + 10h - 56$$
$$0 = (h + 14)(h - 4)$$
$$h + 14 = 0 \text{ or } h - 4 = 0$$
$$h = -14 \text{ or } h = 4$$

15. (continued)

Check. The solutions of the equation are -14 and 4. The height of a triangle cannot have a negative length, so -14 cannot be a solution. Suppose the height is 4 cm. The base is 10 cm greater than the height, so the base is 14 cm and the area is $\frac{1}{2}$ · 14·4, or 28 cm². These numbers check.

State. The height is 4 cm and the base is 14 cm.

16. Height: 2 m, base: 10 m

17. Familiarize. We make a drawing. Let x = the length of a side of the original square. Then x + 3 = the length of a side of the enlarged square.

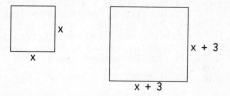

Recall that the area of a square is found by squaring the length of a side.

Translate.
Area of enlarged square is the square of the lengthened side.
$\quad 81 \quad = \quad (x + 3)^2$

Carry out.
$$81 = (x + 3)^2$$
$$81 = x^2 + 6x + 9$$
$$0 = x^2 + 6x - 72$$
$$0 = (x + 12)(x - 6)$$
$$x + 12 = 0 \text{ or } x - 6 = 0$$
$$x = -12 \text{ or } x = 6$$

Check. The solutions of the equation are -12 and 6. The length of a side cannot be negative, so -12 cannot be a solution. Suppose the length of a side of the original square is 6 m. Then the length of a side of the new square is 6 + 3, or 9 m. Its area is 9², or 81 m². The numbers check.

State. The length of a side of the original square is 6 m.

18. 4 km

19. **Familiarize.** Let x = the smaller odd positive integer. Then x + 2 = the larger odd positive integer.
    Translate.

    Square of the smaller odd positive integer + Square of the larger odd positive integer is 74

    $x^2 + (x+2)^2 = 74$

    Carry out.
    $$x^2 + (x+2)^2 = 74$$
    $$x^2 + x^2 + 4x + 4 = 74$$
    $$2x^2 + 4x - 70 = 0$$
    $$2(x^2 + 2x - 35) = 0$$
    $$2(x+7)(x-5) = 0$$
    $$x + 7 = 0 \text{ or } x - 5 = 0$$
    $$x = -7 \text{ or } x = 5$$

    Check. The solutions of the equation are -7 and 5. The problem asks for odd positive integers, so -7 cannot be a solution. When x is 5, x + 2 is 7. The numbers 5 and 7 are consecutive odd positive integers. The sum of their squares, 25 + 49, is 74. The numbers check.
    State. The integers are 5 and 7.

20. 7 and 9

21. **Familiarize.** Reread Example 5 in Section 4.3.
    Translate. Substitute 23 for n.
    $$23^2 - 23 = N$$
    Carry out. We do the computation on the left.
    $$23^2 - 23 = N$$
    $$529 - 23 = N$$
    $$506 = N$$
    Check. We can recheck the computation or we can solve $n^2 - n = 506$. The answer checks.
    State. 506 games will be played.

22. 182

23. **Familiarize.** Reread Example 5 in Section 4.3.
    Translate. Substitute 132 for N.
    $$n^2 - n = 132$$
    Carry out.
    $$n^2 - n = 132$$
    $$n^2 - n - 132 = 0$$
    $$(n-12)(n+11) = 0$$
    $$n - 12 = 0 \text{ or } n + 11 = 0$$
    $$n = 12 \text{ or } n = -11$$
    Check. The solutions of the equation are 12 and -11. Since the number of teams cannot be negative, -11 cannot be a solution. But 12 checks since $12^2 - 12 = 144 - 12 = 132$.
    State. There are 12 teams in the league.

24. 10

25. **Familiarize.** We will use the formula $N = \frac{1}{2}(n^2 - n)$.
    Translate. Substitute 40 for n.
    $$N = \frac{1}{2}(40^2 - 40)$$
    Carry out. We do the computation on the right.
    $$N = \frac{1}{2}(40^2 - 40)$$
    $$N = \frac{1}{2}(1600 - 40)$$
    $$N = \frac{1}{2}(1560)$$
    $$N = 780$$
    Check. We can recheck the computation, or we can solve the equation $780 = \frac{1}{2}(n^2 - n)$. The answer checks.
    State. 780 handshakes are possible.

26. 4950

27. **Familiarize.** We will use the formula $N = \frac{1}{2}(n^2 - n)$.
    Translate. Substitute 190 for N.
    $$190 = \frac{1}{2}(n^2 - n)$$
    Carry out.
    $$190 = \frac{1}{2}(n^2 - n)$$
    $$380 = n^2 - n \quad \text{Multiplying by 2}$$
    $$0 = n^2 - n - 380$$
    $$0 = (n-20)(n+19)$$
    $$n - 20 = 0 \text{ or } n + 19 = 0$$
    $$n = 20 \text{ or } n = -19$$
    Check. The solutions of the equation are 20 and -19. Since the number of people cannot be negative, -19 cannot be a solution. But 20 checks since $\frac{1}{2}(20^2 - 20) = \frac{1}{2}(400 - 20) = \frac{1}{2} \cdot 380 = 190$.
    State. There were 20 people at the party.

28. 25

29. Familiarize. Using the labels shown on the drawing in the text, we let x = the width of the walk. Then the length and width of the rectangle formed by the pool and walk together are 40 + 2x and 20 + 2x, respectively.

Translate.

Area is length times width.
1500 = (40 + 2x) · (20 + 2x)

Carry out.
1500 = (40 + 2x)(20 + 2x)
1500 = 2(20 + x)·2(10 + x)   Factoring 2 out of each factor on the right
1500 = 4·(20 + x)(10 + x)
375 = (20 + x)(10 + x)   Dividing by 4
375 = 200 + 30x + x²
0 = x² + 30x - 175
0 = (x + 35)(x - 5)

x + 35 = 0   or   x - 5 = 0
x = -35   or   x = 5

Check. The solutions of the equation are -35 and 5. Since the width of the walk cannot be negative, -35 is not a solution. When x = 5, 40 + 2x = 40 + 2·5, or 50 and 20 + 2x = 20 + 2·5, or 30. The total area of the pool and walk is 50·30, or 1500 ft². This checks.

State. The width of the walk is 5 ft.

30. 4 sec, $7\frac{1}{4}$ sec

31. a) Familiarize. We will use the formula $h = rt - 4.9t^2$.

    Translate. We substitute 20.6 for r and 21.6 for h.
    $21.6 = 20.6t - 4.9t^2$

    Carry out.
    $21.6 = 20.6t - 4.9t^2$
    $4.9t^2 - 20.6t + 21.6 = 0$
    $49t^2 - 206t + 216 = 0$ Clearing decimals
    $(49t - 108)(t - 2) = 0$

    49t - 108 = 0   or   t - 2 = 0
    49t = 108   or   t = 2
    $t = \frac{108}{49}$ or   t = 2

    Check. The solutions of the equation are $\frac{108}{49}$, or $2\frac{10}{49}$, and 2. Now $20.6\left[\frac{108}{49}\right] - 4.9\left[\frac{108}{49}\right]^2 = 21.6$ and $20.6(2) - 4.9(2)^2 = 21.6$, so both numbers check. This means that after 2 sec the ball (on its way up) will reach a height of 21.6 ft. Also, after $2\frac{10}{49}$ sec the ball (on the way down) will again reach a height of 21.6 ft.

31. (continued)

    State. The ball will first reach a height of 21.6 ft after 2 sec. [Note that it will again reach this height after $2\frac{10}{49}$ sec.]

    b) Familiarize. We will use the formula $h = rt - 4.9t^2$. When the ball hits the ground its height is 0 ft.

    Translate. We substitute 20.6 for r and 0 for h.
    $0 = 20.6t - 4.9t^2$

    Carry out.
    $0 = 20.6t^2 - 4.9t^2$
    $0 = 206t - 49t^2$ Clearing decimals
    $49t^2 - 206t = 0$
    $t(49t - 206) = 0$

    t = 0   or   49t - 206 = 0
    t = 0   or   49t = 206
    t = 0   or $t = \frac{206}{49}$
    t = 0   or   t ≈ 4.2

    Check. The solutions of the equation are 0 and approximately 4.2. Since t = 0 before the ball is thrown, we only need to check 4.2. Now $20.6(4.2) - 4.9(4.2)^2 \approx 0$, so the answer checks.

    State. The ball hits the ground after approximately 4.2 sec.

32. 37

33. Familiarize. We make a drawing. Let x = the length of a side of the base.

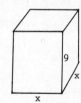

The area of the base of the box is x·x, or x². The area of the top of the box is x·x, or x². The area of each side is 9·x. The total area of the four sides is 4·9x, or 36x. The total surface area, 350 m², is the sum of the areas of the top, the base, and the four sides.

Translate.

Total surface area	=	Area of top	+	Area of base	+	Area of 4 sides
350	=	x²	+	x²	+	36x

33. (continued)

Carry out.
$350 = x^2 + x^2 + 36x$
$0 = 2x^2 + 36x - 350$
$0 = x^2 + 18x - 175$
$0 = (x + 25)(x - 7)$

$x + 25 = 0$ or $x - 7 = 0$
$x = -25$ or $x = 7$

Check. The solutions of the equation are -25 and 7. Since the length of a side cannot be negative, -25 is not a solution.

When $x = 7$, the areas are:
Top: $x^2 = 7^2 = 49$
Base: $x^2 = 7^2 = 49$
4 Sides: $36 = 36 \cdot 7 = 252$

The total surface area is $49 + 49 + 252$, or 350 m^2. These numbers check.

State. The length of a side of the base is 7 m.

34. 30 cm by 15 cm

35. Familiarize. We make a drawing. Let $x =$ the depth of the gutter. Then $20 - 2x =$ the width of the gutter.

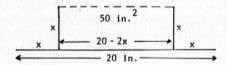

Translate.
Area of cross-section of gutter = Depth · Width
$50 = x \cdot (20 - 2x)$

Carry out.
$50 = x(20 - 2x)$
$50 = 20x - 2x^2$
$2x^2 - 20x + 50 = 0$
$x^2 - 10x + 25 = 0$
$(x - 5)(x - 5) = 0$

$x - 5 = 0$ or $x - 5 = 0$
$x = 5$ or $x = 5$

Check. When $x = 5$, the depth is 5 in. and the width is $20 - 2 \cdot 5$, or 10 in. The area is $5 \cdot 10$, or 50 in^2. This checks.

State. The depth of the gutter is 5 in.

36. 100 cm², 225 cm²

CHAPTER 6   RATIONAL EXPRESSIONS AND EQUATIONS

Exercise Set 6.1

1. $\dfrac{-5}{2x}$

   We find the real number(s) that make the denominator 0. To do so we set the denominator equal to 0 and solve for x:
   $$2x = 0$$
   $$x = 0$$
   The meaningful replacements are all real numbers except 0.

2. All real numbers except 0

3. $\dfrac{a + 7}{a - 8}$

   Set the denominator equal to 0 and solve for a:
   $$a - 8 = 0$$
   $$a = 8$$
   The meaningful replacements are all real numbers except 8.

4. All real numbers except -7

5. $\dfrac{3}{2y + 5}$

   Set the denominator equal to 0 and solve for y:
   $$2y + 5 = 0$$
   $$2y = -5$$
   $$y = -\dfrac{5}{2}$$
   The meaningful replacements are all real numbers except $-\dfrac{5}{2}$.

6. All real numbers except 3

7. $\dfrac{x^2 + 11}{x^2 - 3x - 28}$

   Set the denominator equal to 0 and solve for x:
   $$x^2 - 3x - 28 = 0$$
   $$(x - 7)(x + 4) = 0$$
   $$x - 7 = 0 \text{ or } x + 4 = 0$$
   $$x = 7 \text{ or } x = -4$$
   The meaningful replacements are all real numbers except 7 and -4.

8. All real numbers except 5 and 2

9. $\dfrac{m^3 - 2m}{m^2 - 25}$

   Set the denominator equal to 0 and solve for m:
   $$m^2 - 25 = 0$$
   $$(m + 5)(m - 5) = 0$$
   $$m + 5 = 0 \text{ or } m - 5 = 0$$
   $$m = -5 \text{ or } m = 5$$
   The meaningful replacements are all real numbers except -5 and 5.

10. All real numbers except -7 and 7

11. For $x = 3$, $\dfrac{5x}{7} = \dfrac{5 \cdot 3}{7} = \dfrac{15}{7}$.

    For $x = -2$, $\dfrac{5x}{7} = \dfrac{5(-2)}{7} = \dfrac{-10}{7} = -\dfrac{10}{7}$.

12. $-\dfrac{5}{3}$; $\dfrac{5}{2}$

13. For $x = 3$, $\dfrac{x + 2}{x - 3} = \dfrac{3 + 2}{3 - 3} = \dfrac{5}{0}$. Since division by 0 is not defined, the expression has no value for $x = 3$.

    For $x = -2$, $\dfrac{x + 2}{x - 3} = \dfrac{-2 + 2}{-2 - 3} = \dfrac{0}{-5} = 0$.

14. 0; no value for $x = -2$

15. For $x = 3$, $\dfrac{5x^2}{x^2 + x - 12} = \dfrac{5 \cdot 3^2}{3^2 + 3 - 12} = \dfrac{5 \cdot 9}{9 + 3 - 12} = \dfrac{45}{0}$.

    Since division by 0 is not defined, the expression has no value for $x = 3$.

    For $x = -2$, $\dfrac{5x^2}{x^2 + x - 12} = \dfrac{5(-2)^2}{(-2)^2 + (-2) - 12} = \dfrac{5 \cdot 4}{4 - 2 - 12} = \dfrac{20}{-10} = -2$.

16. $\dfrac{2}{5}$; no value for $x = -2$

17. For $x = 3$, $\dfrac{x^2 + 5}{x^2 - 9} = \dfrac{3^2 + 5}{3^2 - 9} = \dfrac{9 + 5}{9 - 9} = \dfrac{14}{0}$. Since division by 0 is not defined, the expression has no value for $x = 3$.

    For $x = -2$, $\dfrac{x^2 + 5}{x^2 - 9} = \dfrac{(-2)^2 + 5}{(-2)^2 - 9} = \dfrac{4 + 5}{4 - 9} = \dfrac{9}{-5} = -\dfrac{9}{5}$.

18. $\dfrac{2}{5}$; no value for $x = -2$

19. For $x = 3$, $\dfrac{x^2 - x - 6}{x^2 + 5x - 1} = \dfrac{3^2 - 3 - 6}{3^2 + 5 \cdot 3 - 1} = \dfrac{9 - 3 - 6}{9 + 15 - 1} = \dfrac{0}{23} = 0$.

    For $x = -2$, $\dfrac{x^2 - x - 6}{x^2 + 5x - 1} = \dfrac{(-2)^2 - (-2) - 6}{(-2)^2 + 5(-2) - 1} = \dfrac{4 + 2 - 6}{4 - 10 - 1} = \dfrac{0}{-7} = 0$.

20. 0; $\dfrac{25}{12}$

21. For $x = 3$, $\dfrac{2x - 3}{x^2 + 4x + 4} = \dfrac{2 \cdot 3 - 3}{3^2 + 4 \cdot 3 + 4} = \dfrac{6 - 3}{9 + 12 + 4} = \dfrac{3}{25}$.

    For $x = -2$, $\dfrac{2x - 3}{x^2 + 4x + 4} = \dfrac{2(-2) - 3}{(-2)^2 + 4(-2) + 4} = \dfrac{-4 - 3}{4 - 8 + 4} = \dfrac{-7}{0}$.

    Since division by 0 is not defined, the expression has no value for $x = -2$.

Chapter 6   (6.1)

22. No value for x = 3; $-\frac{4}{25}$

23. To find the speed required to make the trip in 5 hr, we evaluate the rational expression for
t = 5: $r = \frac{350}{5} = 70$ mph

    To find the speed required to make the trip in 6 hr, we evaluate the rational expression for
t = 6: $r = \frac{350}{6} = \frac{175}{3}$, or $58\frac{1}{3}$ mph.

24. $6\frac{4}{11}$ hr; $5\frac{5}{13}$ hr

25. $\frac{3x}{2} \cdot \frac{x + 4}{x - 1} = \frac{3x(x + 4)}{2(x - 1)}$

26. $\frac{4x(x - 3)}{5(x + 2)}$

27. $\frac{x - 1}{x + 2} \cdot \frac{x + 1}{x + 2} = \frac{(x - 1)(x + 1)}{(x + 2)(x + 2)}$

28. $\frac{(x - 2)(x - 2)}{(x - 5)(x + 5)}$

29. $\frac{2x + 3}{4} \cdot \frac{x + 1}{x - 5} = \frac{(2x + 3)(x + 1)}{4(x - 5)}$

30. $\frac{(-5)(-6)}{(3x - 4)(5x + 6)}$

31. $\frac{a - 5}{a^2 + 1} \cdot \frac{a + 2}{a^2 - 1} = \frac{(a - 5)(a + 2)}{(a^2 + 1)(a^2 - 1)}$

32. $\frac{(t + 3)(t + 3)}{(t^2 - 2)(t^2 - 2)}$

33. $\frac{x + 1}{2 + x} \cdot \frac{x - 1}{x + 1} = \frac{(x + 1)(x - 1)}{(2 + x)(x + 1)}$

34. $\frac{(m^2 + 5)(m^2 - 4)}{(m + 8)(m^2 - 4)}$

35. $\frac{3a}{3a} \cdot \frac{5a^2}{2c} = \frac{3a \cdot 5a^2}{3a \cdot 2c}$

36. $\frac{5x^2 \cdot 6y^3}{5x^2 \cdot 3z^4}$

37. $\frac{2x}{2x} \cdot \frac{x - 1}{x + 4} = \frac{2x(x - 1)}{2x(x + 4)}$

38. $\frac{(3y - 1)y}{(2y + 1)y}$

39. $\frac{-1}{-1} \cdot \frac{3 - x}{4 - x} = \frac{-1(3 - x)}{-1(4 - x)}$

40. $\frac{-1(x - 5)}{-5(5 - x)}$

41. $\frac{y + 6}{y + 6} \cdot \frac{y - 7}{y + 2} = \frac{(y + 6)(y - 7)}{(y + 6)(y + 2)}$

42. $\frac{(x - 3)(x^2 - 4)}{(x - 3)(x^3 + 1)}$

43. $\frac{4x^2 y}{2xy^3} = \frac{2 \cdot 2 \cdot x \cdot x \cdot y}{2 \cdot x \cdot y \cdot y^2}$ Factoring numerator and denominator

    $= \frac{2xy}{2xy} \cdot \frac{2x}{y^2}$ Factoring the rational expression

    $= 1 \cdot \frac{2x}{y^2}$

    $= \frac{2x}{y^2}$ "Removing" a factor of 1

44. $-\frac{b}{2a^2}$

45. $\frac{48p^7 q^5}{18p^5 q^4} = \frac{8 \cdot 6 \cdot p^5 \cdot p^2 \cdot q^4 \cdot q}{6 \cdot 3 \cdot p^5 \cdot q^4}$ Factoring numerator and denominator

    $= \frac{6p^5 q^4}{6p^5 q^4} \cdot \frac{8p^2 q}{3}$ Factoring the rational expression

    $= 1 \cdot \frac{8p^2 q}{3}$

    $= \frac{8p^2 q}{3}$ "Removing" a factor of 1

46. $\frac{19x^4}{6}$

47. $\frac{4x - 12}{4x} = \frac{4(x - 3)}{4 \cdot x}$

    $= \frac{4}{4} \cdot \frac{x - 3}{x}$

    $= 1 \cdot \frac{x - 3}{x}$

    $= \frac{x - 3}{x}$

48. $\frac{y - 3}{2y}$

49. $\frac{3m^2 + 3m}{6m^2 + 9m} = \frac{3m(m + 1)}{3m(2m + 3)}$

    $= \frac{3m}{3m} \cdot \frac{m + 1}{2m + 3}$

    $= 1 \cdot \frac{m + 1}{2m + 3}$

    $= \frac{m + 1}{2m + 3}$

50. $\frac{2(2y - 1)}{5(y - 1)}$

51. $\frac{a^2 - 9}{a^2 + 5a + 6} = \frac{(a - 3)(a + 3)}{(a + 2)(a + 3)}$

    $= \frac{a - 3}{a + 2} \cdot \frac{a + 3}{a + 3}$

    $= \frac{a - 3}{a + 2} \cdot 1$

    $= \frac{a - 3}{a + 2}$

52. $\frac{t - 5}{t - 4}$

Chapter 6   (6.1)

53. $\dfrac{2t^2 + 6t + 4}{4t^2 - 12t - 16} = \dfrac{2(t^2 + 3t + 2)}{4(t^2 - 3t - 4)}$

$= \dfrac{2(t + 2)(t + 1)}{2 \cdot 2(t - 4)(t + 1)}$

$= \dfrac{2(t + 1)}{2(t + 1)} \cdot \dfrac{t + 2}{2(t - 4)}$

$= 1 \cdot \dfrac{t + 2}{2(t - 4)}$

$= \dfrac{t + 2}{2(t - 4)}$

54. $\dfrac{a - 4}{2(a + 4)}$

55. $\dfrac{x^2 - 25}{x^2 - 10x + 25} = \dfrac{(x - 5)(x + 5)}{(x - 5)(x - 5)}$

$= \dfrac{x - 5}{x - 5} \cdot \dfrac{x + 5}{x - 5}$

$= 1 \cdot \dfrac{x + 5}{x - 5}$

$= \dfrac{x + 5}{x - 5}$

56. $\dfrac{x + 4}{x - 4}$

57. $\dfrac{a^2 - 1}{a - 1} = \dfrac{(a - 1)(a + 1)}{a - 1}$

$= \dfrac{a - 1}{a - 1} \cdot \dfrac{a + 1}{1}$

$= 1 \cdot \dfrac{a + 1}{1}$

$= a + 1$

58. $t - 1$

59. $\dfrac{x^2 + 1}{x + 1}$ cannot be simplified.

Neither the numerator nor the denominator can be factored.

60. $\dfrac{y^2 + 4}{y + 2}$

61. $\dfrac{6x^2 - 54}{4x^2 - 36} = \dfrac{2 \cdot 3(x^2 - 9)}{2 \cdot 2(x^2 - 9)}$

$= \dfrac{2(x^2 - 9)}{2(x^2 - 9)} \cdot \dfrac{3}{2}$

$= 1 \cdot \dfrac{3}{2}$

$= \dfrac{3}{2}$

62. 2

63. $\dfrac{6t + 12}{t^2 - t - 6} = \dfrac{6(t + 2)}{(t - 3)(t + 2)}$

$= \dfrac{6}{t - 3} \cdot \dfrac{t + 2}{t + 2}$

$= \dfrac{6}{t - 3} \cdot 1$

$= \dfrac{6}{t - 3}$

64. $\dfrac{5}{y + 6}$

65. $\dfrac{a^2 - 10a + 21}{a^2 - 11a + 28} = \dfrac{(a - 7)(a - 3)}{(a - 7)(a - 4)}$

$= \dfrac{a - 7}{a - 7} \cdot \dfrac{a - 3}{a - 4}$

$= 1 \cdot \dfrac{a - 3}{a - 4}$

$= \dfrac{a - 3}{a - 4}$

66. $\dfrac{y - 6}{y - 5}$

67. $\dfrac{t^2 - 4}{(t + 2)^2} = \dfrac{(t - 2)(t + 2)}{(t + 2)(t + 2)}$

$= \dfrac{t - 2}{t + 2} \cdot \dfrac{t + 2}{t + 2}$

$= \dfrac{t - 2}{t + 2} \cdot 1$

$= \dfrac{t - 2}{t + 2}$

68. $\dfrac{a - 3}{a + 3}$

69. $\dfrac{6 - x}{x - 6} = \dfrac{-1(-6 + x)}{x - 6}$

$= \dfrac{-1(x - 6)}{x - 6}$

$= -1 \cdot \dfrac{x - 6}{x - 6}$

$= -1 \cdot 1$

$= -1$

70. -1

71. $\dfrac{a - b}{b - a} = \dfrac{-1(-a + b)}{b - a}$

$= \dfrac{-1(b - a)}{b - a}$

$= -1 \cdot \dfrac{b - a}{b - a}$

$= -1 \cdot 1$

$= -1$

72. 1

73. $\dfrac{6t - 12}{2 - t} = \dfrac{-6(-t + 2)}{2 - t}$

$= \dfrac{-6(2 - t)}{2 - t}$

$= -6 \cdot \dfrac{2 - t}{2 - t}$

$= -6 \cdot 1$

$= -6$

74. -5

Chapter 6   (6.1)

75. $\dfrac{a^2 - 1}{1 - a} = \dfrac{(a + 1)(a - 1)}{-1(-1 + a)}$

$= \dfrac{(a + 1)(a - 1)}{-1(a - 1)}$

$= \dfrac{a + 1}{-1} \cdot \dfrac{a - 1}{a - 1}$

$= -(a + 1) \cdot 1$

$= -a - 1$

76. -1

77. $\dfrac{4x^3}{3x} \cdot \dfrac{14}{x} = \dfrac{4x^3 \cdot 14}{3x \cdot x}$ Multiplying the numerators and the denominators

$= \dfrac{4 \cdot x \cdot x \cdot x \cdot 14}{3 \cdot x \cdot x}$ Factoring the numerator and the denominator

$= \dfrac{x \cdot x}{x \cdot x} \cdot \dfrac{4 \cdot x \cdot 14}{3}$ Factoring the rational expression

$= 1 \cdot \dfrac{56x}{3}$

$= \dfrac{56x}{3}$ Removing a factor of 1

78. $\dfrac{12}{b^2}$

79. $\dfrac{3c}{d^2} \cdot \dfrac{4d}{6c^3} = \dfrac{3c \cdot 4d}{d^2 \cdot 6c^3}$ Multiplying the numerators and the denominators

$= \dfrac{3 \cdot c \cdot 2 \cdot 2 \cdot d}{d \cdot d \cdot 3 \cdot 2 \cdot c \cdot c^2}$ Factoring the numerator and the denominator

$= \dfrac{3 \cdot 2 \cdot c \cdot d}{3 \cdot 2 \cdot c \cdot d} \cdot \dfrac{2}{dc^2}$

$= 1 \cdot \dfrac{2}{dc^2}$

$= \dfrac{2}{dc^2}$ Removing a factor of 1

80. $\dfrac{6x}{y^2}$

81. $\dfrac{x^2 - 3x - 10}{(x - 2)^2} \cdot \dfrac{x - 2}{x - 5} = \dfrac{(x^2 - 3x - 10)(x - 2)}{(x - 2)^2(x - 5)}$

$= \dfrac{(x - 5)(x + 2)(x - 2)}{(x - 2)(x - 2)(x - 5)}$

$= \dfrac{(x - 5)(x - 2)}{(x - 5)(x - 2)} \cdot \dfrac{x + 2}{x - 2}$

$= \dfrac{x + 2}{x - 2}$

82. $\dfrac{t}{t + 2}$

83. $\dfrac{a^2 - 9}{a^2} \cdot \dfrac{a^2 - 3a}{a^2 + a - 12} = \dfrac{(a - 3)(a + 3)(a)(a - 3)}{a \cdot a(a + 4)(a - 3)}$

$= \dfrac{a(a - 3)}{a(a - 3)} \cdot \dfrac{(a - 3)(a + 3)}{a(a + 4)}$

$= \dfrac{(a - 3)(a + 3)}{a(a + 4)}$

84. 1

85. $\dfrac{4a^2}{3a^2 - 12a + 12} \cdot \dfrac{3a - 6}{2a} = \dfrac{4a^2(3a - 6)}{(3a^2 - 12a + 12)2a}$

$= \dfrac{2 \cdot 2 \cdot a \cdot a \cdot 3 \cdot (a - 2)}{3 \cdot (a - 2) \cdot (a - 2) \cdot 2 \cdot a}$

$= \dfrac{2 \cdot 3a(a - 2)}{2 \cdot 3a(a - 2)} \cdot \dfrac{2a}{a - 2}$

$= \dfrac{2a}{a - 2}$

86. $\dfrac{5(v - 2)}{v - 1}$

87. $\dfrac{x^4 - 16}{x^4 - 1} \cdot \dfrac{x^2 + 1}{x^2 + 4}$

$= \dfrac{(x^4 - 16)(x^2 + 1)}{(x^4 - 1)(x^2 + 4)}$

$= \dfrac{(x^2 + 4)(x + 2)(x - 2)(x^2 + 1)}{(x^2 + 1)(x + 1)(x - 1)(x^2 + 4)}$

$= \dfrac{(x^2 + 4)(x^2 + 1)}{(x^2 + 4)(x^2 + 1)} \cdot \dfrac{(x + 2)(x - 2)}{(x + 1)(x - 1)}$

$= \dfrac{(x + 2)(x - 2)}{(x + 1)(x - 1)}$

88. $\dfrac{(t + 1)(t - 1)}{(t + 3)(t - 3)}$

89. $\dfrac{(t - 2)^3}{(t - 1)^3} \cdot \dfrac{t^2 - 2t + 1}{t^2 - 4t + 4}$

$= \dfrac{(t - 2)^3(t^2 - 2t + 1)}{(t - 1)^3(t^2 - 4t + 4)}$

$= \dfrac{(t - 2)(t - 2)(t - 2)(t - 1)(t - 1)}{(t - 1)(t - 1)(t - 1)(t - 2)(t - 2)}$

$= \dfrac{(t - 2)(t - 2)(t - 1)(t - 1)}{(t - 2)(t - 2)(t - 1)(t - 1)} \cdot \dfrac{t - 2}{t - 1}$

$= \dfrac{t - 2}{t - 1}$

90. $\dfrac{y + 4}{y + 2}$

91. <u>Familiarize</u>. Let x = the smaller even integer. Then x + 2 = the larger even integer.

<u>Translate</u>. We reword the problem.

Smaller even integer   times   larger even integer   is   360.
    x   ·   (x + 2)   =   360

<u>Carry out</u>.

$x(x + 2) = 360$

$x^2 + 2x = 360$

$x^2 + 2x - 360 = 0$

$(x + 20)(x - 18) = 0$

$x + 20 = 0$ or $x - 18 = 0$

$x = -20$ or $x = 18$

<u>Check</u>. The solutions of the equation are -20 and 18. When x = -20, then x + 2 = -18 and -20(-18) = 360. The numbers -20 and -18 are consecutive even integers which are solutions to the problem. When x = 18, then x + 2 = 20 and 18·20 = 360. The numbers 18 and 20 are also consecutive even integers which are solutions to the problem.

Chapter 6 (6.2)

91. (continued)

   State. We have two solutions each of which consists of a pair of numbers: -20 and -18, and 18 and 20.

92. $(4 + t^2)(2 + t)(2 - t)$

93. We factor by grouping.
   $2y^3 - 10y^2 + y - 5 = 2y^2(y - 5) + 1(y - 5)$
   $ = (2y^2 + 1)(y - 5)$

94. $x^3(x - 7)(x + 5)$

95. $\dfrac{x^4 - 16y^2}{(x^2 + 4y^2)(x - 2y)}$

   $= \dfrac{(x^2 + 4y^2)(x + 2y)(x - 2y)}{(x^2 + 4y^2)(x - 2y)}$

   $= \dfrac{(x^2 + 4y^2)(x - 2y)}{(x^2 + 4y^2)(x - 2y)} \cdot \dfrac{x + 2y}{1}$

   $= x + 2y$

96. $\dfrac{b - a}{a + b}$

97. $\dfrac{t^4 - 1}{t^4 - 81} \cdot \dfrac{t^2 - 9}{t^2 + 1} \cdot \dfrac{(t - 9)^2}{(t + 1)^2}$

   $= \dfrac{(t^2+1)(t+1)(t-1)(t+3)(t-3)(t-9)(t-9)}{(t^2+9)(t+3)(t-3)(t^2+1)(t+1)(t+1)}$

   $= \dfrac{(t^2+1)(t+1)(t+3)(t-3)}{(t^2+1)(t+1)(t+3)(t-3)} \cdot \dfrac{(t-1)(t-9)(t-9)}{(t^2+9)(t+1)}$

   $= \dfrac{(t - 1)(t - 9)(t - 9)}{(t^2 + 9)(t + 1)}$, or $\dfrac{(t - 1)(t - 9)^2}{(t^2 + 9)(t + 1)}$

98. 1

99. $\dfrac{x^2 - y^2}{(x - y)^2} \cdot \dfrac{x^2 - 2xy + y^2}{x^2 - 4xy - 5y^2}$

   $= \dfrac{(x + y)(x - y)(x - y)(x - y)}{(x - y)(x - y)(x - 5y)(x + y)}$

   $= \dfrac{(x + y)(x - y)(x - y)}{(x + y)(x - y)(x - y)} \cdot \dfrac{x - y}{x - 5y}$

   $= \dfrac{x - y}{x - 5y}$

100. $\dfrac{1}{x - 1}$

101. $\dfrac{5(2x + 5) - 25}{10} = \dfrac{10x + 25 - 25}{10}$

   $\phantom{\dfrac{5(2x + 5) - 25}{10}} = \dfrac{10x}{10}$

   $\phantom{\dfrac{5(2x + 5) - 25}{10}} = x$

   You get the same number you selected.

Exercise Set 6.2

1. The reciprocal of $\dfrac{4}{x}$ is $\dfrac{x}{4}$ because $\dfrac{4}{x} \cdot \dfrac{x}{4} = 1$.

2. $\dfrac{a - 1}{a + 3}$

3. The reciprocal of $x^2 - y^2$ is $\dfrac{1}{x^2 - y^2}$ because $\dfrac{x^2 - y^2}{1} \cdot \dfrac{1}{x^2 - y^2} = 1$.

4. $a + b$

5. The reciprocal of $\dfrac{x^2 + 2x - 5}{x^2 - 4x + 7}$ is $\dfrac{x^2 - 4x + 7}{x^2 + 2x - 5}$ because $\dfrac{x^2 + 2x - 5}{x^2 - 4x + 7} \cdot \dfrac{x^2 - 4x + 7}{x^2 + 2x - 5} = 1$.

6. $\dfrac{x^2 + 7xy - y^2}{x^2 - 3xy + y^2}$

7. $\dfrac{2}{5} \div \dfrac{4}{3} = \dfrac{2}{5} \cdot \dfrac{3}{4}$ Multiplying by the reciprocal

   $\phantom{\dfrac{2}{5} \div \dfrac{4}{3}} = \dfrac{2 \cdot 3}{5 \cdot 4}$

   $\phantom{\dfrac{2}{5} \div \dfrac{4}{3}} = \dfrac{2 \cdot 3}{5 \cdot 2 \cdot 2}$ Factoring the denominator

   $\phantom{\dfrac{2}{5} \div \dfrac{4}{3}} = \dfrac{2}{2} \cdot \dfrac{3}{5 \cdot 2}$ Factoring the fractional expression

   $\phantom{\dfrac{2}{5} \div \dfrac{4}{3}} = \dfrac{3}{10}$ Simplifying

8. $\dfrac{5}{4}$

9. $\dfrac{2}{x} \div \dfrac{8}{x} = \dfrac{2}{x} \cdot \dfrac{x}{8}$ Multiplying by the reciprocal

   $\phantom{\dfrac{2}{x} \div \dfrac{8}{x}} = \dfrac{2 \cdot x}{x \cdot 8}$

   $\phantom{\dfrac{2}{x} \div \dfrac{8}{x}} = \dfrac{2 \cdot x \cdot 1}{x \cdot 2 \cdot 4}$ Factoring the numerator and the denominator

   $\phantom{\dfrac{2}{x} \div \dfrac{8}{x}} = \dfrac{2x}{2x} \cdot \dfrac{1}{4}$ Factoring the fractional expression

   $\phantom{\dfrac{2}{x} \div \dfrac{8}{x}} = \dfrac{1}{4}$ Simplifying

10. $\dfrac{x^2}{6}$

11. $\dfrac{x^2}{y} \div \dfrac{x^3}{y^3} = \dfrac{x^2}{y} \cdot \dfrac{y^3}{x^3}$

   $\phantom{\dfrac{x^2}{y} \div \dfrac{x^3}{y^3}} = \dfrac{x^2 \cdot y^3}{y \cdot x^3}$

   $\phantom{\dfrac{x^2}{y} \div \dfrac{x^3}{y^3}} = \dfrac{x^2 \cdot y \cdot y^2}{y \cdot x^2 \cdot x}$

   $\phantom{\dfrac{x^2}{y} \div \dfrac{x^3}{y^3}} = \dfrac{x^2 y}{x^2 y} \cdot \dfrac{y^2}{x}$

   $\phantom{\dfrac{x^2}{y} \div \dfrac{x^3}{y^3}} = \dfrac{y^2}{x}$

12. $\dfrac{b}{a}$

13. $\dfrac{a + 2}{a - 3} \div \dfrac{a - 1}{a + 3} = \dfrac{a + 2}{a - 3} \cdot \dfrac{a + 3}{a - 1}$

   $\phantom{\dfrac{a + 2}{a - 3} \div \dfrac{a - 1}{a + 3}} = \dfrac{(a + 2)(a + 3)}{(a - 3)(a - 1)}$

14. $\dfrac{y + 2}{2y}$

Chapter 6   (6.2)

15. $\dfrac{x^2 - 1}{x} \div \dfrac{x + 1}{x - 1} = \dfrac{x^2 - 1}{x} \cdot \dfrac{x - 1}{x + 1}$
$= \dfrac{(x^2 - 1)(x - 1)}{x(x + 1)}$
$= \dfrac{(x - 1)(x + 1)(x - 1)}{x(x + 1)}$
$= \dfrac{x + 1}{x + 1} \cdot \dfrac{(x - 1)(x - 1)}{x}$
$= \dfrac{(x - 1)^2}{x}$

16. $4(y - 2)$

17. $\dfrac{x + 1}{6} \div \dfrac{x + 1}{3} = \dfrac{x + 1}{6} \cdot \dfrac{3}{x + 1}$
$= \dfrac{(x + 1) \cdot 3}{6(x + 1)}$
$= \dfrac{3(x + 1)}{2 \cdot 3(x + 1)}$
$= \dfrac{3(x + 1)}{3(x + 1)} \cdot \dfrac{1}{2}$
$= \dfrac{1}{2}$

18. $\dfrac{a}{b}$

19. $\dfrac{5x - 5}{16} \div \dfrac{x - 1}{6} = \dfrac{5x - 5}{16} \cdot \dfrac{6}{x - 1}$
$= \dfrac{(5x - 5) \cdot 6}{16(x - 1)}$
$= \dfrac{5(x - 1) \cdot 2 \cdot 3}{2 \cdot 8(x - 1)}$
$= \dfrac{2(x - 1)}{2(x - 1)} \cdot \dfrac{5 \cdot 3}{8}$
$= \dfrac{15}{8}$

20. $\dfrac{1}{2}$

21. $\dfrac{-6 + 3x}{5} \div \dfrac{4x - 8}{25} = \dfrac{-6 + 3x}{5} \cdot \dfrac{25}{4x - 8}$
$= \dfrac{(-6 + 3x) \cdot 25}{5(4x - 8)}$
$= \dfrac{3(x - 2) \cdot 5 \cdot 5}{5 \cdot 4(x - 2)}$
$= \dfrac{5(x - 2)}{5(x - 2)} \cdot \dfrac{3 \cdot 5}{4}$
$= \dfrac{15}{4}$

22. 3

23. $\dfrac{a + 2}{a - 1} \div \dfrac{3a + 6}{a - 5} = \dfrac{a + 2}{a - 1} \cdot \dfrac{a - 5}{3a + 6}$
$= \dfrac{(a + 2)(a - 5)}{(a - 1)(3a + 6)}$
$= \dfrac{(a + 2)(a - 5)}{(a - 1) \cdot 3 \cdot (a + 2)}$
$= \dfrac{a + 2}{a + 2} \cdot \dfrac{a - 5}{3(a - 1)}$
$= \dfrac{a - 5}{3(a - 1)}$

24. $\dfrac{t + 1}{4(t + 2)}$

25. $\dfrac{x^2 - 4}{x} \div \dfrac{x - 2}{x + 2} = \dfrac{x^2 - 4}{x} \cdot \dfrac{x + 2}{x - 2}$
$= \dfrac{(x^2 - 4)(x + 2)}{x(x - 2)}$
$= \dfrac{(x - 2)(x + 2)(x + 2)}{x(x - 2)}$
$= \dfrac{x - 2}{x - 2} \cdot \dfrac{(x + 2)(x + 2)}{x}$
$= \dfrac{(x + 2)^2}{x}$

26. $\dfrac{(x + y)^2}{x^2 + y}$

27. $\dfrac{x^2 - 9}{4x + 12} \div \dfrac{x - 3}{6} = \dfrac{x^2 - 9}{4x + 12} \cdot \dfrac{6}{x - 3}$
$= \dfrac{(x^2 - 9) \cdot 6}{(4x + 12)(x - 3)}$
$= \dfrac{(x - 3)(x + 3) \cdot 3 \cdot 2}{2 \cdot 2(x + 3)(x - 3)}$
$= \dfrac{2(x - 3)(x + 3)}{2(x - 3)(x + 3)} \cdot \dfrac{3}{2}$
$= \dfrac{3}{2}$

28. $\dfrac{5x}{2(x + b)}$

29. $\dfrac{c^2 + 3c}{c^2 + 2c - 3} \div \dfrac{c}{c + 1} = \dfrac{c^2 + 3c}{c^2 + 2c - 3} \cdot \dfrac{c + 1}{c}$
$= \dfrac{(c^2 + 3c)(c + 1)}{(c^2 + 2c - 3)c}$
$= \dfrac{c(c + 3)(c + 1)}{(c + 3)(c - 1)c}$
$= \dfrac{c(c + 3)}{c(c + 3)} \cdot \dfrac{c + 1}{c - 1}$
$= \dfrac{c + 1}{c - 1}$

30. $\dfrac{2x}{x + 5}$

31. $\dfrac{2y^2 - 7y + 3}{2y^2 + 3y - 2} \div \dfrac{6y^2 - 5y + 1}{3y^2 + 5y - 2}$
$= \dfrac{2y^2 - 7y + 3}{2y^2 + 3y - 2} \cdot \dfrac{3y^2 + 5y - 2}{6y^2 - 5y + 1}$
$= \dfrac{(2y^2 - 7y + 3)(3y^2 + 5y - 2)}{(2y^2 + 3y - 2)(6y^2 - 5y + 1)}$
$= \dfrac{(2y - 1)(y - 3)(3y - 1)(y + 2)}{(2y - 1)(y + 2)(3y - 1)(2y - 1)}$
$= \dfrac{(2y - 1)(3y - 1)(y + 2)}{(2y - 1)(3y - 1)(y + 2)} \cdot \dfrac{y - 3}{2y - 1}$
$= \dfrac{y - 3}{2y - 1}$

32. $\dfrac{x + 3}{x - 5}$

Chapter 6 (6.2)

33. $\dfrac{c^2 + 10c + 21}{c^2 - 2c - 15} \div (c^2 + 2c - 35)$

$= \dfrac{c^2 + 10c + 21}{c^2 - 2c - 15} \cdot \dfrac{1}{c^2 + 2c - 35}$

$= \dfrac{(c^2 + 10c + 21) \cdot 1}{(c^2 - 2c - 15)(c^2 + 2c - 35)}$

$= \dfrac{(c + 7)(c + 3)}{(c - 5)(c + 3)(c + 7)(c - 5)}$

$= \dfrac{(c + 7)(c + 3)}{(c + 7)(c + 3)} \cdot \dfrac{1}{(c - 5)(c - 5)}$

$= \dfrac{1}{(c - 5)^2}$

34. $\dfrac{1}{1 + 2z - z^2}$

35. $\dfrac{(t + 5)^3}{(t - 5)^3} \div \dfrac{(t + 5)^2}{(t - 5)^2}$

$= \dfrac{(t + 5)^3}{(t - 5)^3} \cdot \dfrac{(t - 5)^2}{(t + 5)^2}$

$= \dfrac{(t + 5)^3 (t - 5)^2}{(t - 5)^3 (t + 5)^2}$

$= \dfrac{(t + 5)^2 (t - 5)^2}{(t + 5)^2 (t - 5)^2} \cdot \dfrac{t + 5}{t - 5}$

$= \dfrac{t + 5}{t - 5}$

36. $\dfrac{y - 3}{y + 3}$

37. <u>Familiarize</u>. Let x = the number.
    <u>Translate</u>.

    Sixteen <u>more than</u> the square of a number <u>is</u> eight times the number.
    16   + x^2 =   8x

    <u>Carry out</u>.
    $16 + x^2 = 8x$
    $x^2 - 8x + 16 = 0$
    $(x - 4)(x - 4) = 0$
    $x - 4 = 0$ or $x - 4 = 0$
    $x = 4$ or $x = 4$

    <u>Check</u>. The square of 4, which is 16, plus 16 is 32, and eight times 4 is 32. The number checks.
    <u>State</u>. The number is 4.

38. $2x^2 + 16$

39. $(8x^3 - 3x^2 + 7) - (8x^2 + 3x - 5) =$
    $8x^3 - 3x^2 + 7 - 8x^2 - 3x + 5 =$
    $8x^3 - 11x^2 - 3x + 12$

40. $0.06y^3 - 0.09y^2 + 0.01y - 1$

41. $\dfrac{2a^2 - 5ab}{c - 3d} \div (4a^2 - 25b^2)$

$= \dfrac{2a^2 - 5ab}{c - 3d} \cdot \dfrac{1}{4a^2 - 25b^2}$

$= \dfrac{a(2a - 5b)}{(c - 3d)(2a + 5b)(2a - 5b)}$

$= \dfrac{2a - 5b}{2a - 5b} \cdot \dfrac{a}{(c - 3d)(2a + 5b)}$

$= \dfrac{a}{(c - 3d)(2a + 5b)}$

42. 1

43. $\dfrac{3a^2 - 5ab - 12b^2}{3ab + 4b^2} \div (3b^2 - ab)$

$= \dfrac{3a^2 - 5ab - 12b^2}{3ab + 4b^2} \cdot \dfrac{1}{3b^2 - ab}$

$= \dfrac{(3a + 4b)(a - 3b)}{b(3a + 4b) \cdot b(3b - a)}$

$= \dfrac{(3a + 4b)(-1)(3b - a)}{b(3a + 4b) \cdot b(3b - a)}$

$= \dfrac{(3a + 4b)(3b - a)}{(3a + 4b)(3b - a)} \cdot \dfrac{-1}{b \cdot b}$

$= -\dfrac{1}{b^2}$

44. $\dfrac{1}{(x + y)^2}$

45. $xy \cdot \dfrac{y^2 - 4xy}{y - x} \div \dfrac{16x^2y^2 - y^4}{4x^2 - 3xy - y^2}$

$= \dfrac{xy}{1} \cdot \dfrac{y^2 - 4xy}{y - x} \cdot \dfrac{4x^2 - 3xy - y^2}{16x^2y^2 - y^4}$

$= \dfrac{x \cdot y \cdot y \cdot (y - 4x)(4x + y)(x - y)}{(y - x) \cdot y \cdot y \cdot (4x - y)(4x + y)}$

$= \dfrac{x \cdot y \cdot y \cdot (-1)(4x - y)(4x + y)(-1)(y - x)}{(y - x) \cdot y \cdot y \cdot (4x - y)(4x + y)}$

$= \dfrac{y \cdot y (4x - y)(4x + y)(y - x)}{y \cdot y (4x - y)(4x + y)(y - x)} \cdot \dfrac{x(-1)(-1)}{1}$

$= x$

46. $\dfrac{(z + 4)^3}{(z - 4)^3}$

47. $\dfrac{x^2 - x + xy - y}{x^2 + 6x - 7} \div \dfrac{x^2 + 2xy + y^2}{4x + 4y}$

$= \dfrac{x^2 - x + xy - y}{x^2 + 6x - 7} \cdot \dfrac{4x + 4y}{x^2 + 2xy + y^2}$

$= \dfrac{x(x - 1) + y(x - 1)}{x^2 + 6x - 7} \cdot \dfrac{4x + 4y}{x^2 + 2xy + y^2}$

$= \dfrac{(x + y)(x - 1) \cdot 4(x + y)}{(x + 7)(x - 1)(x + y)(x + y)}$

$= \dfrac{(x + y)(x + y)(x - 1)}{(x + y)(x + y)(x - 1)} \cdot \dfrac{4}{x + 7}$

$= \dfrac{4}{x + 7}$

48. $\dfrac{x(x^2 + 1)}{3(x + y - 1)}$

Chapter 6   (6.3)

49. $\left(\dfrac{y^2 + 5y + 6}{y^2} \cdot \dfrac{3y^3 + 6y^2}{y^2 - y - 12}\right) \div \dfrac{y^2 - y}{y^2 - 2y - 8}$

$= \dfrac{y^2 + 5y + 6}{y^2} \cdot \dfrac{3y^3 + 6y^2}{y^2 - y - 12} \cdot \dfrac{y^2 - 2y - 8}{y^2 - y}$

$= \dfrac{(y + 3)(y + 2)(3y^2)(y + 2)(y - 4)(y + 2)}{y^2(y - 4)(y + 3)(y)(y - 1)}$

$= \dfrac{y^2(y - 4)(y + 3)}{y^2(y - 4)(y + 3)} \cdot \dfrac{3(y + 2)(y + 2)(y + 2)}{y(y - 1)}$

$= \dfrac{3(y + 2)^3}{y(y - 1)}$

50. $\dfrac{a - 3b}{c}$

Exercise Set 6.3

1. $\dfrac{5}{12} + \dfrac{7}{12} = \dfrac{5 + 7}{12} = \dfrac{12}{12} = 1$

2. $\dfrac{4}{7}$

3. $\dfrac{1}{3 + x} + \dfrac{5}{3 + x} = \dfrac{1 + 5}{3 + x} = \dfrac{6}{3 + x}$

4. $\dfrac{7x - 6}{6x + 5}$

5. $\dfrac{x^2 + 7x}{x^2 - 5x} + \dfrac{x^2 - 4x}{x^2 - 5x} = \dfrac{(x^2 + 7x) + (x^2 - 4x)}{x^2 - 5x}$

$= \dfrac{2x^2 + 3x}{x^2 - 5x}$

$= \dfrac{x(2x + 3)}{x(x - 5)}$

$= \dfrac{x}{x} \cdot \dfrac{2x + 3}{(x - 5)}$

$= \dfrac{2x + 3}{x - 5}$

6. $\dfrac{a + b}{x + y}$

7. $\dfrac{7}{8} + \dfrac{5}{-8} = \dfrac{7}{8} + \dfrac{-1}{-1} \cdot \dfrac{5}{-8}$

$= \dfrac{7}{8} + \dfrac{-5}{8}$

$= \dfrac{7 + (-5)}{8}$

$= \dfrac{2}{8}$

$= \dfrac{1}{4}$

8. 1

9. $\dfrac{3}{t} + \dfrac{4}{-t} = \dfrac{3}{t} + \dfrac{-1}{-1} \cdot \dfrac{4}{-t}$

$= \dfrac{3}{t} + \dfrac{-4}{t}$

$= \dfrac{3 + (-4)}{t}$

$= \dfrac{-1}{t}$

$= -\dfrac{1}{t}$

10. $\dfrac{3}{a}$

11. $\dfrac{2x + 7}{x - 6} + \dfrac{3x}{6 - x} = \dfrac{2x + 7}{x - 6} + \dfrac{-1}{-1} \cdot \dfrac{3x}{6 - x}$

$= \dfrac{2x + 7}{x - 6} + \dfrac{-3x}{x - 6}$

$= \dfrac{(2x + 7) + (-3x)}{x - 6}$

$= \dfrac{-x + 7}{x - 6}$

12. $\dfrac{x + 3}{4x - 3}$

13. $\dfrac{y^2}{y - 3} + \dfrac{9}{3 - y} = \dfrac{y^2}{y - 3} + \dfrac{-1}{-1} \cdot \dfrac{9}{3 - y}$

$= \dfrac{y^2}{y - 3} + \dfrac{-9}{y - 3}$

$= \dfrac{y^2 + (-9)}{y - 3}$

$= \dfrac{y^2 - 9}{y - 3}$

$= \dfrac{(y + 3)(y - 3)}{y - 3}$

$= \dfrac{y + 3}{1} \cdot \dfrac{y - 3}{y - 3}$

$= y + 3$

14. $t + 2$

15. $\dfrac{b - 7}{b^2 - 16} + \dfrac{7 - b}{16 - b^2} = \dfrac{b - 7}{b^2 - 16} + \dfrac{-1}{-1} \cdot \dfrac{7 - b}{16 - b^2}$

$= \dfrac{b - 7}{b^2 - 16} + \dfrac{b - 7}{b^2 - 16}$

$= \dfrac{(b - 7) + (b - 7)}{b^2 - 16}$

$= \dfrac{2b - 14}{b^2 - 16}$

16. 0

17. $\dfrac{z}{(y+z)(y-z)} + \dfrac{y}{(z+y)(z-y)}$

$= \dfrac{z}{(y+z)(y-z)} + \dfrac{y}{(z+y)(z-y)} \cdot \dfrac{-1}{-1}$

$= \dfrac{z}{(y+z)(y-z)} + \dfrac{-y}{(z+y)(y-z)}$

$= \dfrac{z-y}{(y+z)(y-z)}$

$= \dfrac{-(y-z)}{(y+z)(y-z)}$

$= \dfrac{-1}{y+z} \cdot \dfrac{y-z}{y-z}$

$= -\dfrac{1}{y+z}$

18. $a + b$

19. $\dfrac{x+3}{x-5} + \dfrac{2x-1}{5-x} + \dfrac{2(3x-1)}{x-5}$

$= \dfrac{x+3}{x-5} + \dfrac{-1}{-1} \cdot \dfrac{2x-1}{5-x} + \dfrac{2(3x-1)}{x-5}$

$= \dfrac{x+3}{x-5} + \dfrac{1-2x}{x-5} + \dfrac{2(3x-1)}{x-5}$

$= \dfrac{(x+3) + (1-2x) + (6x-2)}{x-5}$

$= \dfrac{5x+2}{x-5}$

20. $\dfrac{10x-4}{2x-3}$

21. $\dfrac{2(4x+1)}{5x-7} + \dfrac{3(x-2)}{7-5x} + \dfrac{-10x-1}{5x-7}$

$= \dfrac{2(4x+1)}{5x-7} + \dfrac{-1}{-1} \cdot \dfrac{3(x-2)}{7-5x} + \dfrac{-10x-1}{5x-7}$

$= \dfrac{2(4x+1)}{5x-7} + \dfrac{-3(x-2)}{5x-7} + \dfrac{-10x-1}{5x-7}$

$= \dfrac{(8x+2) + (-3x+6) + (-10x-1)}{5x-7}$

$= \dfrac{-5x+7}{5x-7}$

$= \dfrac{-1(5x-7)}{5x-7}$

$= \dfrac{-1}{1} \cdot \dfrac{5x-7}{5x-7}$

$= -1$

22. $\dfrac{12x+7}{4-3x}$

23. $\dfrac{x+1}{(x+3)(x-3)} + \dfrac{4(x-3)}{(x-3)(x+3)} + \dfrac{(x-1)(x-3)}{(3-x)(x+3)}$

$= \dfrac{x+1}{(x+3)(x-3)} + \dfrac{4(x-3)}{(x-3)(x+3)} + \dfrac{-1}{-1} \cdot \dfrac{(x-1)(x-3)}{(3-x)(x+3)}$

$= \dfrac{x+1}{(x+3)(x-3)} + \dfrac{4(x-3)}{(x-3)(x+3)} + \dfrac{-1(x^2-4x+3)}{(x-3)(x+3)}$

$= \dfrac{(x+1) + (4x-12) + (-x^2+4x-3)}{(x+3)(x-3)}$

$= \dfrac{-x^2+9x-14}{(x+3)(x-3)}$

24. $\dfrac{-2x+11}{(2x-3)(x-1)}$

25. $\dfrac{7}{8} - \dfrac{3}{8} = \dfrac{7-3}{8} = \dfrac{4}{8} = \dfrac{1}{2}$

26. $-\dfrac{2}{y}$

27. $\dfrac{x}{x-1} - \dfrac{1}{x-1} = \dfrac{x-1}{x-1} = 1$

28. $x - 4$

29. $\dfrac{x+1}{x^2-2x+1} - \dfrac{5-3x}{x^2-2x+1} = \dfrac{(x+1)-(5-3x)}{x^2-2x+1}$

$= \dfrac{x+1-5+3x}{x^2-2x+1}$

$= \dfrac{4x-4}{x^2-2x+1}$

$= \dfrac{4(x-1)}{(x-1)(x-1)}$

$= \dfrac{4}{x-1} \cdot \dfrac{x-1}{x-1}$

$= \dfrac{4}{x-1}$

30. $\dfrac{1}{x-1}$

31. $\dfrac{11}{6} - \dfrac{5}{-6} = \dfrac{11}{6} - \dfrac{-1}{-1} \cdot \dfrac{5}{-6}$

$= \dfrac{11}{6} - \dfrac{-5}{6}$

$= \dfrac{11-(-5)}{6}$

$= \dfrac{11+5}{6}$

$= \dfrac{16}{6}$

$= \dfrac{8}{3}$

32. $\dfrac{3}{2}$

33. $\dfrac{5}{a} - \dfrac{8}{-a} = \dfrac{5}{a} - \dfrac{-1}{-1} \cdot \dfrac{8}{-a}$

$= \dfrac{5}{a} - \dfrac{-8}{a}$

$= \dfrac{5-(-8)}{a}$

$= \dfrac{5+8}{a}$

$= \dfrac{13}{a}$

34. $\dfrac{7}{t}$

Chapter 6   (6.3)

35. $\dfrac{x}{4} - \dfrac{3x - 5}{-4} = \dfrac{x}{4} - \dfrac{-1}{-1} \cdot \dfrac{3x - 5}{-4}$

$= \dfrac{x}{4} - \dfrac{5 - 3x}{4}$

$= \dfrac{x - (5 - 3x)}{4}$

$= \dfrac{x - 5 + 3x}{4}$

$= \dfrac{4x - 5}{4}$

36. $\dfrac{4}{x - 1}$

37. $\dfrac{3 - x}{x - 7} - \dfrac{2x - 5}{7 - x} = \dfrac{3 - x}{x - 7} - \dfrac{-1}{-1} \cdot \dfrac{2x - 5}{7 - x}$

$= \dfrac{3 - x}{x - 7} - \dfrac{5 - 2x}{x - 7}$

$= \dfrac{(3 - x) - (5 - 2x)}{x - 7}$

$= \dfrac{3 - x - 5 + 2x}{x - 7}$

$= \dfrac{x - 2}{x - 7}$

38. $\dfrac{t^2 + 4}{t - 2}$

39. $\dfrac{x - 8}{x^2 - 16} - \dfrac{x - 8}{16 - x^2} = \dfrac{x - 8}{x^2 - 16} - \dfrac{-1}{-1} \cdot \dfrac{x - 8}{16 - x^2}$

$= \dfrac{x - 8}{x^2 - 16} - \dfrac{8 - x}{x^2 - 16}$

$= \dfrac{(x - 8) - (8 - x)}{x^2 - 16}$

$= \dfrac{x - 8 - 8 + x}{x^2 - 16}$

$= \dfrac{2x - 16}{x^2 - 16}$

40. $\dfrac{4}{x^2 - 25}$

41. $\dfrac{4 - x}{x - 9} - \dfrac{3x - 8}{9 - x} = \dfrac{4 - x}{x - 9} - \dfrac{-1}{-1} \cdot \dfrac{3x - 8}{9 - x}$

$= \dfrac{4 - x}{x - 9} - \dfrac{8 - 3x}{x - 9}$

$= \dfrac{(4 - x) - (8 - 3x)}{x - 9}$

$= \dfrac{4 - x - 8 + 3x}{x - 9}$

$= \dfrac{2x - 4}{x - 9}$

42. $\dfrac{x - 2}{x - 7}$

43. $\dfrac{2(x - 1)}{2x - 3} - \dfrac{3(x + 2)}{2x - 3} - \dfrac{x - 1}{3 - 2x}$

$= \dfrac{2(x - 1)}{2x - 3} - \dfrac{3(x + 2)}{2x - 3} - \dfrac{-1}{-1} \cdot \dfrac{x - 1}{3 - 2x}$

$= \dfrac{2(x - 1)}{2x - 3} - \dfrac{3(x + 2)}{2x - 3} - \dfrac{1 - x}{2x - 3}$

$= \dfrac{(2x - 2) - (3x + 6) - (1 - x)}{2x - 3}$

$= \dfrac{2x - 2 - 3x - 6 - 1 + x}{2x - 3}$

$= \dfrac{-9}{2x - 3}$

44. $\dfrac{-4x - 14}{2x - 3}$

45. $\dfrac{3(2x + 5)}{x - 1} - \dfrac{3(2x - 3)}{1 - x} + \dfrac{6x - 1}{x - 1}$

$= \dfrac{3(2x + 5)}{x - 1} - \dfrac{-1}{-1} \cdot \dfrac{3(2x - 3)}{1 - x} + \dfrac{6x - 1}{x - 1}$

$= \dfrac{3(2x + 5)}{x - 1} - \dfrac{-3(2x - 3)}{x - 1} + \dfrac{6x - 1}{x - 1}$

$= \dfrac{(6x + 15) - (-6x + 9) + (6x - 1)}{x - 1}$

$= \dfrac{6x + 15 + 6x - 9 + 6x - 1}{x - 1}$

$= \dfrac{18x + 5}{x - 1}$

46. $\dfrac{-2x + 4y}{x - y}$

47. $\dfrac{x - y}{x^2 - y^2} + \dfrac{x + y}{x^2 - y^2} - \dfrac{2x}{x^2 - y^2}$

$= \dfrac{x - y + x + y - 2x}{x^2 - y^2}$

$= \dfrac{0}{x^2 - y^2}$

$= 0$

48. $\dfrac{-x + 3y}{x - y}$

49. $\dfrac{10}{2y - 1} - \dfrac{6}{1 - 2y} + \dfrac{y}{2y - 1} + \dfrac{y - 4}{1 - 2y}$

$= \dfrac{10}{2y - 1} - \dfrac{-1}{-1} \cdot \dfrac{6}{1 - 2y} + \dfrac{y}{2y - 1} + \dfrac{-1}{-1} \cdot \dfrac{y - 4}{1 - 2y}$

$= \dfrac{10}{2y - 1} - \dfrac{-6}{2y - 1} + \dfrac{y}{2y - 1} + \dfrac{4 - y}{2y - 1}$

$= \dfrac{10 + 6 + y + 4 - y}{2y - 1}$

$= \dfrac{20}{2y - 1}$

50. $\dfrac{x^2 + 3}{(3 - 2x)(x - 3)}$

Chapter 6   (6.4)

51. <u>Familiarize</u>. Let d = the amount of customs duty to be paid.
<u>Translate</u>. We reword the problem.

Amount of duty	is	5%	of	$2800.
d	=	5%	·	2800

<u>Carry out</u>.
d = 5%·2800
d = 0.05(2800)
d = 140
<u>Check</u>. To check we can repeat the computations.
<u>State</u>. You will pay $140 customs duty.

52. 60

53. $16 = 2·2·2·2$
$18 = 2·3·3$
The LCM is $2·2·2·2·3·3$, or 144.

54. 72

55. $\dfrac{x}{(x-y)(y-z)} - \dfrac{y}{(y-x)(z-y)}$

$= \dfrac{x}{(x-y)(y-z)} - \dfrac{x}{(-1)(x-y)(-1)(y-z)}$

$= \dfrac{x}{(x-y)(y-z)} - \dfrac{x}{(x-y)(y-z)}$

$= \dfrac{x-x}{(x-y)(y-z)}$

$= \dfrac{0}{(x-y)(y-z)}$

$= 0$

56. $\dfrac{x+y}{x-y}$

57. $\dfrac{b-c}{a-(b-c)} - \dfrac{b-a}{(b-a)-c}$

$= \dfrac{b-c}{a-b+c} - \dfrac{b-a}{-a+b-c}$

$= \dfrac{b-c}{a-b+c} - \dfrac{-1}{-1} · \dfrac{b-a}{-a+b-c}$

$= \dfrac{b-c}{a-b+c} - \dfrac{a-b}{a-b+c}$

$= \dfrac{(b-c)-(a-b)}{a-b+c}$

$= \dfrac{b-c-a+b}{a-b+c}$

$= \dfrac{2b-c-a}{a-b+c}$

58. -1

59. $\dfrac{x^2}{3x^2-5x-2} - \dfrac{2x}{3x+1} · \dfrac{1}{x-2}$

$= \dfrac{x^2}{(3x+1)(x-2)} - \dfrac{2x}{(3x+1)(x-2)}$

$= \dfrac{x^2-2x}{(3x+1)(x-2)}$

$= \dfrac{x(x-2)}{(3x+1)(x-2)}$

$= \dfrac{x}{3x+1} · \dfrac{x-2}{x-2}$

$= \dfrac{x}{3x+1}$

60. $\dfrac{30}{(x+4)(x-3)}$

Exercise Set 6.4

1. $12 = 2·2·3$
$27 = 3·3·3$
LCM = $2·2·3·3·3$, or 108

2. 30

3. $8 = 2·2·2$
$9 = 3·3$
LCM = $2·2·2·3·3$, or 72

4. 60

5. $6 = 2·3$
$9 = 3·3$
$21 = 3·7$
LCM = $2·3·3·7$, or 126

6. 360

7. $24 = 2·2·2·3$
$36 = 2·2·3·3$
$40 = 2·2·2·5$
LCM = $2·2·2·3·3·5$, or 360

8. 60

9. $28 = 2·2·7$
$42 = 2·3·7$
$60 = 2·2·3·5$
LCM = $2·2·3·5·7$, or 420

10. 500

Chapter 6   (6.4)

11. $24 = 2 \cdot 2 \cdot 2 \cdot 3$
    $18 = 2 \cdot 3 \cdot 3$
    LCM $= 2 \cdot 2 \cdot 2 \cdot 3 \cdot 3$, or 72
    $\frac{7}{24} + \frac{11}{18} = \frac{7}{2 \cdot 2 \cdot 2 \cdot 3} \cdot \frac{3}{3} + \frac{11}{2 \cdot 3 \cdot 3} \cdot \frac{2 \cdot 2}{2 \cdot 2}$
    $= \frac{21}{2 \cdot 2 \cdot 2 \cdot 3 \cdot 3} + \frac{44}{2 \cdot 2 \cdot 2 \cdot 3 \cdot 3}$
    $= \frac{65}{72}$

12. $\frac{59}{300}$

13. $\frac{1}{6} + \frac{3}{40} + \frac{2}{75}$
    $= \frac{1}{2 \cdot 3} + \frac{3}{2 \cdot 2 \cdot 2 \cdot 5} + \frac{2}{3 \cdot 5 \cdot 5}$
    LCM $= 2 \cdot 2 \cdot 2 \cdot 3 \cdot 5 \cdot 5$, or 600
    $= \frac{1}{2 \cdot 3} \cdot \frac{2 \cdot 2 \cdot 5 \cdot 5}{2 \cdot 2 \cdot 5 \cdot 5} + \frac{3}{2 \cdot 2 \cdot 2 \cdot 5} \cdot \frac{3 \cdot 5}{3 \cdot 5} + \frac{2}{3 \cdot 5 \cdot 5} \cdot \frac{2 \cdot 2 \cdot 2}{2 \cdot 2 \cdot 2}$
    $= \frac{100 + 45 + 16}{2 \cdot 2 \cdot 2 \cdot 3 \cdot 5 \cdot 5}$
    $= \frac{161}{600}$

14. $\frac{71}{120}$

15. $\frac{2}{15} + \frac{5}{9} + \frac{3}{20}$
    $= \frac{2}{3 \cdot 5} + \frac{5}{3 \cdot 3} + \frac{3}{2 \cdot 2 \cdot 5}$
    LCM is $2 \cdot 2 \cdot 3 \cdot 3 \cdot 5$, or 180
    $= \frac{2}{3 \cdot 5} \cdot \frac{2 \cdot 2 \cdot 3}{2 \cdot 2 \cdot 3} + \frac{5}{3 \cdot 3} \cdot \frac{2 \cdot 2 \cdot 5}{2 \cdot 2 \cdot 5} + \frac{3}{2 \cdot 2 \cdot 5} \cdot \frac{3 \cdot 3}{3 \cdot 3}$
    $= \frac{24 + 100 + 27}{2 \cdot 2 \cdot 3 \cdot 3 \cdot 5}$
    $= \frac{151}{180}$

16. $\frac{23}{180}$

17. $6x^2 = 2 \cdot 3 \cdot x \cdot x$
    $12x^3 = 2 \cdot 2 \cdot 3 \cdot x \cdot x \cdot x$
    LCM $= 2 \cdot 2 \cdot 3 \cdot x \cdot x \cdot x$, or $12x^3$

18. $8a^2b^2$

19. $2x^2 = 2 \cdot x \cdot x$
    $6xy = 2 \cdot 3 \cdot x \cdot y$
    $18y^2 = 2 \cdot 3 \cdot 3 \cdot y \cdot y$
    LCM $= 2 \cdot 3 \cdot 3 \cdot x \cdot x \cdot y \cdot y$, or $18x^2y^2$

20. c^3d^2

21. $2(y - 3) = 2 \cdot (y - 3)$
    $6(y - 3) = 2 \cdot 3 \cdot (y - 3)$
    LCM $= 2 \cdot 3 \cdot (y - 3)$, or $6(y - 3)$

22. $8(x - 1)$

23. $t, t + 2, t - 2$
    The expressions are not factorable, so the LCM is their product:
    LCM $= t(t + 2)(t - 2)$

24. $x(x + 3)(x - 3)$

25. $x^2 - 4 = (x + 2)(x - 2)$
    $x^2 + 5x + 6 = (x + 3)(x + 2)$
    LCM $= (x + 2)(x - 2)(x + 3)$

26. $(x + 2)(x + 1)(x - 2)$

27. $t^3 + 4t^2 + 4t = t(t^2 + 4t + 4) = t(t + 2)(t + 2)$
    $t^2 - 4t = t(t - 4)$
    LCM $= t(t + 2)(t + 2)(t - 4) = t(t + 2)^2(t - 4)$

28. $y^2(y + 1)(y - 1)$

29. $a + 1 = a + 1$
    $(a - 1)^2 = (a - 1)(a - 1)$
    $a^2 - 1 = (a + 1)(a - 1)$
    LCM $= (a + 1)(a - 1)(a - 1) = (a + 1)(a - 1)^2$

30. $2(x + y)^2(x - y)$

31. $m^2 - 5m + 6 = (m - 3)(m - 2)$
    $m^2 - 4m + 4 = (m - 2)(m - 2)$
    LCM $= (m - 3)(m - 2)(m - 2) = (m - 3)(m - 2)^2$

32. $(2x + 1)(x + 2)(x - 1)$

33. $2 + 3x = 2 + 3x$
    $4 - 9x^2 = (2 + 3x)(2 - 3x)$
    $2 - 3x = 2 - 3x$
    LCM $= (2 + 3x)(2 - 3x)$

34. $(3 + 2x)(3 - 2x)$

35. $10v^2 + 30v = 10v(v + 3) = 2 \cdot 5 \cdot v(v + 3)$
    $-5v^2 - 35v - 60 = -5(v^2 + 7v + 12)$
    $= -1 \cdot 5(v + 4)(v + 3)$
    LCM $= 2 \cdot 5 \cdot v(v + 3)(v + 4) = 10v(v + 3)(v + 4)$, or
    $2 \cdot (-1) \cdot 5 \cdot v(v + 3)(v + 4) = -10v(v + 3)(v + 4)$

36. $-12a(a + 2)(a + 3)$, or $12a(a + 2)(a + 3)$

37. $9x^3 - 9x^2 - 18x = 9x(x^2 - x - 2)$
    $= 3 \cdot 3 \cdot x(x - 2)(x + 1)$
    $6x^5 - 24x^4 + 24x^3 = 6x^3(x^2 - 4x + 4)$
    $= 2 \cdot 3 \cdot x \cdot x \cdot x(x - 2)(x - 2)$
    LCM $= 2 \cdot 3 \cdot 3 \cdot x \cdot x \cdot x(x - 2)(x - 2)(x + 1) = 18x^3(x - 2)^2(x + 1)$

38. $x^3(x + 2)^2(x - 2)$

Chapter 6   (6.5)

39. $x^5 + 4x^4 + 4x^3 = x^3(x^2 + 4x + 4)$
    $= x \cdot x \cdot x(x + 2)(x + 2)$
    $3x^2 - 12 = 3(x^2 - 4) = 3(x + 2)(x - 2)$
    $2x + 4 = 2(x + 2)$
    LCM $= 2 \cdot 3 \cdot x \cdot x \cdot x(x + 2)(x + 2)(x - 2)$
    $= 6x^3(x + 2)^2(x - 2)$

40. $10x^3(x + 1)^2(x - 1)$

41. The LCM is $2 \cdot 2 \cdot 3 \cdot x \cdot x \cdot x$, or $12x^3$. (See Exercise 17.)
    $\frac{7}{6x^2} = \frac{7}{2 \cdot 3 \cdot x \cdot x} \cdot \frac{2x}{2x} = \frac{7 \cdot 2x}{2 \cdot 2 \cdot 3 \cdot x \cdot x \cdot x}$, or $\frac{14x}{12x^3}$
    $\frac{y}{12x^3}$ already has the LCM as the denominator.

42. $\frac{12bc}{8a^2b^2}, \frac{4ac}{8a^2b^2}$

43. The LCM is $(x + 2)(x - 2)(x + 3)$. (See Exercise 25.)
    $\frac{x + 1}{x^2 - 4} = \frac{x + 1}{(x + 2)(x - 2)} \cdot \frac{x + 3}{x + 3}$
    $= \frac{(x + 1)(x + 3)}{(x + 2)(x - 2)(x + 3)}$
    $\frac{x - 2}{x^2 + 5x + 6} = \frac{x - 2}{(x + 3)(x + 2)} \cdot \frac{x - 2}{x - 2}$
    $= \frac{(x - 2)^2}{(x + 3)(x + 2)(x - 2)}$

44. $\frac{(x - 4)(x + 8)}{(x + 3)(x - 3)(x + 8)}, \frac{(x + 2)(x - 3)}{(x + 3)(x + 8)(x - 3)}$

45. The LCM is $t(t + 2)(t - 2)$. (See Exercise 23.)
    $\frac{3}{t} = \frac{3}{t} \cdot \frac{(t + 2)(t - 2)}{(t + 2)(t - 2)} = \frac{3(t + 2)(t - 2)}{t(t + 2)(t - 2)}$
    $\frac{4}{t + 2} = \frac{4}{t + 2} \cdot \frac{t(t - 2)}{t(t - 2)} = \frac{4t(t - 2)}{t(t + 2)(t - 2)}$
    $\frac{t}{t - 2} = \frac{t}{t - 2} \cdot \frac{t(t + 2)}{t(t + 2)} = \frac{t^2(t + 2)}{t(t - 2)(t + 2)}$

46. $\frac{(x + 3)(x - 3)}{x(x + 3)(x - 3)}, \frac{-2x(x - 3)}{x(x + 3)(x - 3)},$
    $\frac{x^3(x + 3)}{x(x - 3)(x + 3)}$

47. $2x - 3 = 2x - 3$
    $4x^2 - 9 = (2x + 3)(2x - 3)$
    $2x + 3 = 2x + 3$
    LCM $= (2x + 3)(2x - 3)$
    $\frac{x + 1}{2x - 3} = \frac{x + 1}{2x - 3} \cdot \frac{2x + 3}{2x + 3} = \frac{(x + 1)(2x + 3)}{(2x - 3)(2x + 3)}$
    $\frac{x - 2}{4x^2 - 9} = \frac{x - 2}{(2x + 3)(2x - 3)}$ already has the LCM as the denominator.
    $\frac{x + 1}{2x + 3} = \frac{x + 1}{2x + 3} \cdot \frac{2x - 3}{2x - 3} = \frac{(x + 1)(2x - 3)}{(2x + 3)(2x - 3)}$

48. $\frac{3x(x - 2)}{3x^3(x + 2)^2(x - 2)}, \frac{3x^3(x + 2)}{3x^3(x + 2)^2(x - 2)}$

49. $x^2 - 6x + 9 = x^2 - 2 \cdot x \cdot 3 + 3^2$ Trinomial square
    $= (x - 3)^2$

50. $2x(3x + 2)$

51. $x^2 - 9 = x^2 - 3^2$ Difference of squares
    $= (x + 3)(x - 3)$

52. $(x + 7)(x - 3)$

53. $72 = 2 \cdot 2 \cdot 2 \cdot 3 \cdot 3$
    $90 = 2 \cdot 3 \cdot 3 \cdot 5$
    $96 = 2 \cdot 2 \cdot 2 \cdot 2 \cdot 2 \cdot 3$
    LCM $= 2 \cdot 2 \cdot 2 \cdot 2 \cdot 2 \cdot 3 \cdot 3 \cdot 5$, or 1440

54. $120(x + 1)(x - 1)(1 - x)$

55. The time it takes the joggers to meet again at the starting place is the LCM of the times it takes them to complete one round of the course.
    $6 = 2 \cdot 3$
    $8 = 2 \cdot 2 \cdot 2$
    LCM $= 2 \cdot 2 \cdot 2 \cdot 3$, or 24
    It takes 24 min.

56. One expression is a multiple of the other.

Exercise Set 6.5

1. $\frac{2}{x} + \frac{5}{x^2} = \frac{2}{x} + \frac{5}{x \cdot x}$ LCM $= x \cdot x$, or x^2
    $= \frac{2}{x} \cdot \frac{x}{x} + \frac{5}{x \cdot x}$
    $= \frac{2x + 5}{x^2}$

2. $\frac{4x + 8}{x^2}$

3. $6r = 2 \cdot 3 \cdot r$ $\}$ LCM $= 2 \cdot 2 \cdot 2 \cdot 3 \cdot r$, or $24r$
   $8r = 2 \cdot 2 \cdot 2 \cdot r$
    $\frac{5}{6r} + \frac{7}{8r} = \frac{5}{6r} \cdot \frac{4}{4} + \frac{7}{8r} \cdot \frac{3}{3}$
    $= \frac{20 + 21}{24r}$
    $= \frac{41}{24r}$

4. $\frac{37}{18t}$

5. $xy^2 = x \cdot y \cdot y$ $\}$ LCM $= x \cdot x \cdot y \cdot y$, or x^2y^2
   $x^2y = x \cdot x \cdot y$
    $\frac{4}{xy^2} + \frac{6}{x^2y} = \frac{4}{xy^2} \cdot \frac{x}{x} + \frac{6}{x^2y} \cdot \frac{y}{y}$
    $= \frac{4x + 6y}{x^2y^2}$

6. $\frac{2d^2 + 7c}{c^2d^3}$

137

Chapter 6   (6.5)

7. $\left.\begin{array}{l}9t^3 = 3\cdot 3\cdot t\cdot t\cdot t\\ 6t^2 = 2\cdot 3\cdot t\cdot t\end{array}\right\}$ LCM $= 2\cdot 3\cdot 3\cdot t\cdot t\cdot t$, or $18t^3$

$\dfrac{2}{9t^3} + \dfrac{1}{6t^2} = \dfrac{2}{9t^3}\cdot\dfrac{2}{2} + \dfrac{1}{6t^2}\cdot\dfrac{3t}{3t}$

$\qquad = \dfrac{4 + 3t}{18t^3}$

8. $\dfrac{-2xy + 18}{3x^2y^3}$

9. LCM $= x^2y^2$ (See Exercise 5.)

$\dfrac{x + y}{xy^2} + \dfrac{3x + y}{x^2y} = \dfrac{x + y}{xy^2}\cdot\dfrac{x}{x} + \dfrac{3x + y}{x^2y}\cdot\dfrac{y}{y}$

$\qquad = \dfrac{x(x + y) + y(3x + y)}{x^2y^2}$

$\qquad = \dfrac{x^2 + xy + 3xy + y^2}{x^2y^2}$

$\qquad = \dfrac{x^2 + 4xy + y^2}{x^2y^2}$

10. $\dfrac{c^2 + 3cd - d^2}{c^2d^2}$

11. The denominators do not factor, so the LCM is their product, $(x - 2)(x + 2)$.

$\dfrac{3}{x - 2} + \dfrac{3}{x + 2} = \dfrac{3}{x - 2}\cdot\dfrac{x + 2}{x + 2} + \dfrac{3}{x + 2}\cdot\dfrac{x - 2}{x - 2}$

$\qquad = \dfrac{3(x + 2) + 3(x - 2)}{(x - 2)(x + 2)}$

$\qquad = \dfrac{3x + 6 + 3x - 6}{(x - 2)(x + 2)}$

$\qquad = \dfrac{6x}{(x - 2)(x + 2)}$

12. $\dfrac{4x}{(x - 1)(x + 1)}$

13. $\left.\begin{array}{l}3x = 3\cdot x\\ x + 1 = x + 1\end{array}\right\}$ LCM $= 3x(x + 1)$

$\dfrac{3}{x + 1} + \dfrac{2}{3x} = \dfrac{3}{x + 1}\cdot\dfrac{3x}{3x} + \dfrac{2}{3x}\cdot\dfrac{x + 1}{x + 1}$

$\qquad = \dfrac{9x + 2(x + 1)}{3x(x + 1)}$

$\qquad = \dfrac{9x + 2x + 2}{3x(x + 1)}$

$\qquad = \dfrac{11x + 2}{3x(x + 1)}$

14. $\dfrac{11x + 15}{4x(x + 5)}$

15. $\left.\begin{array}{l}x^2 - 16 = (x + 4)(x - 4)\\ x - 4 = x - 4\end{array}\right\}$ LCM $= (x + 4)(x - 4)$

$\dfrac{2x}{x^2 - 16} + \dfrac{x}{x - 4} = \dfrac{2x}{(x + 4)(x - 4)} + \dfrac{x}{x - 4}\cdot\dfrac{x + 4}{x + 4}$

$\qquad = \dfrac{2x + x(x + 4)}{(x + 4)(x - 4)}$

$\qquad = \dfrac{2x + x^2 + 4x}{(x + 4)(x - 4)}$

$\qquad = \dfrac{x^2 + 6x}{(x + 4)(x - 4)}$

16. $\dfrac{x^2 - x}{(x + 5)(x - 5)}$

17. $\dfrac{5}{z + 4} + \dfrac{3}{3z + 12} = \dfrac{5}{z + 4} + \dfrac{3}{3(z + 4)}$ LCM $= 3(z + 4)$

$\qquad = \dfrac{5}{z + 4}\cdot\dfrac{3}{3} + \dfrac{3}{3(z + 4)}$

$\qquad = \dfrac{15 + 3}{3(z + 4)} = \dfrac{18}{3(z + 4)}$

$\qquad = \dfrac{3\cdot 6}{3(z + 4)} = \dfrac{3}{3}\cdot\dfrac{6}{z + 4}$

$\qquad = \dfrac{6}{z + 4}$

18. $\dfrac{4t + 5}{4(t - 3)}$

19. $\dfrac{3}{x - 1} + \dfrac{2}{(x - 1)^2} = \dfrac{3}{x - 1}\cdot\dfrac{x - 1}{x - 1} + \dfrac{2}{(x - 1)^2}$

$\qquad\qquad$ LCM $= (x - 1)^2$

$\qquad = \dfrac{3(x - 1) + 2}{(x - 1)^2}$

$\qquad = \dfrac{3x - 3 + 2}{(x - 1)^2}$

$\qquad = \dfrac{3x - 1}{(x - 1)^2}$

20. $\dfrac{2x + 10}{(x + 3)^2}$

21. $\dfrac{4a}{5a - 10} + \dfrac{3a}{10a - 20} = \dfrac{4a}{5(a - 2)} + \dfrac{3a}{2\cdot 5(a - 2)}$

$\qquad\qquad$ LCM $= 2\cdot 5(a - 2)$

$\qquad = \dfrac{4a}{5(a - 2)}\cdot\dfrac{2}{2} + \dfrac{3a}{2\cdot 5(a - 2)}$

$\qquad = \dfrac{8a + 3a}{10(a - 2)}$

$\qquad = \dfrac{11a}{10(a - 2)}$

22. $\dfrac{9a}{4(a - 5)}$

23. $\dfrac{x + 4}{x} + \dfrac{x}{x + 4} = \dfrac{x + 4}{x}\cdot\dfrac{x + 4}{x + 4} + \dfrac{x}{x + 4}\cdot\dfrac{x}{x}$

$\qquad\qquad$ LCM $= x(x + 4)$

$\qquad = \dfrac{(x + 4)^2 + x^2}{x(x + 4)}$

$\qquad = \dfrac{x^2 + 8x + 16 + x^2}{x(x + 4)}$

$\qquad = \dfrac{2x^2 + 8x + 16}{x(x + 4)}$

24. $\dfrac{2x^2 - 10x + 25}{x(x - 5)}$

25. $\dfrac{x}{x^2 + 2x + 1} + \dfrac{1}{x^2 + 5x + 4}$

$= \dfrac{x}{(x + 1)(x + 1)} + \dfrac{1}{(x + 1)(x + 4)}$

$\qquad\qquad$ LCM $= (x + 1)^2(x + 4)$

$= \dfrac{x}{(x + 1)(x + 1)}\cdot\dfrac{x + 4}{x + 4} + \dfrac{1}{(x + 1)(x + 4)}\cdot\dfrac{x + 1}{x + 1}$

$= \dfrac{x(x + 4) + 1\cdot(x + 1)}{(x + 1)^2(x + 4)} = \dfrac{x^2 + 4x + x + 1}{(x + 1)^2(x + 4)}$

$= \dfrac{x^2 + 5x + 1}{(x + 1)^2(x + 4)}$

Chapter 6   (6.5)

26. $\dfrac{12a - 11}{(a + 2)(a - 1)(a - 3)}$

27. $\dfrac{x + 3}{x - 5} + \dfrac{x - 5}{x + 3} = \dfrac{x + 3}{x - 5} \cdot \dfrac{x + 3}{x + 3} + \dfrac{x - 5}{x + 3} \cdot \dfrac{x - 5}{x - 5}$

    $\quad\quad\quad\quad\quad\quad\quad\quad\quad$ LCM $= (x - 5)(x + 3)$

    $= \dfrac{(x + 3)^2 + (x - 5)^2}{(x - 5)(x + 3)}$

    $= \dfrac{x^2 + 6x + 9 + x^2 - 10x + 25}{(x - 5)(x + 3)}$

    $= \dfrac{2x^2 - 4x + 34}{(x - 5)(x + 3)}$

28. $\dfrac{13xy - 12x}{(2y - 3)(3y - 2)}$

29. $\dfrac{a}{a^2 - 1} + \dfrac{2a}{a^2 - a}$

    $= \dfrac{a}{(a + 1)(a - 1)} + \dfrac{2a}{a(a - 1)}$ LCM $= a(a + 1)(a - 1)$

    $= \dfrac{a}{(a + 1)(a - 1)} \cdot \dfrac{a}{a} + \dfrac{2a}{a(a - 1)} \cdot \dfrac{a + 1}{a + 1}$

    $= \dfrac{a^2 + 2a(a + 1)}{a(a + 1)(a - 1)} = \dfrac{a^2 + 2a^2 + 2a}{a(a + 1)(a - 1)}$

    $= \dfrac{3a^2 + 2a}{a(a + 1)(a - 1)} = \dfrac{a(3a + 2)}{a(a + 1)(a - 1)}$

    $= \dfrac{a}{a} \cdot \dfrac{3a + 2}{(a + 1)(a - 1)} = \dfrac{3a + 2}{(a + 1)(a - 1)}$

30. $\dfrac{3x + 5}{3(x + 2)}$

31. $\dfrac{6}{x - y} + \dfrac{4x}{y^2 - x^2} = \dfrac{6}{x - y} + \dfrac{4x}{(y - x)(y + x)}$

    $= \dfrac{6}{x - y} + \dfrac{-1}{-1} \cdot \dfrac{4x}{(y - x)(y + x)}$

    $= \dfrac{6}{x - y} + \dfrac{-4x}{(x - y)(x + y)}$

    $\quad\quad$ $[-1(y - x) = x - y;$
    $\quad\quad\quad y + x = x + y]$

    $= \dfrac{6}{x - y} \cdot \dfrac{x + y}{x + y} + \dfrac{-4x}{(x - y)(x + y)}$

    $\quad\quad\quad\quad\quad\quad$ LCM $= (x - y)(x + y)$

    $= \dfrac{6(x + y) - 4x}{(x - y)(x + y)}$

    $= \dfrac{6x + 6y - 4x}{(x - y)(x + y)}$

    $= \dfrac{2x + 6y}{(x - y)(x + y)}$

32. $\dfrac{-2a^2 - a + 10}{(a + 3)(a - 3)}$

33. $\dfrac{y + 2}{y - 7} + \dfrac{3 - y}{49 - y^2} = \dfrac{y + 2}{y - 7} + \dfrac{3 - y}{(7 + y)(7 - y)}$

    $= \dfrac{y + 2}{y - 7} + \dfrac{-1}{-1} \cdot \dfrac{3 - y}{(7 + y)(7 - y)}$

    $= \dfrac{y + 2}{y - 7} + \dfrac{-1(3 - y)}{(y + 7)(y - 7)}$

    $\quad\quad$ $[7 + y = y + 7;$
    $\quad\quad\;\; -1(7 - y) = y - 7]$

    $= \dfrac{y + 2}{y - 7} \cdot \dfrac{y + 7}{y + 7} + \dfrac{-1(3 - y)}{(y + 7)(y - 7)}$

    $= \dfrac{(y + 2)(y + 7) - (3 - y)}{(y + 7)(y - 7)}$

    $= \dfrac{y^2 + 9y + 14 - 3 + y}{(y + 7)(y - 7)}$

    $= \dfrac{y^2 + 10y + 11}{(y + 7)(y - 7)}$

34. $\dfrac{p^2 + 7p + 1}{(p + 5)(p - 5)}$

35. $\dfrac{10}{x^2 + x - 6} + \dfrac{3x}{x^2 - 4x + 4}$

    $= \dfrac{10}{(x + 3)(x - 2)} + \dfrac{3x}{(x - 2)(x - 2)}$

    $\quad\quad\quad\quad\quad$ LCM $= (x + 3)(x - 2)^2$

    $= \dfrac{10}{(x + 3)(x - 2)} \cdot \dfrac{x - 2}{x - 2} + \dfrac{3x}{(x - 2)(x - 2)} \cdot \dfrac{x + 3}{x + 3}$

    $= \dfrac{10(x - 2) + 3x(x + 3)}{(x + 3)(x - 2)^2} = \dfrac{10x - 20 + 3x^2 + 9x}{(x + 3)(x - 2)^2}$

    $= \dfrac{3x^2 + 19x - 20}{(x + 3)(x - 2)^2}$

36. $\dfrac{5z + 12}{(z - 3)(z + 2)(z + 3)}$

37. $(x + 2) - (x + 1) = x + 2 - x - 1 = 1$

38. $x^2 - 1$

39. $(4y^3 - 5y^2 + 7y - 24) - (-9y^3 + 9y^2 - 5y + 49)$
    $= 4y^3 - 5y^2 + 7y - 24 + 9y^3 - 9y^2 + 5y - 49$
    $= 13y^3 - 14y^2 + 12y - 73$

40. $25{,}704$ ft^2

41. To find the perimeter we add the lengths of the sides:

    $\dfrac{y + 4}{3} + \dfrac{y + 4}{3} + \dfrac{y - 2}{5} + \dfrac{y - 2}{5}$ LCM $= 3 \cdot 5$

    $= \dfrac{y + 4}{3} \cdot \dfrac{5}{5} + \dfrac{y + 4}{3} \cdot \dfrac{5}{5} + \dfrac{y - 2}{5} \cdot \dfrac{3}{3} + \dfrac{y - 2}{5} \cdot \dfrac{3}{3}$

    $= \dfrac{5y + 20 + 5y + 20 + 3y - 6 + 3y - 6}{3 \cdot 5}$

    $= \dfrac{16y + 28}{15}$

    To find the area we multiply the length and the width:

    $\left(\dfrac{y + 4}{3}\right)\left(\dfrac{y - 2}{5}\right) = \dfrac{(y + 4)(y - 2)}{3 \cdot 5} = \dfrac{y^2 + 2y - 8}{15}$

42. $P = \dfrac{10x - 14}{x^2 - x - 20}$, $A = \dfrac{6}{x^2 - x - 20}$

Chapter 6 (6.6)

43. $\dfrac{5}{z+2} + \dfrac{4z}{z^2-4} + 2$

$= \dfrac{5}{z+2} + \dfrac{4z}{(z+2)(z-2)} + \dfrac{2}{1}$ LCM = $(z+2)(z-2)$

$= \dfrac{5}{z+2} \cdot \dfrac{z-2}{z-2} + \dfrac{4z}{(z+2)(z-2)} + \dfrac{2}{1} \cdot \dfrac{(z+2)(z-2)}{(z+2)(z-2)}$

$= \dfrac{5z - 10 + 4z + 2(z^2 - 4)}{(z+2)(z-2)}$

$= \dfrac{5z - 10 + 4z + 2z^2 - 8}{(z+2)(z-2)} = \dfrac{2z^2 + 9z - 18}{(z+2)(z-2)}$

$= \dfrac{(2z - 3)(z + 6)}{(z+2)(z-2)}$

44. $\dfrac{-2(y^2 - 5y - 30)}{(y+3)(y-3)^2}$

45. $\dfrac{3z^2}{z^4 - 4} + \dfrac{5z^2 - 3}{2z^4 + z^2 - 6}$

$= \dfrac{3z^2}{(z^2+2)(z^2-2)} + \dfrac{5z^2-3}{(2z^2-3)(z^2+2)}$

LCM = $(z^2+2)(z^2-2)(2z^2-3)$

$= \dfrac{3z^2}{(z^2+2)(z^2-2)} \cdot \dfrac{2z^2-3}{2z^2-3} + \dfrac{5z^2-3}{(2z^2-3)(z^2+2)} \cdot \dfrac{z^2-2}{z^2-2}$

$= \dfrac{6z^4 - 9z^2 + 5z^4 - 13z^2 + 6}{(z^2+2)(z^2-2)(2z^2-3)}$

$= \dfrac{11z^4 - 22z^2 + 6}{(z^2+2)(z^2-2)(2z^2-3)}$

46. Answers may vary. $\dfrac{a}{a-b} + \dfrac{3b}{b-a}$

Exercise Set 6.6

1. $\dfrac{x-2}{6} - \dfrac{x+1}{3} = \dfrac{x-2}{6} - \dfrac{x+1}{3} \cdot \dfrac{2}{2}$ LCM = 6

$= \dfrac{x-2}{6} - \dfrac{2x+2}{6}$

$= \dfrac{x - 2 - (2x + 2)}{6}$

$= \dfrac{x - 2 - 2x - 2}{6}$

$= \dfrac{-x - 4}{6}$, or $\dfrac{-(x+4)}{6}$

2. $\dfrac{a+8}{4}$

3. $\dfrac{4z-9}{3z} - \dfrac{3z-8}{4z} = \dfrac{4z-9}{3z} \cdot \dfrac{4}{4} - \dfrac{3z-8}{4z} \cdot \dfrac{3}{3}$

LCM = $3 \cdot 4 \cdot z$, or $12z$

$= \dfrac{16z - 36}{12z} - \dfrac{9z - 24}{12z}$

$= \dfrac{16z - 36 - (9z - 24)}{12z}$

$= \dfrac{16z - 36 - 9z + 24}{12z}$

$= \dfrac{7z - 12}{12z}$

4. $\dfrac{-7x - 13}{4x}$

5. $\dfrac{4x + 2t}{3xt^2} - \dfrac{5x - 3t}{x^2 t}$ LCM = $3x^2 t^2$

$= \dfrac{4x + 2t}{3xt^2} \cdot \dfrac{x}{x} - \dfrac{5x - 3t}{x^2 t} \cdot \dfrac{3t}{3t}$

$= \dfrac{4x^2 + 2tx}{3x^2 t^2} - \dfrac{15xt - 9t^2}{3x^2 t^2}$

$= \dfrac{4x^2 + 2tx - (15xt - 9t^2)}{3x^2 t^2}$

$= \dfrac{4x^2 + 2tx - 15xt + 9t^2}{3x^2 t^2}$

$= \dfrac{4x^2 - 13xt + 9t^2}{3x^2 t^2}$

6. $\dfrac{3y^2 - 3xy - 6x^2}{2x^2 y^2}$

7. $\dfrac{5}{x+5} - \dfrac{3}{x-5}$ LCM = $(x+5)(x-5)$

$= \dfrac{5}{x+5} \cdot \dfrac{x-5}{x-5} - \dfrac{3}{x-5} \cdot \dfrac{x+5}{x+5}$

$= \dfrac{5x - 25}{(x+5)(x-5)} - \dfrac{3x + 15}{(x+5)(x-5)}$

$= \dfrac{5x - 25 - (3x + 15)}{(x+5)(x-5)}$

$= \dfrac{5x - 25 - 3x - 15}{(x+5)(x-5)}$

$= \dfrac{2x - 40}{(x+5)(x-5)}$

8. $\dfrac{-z^2 + 5z}{(z-1)(z+1)}$

9. $\dfrac{3}{2t^2 - 2t} - \dfrac{5}{2t - 2}$

$= \dfrac{3}{2t(t-1)} - \dfrac{5}{2(t-1)}$ LCM = $2t(t-1)$

$= \dfrac{3}{2t(t-1)} - \dfrac{5}{2(t-1)} \cdot \dfrac{t}{t}$

$= \dfrac{3}{2t(t-1)} - \dfrac{5t}{2t(t-1)}$

$= \dfrac{3 - 5t}{2t(t-1)}$

10. $\dfrac{14 - 3x}{(x+2)(x-2)}$

11. $\dfrac{2s}{t^2 - s^2} - \dfrac{s}{t-s}$, LCM = $(t-s)(t+s)$

$= \dfrac{2s}{(t-s)(t+s)} - \dfrac{s}{t-s} \cdot \dfrac{t+s}{t+s}$

$= \dfrac{2s}{(t-s)(t+s)} - \dfrac{st + s^2}{(t-s)(t+s)}$

$= \dfrac{2s - (st + s^2)}{(t-s)(t+s)}$

$= \dfrac{2s - st - s^2}{(t-s)(t+s)}$

12. $\dfrac{5x - 17}{(4-x)(x+3)(x-3)}$, or $\dfrac{17 - 5x}{(x-4)(x+3)(x-3)}$

13. $\dfrac{y-5}{y} - \dfrac{3y-1}{4y} = \dfrac{y-5}{y} \cdot \dfrac{4}{4} - \dfrac{3y-1}{4y}$ LCM = $4y$

$= \dfrac{4y-20}{4y} - \dfrac{3y-1}{4y}$

$= \dfrac{4y-20-(3y-1)}{4y}$

$= \dfrac{4y-20-3y+1}{4y}$

$= \dfrac{y-19}{4y}$

14. $\dfrac{3x-8}{12x}$

15. $\dfrac{a}{x+a} - \dfrac{a}{x-a}$ LCM = $(x+a)(x-a)$

$= \dfrac{a}{x+a} \cdot \dfrac{x-a}{x-a} - \dfrac{a}{x-a} \cdot \dfrac{x+a}{x+a}$

$= \dfrac{ax-a^2}{(x+a)(x-a)} - \dfrac{ax+a^2}{(x+a)(x-a)}$

$= \dfrac{ax-a^2-(ax+a^2)}{(x+a)(x-a)}$

$= \dfrac{ax-a^2-ax-a^2}{(x+a)(x-a)}$

$= \dfrac{-2a^2}{(x+a)(x-a)}$

16. $\dfrac{t^2+2ty-y^2}{(y-t)(y+t)}$

17. $\dfrac{8x}{16-x^2} - \dfrac{5}{x-4}$

$= \dfrac{8x}{(4+x)(4-x)} - \dfrac{5}{x-4}$ $4-x$ and $x-4$ are additive inverses

$= \dfrac{8x}{(4+x)(4-x)} - \dfrac{-1}{-1} \cdot \dfrac{5}{x-4}$

$= \dfrac{8x}{(4+x)(4-x)} - \dfrac{-5}{4-x}$ LCM = $(4-x)(4-x)$

$= \dfrac{8x}{(4+x)(4-x)} - \dfrac{-5}{4-x} \cdot \dfrac{4+x}{4+x}$

$= \dfrac{8x}{(4+x)(4-x)} - \dfrac{-20-5x}{(4-x)(4+x)}$

$= \dfrac{8x-(-20-5x)}{(4+x)(4-x)}$

$= \dfrac{8x+20+5x}{(4+x)(4-x)}$

$= \dfrac{13x+20}{(4+x)(4-x)}$

18. $\dfrac{9x+12}{(x+3)(x-3)}$

19. $\dfrac{t^2}{2t^2-2t} - \dfrac{1}{2t-2}$

$= \dfrac{t^2}{2t(t-1)} - \dfrac{1}{2(t-1)}$ LCM = $2t(t-1)$

$= \dfrac{t^2}{2t(t-1)} - \dfrac{1}{2(t-1)} \cdot \dfrac{t}{t}$

$= \dfrac{t^2}{2t(t-1)} - \dfrac{t}{2t(t-1)}$

$= \dfrac{t^2-t}{2t(t-1)}$

$= \dfrac{t(t-1)}{2t(t-1)}$

$= \dfrac{1}{2} \cdot \dfrac{t(t-1)}{t(t-1)}$

$= \dfrac{1}{2}$

20. $\dfrac{4-3b}{5b(b-1)}$

21. $\dfrac{x}{x^2+5x+6} - \dfrac{2}{x^2+3x+2}$

$= \dfrac{x}{(x+3)(x+2)} - \dfrac{2}{(x+2)(x+1)}$

LCM = $(x+3)(x+2)(x+1)$

$= \dfrac{x}{(x+3)(x+2)} \cdot \dfrac{x+1}{x+1} - \dfrac{2}{(x+2)(x+1)} \cdot \dfrac{x+3}{x+3}$

$= \dfrac{x^2+x}{(x+3)(x+2)(x+1)} - \dfrac{2x+6}{(x+3)(x+2)(x+1)}$

$= \dfrac{x^2+x-(2x+6)}{(x+3)(x+2)(x+1)}$

$= \dfrac{x^2+x-2x-6}{(x+3)(x+2)(x+1)}$

$= \dfrac{x^2-x-6}{(x+3)(x+2)(x+1)}$

$= \dfrac{(x-3)(x+2)}{(x+3)(x+2)(x+1)}$

$= \dfrac{x+2}{x+2} \cdot \dfrac{x-3}{(x+3)(x+1)}$

$= \dfrac{x-3}{(x+3)(x+1)}$

22. $\dfrac{x-6}{(x+6)(x+4)}$

23. $\dfrac{4y}{y^2-1} - \dfrac{2}{y} - \dfrac{2}{y+1}$

$= \dfrac{4y}{(y+1)(y-1)} - \dfrac{2}{y} - \dfrac{2}{y+1}$

LCM = $y(y+1)(y-1)$

$= \dfrac{4y}{(y+1)(y-1)} \cdot \dfrac{y}{y} - \dfrac{2}{y} \cdot \dfrac{(y+1)(y-1)}{(y+1)(y-1)} - \dfrac{2}{y+1} \cdot \dfrac{y(y-1)}{y(y-1)}$

$= \dfrac{4y^2-(2y^2-2)-(2y^2-2y)}{y(y+1)(y-1)}$

$= \dfrac{4y^2-2y^2+2-2y^2+2y}{y(y+1)(y-1)}$

$= \dfrac{2y+2}{y(y+1)(y-1)}$

$= \dfrac{2(y+1)}{y(y+1)(y-1)}$

$= \dfrac{2}{y(y-1)} \cdot \dfrac{y+1}{y+1} = \dfrac{2}{y(y-1)}$

Chapter 6   (6.6)

24. $\dfrac{2x - 3}{2 - x}$

25. $\dfrac{2z}{1 - 2z} + \dfrac{3z}{2z + 1} - \dfrac{3}{4z^2 - 1}$

    $= \dfrac{-1}{-1} \cdot \dfrac{2z}{1 - 2z} + \dfrac{3z}{2z + 1} - \dfrac{3}{4z^2 - 1}$

    $= \dfrac{-2z}{2z - 1} + \dfrac{3z}{2z + 1} - \dfrac{3}{(2z - 1)(2z + 1)}$

    LCM $= (2z - 1)(2z + 1)$

    $= \dfrac{-2z}{2z-1} \cdot \dfrac{2z+1}{2z+1} + \dfrac{3z}{2z+1} \cdot \dfrac{2z-1}{2z-1} - \dfrac{3}{(2z-1)(2z + 1)}$

    $= \dfrac{(-4z^2 - 2z) + (6z^2 - 3z) - 3}{(2z - 1)(2z + 1)}$

    $= \dfrac{2z^2 - 5z - 3}{(2z - 1)(2z + 1)}$

    $= \dfrac{(z - 3)(2z + 1)}{(2z - 1)(2z + 1)}$

    $= \dfrac{z - 3}{2z - 1} \cdot \dfrac{2z + 1}{2z + 1}$

    $= \dfrac{z - 3}{2z - 1}$

26. 0

27. $\dfrac{5}{3 - 2x} + \dfrac{3}{2x - 3} - \dfrac{x - 3}{2x^2 - x - 3}$

    $= \dfrac{-1}{-1} \cdot \dfrac{5}{3 - 2x} + \dfrac{3}{2x - 3} - \dfrac{x - 3}{2x^2 - x - 3}$

    $= \dfrac{-5}{2x - 3} + \dfrac{3}{2x - 3} - \dfrac{x - 3}{(2x - 3)(x + 1)}$

    LCM $= (2x - 3)(x + 1)$

    $= \dfrac{-5}{2x - 3} \cdot \dfrac{x + 1}{x + 1} + \dfrac{3}{2x - 3} \cdot \dfrac{x + 1}{x + 1} - \dfrac{x - 3}{(2x - 3)(x + 1)}$

    $= \dfrac{(-5x - 5) + (3x + 3) - (x - 3)}{(2x - 3)(x + 1)}$

    $= \dfrac{-5x - 5 + 3x + 3 - x + 3}{(2x - 3)(x + 1)}$

    $= \dfrac{-3x + 1}{(2x - 3)(x + 1)}$

28. $\dfrac{2}{r + s}$

29. $\dfrac{3}{2c - 1} - \dfrac{1}{c + 2} - \dfrac{5}{2c^2 + 3c - 2}$

    $= \dfrac{3}{2c - 1} - \dfrac{1}{c + 2} - \dfrac{5}{(2c - 1)(c + 2)}$

    LCM $= (2c - 1)(c + 2)$

    $= \dfrac{3}{2c - 1} \cdot \dfrac{c + 2}{c + 2} - \dfrac{1}{c + 2} \cdot \dfrac{2c - 1}{2c - 1} - \dfrac{5}{(2c - 1)(c + 2)}$

    $= \dfrac{(3c + 6) - (2c - 1) - 5}{(2c - 1)(c + 2)}$

    $= \dfrac{3c + 6 - 2c + 1 - 5}{(2c - 1)(c + 2)}$

    $= \dfrac{c + 2}{(2c - 1)(c + 2)}$

    $= \dfrac{1}{2c - 1} \cdot \dfrac{c + 2}{c + 2}$

    $= \dfrac{1}{2c - 1}$

30. $\dfrac{2y^2 + 2y - 7}{(2y + 3)(y - 1)}$

31. $\dfrac{1}{x + y} - \dfrac{1}{x - y} + \dfrac{2x}{x^2 - y^2}$

    $= \dfrac{1}{x + y} - \dfrac{1}{x - y} + \dfrac{2x}{(x + y)(x - y)}$

    LCM $= (x + y)(x - y)$

    $= \dfrac{1}{x + y} \cdot \dfrac{x - y}{x - y} - \dfrac{1}{x - y} \cdot \dfrac{x + y}{x + y} + \dfrac{2x}{(x + y)(x - y)}$

    $= \dfrac{x - y - (x + y) + 2x}{(x + y)(x - y)}$

    $= \dfrac{x - y - x - y + 2x}{(x + y)(x - y)}$

    $= \dfrac{2x - 2y}{(x + y)(x - y)}$

    $= \dfrac{2(x - y)}{(x + y)(x - y)}$

    $= \dfrac{2}{x + y} \cdot \dfrac{x - y}{x - y}$

    $= \dfrac{2}{x + y}$

32. $\dfrac{4b}{(a + b)(a - b)}$

33. $\dfrac{x^8}{x^3} = x^{8-3} = x^5$

34. $30x^{12}$

35. $(a^2 b^{-5})^{-4} = a^{2(-4)} b^{-5(-4)} = a^{-8} b^{20}$

36. $18x^3$

37. $\dfrac{1}{2xy - 6x + ay - 3a} - \dfrac{ay + xy}{(a^2 - 4x^2)(y^2 - 6y + 9)}$

    $= \dfrac{1}{(2x + a)(y - 3)} - \dfrac{ay + xy}{(a + 2x)(a - 2x)(y - 3)(y - 3)}$

    LCM $= (a + 2x)(a - 2x)(y - 3)^2$

    $= \dfrac{1}{(2x + a)(y - 3)} \cdot \dfrac{(a - 2x)(y - 3)}{(a - 2x)(y - 3)} -$

    $\qquad \dfrac{ay + xy}{(a + 2x)(a - 2x)(y - 3)(y - 3)}$

    $= \dfrac{ay - 3a - 2xy + 6x - (ay + xy)}{(a + 2x)(a - 2x)(y - 3)^2}$

    $= \dfrac{-3a - 3xy + 6x}{(a + 2x)(a - 2x)(y - 3)^2}$

38. $\dfrac{x^2 + xy - x^3 + x^2 y - xy^2 + y^3}{(x^2 + y^2)(x + y)^2(x - y)}$

39. Answers may vary.

    $\dfrac{5x^2}{x^2 + y^2} - \dfrac{2xy}{-x^2 - y^2}$

Chapter 6   (6.7)

Exercise Set 6.7

1. $\frac{3}{8} + \frac{4}{5} = \frac{x}{20}$, LCM = 40   Check:

$40\left(\frac{3}{8} + \frac{4}{5}\right) = 40 \cdot \frac{x}{20}$

$40 \cdot \frac{3}{8} + 40 \cdot \frac{4}{5} = 40 \cdot \frac{x}{20}$

$15 + 32 = 2x$

$47 = 2x$

$\frac{47}{2} = x$

$\frac{3}{8} + \frac{4}{5} = \frac{x}{20}$
$\frac{3}{8} + \frac{4}{5}$ | $\frac{\frac{47}{2}}{20}$
$\frac{15}{40} + \frac{32}{40}$ | $\frac{47}{2} \cdot \frac{1}{20}$
$\frac{47}{40}$ | $\frac{47}{40}$

This checks, so the solution is $\frac{47}{2}$.

2. $\frac{57}{5}$

3. $\frac{2}{3} - \frac{5}{6} = \frac{1}{x}$, LCM = 6x   Check:

$6x\left(\frac{2}{3} - \frac{5}{6}\right) = 6x \cdot \frac{1}{x}$

$6x \cdot \frac{2}{3} - 6x \cdot \frac{5}{6} = 6x \cdot \frac{1}{x}$

$4x - 5x = 6$

$-x = 6$

$x = -6$

$\frac{2}{3} - \frac{5}{6} = \frac{1}{x}$
$\frac{2}{3} - \frac{5}{6}$ | $\frac{1}{-6}$
$\frac{4}{6} - \frac{5}{6}$ | $-\frac{1}{6}$
$-\frac{1}{6}$ |

This checks, so the solution is -6.

4. $-\frac{40}{19}$

5. $\frac{1}{6} + \frac{1}{8} = \frac{1}{t}$, LCM = 24t

$24t\left(\frac{1}{6} + \frac{1}{8}\right) = 24t \cdot \frac{1}{t}$

$24t \cdot \frac{1}{6} + 24t \cdot \frac{1}{8} = 24t \cdot \frac{1}{t}$

$4t + 3t = 24$

$7t = 24$

$t = \frac{24}{7}$

Check:
$\frac{1}{6} + \frac{1}{8} = \frac{1}{t}$
$\frac{1}{6} + \frac{1}{8}$ | $\frac{1}{\frac{24}{7}}$
$\frac{4}{24} + \frac{3}{24}$ | $1 \cdot \frac{7}{24}$
$\frac{7}{24}$ | $\frac{7}{24}$

This checks, so the solution is $\frac{24}{7}$.

6. $\frac{40}{9}$

7. $x + \frac{4}{x} = -5$, LCM = x

$x\left(x + \frac{4}{x}\right) = x(-5)$

$x \cdot x + x \cdot \frac{4}{x} = x(-5)$

$x^2 + 4 = -5x$

$x^2 + 5x + 4 = 0$

$(x + 4)(x + 1) = 0$

$x + 4 = 0$ or $x + 1 = 0$

$x = -4$ or $x = -1$

Check:
$x + \frac{4}{x} = -5$	$x + \frac{4}{x} = -5$
$-4 + \frac{4}{-4}$ | -5	$-1 + \frac{4}{-1}$ | -5
$-4 - 1$	$-1 - 4$
-5	-5

Both of these check, so the two solutions are -4 and -1.

8. -3, -1

9. $\frac{x}{4} - \frac{4}{x} = 0$, LCM = 4x

$4x\left(\frac{x}{4} - \frac{4}{x}\right) = 4x \cdot 0$

$4x \cdot \frac{x}{4} - 4x \cdot \frac{4}{x} = 4x \cdot 0$

$x^2 - 16 = 0$

$(x + 4)(x - 4) = 0$

$x + 4 = 0$ or $x - 4 = 0$

$x = -4$ or $x = 4$

Check:
$\frac{x}{4} - \frac{4}{x} = 0$	$\frac{x}{4} - \frac{4}{x} = 0$
$\frac{-4}{4} - \frac{4}{-4}$ | 0	$\frac{4}{4} - \frac{4}{4}$ | 0
$-1 + 1$	$1 - 1$
0	0

Both of these check, so the two solutions are -4 and 4.

10. -5, 5

11. $\frac{5}{x} = \frac{6}{x} - \frac{1}{3}$, LCM = 3x   Check:

$3x \cdot \frac{5}{x} = 3x\left(\frac{6}{x} - \frac{1}{3}\right)$

$3x \cdot \frac{5}{x} = 3x \cdot \frac{6}{x} - 3x \cdot \frac{1}{3}$

$15 = 18 - x$

$-3 = -x$

$3 = x$

$\frac{5}{x} = \frac{6}{x} - \frac{1}{3}$
$\frac{5}{3}$ | $\frac{6}{3} - \frac{1}{3}$
| $\frac{5}{3}$

This checks, so the solution is 3.

12. 2

13. $\frac{5}{3x} + \frac{3}{x} = 1$, LCM = 3x   Check:

$$3x\left(\frac{5}{3x} + \frac{3}{x}\right) = 3x \cdot 1$$

$$3x \cdot \frac{5}{3x} + 3x \cdot \frac{3}{x} = 3x \cdot 1$$

$$5 + 9 = 3x$$

$$14 = 3x$$

$$\frac{14}{3} = x$$

Check:
$$\frac{5}{3x} + \frac{3}{x} = 1$$
$$\frac{5}{3 \cdot \frac{14}{3}} + \frac{3}{\frac{14}{3}} \bigg| 1$$
$$\frac{5}{14} + \frac{9}{14}$$
$$\frac{14}{14}$$
$$1$$

This checks, so the solution is $\frac{14}{3}$.

14. $\frac{23}{4}$

15. $\frac{x-7}{x+2} = \frac{1}{4}$, LCM = 4(x + 2)   Check:

$$4(x+2) \cdot \frac{x-7}{x+2} = 4(x+2) \cdot \frac{1}{4}$$

$$4(x - 7) = x + 2$$

$$4x - 28 = x + 2$$

$$3x = 30$$

$$x = 10$$

Check:
$$\frac{x-7}{x+2} = \frac{1}{4}$$
$$\frac{10-7}{10+2} \bigg| \frac{1}{4}$$
$$\frac{3}{12}$$
$$\frac{1}{4}$$

16. 5

17. $\frac{2}{x+1} = \frac{1}{x-2}$, LCM = (x+1)(x-2)

$$(x+1)(x-2) \cdot \frac{2}{x+1} = (x+1)(x-2) \cdot \frac{1}{x-2}$$

$$2(x - 2) = x + 1$$

$$2x - 4 = x + 1$$

$$x = 5$$

This checks, so the solution is 5.

18. $-\frac{13}{2}$

19. $\frac{x}{6} - \frac{x}{10} = \frac{1}{6}$, LCM = 30

$$30\left(\frac{x}{6} - \frac{x}{10}\right) = 30 \cdot \frac{1}{6}$$

$$30 \cdot \frac{x}{6} - 30 \cdot \frac{x}{10} = 30 \cdot \frac{1}{6}$$

$$5x - 3x = 5$$

$$2x = 5$$

$$x = \frac{5}{2}$$

This checks, so the solution is $\frac{5}{2}$.

20. 3

21. $\frac{x+1}{3} - \frac{x-1}{2} = 1$, LCM = 6

$$6\left[\frac{x+1}{3} - \frac{x-1}{2}\right] = 6 \cdot 1$$

$$6 \cdot \frac{x+1}{3} - 6 \cdot \frac{x-1}{2} = 6 \cdot 1$$

$$2(x + 1) - 3(x - 1) = 6$$

$$2x + 2 - 3x + 3 = 6$$

$$-x + 5 = 6$$

$$-x = 1$$

$$x = -1$$

This checks, so the solution is -1.

22. -2

23. $\frac{a-3}{3a+2} = \frac{1}{5}$, LCM = 5(3a + 2)

$$5(3a+2) \cdot \frac{a-3}{3a+2} = 5(3a+2) \cdot \frac{1}{5}$$

$$5(a - 3) = 3a + 2$$

$$5a - 15 = 3a + 2$$

$$2a = 17$$

$$a = \frac{17}{2}$$

This checks, so the solution is $\frac{17}{2}$.

24. $\frac{9}{2}$

25. $\frac{x-1}{x-5} = \frac{4}{x-5}$, LCM = x - 5

$$(x-5) \cdot \frac{x-1}{x-5} = (x-5) \cdot \frac{4}{x-5}$$

$$x - 1 = 4$$

$$x = 5$$

Check:
$$\frac{x-1}{x-5} = \frac{4}{x-5}$$
$$\frac{5-1}{5-5} \bigg| \frac{4}{5-5}$$
$$\frac{4}{0} \bigg| \frac{4}{0}$$

The number 5 is not a solution because it makes a denominator zero. Thus, there is no solution.

26. 9

27. $\frac{2}{x+3} = \frac{5}{x}$, LCM = x(x + 3)

$$x(x+3) \cdot \frac{2}{x+3} = x(x+3) \cdot \frac{5}{x}$$

$$2x = 5(x + 3)$$

$$2x = 5x + 15$$

$$-15 = 3x$$

$$-5 = x$$

This checks, so the solution is -5.

28. -16

Chapter 6   (6.7)

29.
$$\frac{x-2}{x-3} = \frac{x-1}{x+1}, \quad LCM = (x-3)(x+1)$$
$$(x-3)(x+1) \cdot \frac{x-2}{x-3} = (x-3)(x+1) \cdot \frac{x-1}{x+1}$$
$$(x+1)(x-2) = (x-3)(x-1)$$
$$x^2 - x - 2 = x^2 - 4x + 3$$
$$-x - 2 = -4x + 3$$
$$3x = 5$$
$$x = \frac{5}{3}$$

This checks, so the solution is $\frac{5}{3}$.

30. $\frac{1}{5}$

31.
$$\frac{1}{x+3} + \frac{1}{x-3} = \frac{1}{x^2-9}, \quad LCM = (x+3)(x-3)$$
$$(x+3)(x-3)\left[\frac{1}{x+3} + \frac{1}{x-3}\right] = (x+3)(x-3) \cdot \frac{1}{(x+3)(x-3)}$$
$$(x-3) + (x+3) = 1$$
$$2x = 1$$
$$x = \frac{1}{2}$$

This checks, so the solution is $\frac{1}{2}$.

32. No solution

33.
$$\frac{x}{x+4} - \frac{4}{x-4} = \frac{x^2+16}{x^2-16}, \quad LCM = (x+4)(x-4)$$
$$(x+4)(x-4)\left[\frac{x}{x+4} - \frac{4}{x-4}\right] = (x+4)(x-4) \cdot \frac{x^2+16}{(x+4)(x-4)}$$
$$x(x-4) - 4(x+4) = x^2 + 16$$
$$x^2 - 4x - 4x - 16 = x^2 + 16$$
$$-8x - 16 = 16$$
$$-8x = 32$$
$$x = -4$$

The number -4 is not a solution because it makes a denominator zero. Thus, there is no solution.

34. 2

35.
$$\frac{-3}{y-7} = \frac{-10-y}{7-y} \quad y-7 \text{ and } 7-y \text{ are additive inverses}$$
$$\frac{-3}{y-7} = \frac{-1}{-1} \cdot \frac{-10-y}{7-y}$$
$$\frac{-3}{y-7} = \frac{10+y}{y-7}, \quad LCM = y-7$$
$$(y-7)\left[\frac{-3}{y-7}\right] = (y-7)\left[\frac{10+y}{y-7}\right]$$
$$-3 = 10 + y$$
$$-13 = y$$

This checks, so the solution is -13.

36. No solution

37. We will do the multiplications and divisions in order from left to right.
$$11 \cdot 6 \div 3 \cdot 2 \div 7 = 66 \div 3 \cdot 2 \div 7$$
$$= 22 \cdot 2 \div 7$$
$$= 44 \div 7$$
$$= \frac{44}{7}$$

38. 13

39. $\frac{37 - 5(4-5)}{2 \cdot 6 + 2} = \frac{37 - 5(-1)}{2 \cdot 6 + 2} = \frac{37 + 5}{12 + 2} = \frac{42}{14} = 3$

40. $\frac{1}{3}$

41.
$$\frac{4}{y-2} - \frac{2y-3}{y^2-4} = \frac{5}{y+2}, \quad LCM = (y+2)(y-2)$$
$$(y+2)(y-2)\left[\frac{4}{y-2} - \frac{2y-3}{y^2-4}\right] = (y+2)(y-2) \cdot \frac{5}{y+2}$$
$$4(y+2) - (2y-3) = 5(y-2)$$
$$4y + 8 - 2y + 3 = 5y - 10$$
$$2y + 11 = 5y - 10$$
$$21 = 3y$$
$$7 = y$$

This checks, so the solution is 7.

42. $-\frac{1}{6}$

43.
$$\frac{y}{y+0.2} - 1.2 = \frac{y-0.2}{y+0.2}, \quad LCM = y+0.2$$
$$(y+0.2)\left[\frac{y}{y+0.2} - 1.2\right] = (y+0.2) \cdot \frac{y-0.2}{y+0.2}$$
$$y - 1.2(y+0.2) = y - 0.2$$
$$y - 1.2y - 0.24 = y - 0.2$$
$$-0.2y - 0.24 = y - 0.2$$
$$-0.04 = 1.2y$$
$$-\frac{0.04}{1.2} = y$$
$$-\frac{4}{120} = y$$
$$-\frac{1}{30} = y$$

This checks, so the solution is $-\frac{1}{30}$.

44. 0, -1

45.
$$4a - 3 = \frac{a+13}{a+1}, \quad LCM = a+1$$
$$(a+1)(4a-3) = (a+1) \cdot \frac{a+13}{a+1}$$
$$4a^2 + a - 3 = a + 13$$
$$4a^2 - 16 = 0$$
$$4(a+2)(a-2) = 0$$
$$a + 2 = 0 \quad \text{or} \quad a - 2 = 0$$
$$a = -2 \quad \text{or} \quad a = 2$$

Both of these check, so the two solutions are -2 and 2.

Chapter 6   (6.8)

46. 2

47. $\dfrac{y^2 - 4}{y + 3} = 2 - \dfrac{y - 2}{y + 3}$, LCM = $y + 3$

$(y + 3) \cdot \dfrac{y^2 - 4}{y + 3} = (y + 3)\left[2 - \dfrac{y - 2}{y + 3}\right]$

$y^2 - 4 = 2(y + 3) - (y - 2)$

$y^2 - 4 = 2y + 6 - y + 2$

$y^2 - 4 = y + 8$

$y^2 - y - 12 = 0$

$(y - 4)(y + 3) = 0$

$y - 4 = 0$ or $y + 3 = 0$

$y = 4$ or $y = -3$

The number 4 is a solution, but -3 is not because it makes a denominator zero.

48. -6

Exercise Set 6.8

1. Familiarize. Let x = the number. Then $\dfrac{1}{x}$ is the reciprocal of the number.

    Translate.

The reciprocal of 4	plus	the reciprocal of 5	is	the reciprocal of the number.
$\dfrac{1}{4}$	+	$\dfrac{1}{5}$	=	$\dfrac{1}{x}$

    Carry out. We solve the equation.

    $\dfrac{1}{4} + \dfrac{1}{5} = \dfrac{1}{x}$, LCM = $20x$

    $20x\left[\dfrac{1}{4} + \dfrac{1}{5}\right] = 20 \cdot \dfrac{1}{x}$

    $5x + 4x = 20$

    $9x = 20$

    $x = \dfrac{20}{9}$

    Check. The reciprocal of $\dfrac{20}{9}$ is $\dfrac{9}{20}$. Also, $\dfrac{1}{4} + \dfrac{1}{5} = \dfrac{5}{20} + \dfrac{4}{20} = \dfrac{9}{20}$, so the value checks.

    State. The number is $\dfrac{20}{9}$.

2. $\dfrac{24}{11}$

3. Familiarize. Let x = the smaller number. Then $x + 5$ = the larger number.

    Translate.

The larger number	divided by	the smaller number	is	$\dfrac{4}{3}$.
$(x + 5)$	÷	x	=	$\dfrac{4}{3}$

3. (continued)

    Carry out. We solve the equation.

    $\dfrac{x + 5}{x} = \dfrac{4}{3}$, LCM = $3x$

    $3x\left[\dfrac{x + 5}{x}\right] = 3x \cdot \dfrac{4}{3}$

    $3(x + 5) = 4x$

    $3x + 15 = 4x$

    $15 = x$

    Check. If the smaller number is 15, then the larger is 15 + 5, or 20. The quotient of 20 divided by 15 is $\dfrac{20}{15}$, or $\dfrac{4}{3}$. The values check.

    State. The numbers are 15 and 20.

4. $2\dfrac{2}{3}$, $6\dfrac{2}{3}$

5. Familiarize. We complete the table shown in the text.

	Distance	Speed	Time	
Slow car	150	r	t	→ 150 = $r(t)$
Fast car	350	$r + 40$	t	→ 350 = $(r + 40)t$

    $d = r \cdot t$

    Translate. We apply the formula $d = rt$ along the rows of the table to obtain two equations:

    $150 = rt$,
    $350 = (r + 40)t$

    Then we solve each equation for t and set the results equal:

    Solving $150 = rt$ for t: $t = \dfrac{150}{r}$

    Solving $350 = (r + 40)t$ for t: $t = \dfrac{350}{r + 40}$

    Thus, we have

    $\dfrac{150}{r} = \dfrac{350}{r + 40}$.

    Carry out. We multiply by the LCM, $r(r + 40)$.

    $r(r + 40) \cdot \dfrac{150}{r} = r(r + 40) \cdot \dfrac{350}{r + 40}$

    $150(r + 40) = 350r$

    $150r + 6000 = 350r$

    $6000 = 200r$

    $30 = r$

    Check. If r is 30 km/h, then $r + 40$ is 70 km/h. The time for the slow car is 150/30, or 5 hr. The time for the fast car is 350/70, or 5 hr. The times are the same. The values check.

    State. The speed of the slow car is 30 km/h, and the speed of the fast car is 70 km/h.

6. 50 km/h, 80 km/h

Chapter 6 (6.8)

7. Familiarize. We let r represent the speed going. Then 2r is the speed returning. We let t represent the time going. Then t - 3 represents the time returning. We organize the information in a table.

d = r · t

	Distance	Speed	Time
Going	120	r	t
Returning	120	2r	t - 3

Translate. The rows of the table give us two equations:
  120 = rt,
  120 = 2r(t - 3)

We can solve each equation for r and set the results equal:

Solving 120 = rt for r: $r = \frac{120}{t}$

Solving 120 = 2r(t - 3) for r: $r = \frac{120}{2(t-3)}$, or $r = \frac{60}{t-3}$

Then $\frac{120}{t} = \frac{60}{t-3}$.

Carry out. We multiply on both sides by the LCM, t(t - 3).

$t(t-3) \cdot \frac{120}{t} = t(t-3) \cdot \frac{60}{t-3}$

  120(t - 3) = 60t
  120t - 360 = 60t
  -360 = -60t
  6 = t

Then substitute 6 for t in either equation to find r, the time going:

  $r = \frac{120}{t}$
  $r = \frac{120}{6}$ Substituting 6 for t
  r = 20

Check. If r = 20 and t = 6, then 2r = 2·20, or 40 mph and t - 3 = 6 - 3, or 3 hr. The distance going is 6·20, or 120 mi. The distance returning is 40·3, or 120 mi. The numbers check.
State. The speed going is 20 mph.

8. 28 mph

9. Familiarize. We complete the table shown in the text.

d = r · t

	Distance	Speed	Time	
Freight	330	r - 14	t	→ 330 = (r-14)t
Passenger	400	r	t	→ 400 = r(t)

9. (continued)

Translate. From the rows of the table we have two equations:
  330 = (r - 14)t,
  400 = rt

We solve each equation for t and set the results equal:

Solving 330 = (r - 14)t for t: $t = \frac{330}{r-14}$

Solving 400 = rt for t: $t = \frac{400}{r}$

Thus, we have

$\frac{330}{r-14} = \frac{400}{r}$.

Carry out. We multiply by the LCM, r(r - 14).

$r(r-14) \cdot \frac{330}{r-14} = r(r-14) \cdot \frac{400}{r}$

  330r = 400(r - 14)
  330r = 400r - 5600
  -70r = -5600
  r = 80

Then substitute 80 for r in either equation to find t:

  $t = \frac{400}{r}$
  $t = \frac{400}{80}$ Substituting 80 for r
  t = 5

Check. If r = 80, then r - 14 = 66. In 5 hr the freight train travels 66·5, or 330 km, and the passenger train travels 80·5, or 400 km. The values check.
State. The speed of the passenger train is 80 km/h. The speed of the freight train is 66 km/h.

10. Passenger: 80 km/h, freight: 65 km/h

11. Familiarize. The job takes David 4 hours working alone and Sierra 5 hours working alone. Then in 1 hour David does $\frac{1}{4}$ of the job and Sierra does $\frac{1}{5}$ of the job. Working together, they can do $\frac{1}{4} + \frac{1}{5}$, or $\frac{9}{20}$ of the job in 1 hour. In two hours, David does $2\left(\frac{1}{4}\right)$ of the job and Sierra does $2\left(\frac{1}{5}\right)$ of the job. Working together they can do $2\left(\frac{1}{4}\right) + 2\left(\frac{1}{5}\right)$, or $\frac{9}{10}$ of the job in 2 hours. In 3 hours they can do $3\left(\frac{1}{4}\right) + 3\left(\frac{1}{5}\right)$, or $\frac{27}{20}$ or $1\frac{7}{20}$ of the job which is more of the job than needs to be done. The answer is somewhere between 2 hr and 3 hr.

Translate. If they work together t hours, then David does $t\left(\frac{1}{4}\right)$ of the job and Sierra does $t\left(\frac{1}{5}\right)$ of the job. We want some number t such that

  $t\left(\frac{1}{4}\right) + t\left(\frac{1}{5}\right) = 1$.

Chapter 6   (6.8)

11. (continued)

   Carry out. We solve the equation.

   $\frac{t}{4} + \frac{t}{5} = 1$, LCM = 20

   $20\left[\frac{t}{4} + \frac{t}{5}\right] = 20 \cdot 1$

   $5t + 4t = 20$

   $9t = 20$

   $t = \frac{20}{9}$, or $2\frac{2}{9}$

   Check. The check can be done by repeating the computations. We also have another check. In the familiarization step we learned the time must be between 2 hr and 3 hr. The answer, $2\frac{2}{9}$ hr, is between 2 hr and 3 hr and is less than 4 hours, the time it takes David alone.

   State. Working together, it takes them $2\frac{2}{9}$ hr to complete the job.

12. $6\frac{6}{7}$ hr

13. Familiarize. The job takes Rory 12 hours working alone and Mira 9 hours working alone. Then in 1 hour Rory does $\frac{1}{12}$ of the job and Mira does $\frac{1}{9}$ of the job. Working together they can do $\frac{1}{12} + \frac{1}{9}$, or $\frac{7}{36}$ of the job in 1 hour. In two hours, Rory does $2\left(\frac{1}{12}\right)$ of the job and Mira does $2\left(\frac{1}{9}\right)$ of the job. Working together they can do $2\left(\frac{1}{12}\right) + 2\left(\frac{1}{9}\right)$, or $\frac{14}{36}$ of the job in two hours. In 3 hours they can do $3\left(\frac{1}{12}\right) + 3\left(\frac{1}{9}\right)$, or $\frac{21}{36}$ of the job. In 4 hours, they can do $\frac{28}{36}$. In 5 hours, they can do $\frac{42}{36}$, or $1\frac{1}{6}$ which is more of the job than needs to be done. The answer is somewhere between 5 hr and 6 hr.

   Translate. If they work together t hours, then Rory does $t\left(\frac{1}{12}\right)$ of the job and Mira does $t\left(\frac{1}{9}\right)$ of the job. We want some number t such that

   $t\left(\frac{1}{12}\right) + t\left(\frac{1}{9}\right) = 1$.

   Carry out. We solve the equation

   $\frac{t}{12} + \frac{t}{9} = 1$, LCM = 36

   $36\left[\frac{t}{12} + \frac{t}{9}\right] = 36 \cdot 1$

   $3t + 4t = 36$

   $7t = 36$

   $t = \frac{36}{7}$, or $5\frac{1}{7}$

13. (continued)

   Check. The check can be done by repeating the computations. We also have another check. In the familiarization step we learned the time must be between 5 hr and 6 hr. The answer, $5\frac{1}{7}$ hr, is between 5 hr and 6 hr and is less than 9 hours, the time it takes Mira alone.

   State. Working together, it takes them $5\frac{1}{7}$ hr to complete the job.

14. $10\frac{2}{7}$ hr

15. $\frac{54 \text{ days}}{6 \text{ days}} = 9$

16. $16 \frac{\text{mi}}{\text{gal}}$

17. $\frac{4.6 \text{ km}}{2 \text{ hr}} = 2.3$ km/h

18. 186,000 mi/sec

19. Familiarize. The coffee beans from 14 trees are required to produce 7.7 kilograms of coffee, and we wish to find how many trees are required to produce 320 kilograms of coffee. We can set up ratios:

   $\frac{T}{320} \qquad \frac{14}{7.7}$

   Translate. Assuming the two ratios are the same, we can translate to a proportion.

   Trees $\longrightarrow \frac{T}{320} = \frac{14}{7.7} \longleftarrow$ Trees
   Kilograms $\longrightarrow \qquad \qquad \longleftarrow$ Kilograms

   Carry out. We solve the equation.
   We multiply by 320 to get T alone.

   $320 \cdot \frac{T}{320} = 320 \cdot \frac{14}{7.7}$

   $T = \frac{4480}{7.7}$, or $\frac{4480}{\frac{77}{10}}$

   $T = 581\frac{9}{11}$

   Check.

   $\frac{581\frac{9}{11}}{320} = \frac{\frac{6400}{11}}{320} = \frac{6400}{11} \cdot \frac{1}{320} = \frac{320 \cdot 20}{11 \cdot 320} = \frac{20}{11}$ and

   $\frac{14}{7.7} = \frac{14}{\frac{77}{10}} = \frac{14}{1} \cdot \frac{10}{77} = \frac{7 \cdot 2 \cdot 10}{7 \cdot 11} = \frac{20}{11}$.

   The ratios are the same.

   State. $581\frac{9}{11}$ trees are required to produce 320 kg of coffee.

20. 200

Chapter 6   (6.8)

21. Familiarize. A student travels 234 kilometers in 14 days, and we wish to find how far the student would travel in 42 days. We can set up ratios:

$$\frac{K}{42} \quad \frac{234}{14}$$

Translate. Assuming the rates are the same, we can translate to a proportion.

Kilometers ⟶ $\frac{K}{42} = \frac{234}{14}$ ⟵ Kilometers
Days ⟶ ⟵ Days

Carry out. We solve the equation.
We multiply by 42 to get K alone.

$$42 \cdot \frac{K}{42} = 42 \cdot \frac{234}{14}$$

$$K = \frac{9828}{14}$$

$$K = 702$$

Check.
$\frac{702}{42} \approx 16.7 \quad \frac{234}{14} \approx 16.7$

The ratios are the same.

State. The student would travel 702 kilometers in 42 days.

22. $3\frac{3}{4}$ cups

23. Familiarize. 10 cm³ of human blood contains 1.2 grams of hemoglobin, and we wish to find how many grams of hemoglobin are contained in 16 cm³ of the same blood. We can set up ratios:

$$\frac{H}{16} \quad \frac{1.2}{10}$$

Translate. Assuming the ratios are the same, we can translate to a proportion.

Grams ⟶ $\frac{H}{16} = \frac{1.2}{10}$ ⟵ Grams
cm³ ⟶ ⟵ cm³

Carry out. We solve the equation.
We multiply by 16 to get H alone.

$$16 \cdot \frac{H}{16} = 16 \cdot \frac{1.2}{10}$$

$$H = \frac{19.2}{10}$$

$$H = 1.92$$

Check.
$\frac{1.92}{16} = 0.12 \quad \frac{1.2}{10} = 0.12$

The ratios are the same.

State. Thus 16 cm³ of the same blood would contain 1.92 grams of hemoglobin.

24. 216

25. Familiarize. The ratio of trout tagged to the total number of trout in the lake, T, is $\frac{112}{T}$. Of the 82 trout caught later, there were 32 trout tagged. The ratio of trout tagged to trout caught is $\frac{32}{82}$.

25. (continued)

Translate. Assuming the two ratios are the same, we can translate to a proportion.

Trout tagged originally ⟶ $\frac{112}{T} = \frac{32}{82}$ ⟵ Tagged trout caught later
Trout in lake ⟶ ⟵ Trout caught later

Carry out. We solve the equation.
We multiply by the LCM, 82T.

$$82T \cdot \frac{112}{T} = 82T \cdot \frac{32}{82}$$

$$82 \cdot 112 = T \cdot 32$$

$$9184 = 32T$$

$$\frac{9184}{32} = T$$

$$287 = T$$

Check.
$\frac{112}{287} \approx 0.39 \quad \frac{32}{82} \approx 0.39$

The ratios are the same.

State. There are 287 trout in the lake.

26. 954

27. Familiarize. The ratio of the weight of an object on the moon to the weight of an object on the earth is 0.16 to 1.

a) We wish to find out how much a 12-ton rocket would weigh on the moon.

b) We wish to find out how much a 180-lb astronaut would weigh on the moon.

We can set up ratios.

$$\frac{0.16}{1} \quad \frac{T}{12} \quad \frac{P}{180}$$

Translate. Assuming the ratios are the same, we can translate to proportions.

a) Wgt. on moon ⟶ $\frac{0.16}{1} = \frac{T}{12}$ ⟵ Wgt. on moon
   Wgt. on earth ⟶ ⟵ Wgt. on earth

b) Wgt. on moon ⟶ $\frac{0.16}{1} = \frac{P}{180}$ ⟵ Wgt. on moon
   Wgt. on earth ⟶ ⟵ Wgt. on earth

Carry out. We solve each proportion.

a) $\frac{0.16}{1} = \frac{T}{12}$ b) $\frac{0.16}{1} = \frac{P}{180}$

$12(0.16) = T \qquad 180(0.16) = P$

$1.92 = T \qquad\quad 28.8 = P$

Check.
$\frac{0.16}{1} = 0.16 \quad \frac{1.92}{12} = 0.16 \quad \frac{28.8}{180} = 0.16$

The ratios are the same.

State.

a) A 12-ton rocket would weigh 1.92 tons on the moon.

b) A 180-lb astronaut would weigh 28.8 lb on the moon.

Chapter 6   (6.8)

28. a) 4.8 tons

    b) 48 lb

29. <u>Familiarize</u>. Let x represent the numerator. Then 104 - x represents the denominator. The ratio is $\frac{x}{104 - x}$.

    <u>Translate</u>. The ratios are equal.
    $$\frac{x}{104 - x} = \frac{9}{17}$$

    <u>Carry out</u>. We solve the proportion.
    We multiply by the LCM, 17(104 - x).
    $$17(104 - x) \cdot \frac{x}{104 - x} = 17(104 - x) \cdot \frac{9}{17}$$
    $$17x = 9(104 - x)$$
    $$17x = 936 - 9x$$
    $$26x = 936$$
    $$x = \frac{936}{26}$$
    $$x = 36$$

    <u>Check</u>. If x = 36, then 104 - x = 68. The ratio is $\frac{36}{68}$. If we multiply $\frac{9}{17}$ by $\frac{4}{4}$, a form of 1, we get $\frac{36}{68}$. The ratios are equal.

    <u>State</u>. The equal ratio is $\frac{36}{68}$.

30. 11

31. <u>Familiarize</u>. Let x represent the numerator and x + 1 represent the denominator of the original fraction. The fraction is $\frac{x}{x + 1}$. If 2 is subtracted from the numerator and the denominator, the resulting fraction is $\frac{x - 2}{x + 1 - 2}$, or $\frac{x - 2}{x - 1}$.

    <u>Translate</u>.
    The resulting fraction is $\frac{1}{2}$.
    $$\frac{x - 2}{x - 1} = \frac{1}{2}$$

    <u>Carry out</u>. We solve the equation.
    $$\frac{x - 2}{x - 1} = \frac{1}{2}, \quad LCM = 2(x - 1)$$
    $$2(x - 1) \cdot \frac{x - 2}{x - 1} = 2(x - 1) \cdot \frac{1}{2}$$
    $$2(x - 2) = x - 1$$
    $$2x - 4 = x - 1$$
    $$x = 3$$

    <u>Check</u>. If x = 3, then x + 1 = 4 and the original fraction is $\frac{3}{4}$. If 2 is subtracted from both numerator and denominator, the resulting fraction is $\frac{3 - 2}{4 - 2}$, or $\frac{1}{2}$. The value checks.

    <u>State</u>. The original fraction was $\frac{3}{4}$.

32. 6 hr, 12 hr

33. <u>Familiarize</u>. We organize the information in a table. Let r = the speed of the current and t = the time it takes to travel upstream.

    d = r · t

	Distance	Speed	Time
Upstream	24	10 - r	t
Downstream	24	10 + r	5 - t

    <u>Translate</u>. From the rows of the table we get two equations:
    $$24 = (10 - r)t$$
    $$24 = (10 + r)(5 - t)$$

    We solve each equation for t and set the results equal:

    Solving 24 = (10 - r)t for t: $t = \frac{24}{10 - r}$

    Solving 24 = (10 + r)(5 - t) for t:
    $$\frac{24}{10 + r} = 5 - t$$
    $$t = 5 - \frac{24}{10 + r}$$

    Then $\frac{24}{10 - r} = 5 - \frac{24}{10 + r}$.

    <u>Carry out</u>. We first multiply on both sides of the equation by the LCM, (10 - r)(10 + r):
    $$(10-r)(10+r) \cdot \frac{24}{10 - r} = (10-r)(10+r)\left[5 - \frac{24}{10 + r}\right]$$
    $$24(10 + r) = 5(10-r)(10+r) - 24(10-r)$$
    $$240 + 24r = 500 - 5r^2 - 240 + 24r$$
    $$240 + 24r = 260 - 5r^2 + 24r$$
    $$5r^2 - 20 = 0$$
    $$5(r^2 - 4) = 0$$
    $$5(r + 2)(r - 2) = 0$$
    $$r + 2 = 0 \quad \text{or} \quad r - 2 = 0$$
    $$r = -2 \quad \text{or} \quad r = 2$$

    <u>Check</u>. We only check 2 since the speed of the current cannot be negative. If r = 2, then the speed upstream is 10 - 2, or 8 mph and the time is $\frac{24}{8}$, or 3 hours. If r = 2, then the speed downstream is 10 + 2, or 12 mph and the time is $\frac{24}{12}$, or 2 hours. The sum of 3 hr and 2 hr is 5 hr. This checks.

    <u>State</u>. The speed of the current is 2 mph.

34. 100 = 75 + 25

Chapter 6   (6.8)

35. 1) Start with the given proportion:
$$\frac{A}{B} = \frac{C}{D}$$
$$BD \cdot \frac{A}{B} = BD \cdot \frac{C}{D}$$
$$AD = BC$$

The product of the means is equal to the product of the extremes.

2) Find a second true proportion:
$$\frac{A}{B} = \frac{C}{D}$$
$$\frac{D}{A} \cdot \frac{A}{B} = \frac{D}{A} \cdot \frac{C}{D}$$
$$\frac{D}{B} = \frac{C}{A}$$

Product of the means: BC
Product of the extremes: AD

3) Find a third true proportion:
$$\frac{A}{B} = \frac{C}{D}$$
$$\frac{B}{C} \cdot \frac{A}{B} = \frac{B}{C} \cdot \frac{C}{D}$$
$$\frac{A}{C} = \frac{B}{D}$$

Product of the means: BC
Product of the extremes: AD

4) Find a fourth true proportion:
$$\frac{A}{B} = \frac{C}{D}$$
$$\frac{DB}{AC} \cdot \frac{A}{B} = \frac{DB}{AC} \cdot \frac{C}{D}$$
$$\frac{D}{C} = \frac{B}{A}$$

Product of the means: BC
Product of the extremes: AD

In each case, the product of the means is equal to the product of the extremes.

36. $\frac{A + B}{B} = \frac{C + D}{D}$

$\frac{A}{B} + \frac{B}{B} = \frac{C}{D} + \frac{D}{D}$

$\frac{A}{B} + 1 = \frac{C}{D} + 1$

$\frac{A}{B} = \frac{C}{D}$

37. <u>Familiarize</u>. The job takes Rosina 8 days working alone and Ng 10 days working alone. Let x represent the number of days it would take Oscar working alone. Then in 1 day Rosina does $\frac{1}{8}$ of the job, Ng does $\frac{1}{10}$ of the job, and Oscar does $\frac{1}{x}$ of the job. In 1 day they would complete $\frac{1}{8} + \frac{1}{10} + \frac{1}{x}$ of the job, and in 3 days they would complete $3\left(\frac{1}{8} + \frac{1}{10} + \frac{1}{x}\right)$, or $\frac{3}{8} + \frac{3}{10} + \frac{3}{x}$.

37. (continued)

<u>Translate</u>. The amount done in 3 days is one entire job, so we have
$$\frac{3}{8} + \frac{3}{10} + \frac{3}{x} = 1.$$

<u>Carry out</u>. We solve the equation.
$$\frac{3}{8} + \frac{3}{10} + \frac{3}{x} = 1, \quad LCM = 40x$$
$$40x\left(\frac{3}{8} + \frac{3}{10} + \frac{3}{x}\right) = 40x \cdot 1$$
$$40x \cdot \frac{3}{8} + 40x \cdot \frac{3}{10} + 40x \cdot \frac{3}{x} = 40x$$
$$15x + 12x + 120 = 40x$$
$$120 = 13x$$
$$\frac{120}{13} = x$$

<u>Check</u>. If it takes Oscar $\frac{120}{13}$, or $9\frac{3}{13}$ days, to complete the job, then in one day Oscar does $\frac{1}{\frac{120}{13}}$, or $\frac{13}{120}$, of the job, and in 3 days he does $3\left(\frac{13}{120}\right)$, or $\frac{13}{40}$, of the job. The portion of the job done by Rosina, Ng, and Oscar in 3 days is $\frac{3}{8} + \frac{3}{10} + \frac{13}{40} = \frac{15}{40} + \frac{12}{40} + \frac{13}{40} = \frac{40}{40} = 1$ entire job. The answer checks.

<u>State</u>. It will take Oscar $9\frac{3}{13}$ days to complete the job working alone.

38. $27\frac{3}{11}$ minutes or $\frac{5}{11}$ hr after 5:00

39. <u>Familiarize</u>. We organize the information in a table. Let r = the speed on the first part of the trip and t = the time driven at that speed.

$$d = r \cdot t$$

	Distance	Speed	Time
First part	30	r	t
Second part	20	r + 15	1 − t

<u>Translate</u>. From the rows of the table we obtain two equations:
$$30 = rt$$
$$20 = (r + 15)(1 - t)$$

We solve each equation for t and set the results equal:

Solving $30 = rt$ for t: $t = \frac{30}{r}$

Solving $20 = (r + 15)(1 - t)$ for t:
$$\frac{20}{r + 15} = 1 - t$$
$$t = 1 - \frac{20}{r + 15}$$

Then $\frac{30}{r} = 1 - \frac{20}{r + 15}.$

Chapter 6   (6.9)

39. (continued)

<u>Carry out</u>. We first multiply the equation by the LCM, $r(r + 15)$:

$$r(r + 15) \cdot \frac{30}{r} = r(r + 15)\left[1 - \frac{20}{r + 15}\right]$$

$$30(r + 15) = r(r + 15) - 20r$$

$$30r + 450 = r^2 + 15r - 20r$$

$$0 = r^2 - 35r - 450$$

$$0 = (r - 45)(r + 10)$$

$r - 45 = 0$ or $r + 10 = 0$

$r = 45$ or $r = -10$

<u>Check</u>. Since the speed cannot be negative, we only check 45. If $r = 45$, then the time for the first part is $\frac{30}{45}$, or $\frac{2}{3}$ hr. If $r = 45$, then $r + 15 = 60$ and the time for the second part is $\frac{20}{60}$, or $\frac{1}{3}$ hr. The total time is $\frac{2}{3} + \frac{1}{3}$, or 1 hour. The value checks.

<u>State</u>. The speed for the first 30 miles was 45 mph.

40. Michelle: 6 hr, Sal: 3 hr, Kristen: 4 hr

Exercise Set 6.9

1. $S = 2\pi rh$

$\frac{S}{2\pi h} = r$ Multiplying by $\frac{1}{2\pi h}$

2. $t = \frac{A - P}{Pr}$

3. $A = \frac{1}{2}bh$

$2A = bh$ Multiplying by 2

$\frac{2A}{h} = b$ Multiplying by $\frac{1}{h}$

4. $g = \frac{2s}{t^2}$

5. $S = 180(n - 2)$

$S = 180n - 360$ Removing parentheses

$S + 360 = 180n$ Adding 360

$\frac{S + 360}{180} = n$ Multiplying by $\frac{1}{180}$

6. $a = \frac{2S - n\ell}{n}$

7. $V = \frac{1}{3}k(B + b + 4M)$

$3V = k(B + b + 4M)$ Multiplying by 3

$3V = kB + kb + 4kM$ Removing parentheses

$3V - kB - 4kM = kb$ Adding $-kb - 4kM$

$\frac{3V - kB - 4kM}{k} = b$ Multiplying by $\frac{1}{k}$

8. $P = \frac{A}{1 + rt}$

9. $S(r - 1) = r\ell - a$

$Sr - S = r\ell - a$ Removing parentheses

$Sr - r\ell = S - a$ Adding $S - r\ell$

$r(S - \ell) = S - a$ Factoring out r

$r = \frac{S - a}{S - \ell}$ Multiplying by $\frac{1}{S - \ell}$

10. $m = \frac{T}{g - f}$

11. $A = \frac{1}{2}h(b_1 + b_2)$

$2A = h(b_1 + b_2)$ Multiplying by 2

$\frac{2A}{b_1 + b_2} = h$ Multiplying by $\frac{1}{b_1 + b_2}$

12. $h = \frac{S}{2\pi r} - r$, or $\frac{S - 2\pi r^2}{2\pi r}$

13. $r = \frac{v^2 pL}{a}$

$ar = v^2 pL$ Multiplying by a

$a = \frac{v^2 pL}{r}$ Multiplying by $\frac{1}{r}$

14. $M = \frac{Lt + g}{t}$

15. $A = \frac{1}{2}h(b_1 + b_2)$

$2A = h(b_1 + b_2)$ Multiplying by 2

$2A = hb_1 + hb_2$ Removing parentheses

$2A - hb_2 = hb_1$ Adding $-hb_2$

$\frac{2A - hb_2}{h} = b_1$ Multiplying by $\frac{1}{h}$

16. $n = \frac{\ell - a + d}{d}$

17. $A = \frac{\pi r^2 E}{180}$

$180A = \pi r^2 E$ Multiplying by 180

$\frac{180A}{\pi r^2} = E$ Multiplying by $\frac{1}{\pi r^2}$

18. $W = \frac{RL + x}{L}$

19. $V = \frac{1}{3}h(B + c + 4M)$

$3V = h(B + c + 4M)$ Multiplying by 3

$3V = hB + hc + 4hM$ Removing parentheses

$3V - hB - hc = 4hM$ Adding $-hB - hc$

$\frac{3V - hB - hc}{4h} = M$ Multiplying by $\frac{1}{4h}$

20. $R = \frac{V}{I^2}$

Chapter 6   (6.9)

21. $y = \dfrac{v^2 pL}{a}$

    $ay = v^2 pL$ Multiplying by a

    $\dfrac{ay}{v^2 p} = L$ Multiplying by $\dfrac{1}{v^2 p}$

22. $b = \dfrac{3V}{h}$

23. $r = \dfrac{v^2 pL}{a}$

    $ar = v^2 pL$ Multiplying by a

    $\dfrac{ar}{v^2 L} = p$ Multiplying by $\dfrac{1}{v^2 L}$

24. $\ell = \dfrac{P - 2w}{2}$

25. $\dfrac{a}{c} = n + bn$

    $\dfrac{a}{c} = n(1 + b)$ Factoring out n

    $\dfrac{a}{c(1 + b)} = n$ Multiplying by $\dfrac{1}{1 + b}$

26. $a = \dfrac{Q}{M(b - c)}$

27. $S = \dfrac{a + 2b}{3b}$

    $3bS = a + 2b$ Multiplying by $3b$

    $3bS - 2b = a$ Adding $-2b$

    $b(3S - 2) = a$ Factoring out b

    $b = \dfrac{a}{3S - 2}$ Multiplying by $\dfrac{1}{3S - 2}$

28. $a = \dfrac{b}{K - C}$

29. $C = \dfrac{5}{9}(F - 32)$

    $9C = 5(F - 32)$ Multiplying by 9

    $9C = 5F - 160$ Removing parentheses

    $9C + 160 = 5F$ Adding 160

    $\dfrac{9C + 160}{5} = F$ Multiplying by $\dfrac{1}{5}$

30. $\pi = \dfrac{3V}{4r^3}$

31. $f = \dfrac{gm - t}{m}$

    $mf = gm - t$ Multiplying by m

    $mf + t = gm$ Adding t

    $\dfrac{mf + t}{m} = g$ Multiplying by $\dfrac{1}{m}$

32. $a = r\ell - S(r - \ell)$

33. $\dfrac{1}{p} + \dfrac{1}{q} = \dfrac{1}{f}$,   LCM $= pqf$

    $pqf\left(\dfrac{1}{p} + \dfrac{1}{q}\right) = pqf \cdot \dfrac{1}{f}$ Multiplying by pqf

    $qf + pf = pq$ Simplifying

    $qf = pq - pf$ Adding $-pf$

    $qf = p(q - f)$ Factoring out p

    $\dfrac{qf}{q - f} = p$ Multiplying by $\dfrac{1}{q - f}$

34. $b = \dfrac{at}{a - t}$

35. $\dfrac{A}{P} = 1 + r$

    $A = P(1 + r)$ Multiplying by P

36. $h = \dfrac{2A}{a + b}$

37. $\dfrac{1}{R} = \dfrac{1}{r_1} + \dfrac{1}{r_2}$,   LCM $= Rr_1 r_2$

    $Rr_1 r_2 \cdot \dfrac{1}{R} = Rr_1 r_2 \left(\dfrac{1}{r_1} + \dfrac{1}{r_2}\right)$ Multiplying by $Rr_1 r_2$

    $r_1 r_2 = Rr_2 + Rr_1$ Simplifying

    $r_1 r_2 = R(r_2 + r_1)$ Factoring out R

    $\dfrac{r_1 r_2}{r_2 + r_1} = R$ Multiplying by $\dfrac{1}{r_2 + r_1}$

38. $r_1 = \dfrac{Rr_2}{r_2 - R}$

39. $\dfrac{A}{B} = \dfrac{C}{D}$,   LCM $= BD$

    $BD \cdot \dfrac{A}{B} = BD \cdot \dfrac{C}{D}$ Multiplying by BD

    $DA = BC$ Simplifying

    $D = \dfrac{BC}{A}$ Multiplying by $\dfrac{1}{A}$

40. $C = \dfrac{AD}{B}$

41. $h_1 = q\left(1 + \dfrac{h_2}{p}\right)$

    $h_1 = q + \dfrac{qh_2}{p}$ Removing parentheses

    $h_1 - q = \dfrac{qh_2}{p}$ Adding $-q$

    $p(h_1 - q) = qh_2$ Multiplying by p

    $\dfrac{p(h_1 - q)}{q} = h_2$ Multiplying by $\dfrac{1}{q}$

42. $a = \dfrac{S(1 - r)}{1 - r^n}$

Chapter 6   (6.10)

43. $\dfrac{A - B}{AB} = \dfrac{1}{C}$,   LCM = ABC

$ABC \cdot \left[\dfrac{A - B}{AB}\right] = ABC \cdot \dfrac{1}{C}$ Multiplying by ABC

$C(A - B) = AB$ Simplifying

$AC - BC = AB$ Removing parentheses

$AC = AB + BC$ Adding $-BC$

$AC = B(A + C)$ Factoring out B

$\dfrac{AC}{A + C} = B$ Multiplying by $\dfrac{1}{A + C}$

44. $A = \dfrac{BC}{C - B}$

45. The exponent is negative, so the number is small. We move the decimal point 5 places to the left.
$3.47 \times 10^{-5} = 0.0000347$

46. 1.209×10^{-4}

47. $100m^2 - 81 = (10m)^2 - 9^2$ Difference of squares
$= (10m + 9)(10m - 9)$

48. $(8t + 5)^2$

49. $V = \dfrac{4}{3}\pi r^3$

When r is doubled,

$V = \dfrac{4}{3}\pi(2r)^3$

$V = \dfrac{4}{3}\pi(8r^3)$

$V = 8\left[\dfrac{4}{3}\pi r^3\right]$

V is multiplied by 8.

50. $-40°$

51. $\dfrac{3B}{A}$ is undefined when $A = 0$.

$\dfrac{4C - 1}{A + 3}$ is undefined when $A + 3 = 0$, or when $A = -3$.

Thus, $\dfrac{3B}{A} = \dfrac{4C - 1}{A + 3}$ is undefined when A equals 0 or -3.

52. N decreases

Exercise Set 6.10

1. $\dfrac{1 + \dfrac{9}{16}}{1 - \dfrac{3}{4}}$ LCM of the denominators is 16

$= \dfrac{1 + \dfrac{9}{16}}{1 - \dfrac{3}{4}} \cdot \dfrac{16}{16}$

$= \dfrac{\left(1 + \dfrac{9}{16}\right)16}{\left(1 - \dfrac{3}{4}\right)16}$

$= \dfrac{1(16) + \dfrac{9}{16}(16)}{1(16) - \dfrac{3}{4}(16)} = \dfrac{16 + 9}{16 - 12} = \dfrac{25}{4}$

2. $\dfrac{5}{2}$

3. $\dfrac{1 - \dfrac{3}{5}}{1 + \dfrac{1}{5}} = \dfrac{1 \cdot \dfrac{5}{5} - \dfrac{3}{5}}{1 \cdot \dfrac{5}{5} + \dfrac{1}{5}} = \dfrac{\dfrac{5 - 3}{5}}{\dfrac{5 + 1}{5}}$

$= \dfrac{\dfrac{2}{5}}{\dfrac{6}{5}}$

$= \dfrac{2}{5} \cdot \dfrac{5}{6} = \dfrac{2}{6} = \dfrac{1}{3}$

4. $-\dfrac{65}{18}$

5. $\dfrac{\dfrac{1}{x} + 3}{\dfrac{1}{x} - 5}$ LCM of the denominators is x

$= \dfrac{\dfrac{1}{x} + 3}{\dfrac{1}{x} - 5} \cdot \dfrac{x}{x}$

$= \dfrac{\left(\dfrac{1}{x} + 3\right)x}{\left(\dfrac{1}{x} - 5\right)x}$

$= \dfrac{\dfrac{1}{x} \cdot x + 3 \cdot x}{\dfrac{1}{x} \cdot x - 5 \cdot x} = \dfrac{1 + 3x}{1 - 5x}$

6. $\dfrac{9 + 3s^2}{4s^2}$

7. $\dfrac{\dfrac{1}{2} + \dfrac{3}{4}}{\dfrac{5}{8} - \dfrac{5}{6}} = \dfrac{\dfrac{1}{2} \cdot \dfrac{2}{2} + \dfrac{3}{4}}{\dfrac{5}{8} \cdot \dfrac{3}{3} - \dfrac{5}{6} \cdot \dfrac{4}{4}} = \dfrac{\dfrac{2 + 3}{4}}{\dfrac{15 - 20}{24}}$

$= \dfrac{\dfrac{5}{4}}{\dfrac{-5}{24}}$

$= \dfrac{5}{4} \cdot \dfrac{24}{-5}$

$= -6$

8. $-\dfrac{4}{39}$

Chapter 6   (6.10)

9. $\dfrac{\dfrac{2}{y} + \dfrac{1}{2y}}{y + \dfrac{y}{2}}$ LCM of the denominators is 2y

$= \dfrac{\dfrac{2}{y} + \dfrac{1}{2y}}{y + \dfrac{y}{2}} \cdot \dfrac{2y}{2y}$

$= \dfrac{\left(\dfrac{2}{y} + \dfrac{1}{2y}\right)2y}{\left(y + \dfrac{y}{2}\right)2y}$

$= \dfrac{\dfrac{2}{y}(2y) + \dfrac{1}{2y}(2y)}{y(2y) + \dfrac{y}{2}(2y)} = \dfrac{4 + 1}{2y^2 + y^2} = \dfrac{5}{3y^2}$

10. $\dfrac{2x + 1}{x}$

11. $\dfrac{8 + \dfrac{8}{d}}{1 + \dfrac{1}{d}} = \dfrac{8 \cdot \dfrac{d}{d} + \dfrac{8}{d}}{1 \cdot \dfrac{d}{d} + \dfrac{1}{d}} = \dfrac{\dfrac{8d + 8}{d}}{\dfrac{d + 1}{d}}$

$= \dfrac{8d + 8}{d} \cdot \dfrac{d}{d + 1}$

$= \dfrac{8(d + 1)}{d} \cdot \dfrac{d}{d + 1}$

$= \dfrac{d(d + 1)}{d(d + 1)} \cdot \dfrac{8}{1}$

$= 8$

12. $\dfrac{3(2b - 3)}{b(6 - b)}$

13. $\dfrac{\dfrac{x}{8} - \dfrac{8}{x}}{\dfrac{1}{8} + \dfrac{1}{x}}$ LCM of the denominators is 8x

$= \dfrac{\dfrac{x}{8} - \dfrac{8}{x}}{\dfrac{1}{8} + \dfrac{1}{x}} \cdot \dfrac{8x}{8x}$

$= \dfrac{\left(\dfrac{x}{8} - \dfrac{8}{x}\right)8x}{\left(\dfrac{1}{8} + \dfrac{1}{x}\right)8x}$

$= \dfrac{\dfrac{x}{8}(8x) - \dfrac{8}{x}(8x)}{\dfrac{1}{8}(8x) + \dfrac{1}{x}(8x)}$

$= \dfrac{x^2 - 64}{x + 8} = \dfrac{(x + 8)(x - 8)}{x + 8} = x - 8$

14. $\dfrac{4 + m^2}{m^2 - 4}$

15. $\dfrac{1 + \dfrac{1}{y}}{1 - \dfrac{1}{y^2}} = \dfrac{1 \cdot \dfrac{y}{y} + \dfrac{1}{y}}{1 \cdot \dfrac{y^2}{y^2} - \dfrac{1}{y^2}} = \dfrac{\dfrac{y + 1}{y}}{\dfrac{y^2 - 1}{y^2}}$

$= \dfrac{y + 1}{y} \cdot \dfrac{y^2}{y^2 - 1}$

$= \dfrac{(y + 1)y \cdot y}{y(y + 1)(y - 1)} = \dfrac{y}{y - 1}$

16. $\dfrac{1 - q}{q}$

17. $\dfrac{\dfrac{1}{5} - \dfrac{1}{a}}{\dfrac{5 - a}{5}}$ LCM of the denominators is 5a

$= \dfrac{\dfrac{1}{5} - \dfrac{1}{a}}{\dfrac{5 - a}{5}} \cdot \dfrac{5a}{5a}$

$= \dfrac{\left(\dfrac{1}{5} - \dfrac{1}{a}\right)5a}{\left(\dfrac{5 - a}{5}\right)5a}$

$= \dfrac{\dfrac{1}{5}(5a) - \dfrac{1}{a}(5a)}{a(5 - a)} = \dfrac{a - 5}{5a - a^2}$

$= \dfrac{a - 5}{-a(-5 + a)} = \dfrac{a - 5}{-a(a - 5)} = -\dfrac{1}{a}$

18. $\dfrac{2x - 1}{2}$

19. $\dfrac{\dfrac{x}{x - y}}{\dfrac{x^2}{x^2 - y^2}} = \dfrac{x}{x - y} \cdot \dfrac{x^2 - y^2}{x^2}$ Multiplying by the reciprocal of the divisor

$= \dfrac{x}{x - y} \cdot \dfrac{(x - y)(x + y)}{x \cdot x}$

$= \dfrac{x(x - y)}{x(x - y)} \cdot \dfrac{x + y}{x}$

$= \dfrac{x + y}{x}$

20. $x - y$

21. $\dfrac{x - 3 + \dfrac{2}{x}}{x - 4 + \dfrac{3}{x}} = \dfrac{x \cdot \dfrac{x}{x} - 3 \cdot \dfrac{x}{x} + \dfrac{2}{x}}{x \cdot \dfrac{x}{x} - 4 \cdot \dfrac{x}{x} + \dfrac{3}{x}}$

$= \dfrac{\dfrac{x^2 - 3x + 2}{x}}{\dfrac{x^2 - 4x + 3}{x}}$

$= \dfrac{x^2 - 3x + 2}{x} \cdot \dfrac{x}{x^2 - 4x + 3}$

$= \dfrac{(x - 2)(x - 1)}{x} \cdot \dfrac{x}{(x - 3)(x - 1)}$

$= \dfrac{x(x - 1)}{x(x - 1)} \cdot \dfrac{x - 2}{x - 3}$

$= \dfrac{x - 2}{x - 3}$

22. $\dfrac{a + b}{a - b}$

155

Chapter 6   (6.10)

23. $(5x^4 - 6x^3 + 23x^2 - 79x + 24) - (-18x^4 - 56x^3 + 84x - 17) = 5x^4 - 6x^3 + 23x^2 - 79x + 24 + 18x^4 + 56x^3 - 84x + 17 = 23x^4 + 50x^3 + 23x^2 - 163x + 41$

24. 14 yd

25. $\dfrac{1}{\dfrac{2}{x-1} - \dfrac{1}{3x-2}} = \dfrac{1}{\dfrac{2}{x-1} - \dfrac{1}{3x-2}} \cdot \dfrac{(x-1)(3x-2)}{(x-1)(3x-2)}$

$= \dfrac{(x-1)(3x-2)}{\left[\dfrac{2}{x-1} - \dfrac{1}{3x-2}\right](x-1)(3x-2)}$

$= \dfrac{(x-1)(3x-2)}{\dfrac{2}{x-1}(x-1)(3x-2) - \dfrac{1}{3x-2}(x-1)(3x-2)}$

$= \dfrac{(x-1)(3x-2)}{2(3x-2) - (x-1)}$

$= \dfrac{(x-1)(3x-2)}{6x - 4 - x + 1}$

$= \dfrac{(x-1)(3x-2)}{5x - 3}$

26. $\dfrac{ac}{bd}$

27. $\dfrac{\dfrac{a}{b} - \dfrac{c}{d}}{\dfrac{b}{a} - \dfrac{d}{c}} = \dfrac{\dfrac{a}{b}\cdot\dfrac{d}{d} - \dfrac{c}{d}\cdot\dfrac{b}{b}}{\dfrac{b}{a}\cdot\dfrac{c}{c} - \dfrac{d}{c}\cdot\dfrac{a}{a}} = \dfrac{\dfrac{ad - bc}{bd}}{\dfrac{bc - ad}{ac}}$

$= \dfrac{ad - bc}{bd} \cdot \dfrac{ac}{bc - ad}$

$= \dfrac{-(bc - ad)}{bd} \cdot \dfrac{ac}{bc - ad}$

$= \dfrac{bc - ad}{bc - ad} \cdot \dfrac{-ac}{bd}$

$= -\dfrac{ac}{bd}$

28. x^5

29. $1 + \dfrac{1}{1 + \dfrac{1}{1 + \dfrac{1}{1 + \dfrac{1}{x}}}} = 1 + \dfrac{1}{1 + \dfrac{1}{1 + \dfrac{1}{\dfrac{x+1}{x}}}}$

$= 1 + \dfrac{1}{1 + \dfrac{1}{1 + \dfrac{x}{x+1}}}$

$= 1 + \dfrac{1}{1 + \dfrac{1}{\dfrac{x+1+x}{x+1}}}$

$= 1 + \dfrac{1}{1 + \dfrac{x+1}{2x+1}}$

$= 1 + \dfrac{1}{\dfrac{2x+1+x+1}{2x+1}}$

$= 1 + \dfrac{2x+1}{3x+2}$

$= \dfrac{3x+2+2x+1}{3x+2}$

$= \dfrac{5x+3}{3x+2}$

30. $\dfrac{-2z(5z-2)}{(2+z)(-13z+6)}$

CHAPTER 7   GRAPHS, LINEAR EQUATIONS, AND SLOPE

Exercise Set 7.1

1. The longest bar represents boredom. Thus this was the reason given most often.

2. Work/Military service

3. We go to the right end of the bar representing grades and then go down to the percent scale. We find that approximately 5% dropped out because of grades.

4. Week 6

5. We locate 2100 on the DJIA scale and then move to the right until we reach the line. At that point, we move down to the Week scale and see that the DJIA closing was about 2100 in week 2.

6. About 250

7. We find the portion of the graph labeled medical care and read that 12% of the income was spent on medical expenses.

8. $220

9. From the top of the bar representing New York we go the left to the Dollars scale and read that the average daily expenses are $270.

10. $210

11. The shortest bar represents San Francisco, so it is the least expensive city.

12. New York

13. Familiarize. Let y = the amount by which the average daily expenses in Washington exceed those in San Francisco. We will use the graph to determine the expenses in each city, and then we will find their difference.

    Translate. We reword the problem.

    Amount of excess   is   Washington amount   minus   San Francisco amount.
       y   = $240   - $190

    Carry out. We do the computation.
    y = $240 - $190 = $50

    Check. We go over the computation. The answer checks.

    State. The average daily expenses in Washington are $50 more than in San Francisco.

14. $30

15. The highest point on the graph is above 1993 on the Year scale, so the estimated sales are greatest in 1993.

16. 1991

17. We go up from 1991 on the Year scale to a point on the graph and then go to the left and read the values on the Estimated sales scale. In 1991 estimated sales are $17 million.

18. $18.5 million

19. Familiarize. Let x = the amount by which the estimated sales in 1993 exceed those in 1995. We will use the graph to find the estimated sales in each year, and then we will find their difference. We keep in mind that the amounts given are in millions.

    Translate.

    Amount of excess   is   1993 amount   minus   1995 amount.
       x   = $19.5   - $18.1

    Carry out. We do the computation.
    x = $19.5 - $18.1 = $1.4

    Check. We go over the computation. The answer checks.

    State. Estimated sales in 1993 are $1.4 million greater than in 1995.

20. $1.5 million

21. We find the portion of the graph labeled Jazz and read that 3.7% of all recordings sold are jazz.

22. 9.0%

23. Familiarize. Let p = the percent of all recordings sold that are either soul or pop/rock. We will use the graph to find the percent for each type of music and then find their sum.

    Translate. We reword the problem.

    Percent together   is   soul percent   plus   pop/rock percent.
       p   =   12.0%   +   58.1%

    Carry out. We do the computation.
    p = 12.0% + 58.1% = 70.1%

    Check. We go over the computation. The answer checks.

    State. Together, 70.1% of all recordings sold are either soul or pop/rock.

24. 10.5%

Chapter 7   (7.1)

25. Familiarize. The graph tells us that 58.1% of all recordings sold are pop/rock, 12.0% are soul, and 9.0% are country. Let p, s, and c represent the number of pop/rock, soul, and country recordings sold, respectively.

    Translate. We reword the problem and write an equation for each type of music.

Number of pop/rock recordings sold	is	58.1%	of	total number sold.
p	=	58.1%	·	3000

Number of soul recordings sold	is	12.0%	of	total number sold.
s	=	12.0%	·	3000

Number of country recordings sold	is	9.0%	of	total number sold.
c	=	9.0%	·	3000

    Carry out. We do the three computations.
    p = 0.581·3000 = 1743
    s = 0.12·3000 = 360
    c = 0.09·3000 = 270

    Check. We go over the computations. The answers check.

    State. The store sells 1743 pop/rock, 360 soul, and 270 country recordings.

26. Pop/rock: 1452, clasical: 170, gospel: 30

27. (2,5) is 2 units right and 5 units up.
    (-1,3) is 1 unit left and 3 units up.
    (3,-2) is 3 units right and 2 units down.
    (-2,-4) is 2 units left and 4 units down.
    (0,4) is 0 units left or right and 4 units up.
    (0,-5) is 0 units left or right and 5 units down.
    (5,0) is 5 units right and 0 units up or down.
    (-5,0) is 5 units left and 0 units up or down.

28.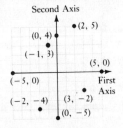

29. Since the first coordinate is negative and the second coordinate positive, the point (-5,3) is located in the second quadrant.

30. II

31. Since the first coordinate is positive and the second coordinate negative, the point (100,-1) is in the fourth quadrant.

32. IV

33. Since both coordinates are negative, the point (-6,-29) is in the third quadrant.

34. III

35. Since both coordinates are positive, the point (3.8,9.2) is the first quadrant.

36. I

37. In quadrant III, first coordinates are always negative and second coordinates are always negative.

38. Second, first

39.

    Point A is 3 units right and 3 units up. The coordinates of A are (3,3).

    Point B is 0 units left or right and 4 units down. The coordinates of B are (0,-4).

    Point C is 5 units left and 0 units up or down. The coordinates of C are (-5,0).

    Point D is 1 unit left and 1 unit down. The coordinates of D are (-1,-1).

    Point E is 2 units right and 0 units up or down. The coordinates of E are (2,0).

40. A: (4,1), B: (0,-5), c: (-4,0), D: (-3,-2), E: (3,0)

41. Familiarize. Let y = the radius of Jupiter.

    Translate.

Radius of Jupiter	is	11 times	radius of earth.
y	=	11 ·	4030

    Carry out. We do the computation.
    y = 11·4030 = 44,330

    Check. We go over the computation. The answer checks.

    State. The radius of Jupiter is about 44,330 mi.

42. 20 in.

Chapter 7   (7.2)

43.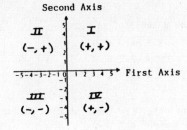

If the first coordinate is positive, then the point must be in either I or IV.

44. III or IV

45. If the first and second coordinates are equal, they must either be both positive or both negative. The point must be in either I (both positive) or III (both negative).

46. II or IV

47.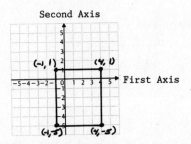

The coordinates of the fourth vertex are (-1,-5).

48. (5,2), (-7,2), or (3,-8)

49. Answers may vary.

We select eight points such that the sum of the coordinates for each point is 6.

(-1,7)	-1 + 7 = 6
(0,6)	0 + 6 = 6
(1,5)	1 + 5 = 6
(2,4)	2 + 4 = 6
(3,3)	3 + 3 = 6
(4,2)	4 + 2 = 6
(5,1)	5 + 1 = 6
(6,0)	6 + 0 = 6

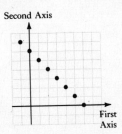

50. Answers may vary

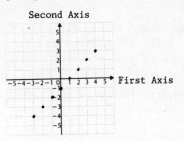

51.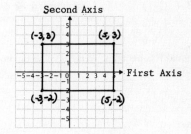

The length is 8, and the width is 5.
P = 2ℓ + 2w
P = 2·8 + 2·5 = 16 + 10 = 26

52.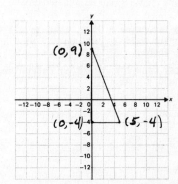

The base is 5 and the height is 13.
$A = \frac{1}{2}bh = \frac{1}{2} \cdot 5 \cdot 13 = \frac{65}{2}$, or $32\frac{1}{2}$

53. Latitude 32.5° north,
Longitude 64.5° west

54. Latitude 27° north,
Longitude 76° west

Exercise Set 7.2

1. y = 3x - 1
   5 | 3·2 - 1   Substituting 2 for x and 5 for y
     | 6 - 1   (alphabetical order of variables)
     | 5

The equation becomes true; (2,5) is a solution.

159

Chapter 7   (7.2)

2. Yes

3.
$$\begin{array}{c|c} 3x - y = 4 \\ \hline 3\cdot 2 - (-3) & 4 \\ 6 + 3 & \\ 9 & \end{array}$$ Substituting 2 for x and -3 for y

The equation becomes false; (2,-3) is not a solution.

4. No

5.
$$\begin{array}{c|c} 2c + 2d = -7 \\ \hline 2(-2) + 2(-1) & -7 \\ -4 - 2 & \\ -6 & \end{array}$$ Substituting -2 for c and -1 for d

The equation becomes false; (-2,-1) is not a solution.

6. No

7. $y = 4x$

We first make a table of values. We choose <u>any</u> number for x and then determine y by substitution.

When $x = 0$, $y = 4\cdot 0 = 0$.
When $x = -1$, $y = 4(-1) = -4$.
When $x = 1$, $y = 4\cdot 1 = 4$.

x	y
0	0
-1	-4
1	4

Since two points determine a line, that is all we really need to graph a line, but you may plot a third point as a check.

Plot these points, draw the line they determine, and label the graph $y = 4x$.

8.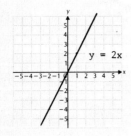

9. $y = -2x$

We first make a table of values.

When $x = 0$, $y = -2\cdot 0 = 0$.
When $x = 2$, $y = -2\cdot 2 = -4$.
When $x = -1$, $y = -2(-1) = 2$.

x	y
0	0
2	-4
-1	2

Plot these points, draw the line they determine, and label the graph $y = -2x$.

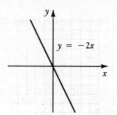

10.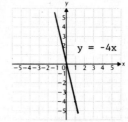

11. $y = \frac{1}{3}x$

We first make a table of values. Using multiples of 3 avoids fractions.

When $x = 0$, $y = \frac{1}{3} \cdot 0 = 0$.
When $x = 6$, $y = \frac{1}{3} \cdot 6 = 2$.
When $x = -3$, $y = \frac{1}{3}(-3) = -1$.

x	y
0	0
6	2
-3	-1

Plot these points, draw the line they determine, and label the graph $y = \frac{1}{3}x$.

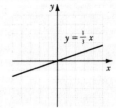

12.

13. $y = -\frac{3}{2}x$

We first make a table of values. Using multiples of 2 avoids fractions.

When $x = 0$, $y = -\frac{3}{2} \cdot 0 = 0$.
When $x = 2$, $y = -\frac{3}{2} \cdot 2 = -3$.
When $x = -2$, $y = -\frac{3}{2}(-2) = 3$.

x	y
0	0
2	-3
-2	3

Plot these points, draw the line they determine, and label the graph $y = -\frac{3}{2}x$.

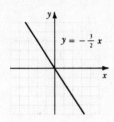

14.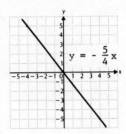

15. $y = x + 1$

We first make a table of values. We choose <u>any</u> number for x and then determine y by substitution.

When $x = 0$, $y = 0 + 1 = 1$.
When $x = 3$, $y = 3 + 1 = 4$.
When $x = -5$, $y = -5 + 1 = -4$.

x	y
0	1
3	4
-5	-4

Plot these points, draw the line they determine, and label the graph $y = x + 1$.

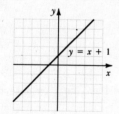

16.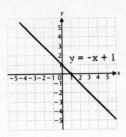

17. $y = 2x + 2$

We first make a table of values.

When $x = 0$, $y = 2 \cdot 0 + 2 = 0 + 2 = 2$.
When $x = -3$, $y = 2(-3) + 2 = -6 + 2 = -4$.
When $x = 1$, $y = 2 \cdot 1 + 2 = 2 + 2 = 4$.

x	y
0	2
-3	-4
1	4

Plot these points, draw the line they determine, and label the graph $y = 2x + 2$.

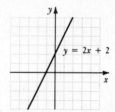

18.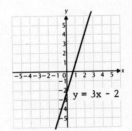

19. $y = \frac{1}{3}x - 1$

We first make a table of values. Using multiples of 3 avoids fractions.

When $x = 0$, $y = \frac{1}{3} \cdot 0 - 1 = 0 - 1 = -1$.
When $x = -6$, $y = \frac{1}{3}(-6) - 1 = -2 - 1 = -3$.
When $x = 3$, $y = \frac{1}{3} \cdot 3 - 1 = 1 - 1 = 0$.

x	y
0	-1
-6	-3
3	0

Plot these points, draw the line they determine, and label the graph $y = \frac{1}{3}x - 1$.

19. (continued)

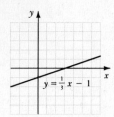

20.

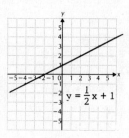

21. $y = -x - 3$

We first make a table of values.
When $x = 0$, $y = -0 - 3 = -3$.
When $x = 1$, $y = -1 - 3 = -4$.
When $x = -5$, $y = -(-5) - 3 = 5 - 3 = 2$.

x	y
0	-3
1	-4
-5	2

Plot these points, draw the line they determine, and label the graph.

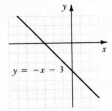

22.

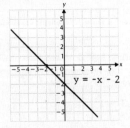

23. $y = \frac{5}{2}x + 3$

We first make a table of values. Using multiples of 2 avoids fractions.
When $x = 0$, $y = \frac{5}{2} \cdot 0 + 3 = 0 + 3 = 3$.
When $x = -2$, $y = \frac{5}{2}(-2) + 3 = -5 + 3 = -2$.
When $x = -4$, $y = \frac{5}{2}(-4) + 3 = -10 + 3 = -7$.

x	y
0	3
-2	-2
-4	-7

Plot these points, draw the line they determine, and label the graph.

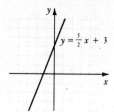

24.

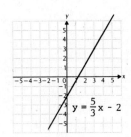

25. $y = -\frac{5}{2}x - 2$

We first make a table of values. Using multiples of 2 avoids fractions.
When $x = 0$, $y = -\frac{5}{2} \cdot 0 - 2 = 0 - 2 = -2$.
When $x = -2$, $y = -\frac{5}{2}(-2) - 2 = 5 - 2 = 3$.
When $x = 2$, $y = -\frac{5}{2}(2) - 2 = -5 - 2 = -7$.

x	y
0	-2
-2	3
2	-7

Plot these points, draw the line they determine, and label the graph.

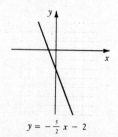

Chapter 7   (7.2)

26.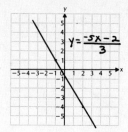

27. $y = x$

We first make a table of values.

When $x = -4$, $y = -4$.
When $x = -1$, $y = -1$.
When $x = 3$, $y = 3$.

x	y
-4	-4
-1	-1
3	3

Plot these points, draw the line they determine, and label the graph.

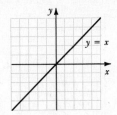

28.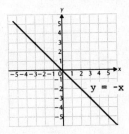

29. $y = 3 - 2x$

We first make a table of values.

When $x = 1$, $y = 3 - 2 \cdot 1 = 3 - 2 = 1$.
When $x = -1$, $y = 3 - 2(-1) = 3 + 2 = 5$.
When $x = 3$, $y = 3 - 2 \cdot 3 = 3 - 6 = -3$.

x	y
1	1
-1	5
3	-3

Plot these points, draw the line they determine, and label the graph.

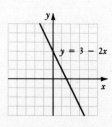

30.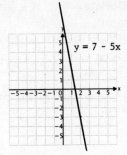

31. $y = \frac{4}{3} - \frac{1}{3}x$

We first make a table of values.

When $x = 1$, $y = \frac{4}{3} - \frac{1}{3} \cdot 1 = \frac{4}{3} - \frac{1}{3} = \frac{3}{3} = 1$.
When $x = -2$, $y = \frac{4}{3} - \frac{1}{3}(-2) = \frac{4}{3} + \frac{2}{3} = \frac{6}{3} = 2$.
When $x = 4$, $y = \frac{4}{3} - \frac{1}{3}(4) = \frac{4}{3} - \frac{4}{3} = 0$.
When $x = 0$, $y = \frac{4}{3} - \frac{1}{3} \cdot 0 = \frac{4}{3} - 0 = \frac{4}{3}$.

x	y
1	1
-2	2
4	0
0	$\frac{4}{3}$

Plot these points, draw the line they determine, and label the graph.

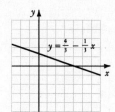

32.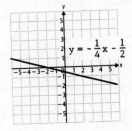

33. a) We substitute and calculate.
   1 hr: d = 55·1 = 55 mi
   2 hr: d = 55·2 = 110 mi
   5 hr: d = 55·5 = 275 mi
   10 hr: d = 55·10 = 550 mi

   b) We plot the points found in part a) and draw the line they determine.

   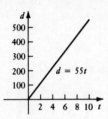

34. a) 1 in., 2.5 in., 4 in., 6 in.

   b)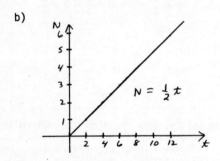

35. a) We substitute and calculate.
   Size 4: y = 4 - 2 = 2
   Size 5: y = 5 - 2 = 3
   Size 6: y = 6 - 2 = 4
   Size 7: y = 7 - 2 = 5
   Size 8: y = 8 - 2 = 6

   b) We plot the points found in part a) and draw the line they determine.

   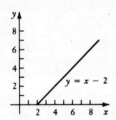

36. a) $0.46, $2.89, $4.24

   b)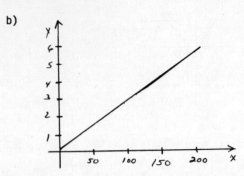

37. Familiarize. We first make a drawing. Let x represent the length of the post. Then $\frac{1}{2}x$ and $\frac{1}{3}x$ represent the length of the post in the mud and the length of the post in the water, respectively.

   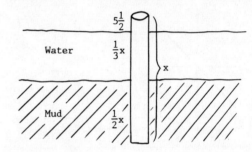

   Translate.

   $$\underbrace{\text{Length in mud}}_{\frac{1}{2}x} + \underbrace{\text{Length in water}}_{\frac{1}{3}x} + \underbrace{\text{Length above the water}}_{5\frac{1}{2}} = \underbrace{\text{Entire length of post}}_{x}$$

   Carry out.

   $$\frac{1}{2}x + \frac{1}{3}x + \frac{11}{2} = x, \quad \text{LCM} = 6$$

   $$6\left(\frac{1}{2}x + \frac{1}{3}x + \frac{11}{2}\right) = 6 \cdot x$$

   $$6 \cdot \frac{1}{2}x + 6 \cdot \frac{1}{3}x + 6 \cdot \frac{11}{2} = 6x$$

   $$3x + 2x + 33 = 6x$$

   $$5x + 33 = 6x$$

   $$33 = x$$

   Check. If the post is 33 ft long, then $\frac{1}{2} \cdot 33$, or $16\frac{1}{2}$ ft are in the mud, $\frac{1}{3} \cdot 33$, or 11 ft, are in the water. The sum of the three lengths is $16\frac{1}{2} + 11 + 5\frac{1}{2}$, or 33 ft. The numbers check.

   State. The post is 33 ft long.

38. 4224

Chapter 7   (7.2)

39. $25t^2 - 49 = 0$
    $(5t + 7)(5t - 7) = 0$
    $5t + 7 = 0$ or $5t - 7 = 0$
        $5t = -7$ or $5t = 7$
         $t = -\frac{7}{5}$ or $t = \frac{7}{5}$

    The solutions are $-\frac{7}{5}$ and $\frac{7}{5}$.

40.   0, 4

41. $y = x^2 + 1$
    When $x = 0$, $y = 0^2 + 1 = 0 + 1 = 1$.
    When $x = -1$, $y = (-1)^2 + 1 = 1 + 1 = 2$.
    When $x = 1$, $y = 1^2 + 1 = 1 + 1 = 2$.
    When $x = -2$, $y = (-2)^2 + 1 = 4 + 1 = 5$.
    When $x = 2$, $y = 2^2 + 1 = 4 + 1 = 5$.
    When $x = -3$, $y = (-3)^2 + 1 = 9 + 1 = 10$.
    When $x = 3$, $y = 3^2 + 1 = 9 + 1 = 10$.

x	0	-1	1	-2	2	-3	3
y	1	2	2	5	5	10	10

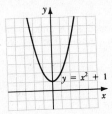

42.   (0,6), (1,5), (2,4), (3,3), (4,2), (5,1), (6,0)

43. $x + 3y = 15$
    Since $3y$ and 15 are both multiples of 3, x must also be a multiple of 3.
    When $x = 0$, $0 + 3y = 15$
                  $3y = 15$
                   $y = 5$
    When $x = 3$, $3 + 3y = 15$
                  $3y = 12$
                   $y = 4$
    When $x = 6$, $6 + 3y = 15$
                  $3y = 9$
                   $y = 3$
    When $x = 9$, $9 + 3y = 15$
                  $3y = 6$
                   $y = 2$
    When $x = 12$, $12 + 3y = 15$
                  $3y = 3$
                   $y = 1$

43.   (continued)
    When $x = 15$, $15 + 3y = 15$
                  $3y = 0$
                   $y = 0$

    The whole number solutions are (0,5), (3,4), (6,3), (9,2), (12,1), (15,0).

44. $5n + 10d = 195$, ordered pairs of the form (d,n), (10,19), (0,39), (15,9). Answers may vary.

45.   The value of n nickels is 0.05n.
    The value of q quarters is 0.25q.
    Thus, $0.05n + 0.25q = 2.35$
       or $5n + 25q = 235$ Clearing decimals

    When $n = 42$, $5 \cdot 42 + 25q = 235$
                  $210 + 25q = 235$
                   $25q = 25$
                   $q = 1$

    When $n = 7$, $5 \cdot 7 + 25q = 235$
                 $35 + 25q = 235$
                 $25q = 200$
                 $q = 8$

    When $n = 27$, $5 \cdot 27 + 25q = 235$
                  $135 + 25q = 235$
                   $25q = 100$
                   $q = 4$

    Solutions are ordered pairs of the form (n,q). Three are (42,1), (7,8), and (27,4). Answers may vary.

46.   Answers may vary. (-3,3), (2,2), (0,0)

Chapter 7 (7.3)

Exercise Set 7.3

1. $x + 3y = 6$

   To find the x-intercept, let $y = 0$.
   $$x + 3y = 6$$
   $$x + 3 \cdot 0 = 6$$
   $$x = 6$$
   Thus, $(6,0)$ is the x-intercept.

   To find the y-intercept, let $x = 0$.
   $$x + 3y = 6$$
   $$0 + 3y = 6$$
   $$3y = 6$$
   $$y = 2$$
   Thus, $(0,2)$ is the y-intercept.

   Plot these points and draw the line.

   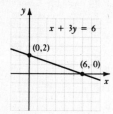

   A third point should be used as a check. We substitute any value for x and solve for y.
   We let $x = 3$. Then
   $$x + 3y = 6$$
   $$3 + 3y = 6$$
   $$3y = 3$$
   $$y = 1$$
   The point $(3,1)$ is on the graph, so the graph is probably correct.

2.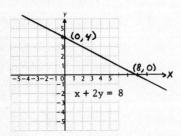

3. $-x + 2y = 4$

   To find the x-intercept, let $y = 0$.
   $$-x + 2y = 4$$
   $$-x + 2 \cdot 0 = 4$$
   $$-x = 4$$
   $$x = -4$$
   Thus, $(-4,0)$ is the x-intercept.

   To find the y-intercept, let $x = 0$.
   $$-x + 2y = 4$$
   $$-0 + 2y = 4$$
   $$2y = 4$$
   $$y = 2$$
   Thus, $(0,2)$ is the y-intercept.

   Plot these points and draw the line.

   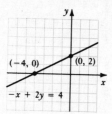

   A third point should be used as a check. We substitute any value for x and solve for y.
   We let $x = 4$. Then
   $$-x + 2y = 4$$
   $$-4 + 2y = 4$$
   $$2y = 8$$
   $$y = 4$$
   The point $(4,4)$ is on the graph, so the graph is probably correct.

4.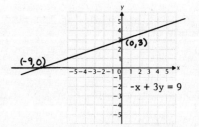

5. $3x + y = 9$

To find the x-intercept, let $y = 0$.

$3x + y = 9$
$3x + 0 = 9$
$3x = 9$
$x = 3$

Thus, $(3,0)$ is the x-intercept.

To find the y-intercept, let $x = 0$.

$3x + y = 9$
$3 \cdot 0 + y = 9$
$y = 9$

Thus, $(0,9)$ is the y-intercept.

Plot these points and draw the line.

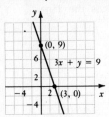

A third point should be used as a check. We substitute any value for x and solve for y.

We let $x = 2$. Then

$3x + y = 9$
$3 \cdot 2 + y = 9$
$6 + y = 9$
$y = 3$

The point $(2,3)$ is on the graph, so the graph is probably correct.

6.

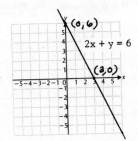

7. $2y - 2 = 6x$

To find the x-intercept, let $y = 0$.

$2y - 2 = 6x$
$2 \cdot 0 - 2 = 6x$
$-2 = 6x$
$-\frac{1}{3} = x$

Thus, $\left(-\frac{1}{3}, 0\right)$ is the x-intercept.

To find the y-intercept, let $x = 0$.

$2y - 2 = 6x$
$2y - 2 = 6 \cdot 0$
$2y - 2 = 0$
$2y = 2$
$y = 1$

Thus, $(0,1)$ is the y-intercept.

It is helpful to plot another point since the intercepts are so close together. This point can also serve as a check.

We let $x = 1$. Then

$2y - 2 = 6x$
$2y - 2 = 6 \cdot 1$
$2y - 2 = 6$
$2y = 8$
$y = 4$

Plot the point $(1,4)$ and the intercepts and draw the line.

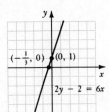

8.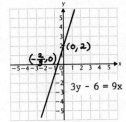

9. $3x - 9 = 3y$

To find the x-intercept, let $y = 0$.
$$3x - 9 = 3y$$
$$3x - 9 = 3 \cdot 0$$
$$3x - 9 = 0$$
$$3x = 9$$
$$x = 3$$

Thus, $(3,0)$ is the x-intercept.

To find the y-intercept, let $x = 0$.
$$3x - 9 = 3y$$
$$3 \cdot 0 - 9 = 3y$$
$$-9 = 3y$$
$$-3 = y$$

Thus, $(0,-3)$ is the y-intercept.

Plot these points and draw the line.

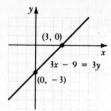

A third point should be used as a check. We substitute any value for x and solve for y.

We let $x = 1$. Then
$$3x - 9 = 3y$$
$$3 \cdot 1 - 9 = 3y$$
$$3 - 9 = 3y$$
$$-6 = 3y$$
$$-2 = y$$

The point $(1,-2)$ is on the graph, so the graph is probably correct.

10.

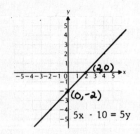

11. $2x - 3y = 6$

To find the x-intercept, let $y = 0$.
$$2x - 3y = 6$$
$$2x - 3 \cdot 0 = 6$$
$$2x = 6$$
$$x = 3$$

Thus, $(3,0)$ is the x-intercept.

To find the y-intercept, let $x = 0$.
$$2x - 3y = 6$$
$$2 \cdot 0 - 3y = 6$$
$$-3y = 6$$
$$y = -2$$

Thus, $(0,-2)$ is the y-intercept.

Plot these points and draw the line.

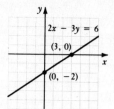

A third point should be used as a check. We substitute any value for x and solve for y.

We let $x = -3$.
$$2x - 3y = 6$$
$$2(-3) - 3y = 6$$
$$-6 - 3y = 6$$
$$-3y = 12$$
$$y = -4$$

The point $(-3,-4)$ is on the graph, so the graph is probably correct.

12.

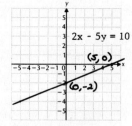

13. $4x + 5y = 20$

To find the x-intercept, let $y = 0$.

$4x + 5y = 20$
$4x + 5 \cdot 0 = 20$
$4x = 20$
$x = 5$

Thus, $(5,0)$ is the x-intercept.

To find the y-intercept, let $x = 0$.

$4x + 5y = 20$
$4 \cdot 0 + 5y = 20$
$5y = 20$
$y = 4$

Thus, $(0,4)$ is the y-intercept.

Plot these points and draw the graph.

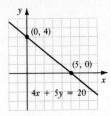

A third point should be used as a check. We substitute any value for x and solve for y.

We let $x = 4$. Then
$4x + 5y = 20$
$4 \cdot 4 + 5y = 20$
$16 + 5y = 20$
$5y = 4$
$y = \frac{4}{5}$

The point $\left(4, \frac{4}{5}\right)$ is on the graph, so the graph is probably correct.

14.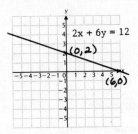

15. $2x + 3y = 8$

To find the x-intercept, let $y = 0$.

$2x + 3y = 8$
$2x + 3 \cdot 0 = 8$
$2x = 8$
$x = 4$

Thus, $(4,0)$ is the x-intercept.

To find the y-intercept, let $x = 0$.

$2x + 3y = 8$
$2 \cdot 0 + 3y = 8$
$3y = 8$
$y = \frac{8}{3}$

Thus, $\left(0, \frac{8}{3}\right)$ is the y-intercept.

Plot these points and draw the graph.

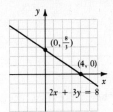

A third point should be used as a check.

We let $x = 1$. Then
$2x + 3y = 8$
$2 \cdot 1 + 3y = 8$
$2 + 3y = 8$
$3y = 6$
$y = 2$

The point $(1,2)$ is on the graph, so the graph is probably correct.

16.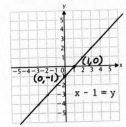

17. $x - 3 = y$

To find the x-intercept, let $y = 0$.
$$x - 3 = y$$
$$x - 3 = 0$$
$$x = 3$$
Thus, $(3,0)$ is the x-intercept.

To find the y-intercept, let $x = 0$.
$$x - 3 = y$$
$$0 - 3 = y$$
$$-3 = y$$
Thus, $(0,-3)$ is the y-intercept.

Plot these points and draw the line.

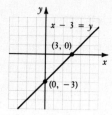

A third point should be used as a check.

We let $x = -2$. Then
$$x - 3 = y$$
$$-2 - 3 = y$$
$$-5 = y$$
The point $(-2,-5)$ is on the graph, so the graph is probably correct.

18.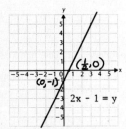

19. $3x - 2 = y$

To find the x-intercept, let $y = 0$.
$$3x - 2 = y$$
$$3x - 2 = 0$$
$$3x = 2$$
$$x = \frac{2}{3}$$
Thus, $\left(\frac{2}{3}, 0\right)$ is the x-intercept.

To find the y-intercept, let $x = 0$.
$$3x - 2 = y$$
$$3 \cdot 0 - 2 = y$$
$$-2 = y$$
Thus, $(0,-2)$ is the y-intercept.

19. (continued)

Plot these points and draw the line.

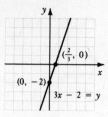

A third point should be used as a check.

We let $x = 2$. Then
$$3x - 2 = y$$
$$3 \cdot 2 - 2 = y$$
$$6 - 2 = y$$
$$4 = y$$
The point $(2,4)$ is on the graph, so the graph is probably correct.

20.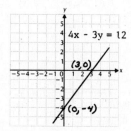

21. $6x - 2y = 18$

To find the x-intercept, let $y = 0$.
$$6x - 2y = 18$$
$$6x - 2 \cdot 0 = 18$$
$$6x = 18$$
$$x = 3$$
Thus, $(3,0)$ is the x-intercept.

To find the y-intercept, let $x = 0$.
$$6x - 2y = 18$$
$$6 \cdot 0 - 2y = 18$$
$$-2y = 18$$
$$y = -9$$
Thus, $(0,-9)$ is the y-intercept.

Plot these points and draw the line.

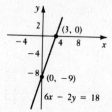

We use a third point as a check.

Chapter 7   (7.3)

21. (continued)

We let x = 1. Then

$6x - 2y = 18$

$6 \cdot 1 - 2y = 18$

$6 - 2y = 18$

$-2y = 12$

$y = -6$

The point (1,-6) is on the graph, so the graph is probably correct.

22.

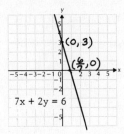

23. $3x + 4y = 5$

To find the x-intercept, let y = 0.

$3x + 4y = 5$

$3x + 4 \cdot 0 = 5$

$3x = 5$

$x = \frac{5}{3}$

Thus, $\left(\frac{5}{3}, 0\right)$ is the x-intercept.

To find the y-intercept, let x = 0.

$3x + 4y = 5$

$3 \cdot 0 + 4y = 5$

$4y = 5$

$y = \frac{5}{4}$

Thus, $\left(0, \frac{5}{4}\right)$ is the y-intercept.

It is helpful to plot another point since the intercepts are so close together. This point can also serve as a check.

We let x = 3. Then

$3x + 4y = 5$

$3 \cdot 3 + 4y = 5$

$9 + 4y = 5$

$4y = -4$

$y = -1$

Plot the point (3,-1) and the intercepts and draw the line.

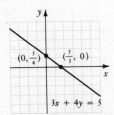

24.

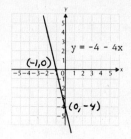

25. $y = -3 - 3x$

To find the x-intercept, let y = 0.

$y = -3 - 3x$

$0 = -3 - 3x$

$3x = -3$

$x = -1$

Thus, (-1,0) is the x-intercept.

To find the y-intercept, let x = 0.

$y = -3 - 3x$

$y = -3 - 3 \cdot 0$

$y = -3$

Thus, (0,-3) is the y-intercept.

Plot these points and draw the graph.

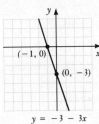

We use a third point as a check.

We let x = -2. Then

$y = -3 - 3x$

$y = -3 - 3 \cdot (-2)$

$y = -3 + 6$

$y = 3$

The point (-2,3) is on the graph, so the graph is probably correct.

26.

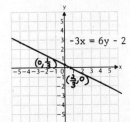

171

27. $-4x = 8y - 5$

To find the x-intercept, let $y = 0$.
$$-4x = 8 \cdot 0 - 5$$
$$-4x = -5$$
$$x = \frac{5}{4}$$

Thus, $\left(\frac{5}{4}, 0\right)$ is the x-intercept.

To find the y-intercept, let $x = 0$.
$$-4x = 8y - 5$$
$$-4 \cdot 0 = 8y - 5$$
$$0 = 8y - 5$$
$$5 = 8y$$
$$\frac{5}{8} = y$$

Thus, $\left(0, \frac{5}{8}\right)$ is the y-intercept.

It is helpful to plot another point since the intercepts are so close together. This point can also serve as a check.

We let $x = -5$. Then
$$-4x = 8y - 5$$
$$-4(-5) = 8y - 5$$
$$20 = 8y - 5$$
$$25 = 8y$$
$$\frac{25}{8} = y$$

Plot the point $\left(-5, \frac{25}{8}\right)$ and the intercepts and draw the line.

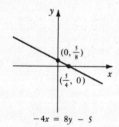

28.

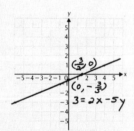

29. $y - 3x = 0$

To find the x-intercept, let $y = 0$.
$$0 - 3x = 0$$
$$-3x = 0$$
$$x = 0$$

Thus, $(0,0)$ is the x-intercept. Note that this is also the y-intercept.

In order to graph the line, we will find a second point.

When $x = 1$, $y - 3 \cdot 1 = 0$
$$y - 3 = 0$$
$$y = 3$$

Plot these points and draw the graph.

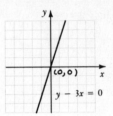

We use a third point as a check.

We let $x = -1$. Then
$$y - 3(-1) = 0$$
$$y + 3 = 0$$
$$y = -3$$

The point $(-1,-3)$ is on the graph, so the graph is probably correct.

30.

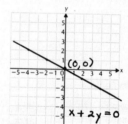

31. $x = -2$

Any ordered pair $(-2, y)$ is a solution. The variable x must be -2, but the y variable can be any number we choose. A few solutions are listed below. Plot these points and draw the line.

x	y
-2	-2
-2	0
-2	4

32.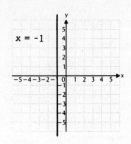

33. y = 2

Any ordered pair (x,2) is a solution. The variable y must be 2, but the x variable can be any number we choose. A few solutions are listed below. Plot these points and draw the line.

x	y
-3	2
0	2
2	2

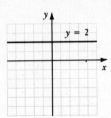

34.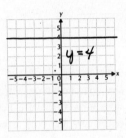

35. x = 2

Any ordered pair (2,y) is a solution. The variable x must be 2, but the y variable can be any number we choose. A few solutions are listed below. Plot these points and draw the line.

x	y
2	-1
2	4
2	5

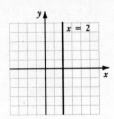

36.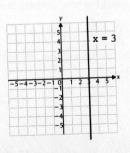

37. y = 0

Any ordered pair (x,0) is a solution. The variable y must be 0, but the x variable can be any number we choose. A few solutions are listed below. Plot these points and draw the line.

x	y
-5	0
-1	0
3	0

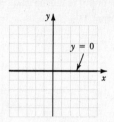

38.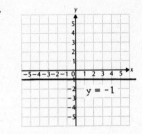

39. $x = \frac{3}{2}$

Any ordered pair $\left(\frac{3}{2}, y\right)$ is a solution. The variable x must be $\frac{3}{2}$, but the y variable can be any number we choose. A few solutions are listed below. Plot these points and draw the line.

x	y
$\frac{3}{2}$	-2
$\frac{3}{2}$	0
$\frac{3}{2}$	4

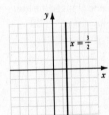

40.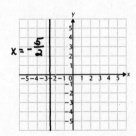

Chapter 7   (7.3)

41. $3y = -5$

$y = -\dfrac{5}{3}$ Solving for y

Any ordered pair $\left(x, -\dfrac{5}{3}\right)$ is a solution. A few solutions are listed below. Plot these points and draw the line.

x	y
-3	$-\frac{5}{3}$
0	$-\frac{5}{3}$
2	$-\frac{5}{3}$

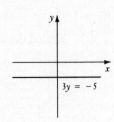

42.

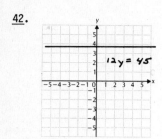

43. $4x + 3 = 0$

$4x = -3$

$x = -\dfrac{3}{4}$ Solving for x

Any ordered pair $\left(-\dfrac{3}{4}, y\right)$ is a solution. A few solutions are listed below. Plot these points and draw the line.

x	y
$-\frac{3}{4}$	-2
$-\frac{3}{4}$	0
$-\frac{3}{4}$	3

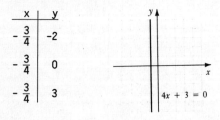

44.

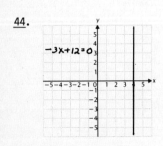

45. $18 - 3y = 0$

$-3y = -18$

$y = 6$ Solving for y

Any ordered pair $(x, 6)$ is a solution. A few solutions are listed below. Plot these points and draw the line.

x	y
-4	6
0	6
2	6

 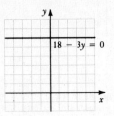

46.

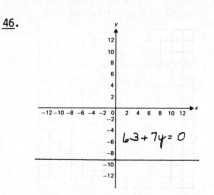

47. <u>Familiarize</u>. In 1 hr Crew A does $\dfrac{1}{10}$ of the job and Crew B does $\dfrac{1}{12}$ of the job, and together they do $\dfrac{1}{10} + \dfrac{1}{12}$ of the job. We want to find a number t such that the fraction of the job completed by the crews together is 1.

<u>Translate</u>. From the familiarization step we see that we want some number t for which

$$t\left(\dfrac{1}{10}\right) + t\left(\dfrac{1}{12}\right) = 1.$$

<u>Carry out</u>. We solve the equation.

$\dfrac{t}{10} + \dfrac{t}{12} = 1$,   LCM = 2·2·3·5, or 60

$60\left(\dfrac{t}{10} + \dfrac{t}{12}\right) = 60 \cdot 1$

$60 \cdot \dfrac{t}{10} + 60 \cdot \dfrac{t}{12} = 60$

$6t + 5t = 60$

$11t = 60$

$t = \dfrac{60}{11}$, or $5\dfrac{5}{11}$

<u>Check</u>. We check by repeating the computations.

$\dfrac{60}{11}\left(\dfrac{1}{10}\right) + \dfrac{60}{11}\left(\dfrac{1}{12}\right) = \dfrac{6}{11} + \dfrac{5}{11} = \dfrac{11}{11} = 1$

<u>State</u>. It takes $5\dfrac{5}{11}$ hr for the crews to paint the house working together.

Chapter 7   (7.4)

48. $25\frac{5}{11}$ km/h, $45\frac{5}{11}$ km/h

49.
$$\frac{7}{1+x} - 1 = \frac{5x}{x^2 + 3x + 2},$$
$$LCM = (x+1)(x+2)$$

$$(x+1)(x+2)\left[\frac{7}{1+x} - 1\right] =$$
$$(x+1)(x+2) \cdot \frac{5x}{(x+1)(x+2)}$$

$$(x+1)(x+2) \cdot \frac{7}{1+x} + (x+1)(x+2)(-1) = 5x$$
$$7(x+2) + (x^2 + 3x + 2)(-1) = 5x$$
$$7x + 14 - x^2 - 3x - 2 = 5x$$
$$-x^2 + 4x + 12 = 5x$$
$$0 = x^2 + x - 12$$
$$0 = (x+4)(x-3)$$

$x + 4 = 0$ or $x - 3 = 0$
$x = -4$ or $x = 3$

Both numbers check. The solutions are -4 and 3.

50. $\frac{2}{3}$

51. The y-axis is a vertical line, so it is of the form $x = a$. All points on the y-axis are of the form $(0,y)$, so a must be 0 and the equation is $x = 0$.

52. $y = 0$

53. The x-coordinate must be -3, and the y-coordinate must be 6. The point of intersection is (-3,6).

54. $y = -5$

55. A line parallel to the y-axis has an equation of the form $x = a$. Since the line is 13 units to the right of the y-axis, all points on the line are of the form $(13,y)$. Thus, a is 13, and the equation is $x = 13$.

56. $y = 2.8$

57. We substitute 2 for x and 0 for y and solve for m:
$$y = mx + 3$$
$$0 = m(2) + 3$$
$$-3 = 2m$$
$$-\frac{3}{2} = m$$

58. -8

Exercise Set 7.4

1. (3,2) and (-1,5)
$m = \frac{5 - 2}{-1 - 3} = \frac{3}{-4} = -\frac{3}{4}$

2. $\frac{2}{3}$

3. (-2,4) and (3,0)
$m = \frac{4 - 0}{-2 - 3} = \frac{4}{-5} = -\frac{4}{5}$

4. $-\frac{5}{6}$

5. (4,0) and (5,7)
$m = \frac{0 - 7}{4 - 5} = \frac{-7}{-1} = 7$

6. $\frac{2}{3}$

7. (0,8) and (-3,10)
$m = \frac{8 - 10}{0 - (-3)} = \frac{8 - 10}{0 + 3} = \frac{-2}{3} = -\frac{2}{3}$

8. $-\frac{1}{2}$

9. (3,-2) and (5,-6)
$m = \frac{-2 - (-6)}{3 - 5} = \frac{-2 + 6}{3 - 5} = \frac{4}{-2} = -2$

10. $-\frac{11}{8}$

11. $\left(-2, \frac{1}{2}\right)$ and $\left(-5, \frac{1}{2}\right)$
$m = \frac{\frac{1}{2} - \frac{1}{2}}{-2 - (-5)} = \frac{\frac{1}{2} - \frac{1}{2}}{-2 + 5} = \frac{0}{3} = 0$

12. 0

13. (9,-4) and (9,-7)
$m = \frac{-4 - (-7)}{9 - 9} = \frac{-4 + 7}{9 - 9} = \frac{3}{0}$

Since division by 0 is not defined, this line has no slope.

14. No slope

15. The line $x = -8$ is a vertical line. A vertical line has no slope.

16. No slope

17. The line $y = 2$ is a horizontal line. A horizontal line has slope 0.

18. 0

19. The line $x = 9$ is a vertical line. A vertical line has no slope.

Chapter 7   (7.4)

20. No slope

21. The line $y = -9$ is a horizontal line. A horizontal line has slope 0.

22. 0

23. $y = 2.3x$
    Think of this as $y = 2.3x + 0$. Then the slope is 2.3.

24. $-\frac{3}{5}$

25. $y = -5x + 7$
    This equation is already in the form $y = mx + b$. The slope is -5.

26. 3.8

27. We solve for y.
    $3x + 2y = 6$
    $2y = -3x + 6$
    $y = -\frac{3}{2}x + 3$
    The slope is $-\frac{3}{2}$.

28. 4

29. We solve for y.
    $x + 4y = 8$
    $4y = -x + 8$
    $y = -\frac{1}{4}x + 2$
    The slope is $-\frac{1}{4}$.

30. $-\frac{1}{3}$

31. We solve for y.
    $-2x + y = 4$
    $y = 2x + 4$
    The slope is 2.

32. 5

33. We solve for y.
    $4x - 3y = -12$
    $-3y = -4x - 12$
    $y = -\frac{1}{3}(-4x - 12)$
    $y = \frac{4}{3}x + 4$
    The slope is $\frac{4}{3}$.

34. $\frac{1}{2}$

35. We solve for y.
    $x - 3y = -2$
    $-3y = -x - 2$
    $y = -\frac{1}{3}(-x - 2)$
    $y = \frac{1}{3}x + \frac{2}{3}$
    The slope is $\frac{1}{3}$.

36. -1

37. We solve for y.
    $-2x + 4y = 8$
    $4y = 2x + 8$
    $y = \frac{1}{4}(2x + 8)$
    $y = \frac{1}{2}x + 2$
    The slope is $\frac{1}{2}$.

38. $\frac{5}{7}$

39. $y = -4x - 9$
    The equation is already in the form $y = mx + b$. The slope is -4 and the y-intercept is $(0,-9)$.

40. -3, $(0,-5)$

41. $y = 1.8x$ (Think: $y = 1.8x + 0$)
    The slope is 1.8 and the y-intercept is $(0,0)$.

42. -27.4, $(0,0)$

43. We solve for y.
    $2x + 3y = 9$
    $3y = -2x + 9$
    $y = \frac{1}{3}(-2x + 9)$
    $y = -\frac{2}{3}x + 3$
    The slope is $-\frac{2}{3}$ and the y-intercept is $(0,3)$.

44. $-\frac{5}{4}$, $(0,3)$

45. We solve for y.
    $-8x - 7y = 21$
    $-7y = 8x + 21$
    $y = -\frac{1}{7}(8x + 21)$
    $y = -\frac{8}{7}x - 3$
    The slope is $-\frac{8}{7}$ and the y-intercept is $(0,-3)$.

46. $-\frac{2}{9}$, $\left(0, -\frac{13}{9}\right)$

Chapter 7   (7.4)

47. We solve for y.
$$9x = 3y + 5$$
$$9x - 5 = 3y$$
$$\tfrac{1}{3}(9x - 5) = y$$
$$3x - \tfrac{5}{3} = y$$
The slope is 3 and the y-intercept is $\left(0, -\tfrac{5}{3}\right)$.

48. $\tfrac{4}{9}$, $\left(0, -\tfrac{7}{9}\right)$

49. We solve for y.
$$-6x = 4y + 2$$
$$-6x - 2 = 4y$$
$$\tfrac{1}{4}(-6x - 2) = y$$
$$-\tfrac{3}{2}x - \tfrac{1}{2} = y$$
The slope is $-\tfrac{3}{2}$ and the y-intercept is $\left(0, -\tfrac{1}{2}\right)$.

50. 0, (0,-17)

51. We use the slope-intercept equation and substitute 5 for m and 6 for b:
$$y = mx + b$$
$$y = 5x + 6$$

52. $y = -4x - 2$

53. We use the slope-intercept equation and substitute 4.3 for m and -1 for b:
$$y = mx + b$$
$$y = 4.3x - 1$$

54. $y = -2.8x + 4$

55. We use the slope-intercept equation and substitute $-\tfrac{5}{3}$ for m and -8 for b:
$$y = -\tfrac{5}{3}x - 8$$

56. $y = \tfrac{3}{4}x + 23$

57. $y = \tfrac{3}{5}x + 2$

First we plot the y-intercept (0,2). We can start at the y-intercept and use the slope, $\tfrac{3}{5}$, to find another point. We move up 3 units and right 5 units to get a new point (5,5). Thinking of the slope as $\tfrac{-3}{-5}$ we can start at (0,2) and move down 3 units and left 5 units to get another point (-5,-1).

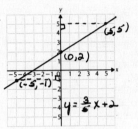

58.

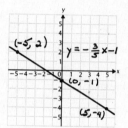

59. $y = -\tfrac{3}{5}x + 4$

First we plot the y-intercept (0,4). We can start at the y-intercept and, thinking of the slope as $\tfrac{-3}{5}$, find another point by moving down 3 units and right 5 units to the point (5,1). Thinking of the slope as $\tfrac{3}{-5}$ we can start at (0,4) and move up 3 units and left 5 units to get another point (-5,7).

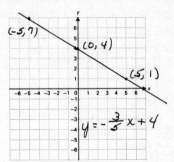

60.

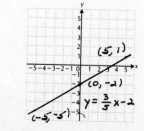

Chapter 7 (7.4)

61. $y = \frac{5}{3}x + 3$

First we plot the y-intercept (0,3). We can start at the y-intercept and use the slope, $\frac{5}{3}$, to find another point. We move up 5 units and right 3 units to the point (3,8). Thinking of the slope as $\frac{-5}{-3}$ we can start at (0,3) and move down 5 units and left 3 units to get another point (-3,-2).

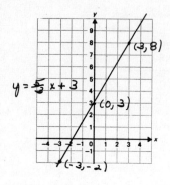

62.
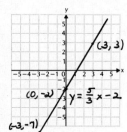

63. $y = -\frac{3}{2}x - 2$

First we plot the y-intercept (0,-2). We can start at the y-intercept and, thinking of the slope as $\frac{-3}{2}$, find another point by moving down 3 units and right 2 units to the point (2,-5). Thinking of the slope as $\frac{3}{-2}$ we can start at (0,-2) and move up 3 units and left 2 units to get another point (-2,1).

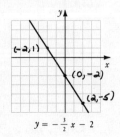

64.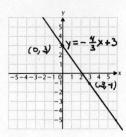

65. We can use the slope and y-intercept to graph each equation.

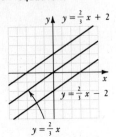

66.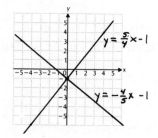

67. Grade = slope = $\frac{\text{vertical change}}{\text{horizontal change}} = \frac{920.58}{13,740} =$ 0.067 = 6.7%

68. $\frac{12}{41}$

69. Grade = slope = $\frac{\text{vertical change}}{\text{horizontal change}} = \frac{0.4}{5} =$ 0.08 = 8%

70. $-\frac{28}{129}$

71. Grade = $\frac{\text{vertical change}}{\text{horizontal change}} = \frac{-158.4}{5280} = -0.03 = -3\%$

72. -0.045, or $-\frac{9}{200}$

73. Grade = $\frac{\text{vertical change}}{\text{horizontal change}}$

$0.12 = \frac{v}{5}$

$0.6 = v$ Multiplying by 5

The end of the treadmill is set at 0.6 ft vertically.

74. 30 ft

Chapter 7 (7.5)

Exercise Set 7.5

1. $y - y_1 = m(x - x_1)$
   We substitute 5 for m, 2 for x_1, and 5 for y_1.
   $y - 5 = 5(x - 2)$
   $y - 5 = 5x - 10$
   $y = 5x - 5$

2. $y = -2x - 6$

3. $y - y_1 = m(x - x_1)$
   We substitute $\frac{3}{4}$ for m, 2 for x_1, and 4 for y_1.
   $y - 4 = \frac{3}{4}(x - 2)$
   $y - 4 = \frac{3}{4}x - \frac{3}{2}$
   $y = \frac{3}{4}x + \frac{5}{2}$

4. $y = -x + \frac{5}{2}$

5. $y - y_1 = m(x - x_1)$
   We substitute 1 for m, 2 for x_1, and -6 for y_1.
   $y - (-6) = 1(x - 2)$
   $y + 6 = x - 2$
   $y = x - 8$

6. $y = 6x - 26$

7. $y - y_1 = m(x - x_1)$
   We substitute -3 for m, -3 for x_1, and 0 for y_1.
   $y - 0 = -3[x - (-3)]$
   $y = -3(x + 3)$
   $y = -3x - 9$

8. $y = -3x + 3$

9. $y - y_1 = m(x - x_1)$
   We substitute $\frac{2}{3}$ for m, 5 for x_1, and 6 for y_1.
   $y - 6 = \frac{2}{3}(x - 5)$
   $y - 6 = \frac{2}{3}x - \frac{10}{3}$
   $y = \frac{2}{3}x + \frac{8}{3}$

10. $y = \frac{5}{6}x + \frac{16}{3}$

11. (-6,1) and (2,3)
    First we find the slope.
    $m = \frac{1 - 3}{-6 - 2} = \frac{-2}{-8} = \frac{1}{4}$
    Then we use the point-slope equation.
    $y - y_1 = m(x - x_1)$
    We substitute $\frac{1}{4}$ for m, -6 for x_1, and 1 for y_1.
    $y - 1 = \frac{1}{4}[x - (-6)]$
    $y - 1 = \frac{1}{4}(x + 6)$
    $y - 1 = \frac{1}{4}x + \frac{3}{2}$
    $y = \frac{1}{4}x + \frac{5}{2}$
    We also could substitute $\frac{1}{4}$ for m, 2 for x_1, and 3 for y_1.
    $y - 3 = \frac{1}{4}(x - 2)$
    $y - 3 = \frac{1}{4}x - \frac{1}{2}$
    $y = \frac{1}{4}x + \frac{5}{2}$

12. $y = x + 4$

13. (0,4) and (4,2)
    First we find the slope.
    $m = \frac{4 - 2}{0 - 4} = \frac{2}{-4} = -\frac{1}{2}$
    Then we use the point-slope equation.
    $y - y_1 = m(x - x_1)$
    We substitute $-\frac{1}{2}$ for m, 0 for x_1, and 4 for y_1.
    $y - 4 = -\frac{1}{2}(x - 0)$
    $y - 4 = -\frac{1}{2}x$
    $y = -\frac{1}{2}x + 4$

14. $y = \frac{1}{2}x$

15. (3,2) and (1,5)
    First we find the slope.
    $m = \frac{2 - 5}{3 - 1} = \frac{-3}{2} = -\frac{3}{2}$
    Then we use the point-slope equation.
    $y - y_1 = m(x - x_1)$
    We substitute $-\frac{3}{2}$ for m, 3 for x_1, and 2 for y_1.
    $y - 2 = -\frac{3}{2}(x - 3)$
    $y - 2 = -\frac{3}{2}x + \frac{9}{2}$
    $y = -\frac{3}{2}x + \frac{13}{2}$

Chapter 7   (7.6)

16. $y = x + 5$

17. $(5,0)$ and $(0,-2)$
We first find the slope.
$$m = \frac{0 - (-2)}{5 - 0} = \frac{2}{5}$$
Then we use the point-slope equation.
$$y - y_1 = m(x - x_1)$$
We substitute $\frac{2}{5}$ for m, 5 for x_1, and 0 for y_1.
$$y - 0 = \frac{2}{5}(x - 5)$$
$$y = \frac{2}{5}x - 2$$

18. $y = \frac{5}{3}x + \frac{4}{3}$

19. $(-2,-4)$ and $(2,-1)$
We first find the slope.
$$m = \frac{-4 - (-1)}{-2 - 2} = \frac{-4 + 1}{-2 - 2} = \frac{-3}{-4} = \frac{3}{4}$$
Then we use the point-slope equation.
$$y - y_1 = m(x - x_1)$$
We substitute $\frac{3}{4}$ for m, -2 for x_1, and -4 for y_1.
$$y - (-4) = \frac{3}{4}[x - (-2)]$$
$$y + 4 = \frac{3}{4}(x + 2)$$
$$y + 4 = \frac{3}{4}x + \frac{3}{2}$$
$$y = \frac{3}{4}x - \frac{5}{2}$$

20. $y = -4x - 7$

21. $2x^2 + 6x = 0$
$2x(x + 3) = 0$
$2x = 0$ or $x + 3 = 0$
$x = 0$ or $x = -3$
The solutions are 0 and -3.

22. $x(x + 7)(x - 2)$

23. Familiarize. Let y = the smaller odd integer. Then $y + 2$ = the larger odd integer.
Translate. We reword the problem.

Smaller odd integer	times	larger odd integer	is	195.
y	$\cdot$	$(y + 2)$	=	195

23. (continued)

Carry out. We solve the equation.
$$y(y + 2) = 195$$
$$y^2 + 2y = 195$$
$$y^2 + 2y - 195 = 0$$
$$(y + 15)(y - 13) = 0$$
$y + 15 = 0$ or $y - 13 = 0$
$y = -15$ or $y = 13$

Check. If $y = -15$, then $y + 2 = -13$. These are consecutive odd integers and their product is $(-15)(-13)$, or 195. This pair checks. If $y = 13$ then $y + 2 = 15$. These are consecutive odd integers and their product is $13 \cdot 15$, or 195. This pair checks also.

State. The integers are -15 and -13 or 13 and 15.

24. 11, -1

25. First find the slope of $3x - y + 4 = 0$.
$3x - y + 4 = 0$
$3x + 4 = y$
The slope is 3.

Then find an equation of the line containing $(2,-3)$ and having slope 3.
$$y - y_1 = m(x - x_1)$$
We substitute 3 for m, 2 for x_1, and -3 for y_1.
$$y - (-3) = 3(x - 2)$$
$$y + 3 = 3x - 6$$
$$y = 3x - 9$$

26. $y = \frac{1}{5}x - 2$

Exercise Set 7.6

1. 1. The first equation is already solved for y:
$$y = x + 4$$

2. We solve the second equation for y:
$$y - x = -3$$
$$y = x - 3$$

The slope of each line is 1. The y-intercepts, $(0,4)$ and $(0,-3)$, are different. The lines are parallel.

2. Yes

3. We solve each equation for y:

1. $y + 3 = 6x$ 2. $-6x - y = 2$
   $y = 6x - 3$ $-y = 6x + 2$
                            $y = -6x - 2$

The slope of the first line is 6 and of the second is -6. Since the slopes are different, the lines are not parallel.

Chapter 7   (7.6)

4. No

5. We solve each equation for y:
   1. $y + 3.5 = 0.3125x$
      $y = 0.3125x - 3.5$
   2. $5y = -32x + 23.5$
      $y = -6.4x + 4.7$

   The slope of the first line is 0.3125 and of the second is -6.4. Since the slopes are different, the lines are not parallel.

6. Yes

7. 1. The first equation is already solved for y:
      $y = 2x + 7$
   2. We solve the second equation for y:
      $5y + 10x = 20$
      $5y = -10x + 20$
      $y = -2x + 4$

   The slope of the first line is 2 and of the second is -2. Since the slopes are different, the lines are not parallel.

8. No

9. We solve each equation for y:
   1. $3x - y = -9$ 2. $2y - 6x = -2$
      $3x + 9 = y$ $2y = 6x - 2$
                               $y = 3x - 1$

   The slope of each line is 3. The y-intercepts, (0,9) and (0,-1) are different. The lines are parallel.

10. No

11. $x = 3$,
    $x = 4$

    These are vertical lines with equations of the form $x = p$ and $x = q$, where $p \neq q$. Thus, they are parallel.

12. No

13. 1. The first equation is already solved for y:
       $y = -4x + 3$
    2. We solve the second equation for y:
       $4y + x = -1$
       $4y = -x - 1$
       $y = -\frac{1}{4}x - \frac{1}{4}$

    The slopes are -4 and $-\frac{1}{4}$. Their product is
    $-4\left(-\frac{1}{4}\right) = 1.$

    Since the product of the slopes is not -1, the lines are not perpendicular.

14. No

15. We solve each equation for y:
    1. $x + y = 6$ 2. $4y - 4x = 12$
       $y = -x + 6$ $4y = 4x + 12$
                              $y = x + 3$

    The slopes are -1 and 1. Their product is
    $-1 \cdot 1 = -1.$
    Thus the lines are perpendicular.

16. Yes

17. 1. The first equation is already solved for y:
       $y = -6.4x - 7$
    2. We solve the second equation for y:
       $64y - 5x = 32$
       $64y = 5x + 32$
       $y = \frac{5}{64}x + \frac{1}{2}$

    The slopes are -6.4 and $\frac{5}{64}$, or $-\frac{64}{10}$ and $\frac{5}{64}$.

    Their product is
    $-\frac{64}{10}\left(\frac{5}{64}\right) = -\frac{1}{2}.$

    Since the product of the slopes is not -1, the lines are not perpendicular.

18. Yes

19. 1. The first equation is already solved for y:
       $y = -x + 8$
    2. We solve the second equation for y:
       $x - y = -1$
       $x + 1 = y$

    The slopes are -1 and 1. Their product is
    $-1 \cdot 1 = -1.$
    Thus the lines are perpendicular.

20. No

Chapter 7   (7.6)

21. We solve each equation for y:

1. $$\frac{3}{8}x - \frac{y}{2} = 1$$
$$8\left(\frac{3}{8}x - \frac{y}{2}\right) = 8 \cdot 1$$
$$8 \cdot \frac{3}{8}x - 8 \cdot \frac{y}{2} = 8$$
$$3x - 4y = 8$$
$$-4y = -3x + 8$$
$$y = \frac{3}{4}x - 2$$

2. $$\frac{4}{3}x - y + 1 = 0$$
$$\frac{4}{3}x + 1 = y$$

The slopes are $\frac{3}{4}$ and $\frac{4}{3}$. Their product is
$$\frac{3}{4}\left(\frac{4}{3}\right) = 1.$$
Since the product of the slopes is not −1, the lines are not perpendicular.

22. Yes

23. Familiarize. We organize the information in a table.

	Distance	Speed	Time
Slow car	130	r	t
Fast car	140	r + 10	t

$d = r \cdot t$

Translate. From the rows of the table we get two equations:
$$130 = rt$$
$$140 = (r + 10)t$$

We solve each equation for t and set the results equal.

Solving $130 = rt$ for t: $\;t = \frac{130}{r}$

Solving $140 = (r + 10)t$ for t: $\;t = \frac{140}{r + 10}$

Then $\frac{130}{r} = \frac{140}{r + 10}$.

Carry out. We first multiply by the LCM, $r(r + 10)$.
$$r(r+10) \cdot \frac{130}{r} = r(r+10) \cdot \frac{140}{r+10}$$
$$130(r + 10) = 140r$$
$$130r + 1300 = 140r$$
$$1300 = 10r$$
$$130 = r$$

Check. If the speed of the slow car is 130 km/h, then it travels 130 km in $\frac{130}{130}$, or 1 hr. If the speed of the fast car is 130 + 10, or 140 km/h, then it travels 140 km in $\frac{140}{140}$, or 1 hr. The times are the same, so the answer checks.

23. (continued)

State. The speed of the slow car is 130 km/h, and the speed of the fast car is 140 km/h.

24. 45 oz

25. $(5t^2 - 6m)(5t^2 + m)$
$= 25t^4 + 5mt^2 - 30mt^2 - 6m^2$ Using FOIL
$= 25t^4 - 25mt^2 - 6m^2$

26. 5

27. We find the product of the slopes:
$m_1 m_2 = 1.25(-0.79) = -0.9875$
Since the product of the slopes is not −1, the lines are not perpendicular.

28. Yes

29. First we find the slope of the given line:
$$y - 3x = 4$$
$$y = 3x + 4$$
The slope is 3.

Then we use the slope-intercept equation to write the equation of a line with slope 3 and y-intercept (0,6):
$$y = mx + b$$
$$y = 3x + 6 \quad \text{Substituting 3 for m and 6 for b}$$

30. $y = 2x + 8$

31. First we find the slope of the given line:
$$3y - x = 0$$
$$3y = x$$
$$y = \frac{1}{3}x$$
The slope is $\frac{1}{3}$.

We can find the slope of the line perpendicular to the given line by taking the reciprocal of $\frac{1}{3}$ and changing the sign. We get −3.

Then we use the slope-intercept equation to write the equation of a line with slope −3 and y-intercept (0,2):
$$y = mx + b$$
$$y = -3x + 2 \quad \text{Substituting −3 for m and 2 for b}$$

32. $y = \frac{1}{2}x - \frac{1}{2}$

33. First we find the slope of the given line:
$$4x - 8y = 12$$
$$-8y = -4x + 12$$
$$y = \frac{1}{2}x - \frac{3}{2}$$
The slope is $\frac{1}{2}$.

Then we use the point-slope equation to find the equation of a line with slope $\frac{1}{2}$ containing the point $(-2,0)$:
$$y - y_1 = m(x - x_1)$$
$$y - 0 = \frac{1}{2}[x - (-2)]$$
$$y = \frac{1}{2}(x + 2)$$
$$y = \frac{1}{2}x + 1$$

34. -1

35. We find the slope of each line:

1. $4y = kx - 6$
$$y = \frac{k}{4}x - \frac{3}{2}$$

2. $5x + 20y = 12$
$$20y = -5x + 12$$
$$y = -\frac{1}{4}x + \frac{3}{5}$$

The slopes are $\frac{k}{4}$ and $-\frac{1}{4}$. If the lines are perpendicular, the product of their slopes is -1.
$$\frac{k}{4}\left(-\frac{1}{4}\right) = -1$$
$$-\frac{k}{16} = -1$$
$$k = 16$$

36. A: $y = \frac{3}{2}x + \frac{1}{2}$, B: $y = -\frac{2}{3}x - \frac{5}{3}$

37. First we find the equation of A, a line containing the points $(1,-1)$ and $(4,3)$:

The slope is $\frac{3 - (-1)}{4 - 1} = \frac{4}{3}$.

Use the point-slope equation:
$$y - y_1 = m(x - x_1)$$
$$y - 3 = \frac{4}{3}(x - 4)$$
$$y - 3 = \frac{4}{3}x - \frac{16}{3}$$
$$y = \frac{4}{3}x - \frac{7}{3}$$

The slope of A is $\frac{4}{3}$. Since A and B are perpendicular we find the slope of B by taking the reciprocal of $\frac{4}{3}$ and changing the sign. We get $-\frac{3}{4}$. Then we use the point-slope equation to find the equation of B, a line with slope $-\frac{3}{4}$ and containing the point $(1,-1)$:

37. (continued)
$$y - y_1 = m(x - x_1)$$
$$y - (-1) = -\frac{3}{4}(x - 1)$$
$$y + 1 = -\frac{3}{4}x + \frac{3}{4}$$
$$y = -\frac{3}{4}x - \frac{1}{4}$$

Exercise Set 7.7

1. We substitute to find k.
$$y = kx$$
$$28 = k \cdot 7 \quad \text{Substituting 28 for y and 7 for x}$$
$$\frac{28}{7} = k$$
$$4 = k \quad \text{k is the constant of variation}$$
The equation of variation is $y = 4x$.

2. $y = 3.75x$

3. We substitute to find k.
$$y = kx$$
$$0.7 = k \cdot 0.4 \quad \text{Substituting 0.7 for y and 0.4 for x}$$
$$\frac{0.7}{0.4} = k$$
$$\frac{7}{4} = k, \text{ or } k = 1.75$$
The equation of variation is $y = 1.75x$.

4. $y = 1.6x$

5. We substitute to find k.
$$y = kx$$
$$400 = k \cdot 125 \quad \text{Substituting 400 for y and 125 for x}$$
$$\frac{400}{125} = k$$
$$\frac{16}{5} = k, \text{ or } k = 3.2$$
The equation of variation is $y = 3.2x$.

6. $y = 3.6x$

7. We substitute to find k.
$$y = kx$$
$$200 = k \cdot 300 \quad \text{Substituting 200 for y and 300 for x}$$
$$\frac{200}{300} = k$$
$$\frac{2}{3} = k$$
The equation of variation is $y = \frac{2}{3}x$.

Chapter 7   (7.7)

8. $y = \frac{25}{3}x$

9. <u>Familiarize</u>. The problem states that we have direct variation between the variables P and H. Thus, an equation $P = kH$, $k > 0$, applies. As the number of hours increases, the paycheck increases.

    <u>Translate</u>. We write an equation of variation.
    Paycheck varies directly as number of hours.
    This translates to $P = kH$.

    <u>Carry out</u>.
    a) First find an equation of variation.

    $P = kH$
    $78.75 = k \cdot 15$ Substituting 78.75 for P and 15 for H
    $\frac{78.75}{15} = k$
    $5.25 = k$

    The equation of variation is $P = 5.25H$.

    b) Use the equation to find the pay for 35 hours work.

    $P = 5.25H$
    $P = 5.25(35)$ Substituting 35 for H
    $P = 183.75$

    <u>Check</u>. This check might be done by repeating the computations. We might also do some reasoning about the answer. The paycheck increased from $78.75 to $183.75. Similarly, the hours increased from 15 to 35.

    <u>State</u>. For 35 hours work, the paycheck is $183.75.

10. 16,445

11. <u>Familiarize</u>. This problem states that we have direct variation between S and W. Thus, an equation $S = kW$, $k > 0$, applies. As the weight increases, the number of servings increases.

    <u>Translate</u>. We write an equation of variation.
    Number of servings varies directly as weight.
    This translates to $S = kW$.

    <u>Carry out</u>.
    a) First find an equation of variation.

    $S = kW$
    $40 = k \cdot 14$ Substituting 40 for S and 14 for W
    $\frac{40}{14} = k$
    $\frac{20}{7} = k$

    The equation of variation is $S = \frac{20}{7}W$.

11. (continued)

    b) Use the equation to find the number of servings from an 8-kg turkey.

    $S = \frac{20}{7}W$
    $S = \frac{20}{7} \cdot 8$ Substituting 8 for W
    $S = \frac{160}{7}$, or $22\frac{6}{7}$

    <u>Check</u>. A check can always be done by repeating the computations. We can also do some reasoning about the answer. The number of servings decreased from 40 to $22\frac{6}{7}$. Similarly, the weight decreased from 14 kg to 8 kg.

    <u>State</u>. $22\frac{6}{7}$ servings can be obtained from an 8-kg turkey.

12. $93\frac{1}{3}$

13. <u>Familiarize</u>. The problem states that we have direct variation between the variables M and E. Thus, an equation $M = kE$, $k > 0$, applies. As the weight on earth increases, the weight on moon increases.

    <u>Translate</u>. We write an equation of variation.
    Weight on moon varies directly as weight on earth.
    This translates to $M = kE$.

    <u>Carry out</u>.
    a) First find an equation of variation.

    $M = kE$
    $28.6 = k \cdot 171.6$ Substituting 28.6 for M and 171.6 for E
    $286 = 1716k$ Clearing decimals
    $\frac{286}{1716} = k$
    $\frac{1}{6} = k$

    The equation of variation is $M = \frac{1}{6}E$.

    b) Use the equation to find how much a 220-lb person would weigh on the moon.

    $M = \frac{1}{6}E$
    $M = \frac{1}{6} \cdot 220$ Substituting 220 for E
    $M = \frac{220}{6}$, or $36\frac{2}{3}$

    <u>Check</u>. In addition to repeating the computations we can do some reasoning. The weight on the earth increased from 171.6 lb to 220 lb. Similarly, the weight on the moon increased from 28.6 lb to $36\frac{2}{3}$ lb.

    <u>State</u>. A 220-lb person would weigh $36\frac{2}{3}$ lb on the moon.

14. 66.88 lb

15. Familiarize. This problem states that we have direct variation between the variables W and B. Thus, an equation W = kB, k > 0, applies. As the body weight increases, the water weight increases.

   Translate. We write an equation of variation. Water weight varies directly as the total body weight.

   This translates to W = kB.

   Carry out.
   a) First find an equation of variation.

   W = kB

   54 = k·75   Substituting 54 for W and 75 for B

   $\frac{54}{75}$ = k

   $\frac{18}{25}$ = k, or k = 0.72

   The equation of variation is W = 0.72B.

   b) Use the equation to find how many kilograms of water are in a person weighing 95 kg.

   W = 0.72B
   W = 0.72(95)   Substituting 95 for B
   W = 68.4

   Check. This check might be done by repeating computations. We might also do some reasoning about the answer. The water weight increased from 54 kg to 68.4 kg. Similarly, the body weight increased from 75 kg to 95 kg.

   State. There are 68.4 kg of water in a person weighing 95 kg.

16. $4500

17. Familiarize. This problem states that we have direct variation between the variables c and n. Thus, an equation c = kn, k > 0, applies. As the number of operating hours increases, the operating cost increases. Note that 30 days = 30·24, or 720 hr.

   Translate. We write an equation of variation. Operating cost varies directly as operating hours.

   This translates to n = kc.

   Carry out.
   a) First we find an equation of variation.

   c = kn

   14 = k·720   Substituting 14 for c and 720 (the number of hours in 30 days) for n

   $\frac{14}{720}$ = k

   $\frac{7}{360}$ = k

   The equation of variation is c = $\frac{7}{360}$n.

17. (continued)

   b) Use the equation of variation to find the operating cost for 1 day and 1 hr.

   For 1 day: 1 day is equivalent to 24 hr, so we substitute 24 for n.

   c = $\frac{7}{360}$ · 24

   c = $\frac{168}{360}$, or c ≈ 0.467

   For 1 hr: c = $\frac{7}{360}$n

   c = $\frac{7}{360}$ · 1   Substituting 1 for n

   c ≈ 0.019

   Check. In addition to repeating the computations we can do some reasoning. The hours decreased from 720 to 24 and from 720 to 1. Similarly, the cost decreased from $14 to $0.467 and from $14 to $0.019.

   State. It cost $0.467, or 46.7¢, to operate a television for 1 day. It cost $0.019, or 1.9¢, to operate a television for 1 hr.

18. 174.24 lb

19. We substitute to find k.

   y = $\frac{k}{x}$

   25 = $\frac{k}{3}$ Substituting 25 for y and 3 for x

   75 = k   k is the constant of variation

   The equation of variation is y = $\frac{75}{x}$.

20. y = $\frac{90}{x}$

21. We substitute to find k.

   y = $\frac{k}{x}$

   8 = $\frac{k}{10}$ Substituting 8 for y and 10 for x

   80 = k

   The equation of variation is y = $\frac{80}{x}$.

22. y = $\frac{70}{x}$

23. We substitute to find k.

   y = $\frac{k}{x}$

   0.125 = $\frac{k}{8}$ Substituting 0.125 for y and 8 for x

   1 = k

   The equation of variation is y = $\frac{1}{x}$.

24. y = $\frac{1}{x}$

25. We substitute to find k.

$y = \dfrac{k}{x}$

$42 = \dfrac{k}{25}$ Substituting 42 for y and 25 for x

$1050 = k$

The equation of variation is $y = \dfrac{1050}{x}$.

26. $y = \dfrac{2100}{x}$

27. We substitute to find k.

$y = \dfrac{k}{x}$

$0.2 = \dfrac{k}{0.3}$ Substituting 0.2 for y and 0.3 for x

$0.06 = k$

The equation of variation is $y = \dfrac{0.06}{x}$.

28. $y = \dfrac{0.24}{x}$

29. a) It seems reasonable that, as the number of workers increases, the number of hours required to do the job decreases, so inverse variation might apply.

   b) <u>Familiarize</u>. We let T = the time to do the job and N = the number of workers. An equation T = k/N, k > 0, applies. (See part a)).

   <u>Translate</u>. We write an equation of variation.
   Time varies inversely as number of people.
   This translates to $T = \dfrac{k}{N}$.

   <u>Carry out</u>.
   a) First find an equation of variation.

   $T = \dfrac{k}{N}$

   $16 = \dfrac{k}{2}$ Substituting 16 for T and 2 for N

   $32 = k$

   The equation of variation is $T = \dfrac{32}{N}$.

   b) Use the equation to find the amount of time it takes 6 people to do the job.

   $T = \dfrac{32}{N}$

   $T = \dfrac{32}{6}$ Substituting 6 for N

   $T = 5\dfrac{1}{3}$

   <u>Check</u>. The check might be done by repeating the computations. We might also analyze the results. The number of people increased from 2 to 6. The time decreased from 16 hours to $5\dfrac{1}{3}$ hours. This is what we would expect with inverse variation.

   <u>State</u>. It would take 6 people $5\dfrac{1}{3}$ hours to do the job.

30. a) Inverse

   b) $4\dfrac{1}{2}$ hr

31. a) It seems reasonable that, as the number of hours of production increases, the number of compact-disc players produced will increase, so direct variation might apply.

   b) <u>Familiarize</u>. Let H = the number of hours the production line is working, and let P = the number of compact-disc players produced. An equation P = kH, k > 0, applies. (See part a)).

   <u>Translate</u>. We write an equation of variation.
   Number of players produced varies directly as hours of production. This translates to P = kH.

   <u>Carry out</u>.
   a) First we find an equation of variation.

   $P = kH$

   $15 = k \cdot 8$ Substituting 8 for H and 15 for P

   $\dfrac{15}{8} = k$

   The equation of variation is $P = \dfrac{15}{8}H$.

   b) Use the equation to find the number of players produced in 37 hr.

   $P = \dfrac{15}{8}H$

   $P = \dfrac{15}{8} \cdot 37$ Substituting 37 for H

   $P = \dfrac{555}{8}$

   $P \approx 69$

   <u>Check</u>. In addition to repeating the computations, we can do some reasoning. The number of hours increased from 8 to 37. Similarly, the number of compact disc players produced increased from 15 to about 69.

   <u>State</u>. About 69 compact-disc players can be produced in 37 hr.

32. a) Direct

   b) $218.75

33. <u>Familiarize</u>. The problem states that we have inverse variation between the variables V and P. Thus, an equation $V = \dfrac{k}{P}$, k > 0, applies. As the pressure increases, the volume decreases.

   <u>Translate</u>. We write an equation of variation.
   Volume varies inversely as pressure.
   This translates to $V = \dfrac{k}{P}$.

33. (continued)

    Carry out.
    a) First find an equation of variation.

    $V = \dfrac{k}{P}$

    $200 = \dfrac{k}{32}$ Substituting 200 for V and 32 for P

    $6400 = k$

    The equation of variation is $V = \dfrac{6400}{P}$.

    b) Use the equation to find the volume of a gas under a pressure of 20 kg/cm².

    $V = \dfrac{6400}{P}$

    $V = \dfrac{6400}{20}$ Substituting 20 for P

    $V = 320$

    Check. Checking can be done by repeating the computations. We can also analyze the results. The pressure decreased from 32 km/cm² to 20 km/cm². The volume increased from 200 cm³ to 320 cm³. This is what we would expect with inverse variation.

    State. The volume is 320 cm³ under a pressure of 20 km/cm².

34. $3\dfrac{5}{9}$ amperes

35. Familiarize. The problem states that we have inverse variation between the variables, t and r. Thus, an equation $t = \dfrac{k}{r}$, $k > 0$, applies. As the rate increases, the time decreases.

    Translate. We write an equation of variation. Time varies inversely as the rate.

    This translates to $t = \dfrac{k}{r}$.

    Carry out.
    a) First find an equation of variation.

    $t = \dfrac{k}{r}$

    $90 = \dfrac{k}{1200}$ Substituting 90 for t and 1200 for r

    $108{,}000 = k$

    The equation of variation is $t = \dfrac{108{,}000}{r}$.

    b) Use the equation to find the time it will take the pump to empty the tank at 2000 L/min.

    $t = \dfrac{108{,}000}{r}$

    $t = \dfrac{108{,}000}{2000}$ Substituting 2000 for r

    $t = 54$

    Check. The check might be done by repeating the computations. Let us also analyze the results. The rate increased from 1200 L/min to 2000 L/min. The time decreased from 90 min to 54 min. This is what we would expect with inverse variation.

35. (continued)

    State. It will take 54 min for the pump to empty the tank at 2000 L/min.

36. 250 cm, 1000 cm²

37. Familiarize. The problem states that we have inverse variation between the variables t and r. Thus, an equation $t = \dfrac{k}{r}$, $k > 0$, applies. As the speed increases, the time decreases.

    Translate.
    Time varies inversely as speed. This translates to $t = \dfrac{k}{r}$.

    Carry out.
    a) First find an equation of variation.

    $t = \dfrac{k}{r}$

    $5 = \dfrac{k}{60}$ Substituting 5 for t and 60 for r

    $300 = k$

    The equation of variation is $t = \dfrac{300}{r}$.

    b) Use the equation to find the time when the speed is 40 km/h.

    $t = \dfrac{300}{r}$

    $t = \dfrac{300}{40}$ Substituting 40 for r

    $t = \dfrac{15}{2}$, or $7\dfrac{1}{2}$

    Check. We can check by repeating the computations. Let us also analyze the results. The speed decreased from 60 km/h to 40 km/h. The time increased from 5 hr to $7\dfrac{1}{2}$ hr. We would expect these results with inverse variation.

    State. It would take $7\dfrac{1}{2}$ hr to drive the fixed distance at 40 km/h.

38. 2.4 ft

39. $P = kS$, where k = the number of sides

40. $C = kr$ ($k = 2\pi \approx 6.28$)

    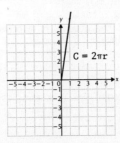

41. $B = kN$

Chapter 7   (7.8)

42. $C = kA$

43. $p = kq$

$\frac{1}{k} \cdot p = \frac{1}{k} \cdot kq$

$\frac{1}{k} \cdot p = q$, or $q = \frac{1}{k}p$

Since k is a constant, so is $\frac{1}{k}$, and q varies directly as p.

44. π

45. $S = kV^2$

46. $P^2 = kt$

47. $V = kr^3$

48. $P = kV^3$

49. $C = \frac{k}{N}$

50. $N = \frac{k}{C}$

51. $I = \frac{k}{R}$

52. $D = \frac{k}{V}$

53. $I = \frac{k}{d^2}$

54. No

55. Yes. As the runner's speed increases, the time decreases.

56. Yes

57. No. As the weight increases, the time increases.

Exercise Set 7.8

1. Yes, since each number of the domain is matched to only one member of the range.

2. Yes

3. Yes, since each member of the domain is matched to only one member of the range.

4. No

5. No, the members of the domain New York and Los Angeles are each matched to more than one member of the range.

6. Yes

7. Yes, since each member of the domain, or student, is matched to only one member of the range, or student ID number.

8. Yes

9. $f(x) = x + 5$
$f(3) = 3 + 5 = 8$
$f(7) = 7 + 5 = 12$
$f(-9) = -9 + 5 = -4$

10. $-6, 0, 12$

11. $h(p) = 3p$
$h(-2) = 3(-2) = -6$
$h(5) = 3 \cdot 5 = 15$
$h(24) = 3 \cdot 24 = 72$

12. $-24, 2, -80$

13. $g(s) = 2s + 4$
$g(1) = 2 \cdot 1 + 4 = 2 + 4 = 6$
$g(-7) = 2(-7) + 4 = -14 + 4 = -10$
$g(6.7) = 2(6.7) + 4 = 13.4 + 4 = 17.4$

14. $19, 19, 19$

15. $F(x) = 2x^2 - 3x + 2$
$F(0) = 2 \cdot 0^2 - 3 \cdot 0 + 2 = 0 - 0 + 2 = 2$
$F(-1) = 2(-1)^2 - 3(-1) + 2 = 2 + 3 + 2 = 7$
$F(2) = 2 \cdot 2^2 - 3 \cdot 2 + 2 = 8 - 6 + 2 = 4$

16. $5, 21, 26$

17. $h(x) = |x|$
$h(-4) = |-4| = 4$
$h\left(\frac{2}{3}\right) = \left|\frac{2}{3}\right| = \frac{2}{3}$
$h(-3.8) = |-3.8| = 3.8$

18. $6, 1, \frac{13}{4}$

19. $f(x) = |x| - 2$
$f(-3) = |-3| - 2 = 3 - 2 = 1$
$f(93) = |93| - 2 = 93 - 2 = 91$
$f(-100) = |-100| - 2 = 100 - 2 = 98$

20. $4, -122, 3$

21. $h(x) = x^4 - 3$
$h(0) = 0^4 - 3 = 0 - 3 = -3$
$h(-1) = (-1)^4 - 3 = 1 - 3 = -2$
$h(3) = 3^4 - 3 = 81 - 3 = 78$

22. $-\frac{2}{3}, -1, -2, 4, 2, 1, \frac{2}{3}, \frac{1}{5}$

Chapter 7 (7.8)

23. $M = \frac{1}{5}n$

    To complete the table, we successively substitute values of n and compute M.

    For n = 15: $M = \frac{1}{5}(15) = 3$

    For n = 7: $M = \frac{1}{5}(7) = \frac{7}{5}$

    For n = 8: $M = \frac{1}{5}(8) = \frac{8}{5}$

    For n = 10: $M = \frac{1}{5}(10) = 2$

n (sec)	15	7	8	10
M(mi)	3	$\frac{7}{5}$	$\frac{8}{5}$	2

24. a) 209.36 cm

    b) 203.48 cm

25. $P(d) = 1 + \frac{d}{33}$

    $P(20) = 1 + \frac{20}{33} = 1\frac{20}{33} \approx 1.606$ atmospheres

    $P(30) = 1 + \frac{30}{33} = 1\frac{30}{33} \approx 1.909$ atmospheres

    $P(100) = 1 + \frac{100}{33} = 1 + 3\frac{1}{33} = 4\frac{1}{33} \approx 4.03$ atmospheres

26. $166\frac{2}{3}, 666\frac{2}{3}, 833\frac{1}{3}$

27. $T(d) = 10d + 20$

    $T(5) = 10 \cdot 5 + 20 = 50 + 20 = 70°$ C

    $T(20) = 10 \cdot 20 + 20 = 200 + 20 = 220°$ C

    $T(1000) = 10 \cdot 1000 + 20 = 10,000 + 20 = 10,020°$ C

28. 1.792 cm, 2.8 cm, 11.2 cm

29. $D(p) = -2.7p + 16.3$

    $D(1) = -2.7(1) + 16.3 = -2.7 + 16.3 = 13.6$ million

    $D(2) = -2.7(2) + 16.3 = -5.4 + 16.3 = 10.9$ million

    $D(3) = -2.7(3) + 16.3 = -8.1 + 16.3 = 8.2$ million

    $D(4) = -2.7(4) + 16.3 = -10.8 + 16.3 = 5.5$ million

    $D(5) = -2.7(5) + 16.3 = -13.5 + 16.3 = 2.8$ million

30. 1.1 million, 8.1 million, 15.1 million, 22.1 million, 29.1 million

31. Graph $f(x) = x + 4$.

    Make a list of function values in a table.

    When $x = -2$, $f(-2) = -2 + 4 = 2$.
    When $x = 0$, $f(0) = 0 + 4 = 4$.
    When $x = 1$, $f(1) = 1 + 4 = 5$.

x	f(x)
-2	2
0	4
1	5

    We plot these points and connect them.

    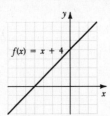

32.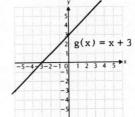

33. Graph $h(x) = 2x - 3$

    Make a list of function values in a table.

    When $x = -1$, $h(-1) = 2(-1) - 3 = -2 - 3 = -5$.
    When $x = 0$, $h(0) = 2 \cdot 0 - 3 = 0 - 3 = -3$.
    When $x = 3$, $h(3) = 2 \cdot 3 - 3 = 6 - 3 = 3$.

x	h(x)
-1	-5
0	-3
3	3

    We plot these points and connect them.

    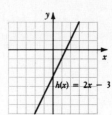

34.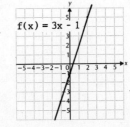

35. Graph g(x) = x - 6.

Make a list of function values in a table.

When x = 0, g(0) = 0 - 6 = -6.
When x = 2, g(2) = 2 - 6 = -4.
When x = 6, g(6) = 6 - 6 = 0.

x	g(x)
0	-6
2	-4
6	0

We plot these points and connect them.

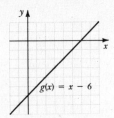

36.

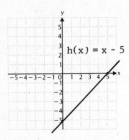

37. Graph f(x) = 2x - 7.

Make a list of function values in a table.
When x = 1, f(1) = 2·1 - 7 = 2 - 7 = -5.
When x = 2, f(2) = 2·2 - 7 = 4 - 7 = -3.
When x = 5, f(5) = 2·5 - 7 = 10 - 7 = 3.

x	f(x)
1	-5
2	-3
5	3

Plot these points and connect them.

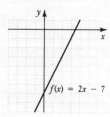

38.

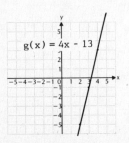

39. Graph f(x) = $\frac{1}{2}$x + 1.

Make a list of function values in a table.

When x = -2, f(-2) = $\frac{1}{2}$(-2) + 1 = -1 + 1 = 0.

When x = 0, f(0) = $\frac{1}{2}$ · 0 + 1 = 0 + 1 = 1.

When x = 4, f(4) = $\frac{1}{2}$ · 4 + 1 = 2 + 1 = 3.

x	f(x)
-2	0
0	1
4	3

Plot these points and connect them.

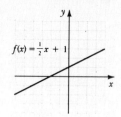

40.

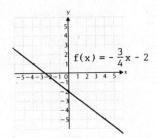

41. Graph g(x) = 2|x|.

Make a list of function values in a table.

When x = -3, g(-3) = 2|-3| = 2·3 = 6.
When x = -1, g(-1) = 2|-1| = 2·1 = 2.
When x = 0, g(0) = 2|0| = 2·0 = 0.
When x = 1, g(1) = 2|1| = 2·1 = 2.
When x = 3, g(3) = 2|3| = 2·3 = 6.

x	g(x)
-3	6
-1	2
0	0
1	2
3	6

Plot these points and connect them.

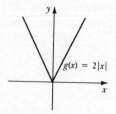

42.

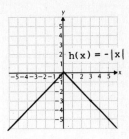

43. Graph $g(x) = x^2$.

Make a list of function values in a table.

When $x = -2$, $g(-2) = (-2)^2 = 4$.
When $x = -1$, $g(-1) = (-1)^2 = 1$.
When $x = 0$, $g(0) = 0^2 = 0$.
When $x = 1$, $g(1) = 1^2 = 1$.
When $x = 2$, $g(2) = 2^2 = 4$.

x	g(x)
-2	4
-1	1
0	0
1	1
2	4

Plot these points and connect them.

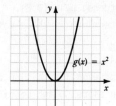

44.

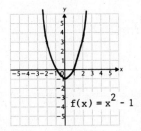

45. Graph $f(x) = \frac{2}{x}$.

Make a list of function values in a table.

When $x = -4$, $f(-4) = \frac{2}{-4} = -\frac{1}{2}$.

When $x = -2$, $f(-2) = \frac{2}{-2} = -1$.

When $x = -1$, $f(-1) = \frac{2}{-1} = -2$.

When $x = -\frac{1}{2}$, $f\left(-\frac{1}{2}\right) = \frac{2}{-\frac{1}{2}} = 2 \cdot (-2) = -4$.

When $x = \frac{1}{2}$, $f\left(\frac{1}{2}\right) = \frac{2}{\frac{1}{2}} = 2 \cdot 2 = 4$.

When $x = 1$, $f(1) = \frac{2}{1} = 2$.

When $x = 2$, $f(2) = \frac{2}{2} = 1$.

When $x = 4$, $f(4) = \frac{2}{4} = \frac{1}{2}$.

x	f(x)
-4	$-\frac{1}{2}$
-2	-1
-1	-2
$-\frac{1}{2}$	-4
$\frac{1}{2}$	4
1	2
2	1
4	$\frac{1}{2}$

Plot these points and draw the two parts of the graph. Since division by 0 is undefined, 0 cannot be an element of the domain.

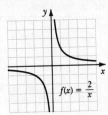

46.

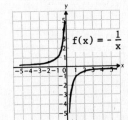

47. Graph $f(x) = 3 - x^2$.

Make a list of function values in a table.

When $x = -2$, $f(-2) = 3 - (-2)^2 = 3 - 4 = -1$
When $x = -1$, $f(-1) = 3 - (-1)^2 = 3 - 1 = 2$
When $x = 0$, $f(0) = 3 - 0^2 = 3 - 0 = 3$
When $x = 1$, $f(1) = 3 - 1^2 = 3 - 1 = 2$
When $x = 2$, $f(2) = 3 - 2^2 = 3 - 4 = -1$

x	f(x)
-2	-1
-1	2
0	3
1	2
2	-1

Plot these points and connect them.

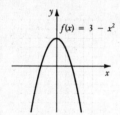

48.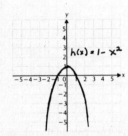

49. This graph is not a graph of a function. It is possible for a vertical line to cross the graph at more than one point.

50. Yes

51. This graph is a graph of a function. No vertical line can cross the graph at more than one point.

52. No

53. $f(x) = 3x + 5$

The domain is the set $\{0, 1, 2, 3\}$.

$f(0) = 3 \cdot 0 + 5 = 0 + 5 = 5$
$f(1) = 3 \cdot 1 + 5 = 3 + 5 = 8$
$f(2) = 3 \cdot 2 + 5 = 6 + 5 = 11$
$f(3) = 3 \cdot 3 + 5 = 9 + 5 = 14$

The range is the set $\{5, 8, 11, 14\}$.

54. $\{-5, -4, -1, 4\}$

55. $h(x) = |x| - x$

The domain is the set $\{-1, 0, 1, 2, 3, 4, 5, 6, 7, 8, 9, 10, 11, 12, 13, 14, 15, 16, 17, 18, 19\}$.

$h(-1) = |-1| - (-1) = 1 + 1 = 2$
$h(0) = |0| - 0 = 0 - 0 = 0$
$h(1) = |1| - 1 = 1 - 1 = 0$
$h(2) = |2| - 2 = 2 - 2 = 0$
$h(3) = |3| - 3 = 3 - 3 = 0$
.
.
.
$h(19) = |19| - 19 = 19 - 19 = 0$

The range is the set $\{0, 2\}$.

56. $\{-7, 0, 1, 2, 9\}$

57. $f(-1) = -7$ gives us the ordered pair $(-1, -7)$.
$f(3) = 8$ gives us the ordered pair $(3, 8)$.

The slope of the line determined by these points is

$$m = \frac{-7 - 8}{-1 - 3} = \frac{-15}{-4} = \frac{15}{4}.$$

We use the slope-intercept equation to find the equation of the line with slope $\frac{15}{4}$ and containing the point $(3, 8)$.

$$y - y_1 = m(x - x_1)$$
$$y - 8 = \frac{15}{4}(x - 3)$$
$$y - 8 = \frac{15}{4}x - \frac{45}{4}$$
$$y = \frac{15}{4}x - \frac{13}{4}$$

58. $-1, 1, -1, 1$, undefined

59. $g(x) = |x| + x$

$g(-3) = |-3| + (-3) = 3 - 3 = 0$
$g(3) = |3| + 3 = 3 + 3 = 6$
$g(-2) = |-2| + (-2) = 2 - 2 = 0$
$g(2) = |2| + 2 = 2 + 2 = 4$
$g(0) = |0| + 0 = 0 + 0 = 0$

60. $6, 0, 4, 0, 0$

61. Graph $f(x) = \frac{|x|}{x}$.

We list the function values found in Exercise 58 in a table.

x	f(x)
-3	-1
-2	-1
0	undefined
2	1
3	1

We plot these points and draw the graph. We use open circles on the y-axis to indicate that the function is undefined for x = 0.

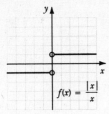

62.

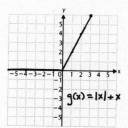

63. Graph $h(x) = |x| - x$.

We list the function values found in Exercise 60 in a table.

x	h(x)
-3	6
-2	4
0	0
2	0
3	0

We plot these points and draw the graph.

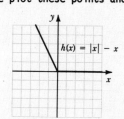

64. Answers may vary. Here are two possible answers.

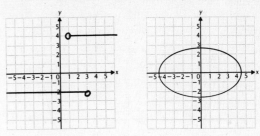

65. Graph $|y| = x$.

We list some function values in a table. For this function it is perhaps best to choose any value for y and then find the corresponding x value. Note that the ordered pairs are still of the form (x,y) even though we choose y first.

x	y
3	-3
1	-1
0	0
1	1
3	3

When y = -3, x = |-3| = 3.
When y = -1, x = |-1| = 1.
When y = 0, x = |0| = 0.
When y = 1, x = |1| = 1.
When y = 3, x = |3| = 3.

We plot these points and draw the graph.

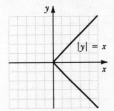

This is not the graph of a function since it is possible for a vertical line to intersect the graph at more than one point.

66.

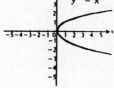

No

Chapter 7   (7.8)

67.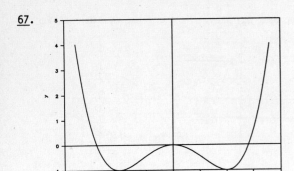

68.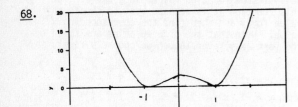

69.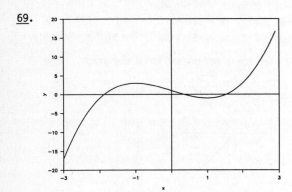

70.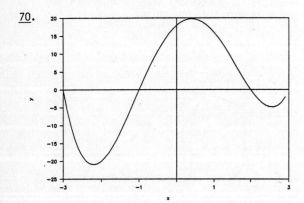

194

CHAPTER 8   SYSTEMS OF EQUATIONS AND PROBLEM SOLVING

Exercise Set 8.1

1. <u>Familiarize</u>. Let x = one number and y = the other.

   <u>Translate</u>. We translate the first statement.

   The sum of two numbers, is 58.
   x + y   = 58

   Now we translate the second statement.

   The difference of two numbers, is 16.
   x - y   = 16

   (The second statement could also be translated as y - x = 16.)

   We have a system of equations:
   x + y = 58,
   x - y = 16

2. x + y = 26.4,
   x = 5y;
   x is one number, y is the other

3. <u>Familiarize</u>. From the drawing in the text we see that we have a rectangle with length L and width W. The perimeter is L + L + W + W, or 2L + 2W.

   <u>Translate</u>. We translate the first statement.

   The perimeter, is 1280 mi.
   2L + 2W   =   1280

   Now we reword and translate the second statement.

   The width, is the length, less 90 mi.
   W   =   L   -   90

   We have a system of equations:
   2L + 2W = 1280,
   W = L - 90

4. 2L + 2W = 42,
   L = W + 3;
   L = length, W = width

5. <u>Familiarize</u>. We let m = the number of miles driven and c = the total cost of the car rental.

   <u>Translate</u>. We reword and translate the first statement, using $0.30 for 30¢.

   $53.95 plus 30¢ times the number of miles driven is cost.
   53.95 + 0.30 ·   m   = c

   We reword and translate the second statement using $0.20 for 20¢.

   $54.95 plus 20¢ times the number of miles driven is cost.
   54.95 + 0.20 ·   m   = c

   We have a system of equations:
   53.95 + 0.30m = c,
   54.95 + 0.20m = c

6. 45.95 + 0.40m = c,
   46.95 + 0.20m = c;
   m = mileage, c = cost

7. <u>Familiarize</u>. We let x = the larger number and y = the smaller number.

   <u>Translate</u>. We translate the first statement:

   The difference between two numbers, is 16.
   x - y   = 16

   Now we translate the second statement.

   Three times the is seven times the
   larger number   smaller number.
        3x   =   7y

   We have a system of equations:
   x - y = 16,
   3x = 7y

8. x - y = 18,
   2y + 3x = 74;
   x is the larger number, y is the smaller number.

9. <u>Familiarize</u>. We let x = the smaller angle and y = the larger angle.

   <u>Translate</u>. We reword and translate the first statement.

   The sum of two angles, is 180°.
   x + y   = 180

   We reword and translate the second statement.

   One   is three times less 8°.
   angle   the other
    y   =   3x   -   8

   We have a system of equations.
   x + y = 180,
   y = 3x - 8

10. x + y = 180,
    y = 2x + 30;
    x = the smaller angle, y = the larger angle

11. <u>Familiarize</u>. We let x = the larger angle and y = the smaller angle.

    <u>Translate</u>. We reword and translate the first statement.

    The sum of two angles, is 90°.
    x + y   = 90

    We reword and translate the second statement.

    The difference of two angles, is 34°.
    x - y   = 34

    We have a system of equations:
    x + y = 90,
    x - y = 34

Chapter 8 (8.1)

12. $x + y = 90$,
$y = \frac{1}{2}x + 42$;
x and y are the angles

13. Familiarize. We let x = the number of hectares of Riesling grapes that should be planted and y = the number of hectares of Chardonnay grapes that should be planted.

Translate. We reword and translate the first statement.

$\underline{\text{Total number of hectares}}$ is $\underline{820}$.
$\quad\quad x + y \quad\quad\quad\quad\quad = \quad 820$

Now we reword and translate the second statement.

$\underline{\text{Hectares of Chardonnay grapes}}$ is $\underline{140\text{ hectares}}$ $\underline{\text{more than}}$ $\underline{\text{hectares of Riesling grapes}}$.
$\quad\quad y \quad\quad\quad = \quad 140 \quad + \quad x$

We have a system of equations:
$x + y = 820$,
$y = 140 + x$

14. $x + y = 650$,
$x = y + 180$;
x = hectares of hay, y = hectares of oats

15. Familiarize. We make a drawing. We let ℓ = the length and w = the width.

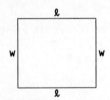

The perimeter is $\ell + \ell + w + w$, or $2\ell + 2w$.
Translate. We translate the first statement.

$\underline{\text{The perimeter}}$ is $\underline{400\text{ m.}}$
$\quad 2\ell + 2w \quad\quad = \quad 400$

We translate the second statement.

$\underline{\text{The length}}$ is $\underline{3m}$ $\underline{\text{more than}}$ $\underline{\text{twice the width}}$.
$\quad \ell \quad\quad = \quad 3 \quad + \quad 2w$

We have a system of equations:
$2\ell + 2w = 400$,
$\ell = 3 + 2w$

16. $2\ell + 2w = 76$,
$w = \ell - 17$;
ℓ = length, w = width

17. Familiarize. We make a drawing. We let ℓ = the length and w = the width.

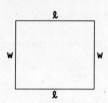

The perimeter is $\ell + \ell + w + w$, or $2\ell + 2w$.
Translate. We translate the first statement.

$\underline{\text{The perimeter}}$ is $\underline{876\text{ cm.}}$
$\quad 2\ell + 2w \quad\quad = \quad 876$

We reword and translate the second statement.

$\underline{\text{The length}}$ is $\underline{\text{three times the width}}$ $\underline{\text{less}}$ $\underline{1\text{ cm.}}$
$\quad \ell \quad\quad = \quad 3w \quad - \quad 1$

We have a system of equations:
$2\ell + 2w = 876$,
$\ell = 3w - 1$

18. $2\ell + 2w = 306\frac{2}{3}$,
$\ell = w + 46\frac{2}{3}$;
ℓ = length, w = width

19. $(9x^{-5})(12x^{-8}) = 9 \cdot 12 x^{-5+(-8)} = 108x^{-13}$

20. $3x^3$

21. $(5x^{-2}y^5)^{-4} = 5^{-4}x^{-2(-4)}y^{5(-4)}$
$= \frac{1}{5^4}x^8y^{-20}$
$= \frac{1}{625}x^8y^{-20}$, or $\frac{x^8}{625y^{20}}$

22. 3.24×10^{-5}

23. Familiarize. We let x = Patrick's age now and y = his father's age now. Twenty years from now Patrick's age will be $x + 20$, and his father's age will be $y + 20$.

Translate. We translate the first statement.

$\underline{\text{Patrick's age now}}$ is $\underline{20\%}$ of $\underline{\text{his father's age now.}}$
$\quad x \quad\quad\quad = \quad 20\% \quad \cdot \quad\quad y$

Now we translate the second statement.

$\underline{\text{Patrick's age twenty years from now}}$ $\underline{\text{will be}}$ $\underline{52\%}$ of $\underline{\text{his father's age twenty years from now.}}$
$\quad x + 20 \quad\quad\quad = \quad 52\% \quad \cdot \quad (y + 20)$

We have a system of equations:
$x = 0.2y$,
$x + 20 = 0.52(y + 20)$

Chapter 8 (8.2)

24. $\frac{x+5}{5} = y$,

    $x - 5 = 8(y - 5)$;

    x = the man's age now, y = his daughter's age now

25. <u>Familiarize</u>. We first make a drawing. We have used h for the height and b for the base in the original triangle. Then $h - 1$ and $b + 2$ represent the height and base in the new triangle.

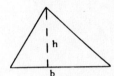
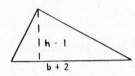

The area of a triangle is $\frac{1}{2} \cdot$ base $\cdot$ height.

<u>Translate</u>. We reword and translate the first statement.

The height of the new triangle	is	$\frac{1}{3}$	of	the base of the new triangle.
$h - 1$	=	$\frac{1}{3}$	$\cdot$	$(b + 2)$

Now we reword and translate the second statement.

The area of the new triangle	is	$\frac{1}{2}$	$\cdot$	base	$\cdot$	height
24	=	$\frac{1}{2}$	$\cdot$	$(b + 2)$	$\cdot$	$(h - 1)$

We have a system of equations:

$h - 1 = \frac{1}{3}(b + 2)$,

$24 = \frac{1}{2}(b + 2)(h - 1)$

Exercise Set 8.2

1. We check by substituting alphabetically 3 for x and 2 for y.

$2x + 3y = 12$		$x - 4y = -5$	
$2 \cdot 3 + 3 \cdot 2$	12	$3 - 4 \cdot 2$	-5
$6 + 6$		$3 - 8$	
12		-5	

    The ordered pair (3,2) is a solution of each equation. Therefore it is a solution of the system of equations.

2. Yes

3. We check by substituting alphabetically 3 for s and 2 for t.

$3t - 2s = 0$		$t + 2s = 15$	
$3 \cdot 2 - 2 \cdot 3$	0	$2 + 2 \cdot 3$	15
$6 - 6$		$2 + 6$	
0		8	

    The ordered pair (3,2) is not a solution of $t + 2s = 15$. Therefore it is not a solution of the system of equations.

4. Yes

5. We check by substituting alphabetically 15 for x and 20 for y.

$3x - 2y = 5$		$6x - 5y = -10$	
$3 \cdot 15 - 2 \cdot 20$	5	$6 \cdot 15 - 5 \cdot 20$	-10
$45 - 40$		$90 - 100$	
5		-10	

    The ordered pair (15,20) is a solution of each equation. Therefore it is a solution of the system of equations.

6. Yes

7. We check by substituting alphabetically -1 for x and 1 for y.

$x = -1$		$x - y = -2$	
-1	-1	$-1 - 1$	-2
		-2	

    The ordered pair $(-1,1)$ is a solution of each equation. Therefore it is a solution of the system of equations.

8. No

9. We check by substituting alphabetically 12 for x and 3 for y.

$y = \frac{1}{4}x$		$3x - y = 33$	
3	$\frac{1}{4} \cdot 12$	$3 \cdot 12 - 3$	33
	3	$36 - 3$	
		33	

    The ordered pair (12,3) is a solution of each equation. Therefore it is a solution of the system of equations.

10. Yes

Chapter 8   (8.2)

11. We graph the equations.

    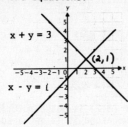

    The point of intersection looks as if it has coordinates (2,1).
    Check:

$x + y = 3$	$x - y = 1$
$2 + 1$ | 3	$2 - 1$ | 1
3	1

    The solution is (2,1).

12. (4,2)

13. We graph the equations.

    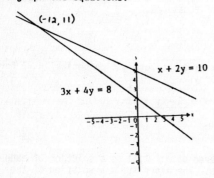

    The point of intersection looks as if it has coordinates (-12,11).
    Check:

$x + 2y = 10$	$3x + 4y = 8$
$-12 + 2 \cdot 11$ | 10	$3(-12) + 4(11)$ | 8
$-12 + 22$	$-36 + 44$
10	8

    The solution is (-12,11).

14. (-8,-7)

15. We graph the equations.

    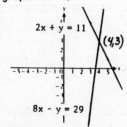

    The point of intersection looks as if it has coordinates (4,3).

15. (continued)
    Check:

$8x - y = 29$	$2x + y = 11$
$8 \cdot 4 - 3$ | 29	$2 \cdot 4 + 3$ | 11
$32 - 3$	$8 + 3$
29	11

    The solution is (4,3).

16. (3,2)

17. We graph the equations.

    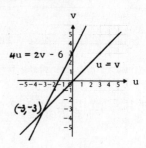

    The point of intersection looks as if it has coordinates (-3,-3).
    Check:

$u = v$	$4u = 2v - 6$
-3 | -3	$4(-3)$ | $2(-3) - 6$
	-12 | $-6 - 6$
	-12

    The solution is (-3,-3).

18. (-6,-2)

19. We graph the equations.

    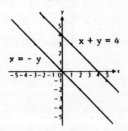

    The lines are parallel. There is no solution.

20. Infinite number of solutions

21. We graph the equations.

    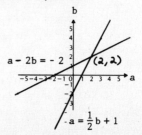

198

Chapter 8   (8.2)

21. (continued)

    The point of intersection looks as if it has coordinates (2,2).
    Check:

    $$\begin{array}{c|c} a = \frac{1}{2}b + 1 \\ \hline 2 & \frac{1}{2} \cdot 2 + 1 \\ & 1 + 1 \\ & 2 \end{array} \qquad \begin{array}{c|c} a - 2b = -2 \\ \hline 2 - 2\cdot 2 & -2 \\ 2 - 4 & \\ & -2 \end{array}$$

    The solution is (2,2).

22. (1,-3)

23. We graph the equations.

    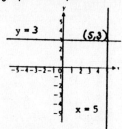

    The point of intersection looks as if it has coordinates (5,3).
    Check:

    $$\begin{array}{c|c} y = 3 \\ \hline 3 & 3 \end{array} \qquad \begin{array}{c|c} x = 5 \\ \hline 5 & 5 \end{array}$$

    The solution is (5,3).

24. $\left(\frac{1}{3}, 1\right)$

25. We graph the equations.

    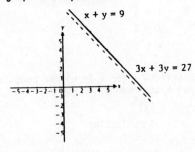

    The lines coincide. The system has an infinite number of solutions.

26. No solution

27. We graph the equations.

    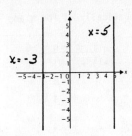

    The lines are parallel. The system has no solution.

28. (5,-3)

29. $\frac{1}{x} - \frac{1}{x^2} + \frac{1}{x+1}$,   LCM is $x^2(x+1)$

    $= \frac{1}{x} \cdot \frac{x(x+1)}{x(x+1)} - \frac{1}{x^2} \cdot \frac{x+1}{x+1} + \frac{1}{x+1} \cdot \frac{x^2}{x^2}$

    $= \frac{x(x+1) - (x+1) + x^2}{x^2(x+1)}$

    $= \frac{x^2 + x - x - 1 + x^2}{x^2(x+1)}$

    $= \frac{2x^2 - 1}{x^2(x+1)}$

30. $\frac{-4}{x-2}$

31. $\frac{x+2}{x-4} - \frac{x+1}{x+4}$,   LCM is $(x-4)(x+4)$

    $= \frac{x+2}{x-4} \cdot \frac{x+4}{x+4} - \frac{x+1}{x+4} \cdot \frac{x-4}{x-4}$

    $= \frac{(x+2)(x+4) - (x+1)(x-4)}{(x-4)(x+4)}$

    $= \frac{x^2 + 6x + 8 - (x^2 - 3x - 4)}{(x-4)(x+4)}$

    $= \frac{x^2 + 6x + 8 - x^2 + 3x + 4}{(x-4)(x+4)}$

    $= \frac{9x + 12}{(x-4)(x+4)}$

32. $\frac{2x+5}{x+3}$

33. (2,-3) is a solution of $Ax - 3y = 13$.
    Substitute 2 for x and -3 for y and solve for A.

    $$Ax - 3y = 13$$
    $$A \cdot 2 - 3(-3) = 13$$
    $$2A + 9 = 13$$
    $$2A = 4$$
    $$A = 2$$

    (2,-3) is a solution of $x - By = 8$. Substitute 2 for x and -3 for y and solve for B.

    $$x - By = 8$$
    $$2 - B(-3) = 8$$
    $$2 + 3B = 8$$
    $$3B = 6$$
    $$B = 2$$

199

Chapter 8   (8.3)

34. No

35. We substitute alphabetically -1 for a and -5 for b.

$4a - b = 1$		$-a + b = -4$	
$4(-1) - (-5)$	1	$-(-1) + (-5)$	-4
$-4 + 5$		$1 - 5$	
1		-4	

$2a + 3b = -17$	
$2(-1) + 3(-5)$	-17
$-2 - 15$	
-17	

    The ordered pair $(-1,-5)$ is a solution of each equation. Therefore it is a solution of the system of equations.

36. Three lines intersecting in one point. The solution is one point.

37. We graph the equations.

    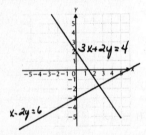

    It is difficult to determine the solution since it appears that fractions are involved. The solution is $\left(\frac{5}{2}, -\frac{7}{4}\right)$. The check would tell you whether the possible pair you found by graphing is the solution.

38. 11-18, 20-25, 28

39. Systems in which the graphs of the equations coincide are dependent. This is the case in Exercises 20 and 25.

40. 19, 26, 27

41. Systems in which the graphs of the equations intersect in exactly one point or are parallel are independent. This is the case in Exercises 11-19, 21-24, and 26-28.

42. Answers may vary.
    $2x - y = 8$,
    $x + 3y = -10$

43. Answers may vary. Any equation with the solution $(3,-2)$ will do. One possibility is $2x - y = 8$.

Exercise Set 8.3

1. $x + y = 4$,   (1)
   $y = 2x + 1$ (2)

   We substitute $2x + 1$ for y in equation (1) and solve for x.

   $x + y = 4$ (1)
   $x + (2x + 1) = 4$ Substituting
   $3x + 1 = 4$
   $3x = 3$
   $x = 1$

   Next we substitute 1 for x in either equation of the original system and solve for y.

   $x + y = 4$ (1)
   $1 + y = 4$ Substituting
   $y = 3$

   We check the orderd pair (1,3).

$x + y = 4$		$y = 2x + 1$	
$1 + 3$	4	3	$2 \cdot 1 + 1$
4			$2 + 1$
			3

   Since (1,3) checks in both equations, it is the solution.

2. (1,9)

3. $y = x + 1$,   (1)
   $2x + y = 4$ (2)

   We substitute $x + 1$ for y in equation (2) and solve for x.

   $2x + y = 4$ (2)
   $2x + (x + 1) = 4$ Substituting
   $3x + 1 = 4$
   $3x = 3$
   $x = 1$

   Next we substitute 1 for x in either equation of the original system and solve for y.

   $y = x + 1$ (1)
   $y = 1 + 1$ Substituting
   $y = 2$

   We check the ordered pair (1,2).

$y = x + 1$		$2x + y = 4$	
2	$1 + 1$	$2 \cdot 1 + 2$	4
	2	$2 + 2$	
		4	

   Since (1,2) checks in both equations, it is the solution.

4. (2,-4)

200

Chapter 8 (8.3)

5. $y = 2x - 5$, (1)
    $3y - x = 5$ (2)

    We substitute $2x - 5$ for y in equation (2) and solve for x.
    $$3y - x = 5 \quad (2)$$
    $$3(2x - 5) - x = 5 \quad \text{Substituting}$$
    $$6x - 15 - x = 5$$
    $$5x - 15 = 5$$
    $$5x = 20$$
    $$x = 4$$

    Next we substitute 4 for x in either equation of the original system and solve for y.
    $$y = 2x - 5 \quad (1)$$
    $$y = 2 \cdot 4 - 5 \quad \text{Substituting}$$
    $$y = 8 - 5$$
    $$y = 3$$

    We check the ordered pair (4,3).

$y = 2x - 5$	$3y - x = 5$
3 | 2·4 - 5	3·3 - 4 | 5
8 - 5	9 - 4
3	5

    Since (4,3) checks in both equations, it is the solution.

6. $(-1,-1)$

7. $x = -2y$, (1)
    $x + 4y = 2$ (2)

    We substitute $-2y$ for x in equation (2) and solve for y.
    $$x + 4y = 2 \quad (2)$$
    $$-2y + 4y = 2$$
    $$2y = 2$$
    $$y = 1$$

    Next we substitute 1 for y in either equation of the original system and solve for x.
    $$x = -2y \quad (1)$$
    $$x = -2 \cdot 1$$
    $$x = -2$$

    We check the ordered pair (-2,1).

$x = -2y$	$3y - x = 5$
-2 | -2·1	3·1 - (-2) | 5
-2	3 + 2
	5

    Since (-2,1) checks in both equations, it is the solution.

8. $(-30,10)$

9. $s + t = -4$, (1)
    $s - t = 2$ (2)

    We solve equation (2) for s.
    $$s - t = 2 \quad (2)$$
    $$s = t + 2 \quad (3)$$

    We substitute $t + 2$ for s in equation (1) and solve for t.
    $$s + t = -4 \quad (1)$$
    $$(t + 2) + t = -4 \quad \text{Substituting}$$
    $$2t + 2 = -4$$
    $$2t = -6$$
    $$t = -3$$

    Now we substitute -3 for t in either of the original equations or in equation (3) and solve for s. It is easiest to use (3).
    $$s = t + 2 = -3 + 2 = -1$$

    We check the ordered pair (-1,-3).

$s + t = -4$	$s - t = 2$
-1 + (-3) | -4	-1 - (-3) | 2
-4	-1 + 3
	2

    Since (-1,-3) checks in both equations, it is the solution.

10. $(2,-4)$

11. $y - 2x = -6$, (1)
    $2y - x = 5$ (2)

    We solve equation (1) for y.
    $$y - 2x = -6 \quad (1)$$
    $$y = 2x - 6 \quad (3)$$

    We substitute $2x - 6$ for y in equation (2) and solve for x.
    $$2y - x = 5 \quad (2)$$
    $$2(2x - 6) - x = 5 \quad \text{Substituting}$$
    $$4x - 12 - x = 5$$
    $$3x - 12 = 5$$
    $$3x = 17$$
    $$x = \frac{17}{3}$$

    We substitute $\frac{17}{3}$ for x in equation (3) and compute y.
    $$y = 2x - 6 = 2\left(\frac{17}{3}\right) - 6 = \frac{34}{3} - \frac{18}{3} = \frac{16}{3}.$$

    The ordered pair $\left(\frac{17}{3}, \frac{16}{3}\right)$ checks in both equations. It is the solution.

12. $\left(\frac{17}{3}, \frac{2}{3}\right)$

Chapter 8   (8.3)

13. $2x + 3y = -2$,   (1)
    $2x - y = 9$ (2)

    We solve equation (2) for y.
    $2x - y = 9$ (2)
    $2x - 9 = y$ (3)

    We substitute $2x - 9$ for y in equation (1) and solve for x.
    $2x + 3y = -2$ (1)
    $2x + 3(2x - 9) = -2$
    $2x + 6x - 27 = -2$
    $8x - 27 = -2$
    $8x = 25$
    $x = \frac{25}{8}$

    Now we substitute $\frac{25}{8}$ for x in equation (3) and compute y.
    $y = 2x - 9 = 2\left[\frac{25}{8}\right] - 9 = \frac{25}{4} - \frac{36}{4} = -\frac{11}{4}$

    The ordered pair $\left(\frac{25}{8}, -\frac{11}{4}\right)$ checks in both equations. It is the solution.

14. $(-12, 11)$

15. $x - y = -3$,   (1)
    $2x + 3y = -6$ (2)

    We solve equation (1) for x.
    $x - y = -3$ (1)
    $x = y - 3$ (3)

    We substitute $y - 3$ for x in equation (2) and solve for y.
    $2x + 3y = -6$ (2)
    $2(y - 3) + 3y = -6$ Substituting
    $2y - 6 + 3y = -6$
    $5y - 6 = -6$
    $5y = 0$
    $y = 0$

    Now we substitute 0 for y in equation (3) and compute x.
    $x = y - 3 = 0 - 3 = -3$

    The ordered pair $(-3, 0)$ checks in both equations. It is the solution.

16. $(4, -2)$

17. $r - 2s = 0$,   (1)
    $4r - 3s = 15$ (2)

    We solve equation (1) for r.
    $r - 2s = 0$ (1)
    $r = 2s$ (3)

    We substitute $2s$ for r in equation (2) and solve for s.
    $4r - 3s = 15$ (2)
    $4(2s) - 3s = 15$ Substituting
    $8s - 3s = 15$
    $5s = 15$
    $s = 3$

    Now we substitute 3 for s in equation (3) and compute r.
    $r = 2s = 2 \cdot 3 = 6$

    The ordered pair $(6, 3)$ checks in both equations. It is the solution.

18. $(1, 2)$

19. $x - 3y = 7$,   (1)
    $-4x + 12y = 28$ (2)

    We solve equation (1) for x.
    $x - 3y = 7$ (1)
    $x = 3y + 7$ (3)

    We substitute $3y + 7$ for x in equation (2) and solve for y.
    $-4x + 12y = 28$ (2)
    $-4(3y + 7) + 12y = 28$ Substituting
    $-12y - 28 + 12y = 28$
    $-28 = 28$

    We obtain a false equation, $-28 = 28$, so the system has no solution.

20. Infinitely many solutions

21. $y = 2x + 5$,   (1)
    $y = 2x - 4$ (2)

    We substitute $2x + 5$ for y in equation (2) and solve for x.
    $(2x + 5) = 2x - 4$
    $5 = -4$

    We obtain a false equation, so the system has no solution.

22. $\left(-\frac{3}{2}, -\frac{11}{6}\right)$

Chapter 8   (8.3)

23. $2x = y - 3$,   (1)
    $2x = y + 5$ (2)

    We solve equation (1) for y.
    $2x = y - 3$ (1)
    $2x + 3 = y$ (3)

    We substitute $2x + 3$ for y in equation (2) and solve for x.
    $2x = (2x + 3) + 5$
    $2x = 2x + 8$
    $0 = 8$

    We obtain a false equation, so the system has no solution.

24. No solution

25. **Familiarize.** We let x = the larger number and y = the smaller number.
    Translate.

    The sum of two numbers, is 27.
    $x + y$ = 27

    One number, is 3 more than, the other.
    x = 3 + y

    The resulting system is
    $x + y = 27$,   (1)
    $x = 3 + y$.   (2)

    Carry out. We solve the system of equations.
    We substitute $3 + y$ for x in equation (1) and solve for y.
    $x + y = 27$ (1)
    $(3 + y) + y = 27$ Substituting
    $3 + 2y = 27$
    $2y = 24$
    $y = 12$

    Next we substitute 12 for y in either equation of the original system and solve for x.
    $x + y = 27$ (1)
    $x + 12 = 27$ Substituting
    $x = 15$

    Check. The sum of 12 and 15 is 27. The number 15 is 3 more than 12. These numbers check.
    State. The numbers are 15 and 12.

26. 19, 17

27. The Familiarize and Translate steps were done in Exercise 1 of Exercise Set 8.1. The resulting system of equations is
    $x + y = 58$,   (1)
    $x - y = 16$,   (2)
    where x and y are the numbers.
    Carry out. We solve the system.
    We solve equation (2) for x.
    $x - y = 16$ (2)
    $x = y + 16$ (3)

    We substitute $y + 16$ for x in equation (1) and solve for y.
    $x + y = 58$ (1)
    $(y + 16) + y = 58$ Substituting
    $2y + 16 = 58$
    $2y = 42$
    $y = 21$

    Now we substitute 21 for y in equation (3) and compute x.
    $x = y + 16 = 21 + 16 = 37$

    Check. The sum of 37 and 21 is 58. The difference between 37 and 21, $37 - 21$, is 16. The numbers check.
    State. The numbers are 37 and 21.

28. 37, 29

29. The Familiarize and Translate steps were done in Exercise 7 of Exercise Set 8.1. The resulting system is
    $x - y = 16$,   (1)
    $3x = 7y$,   (2)
    where x represents the larger number and y the smaller.
    Carry out. We solve the system.
    We solve equation (1) for x.
    $x - y = 16$ (1)
    $x = y + 16$ (3)

    We substitute $y + 16$ for x in equation (2) and solve for y.
    $3x = 7y$ (2)
    $3(y + 16) = 7y$ Substituting
    $3y + 48 = 7y$
    $48 = 4y$
    $12 = y$

    Next we substitute 12 for y in equation (3) and compute x.
    $x = y + 16 = 12 + 16 = 28$

    Check. The difference between 28 and 12, $28 - 12$, is 16. Three times the larger, $3 \cdot 28$ or 84, is seven times the smaller, $7 \cdot 12 = 84$. The numbers check.
    State. The numbers are 28 and 12.

30. 22, 4

31. The Familiarize and Translate steps were done in Exercise 3 of Exercise Set 8.1. The resulting system is

$$2L + 2W = 1280, \quad (1)$$
$$W = L - 90, \quad (2)$$

where L represents the length and W the width.

Carry out. We solve the system.

We substitute $L - 90$ for W in equation (1) and solve for L.

$$2L + 2W = 1280 \quad (1)$$
$$2L + 2(L - 90) = 1280 \quad \text{Substituting}$$
$$2L + 2L - 180 = 1280$$
$$4L - 180 = 1280$$
$$4L = 1460$$
$$L = 365$$

Now we substitute 365 for L in equation (2).

$$W = L - 90 \quad (2)$$
$$W = 365 - 90 \quad \text{Substituting}$$
$$W = 275$$

Check. A possible solution is a length of 365 mi and a width of 275 mi. The perimeter would be $2(365) + 2(275)$, or $730 + 550$, or 1280. Also, the width is 90 mi less than the length. These numbers check.

State. The length is 365 mi, and the width is 275 mi.

32. Length: 12 ft, width: 9 ft

33. The Familiarize and Translate steps were done in Exercise 15 of Exercise Set 8.1. The resulting system is

$$2\ell + 2w = 400, \quad (1)$$
$$\ell = 3 + 2w \quad (2)$$

where ℓ represents the length and w the width.

Carry out. We solve the system.

We substitute $3 + 2w$ for ℓ in equation (1) and solve for w.

$$2\ell + 2w = 400 \quad (1)$$
$$2(3 + 2w) + 2w = 400 \quad \text{Substituting}$$
$$6 + 4w + 2w = 400$$
$$6w = 394$$
$$w = \frac{394}{6} = \frac{197}{3}, \text{ or } 65\frac{2}{3}$$

Now we substitute $\frac{197}{3}$ for w in equation (2).

$$\ell = 3 + 2w$$
$$\ell = 3 + 2\left[\frac{197}{3}\right]$$
$$\ell = \frac{9}{3} + \frac{394}{3}$$
$$\ell = \frac{403}{3}, \text{ or } 134\frac{1}{3}$$

33. (continued)

Check. A possible solution is a length of $134\frac{1}{3}$ m and a width of $65\frac{2}{3}$ m. The perimeter would be $2\left[134\frac{1}{3}\right] + 2\left[65\frac{2}{3}\right]$, or $2\left[\frac{403}{3}\right] + 2\left[\frac{197}{3}\right]$, or $\frac{806}{3} + \frac{394}{3}$, or $\frac{1200}{3}$, or 400. Also, 3 more than twice the width is $3 + 2\left[65\frac{2}{3}\right]$, or $3 + 2\left[\frac{197}{3}\right]$, or $\frac{9}{3} + \frac{394}{3}$, or $\frac{403}{3}$, or $134\frac{1}{3}$, which is the length. These numbers check.

State. The length is $134\frac{1}{3}$ m, and the width is $65\frac{2}{3}$ m.

34. Length: $328\frac{1}{4}$ cm, width: $109\frac{3}{4}$ cm

35. The Familiarize and Translate steps were done in Exercise 18 of Exercise Set 8.1. The resulting system is

$$2\ell + 2w = 306\frac{2}{3}, \quad (1)$$
$$\ell = 46\frac{2}{3} + w, \quad (2)$$

where ℓ represents the length and w the width.

Carry out. We solve the system.

We substitute $46\frac{2}{3} + w$ for ℓ in equation (1) and solve for w.

$$2\ell + 2w = 306\frac{2}{3} \quad (1)$$
$$2\left[46\frac{2}{3} + w\right] + 2w = 306\frac{2}{3} \quad \text{Substituting}$$
$$2\left[\frac{140}{3} + w\right] + 2w = \frac{920}{3}$$
$$\frac{280}{3} + 2w + 2w = \frac{920}{3}$$
$$\frac{280}{3} + 4w = \frac{920}{3}$$
$$4w = \frac{640}{3}$$
$$w = \frac{1}{4} \cdot \frac{640}{3}$$
$$w = \frac{160}{3}, \text{ or } 53\frac{1}{3}$$

Next we substitute $53\frac{1}{3}$ for w in equation (2) and solve for ℓ.

$$\ell = 46\frac{2}{3} + w \quad (2)$$
$$\ell = 46\frac{2}{3} + 53\frac{1}{3} \quad \text{Substituting}$$
$$\ell = 100$$

Check. A possible solution is a length of 100 yd and a width of $53\frac{1}{3}$ yd. The perimeter would be $2 \cdot 100 + 2 \cdot 53\frac{1}{3}$, or $306\frac{2}{3}$ yd. The length, 100 yd, is $46\frac{2}{3}$ yd more than the width. These numbers check.

State. The length is 100 yd, and the width is $53\frac{1}{3}$ yd.

Chapter 8   (8.3)

36.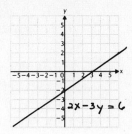

37. Graph: $2x + 3y = 6$

To find the x-intercept let $y = 0$:
$$2x + 3 \cdot 0 = 6$$
$$2x = 6$$
$$x = 3$$
The x-intercept is $(3,0)$.

To find the y-intercept, let $x = 0$.
$$2 \cdot 0 + 3y = 6$$
$$3y = 6$$
$$y = 2$$
The y-intercept is $(0,2)$.

We plot these points and draw the line.

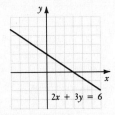

A third point should be used as a check. If we let $x = -3$, then
$$2(-3) + 3y = 6$$
$$-6 + 3y = 6$$
$$3y = 12$$
$$y = 4$$
The point $(-3,4)$ is on the graph, so our graph is probably correct.

38.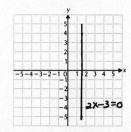

39. Graph: $y = 2x - 5$
We make a table of values.

x	y
-1	-7
2	-1
4	3

We plot these points and draw the line they determine.

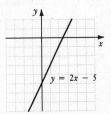

40. $(4.382, 4.328)$

41. $\frac{1}{4}(a - b) = 2$ (1)

$\frac{1}{6}(a + b) = 1$ (2)

We first clear the fractions.
$a - b = 8$ (1a)   Multiplying equation (1) by 4
$a + b = 6$ (2a)   Multiplying equation (2) by 6

We solve equation (1a) for a.
$$a - b = 8 \quad (1a)$$
$$a = b + 8$$

We substitute $b + 8$ for a in equation (2a) and solve for b.
$$a + b = 6 \quad (2a)$$
$$(b + 8) + b = 6 \quad \text{Substituting}$$
$$2b + 8 = 6$$
$$2b = -2$$
$$b = -1$$

Next we substitute -1 for b in equation (1a) and solve for a.
$$a - b = 8$$
$$a - (-1) = 8$$
$$a + 1 = 8$$
$$a = 7$$

We check the ordered pair $(7,-1)$.

$\frac{1}{4}(a - b) = 2$	$\frac{1}{6}(a + b) = 1$
$\frac{1}{4}[7 - (-1)]$? 2	$\frac{1}{6}[7 + (-1)]$? 1
$\frac{1}{4} \cdot 8$	$\frac{1}{6} \cdot 6$
2	1

Since $(7,-1)$ checks in both equations, it is the solution.

42. $(10,-2)$

205

Chapter 8 (8.3)

43. $0.4x + 0.7y = 0.1$ (1)
    $0.5x - 0.1y = 1.1$ (2)

We first multiply each equation by 10 to clear the decimals.

$4x + 7y = 1$ (1a)
$5x - y = 11$ (2a)

We solve equation (2a) for y.

$5x - y = 11$ (2a)
$5x = y + 11$
$5x - 11 = y$ (3)

Substitute $5x - 11$ for y in equation (1a) and solve for x.

$4x + 7y = 1$ (1a)
$4x + 7(5x - 11) = 1$ Substituting
$4x + 35x - 77 = 1$
$39x = 78$
$x = 2$

Next we substitute 2 for x in equation (3) and compute y.

$y = 5x - 11 = 5 \cdot 2 - 11 = 10 - 11 = -1$

We check the ordered pair (2,-1).

$0.4x + 0.7y = 0.1$		$0.5x - 0.1y = 1.1$	
$0.4(2) + 0.7(-1)$	0.1	$0.5(2) - 0.1(-1)$	1.1
$0.8 - 0.7$		$1.0 + 0.1$	
0.1			1.1

Since (2,-1) is a solution of both equations, it is the solution.

44. No

45.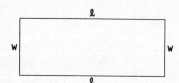

$P = 2\ell + 2w$, (1)
$w = \ell - 5$ (2)

Substitute $\ell - 5$ for w in equation (1) and solve for ℓ.

$P = 2\ell + 2w$
$P = 2\ell + 2(\ell - 5)$
$P = 2\ell + 2\ell - 10$
$P = 4\ell - 10$
$P + 10 = 4\ell$
$\frac{P + 10}{4} = \ell$

46. $w = \frac{P - 16}{4}$

47. $x + y + z = 4$, (1)
    $x - 2y - z = 1$, (2)
    $y = -1$ (3)

Substitute -1 for y in equations (1) and (2).

$x + y + z = 4$ (1)	$x - 2y - z = 1$ (2)
$x + (-1) + z = 4$	$x - 2(-1) - z = 1$
$x + z = 5$	$x + 2 - z = 1$
	$x - z = -1$

We now have a system of two equations in two variables.

$x + z = 5$, (4)
$x - z = -1$ (5)

We solve equation (5) for x.

$x - z = -1$ (5)
$x = z - 1$ (6)

We substitute $z - 1$ for x in equation (4) and solve for z.

$x + z = 5$ (4)
$(z - 1) + z = 5$ Substituting
$2z - 1 = 5$
$2z = 6$
$z = 3$

Next we substitute 3 for z in equation (6) and compute x.

$x = z - 1 = 3 - 1 = 2$

We check the ordered triple (2,-1,3).

$x + y + z = 4$		$x - 2y - z = 1$		$y = -1$	
$2 + (-1) + 3$	4	$2 - 2(-1) - 3$	1	-1	-1
4		$2 + 2 - 3$			
		1			

Since (2,-1,3) checks in all three equations, it is the solution.

48. (30,50,100)

49. $3y + 3x = 14$, (1)
    $y = -x + 4$ (2)

Substitute $-x + 4$ for y in equation (1) and solve for x.

$3y + 3x = 14$ (1)
$3(-x + 4) + 3x = 14$
$-3x + 12 + 3x = 14$
$12 = 14$

The x-terms drop out and leave 12 = 14, a false equation. There is no solution; the graphs of the equations are parallel lines.

50. We obtain 15 = 15, an equation that is true for all values of x and y. The graphs of the equations coincide so the system has infinitely many solutions.

Chapter 8   (8.4)

Exercise Set 8.4

1. $x + y = 10$ (1)
    $\underline{x - y = 8}$ (2)
    $2x = 18$ Adding
    $x = 9$

    Substitute 9 for x in one of the original equations and solve for y.
    $x + y = 10$ (1)
    $9 + y = 10$ Substituting
    $y = 1$

    Check:
$x + y = 10$		$x - y = 8$	
$9 + 1$	10	$9 - 1$	8
10		8	

    Since (9,1) checks, it is the solution.

2. (5,-2)

3. $x + y = 8$ (1)
    $\underline{-x + 2y = 7}$ (2)
    $3y = 15$ Adding
    $y = 5$

    Substitute 5 for y in one of the original equations and solve for x.
    $x + y = 8$ (1)
    $x + 5 = 8$ Substituting
    $x = 3$

    Check:
$x + y = 8$		$-x + 2y = 7$	
$3 + 5$	8	$-3 + 2 \cdot 5$	7
8		$-3 + 10$	
		7	

    Since (3,5) checks, it is the solution.

4. (5,1)

5. $3x - y = 9$ (1)
    $\underline{2x + y = 6}$ (2)
    $5x = 15$ Adding
    $x = 3$

    Substitute 3 for x in one of the original equations and solve for y.
    $2x + y = 6$ (2)
    $2 \cdot 3 + y = 6$ Substituting
    $6 + y = 6$
    $y = 0$

5. (continued)

    Check:
$3x - y = 9$		$2x + y = 6$	
$3 \cdot 3 - 0$	9	$2 \cdot 3 + 0$	6
$9 - 0$		$6 + 0$	
9		6	

    Since (3,0) checks, it is the solution.

6. (2,7)

7. $4a + 3b = 7$ (1)
    $\underline{-4a + b = 5}$ (2)
    $4b = 12$ Adding
    $b = 3$

    Substitute 3 for b in one of the original equations and solve for a.
    $4a + 3b = 7$ (1)
    $4a + 3 \cdot 3 = 7$ Substituting
    $4a + 9 = 7$
    $4a = -2$
    $a = -\frac{1}{2}$

    Check:
$4a + 3b = 7$		$-4a + b = 5$	
$4\left(-\frac{1}{2}\right) + 3 \cdot 3$	7	$-4\left(-\frac{1}{2}\right) + 3$	5
$-2 + 9$		$2 + 3$	
7		5	

    Since $\left(-\frac{1}{2}, 3\right)$ checks, it is the solution.

8. $\left(2, \frac{4}{5}\right)$

9. $8x - 5y = -9$ (1)
    $\underline{3x + 5y = -2}$ (2)
    $11x = -11$ Adding
    $x = -1$

    Substitute -1 for x in one of the original equations and solve for y.
    $3x + 5y = -2$ (2)
    $3(-1) + 5y = -2$ Substituting
    $-3 + 5y = -2$
    $5y = 1$
    $y = \frac{1}{5}$

    Check:
$8x - 5y = -9$		$3x + 5y = -2$	
$8(-1) - 5\left(\frac{1}{5}\right)$	-9	$3(-1) + 5\left(\frac{1}{5}\right)$	-2
$-8 - 1$		$-3 + 1$	
-9		-2	

    Since $\left(-1, \frac{1}{5}\right)$ checks, it is the solution.

10. (-2,3)

11. $4x - 5y = 7$
    $-4x + 5y = 7$
    $ 0 = 14$ Adding

    We obtain a false equation, 0 = 14, so there is no solution.

12. Infinitely many solutions

13. $-x - y = 8$,   (1)
    $2x - y = -1$ (2)

    We multiply by -1 on both sides of equation (1) and then add.

    $x + y = -8$ Multiplying by -1
    $2x - y = -1$
    $3x = -9$ Adding
    $x = -3$

    Substitute -3 for x in one of the original equations and solve for y.

    $2x - y = -1$ (2)
    $2(-3) - y = -1$ Substituting
    $-6 - y = -1$
    $-y = 5$
    $y = -5$

    Check:

$-x - y = 8$		$2x - y = -1$	
$-(-3) - (-5)$	8	$2(-3) - (-5)$	-1
$3 + 5$		$-6 + 5$	
8			-1

    Since (-3,-5) checks, it is the solution.

14. (-1,-6)

15. $x + 3y = 19$,
    $x - y = -1$

    We multiply by -1 on both sides of equation (2) and then add.

    $x + 3y = 19$
    $-x + y = 1$ Multiplying by -1
    $ 4y = 20$ Adding
    $y = 5$

    Substitute 5 for y in one of the original equations and solve for x.

    $x - y = -1$ (2)
    $x - 5 = -1$ Substituting
    $x = 4$

15. (continued)

    Check:

$x + 3y = 19$		$x - y = -1$	
$4 + 3 \cdot 5$	19	$4 - 5$	-1
$4 + 15$			-1
19			

    Since (4,5) checks, it is the solution.

16. (3,1)

17. $x + y = 5$,   (1)
    $5x - 3y = 17$ (2)

    We multiply by 3 on both sides of equation (1) and then add.

    $3x + 3y = 15$ Multiplying by 3
    $5x - 3y = 17$
    $8x = 32$
    $x = 4$

    Substitute 4 for x in one of the original equations and solve for y.

    $x + y = 5$ (1)
    $4 + y = 5$ Substituting
    $y = 1$

    Check:

$x + y = 5$		$5x - 3y = 17$	
$4 + 1$	5	$5 \cdot 4 - 3 \cdot 1$	17
5		$20 - 3$	
		17	

    Since (4,1) checks, it is the solution.

18. (10,3)

19. $2w - 3z = -1$,   (1)
    $3w + 4z = 24$ (2)

    We use the multiplication principle with both equations and then add.

    $8w - 12z = -4$ Multiplying (1) by 4
    $9w + 12z = 72$ Multiplying (2) by 3
    $17w = 68$ Adding
    $w = 4$

    Substitute 4 for w in one of the original equations and solve for z.

    $3w + 4z = 24$ (2)
    $3 \cdot 4 + 4z = 24$ Substituting
    $12 + 4z = 24$
    $4z = 12$
    $z = 3$

Chapter 8   (8.4)

19. (continued)

    Check:

$2w - 3z = -1$		$3w + 4z = 24$	
$2 \cdot 4 - 3 \cdot 3$	-1	$3 \cdot 4 + 4 \cdot 3$	24
$8 - 9$		$12 + 12$	
-1		24	

    Since (4,3) checks, it is the solution.

20. (1,-1)

21. $2a + 3b = -1$,   (1)
    $3a + 5b = -2$ (2)

    We use the multiplication principle with both equations and then add.

    $-10a - 15b = 5$ Multiplying (1) by -5
    $\underline{9a + 15b = -6}$ Multiplying (2) by 3
    $-a \quad\quad = -1$ Adding
    $a = 1$

    Substitute 1 for a in one of the original equations and solve for b.

    $2a + 3b = -1$ (1)
    $2 \cdot 1 + 3b = -1$ Substituting
    $3b = -3$
    $b = -1$

    Check:

$2a + 3b = -1$		$3a + 5b = -2$	
$2 \cdot 1 + 3(-1)$	-1	$3 \cdot 1 + 5(-1)$	-2
$2 - 3$		$3 - 5$	
-1		-2	

    Since (1,-1) checks, it is the solution.

22. (4,-1)

23. $x = 3y$,
    $5x + 14 = y$

    We first get each equation in the form $Ax + By = C$.

    $x - 3y = 0$,   (1) Adding $-3y$
    $5x - y = -14$ (2) Adding $-y - 14$

    We multiply by -5 on both sides of equation (1) and add.

    $-5x + 15y = 0$ Multiplying by -5
    $\underline{5x - y = -14}$
    $14y = -14$ Adding
    $y = -1$

    Substitute -1 for y in equation (1) and solve for x.

    $x - 3y = 0$
    $x - 3(-1) = 0$ Substituting
    $x + 3 = 0$
    $x = -3$

23. (continued)

    Check:

$x - 3y = 0$		$5x - y = -14$	
$-3 - 3(-1)$	0	$5(-3) - (-1)$	-14
$-3 + 3$		$-15 + 1$	
0		-14	

    Since (-3,-1) checks, it is the solution.

24. (2,5)

25. $3x - 2y = 10$,   (1)
    $5x + 3y = 4$ (2)

    We use the multiplication principle with both equations and add.

    $9x - 6y = 30$ Multiplying (1) by 3
    $\underline{10x + 6y = 8}$ Multiplying (2) by 2
    $19x \quad\quad = 38$ Adding
    $x = 2$

    Substitute 2 for x in one of the original equations and solve for y.

    $5x + 3y = 4$ (2)
    $5 \cdot 2 + 3y = 4$ Substituting
    $10 + 3y = 4$
    $3y = -6$
    $y = -2$

    Check:

$3x - 2y = 10$		$5x + 3y = 4$	
$3 \cdot 2 - 2(-2)$	10	$5 \cdot 2 + 3(-2)$	4
$6 + 4$		$10 - 6$	
10		4	

    Since (2,-2) checks, it is the solution.

26. (2,1)

27. $3x = 8y + 11$,
    $x + 6y - 8 = 0$

    We first get each equation in the form $Ax + By = C$.

    $3x - 8y = 11$,   (1) Adding $-8y$
    $x + 6y = 8$ (2) Adding 8

    We multiply by -3 on both sides of equation (2) and add.

    $3x - 8y = 11$
    $\underline{-3x - 18y = -24}$ Multiplying by -3
    $-26y = -13$ Adding
    $y = \frac{1}{2}$

Chapter 8   (8.4)

27. (continued)

   Substitute $\frac{1}{2}$ for y in equation (1) and solve for x.

   $3x - 8y = 11$

   $3x - 8 \cdot \frac{1}{2} = 11$ Substituting

   $3x - 4 = 11$

   $3x = 15$

   $x = 5$

   Check:

$3x - 8y = 11$		$x + 6y = 0$	
$3 \cdot 5 - 8 \cdot \frac{1}{2}$	11	$5 + 6 \cdot \frac{1}{2}$	8
$15 - 4$		$5 + 3$	
11		8	

   Since $\left(5, \frac{1}{2}\right)$ checks, it is the solution.

28. (50, 18)

29. $3x - 2y = 10,$ (1)
    $-6x + 4y = -20$ (2)

   We multiply by 2 on both sides of equation (1) and add.

   $6x - 4y = 20$
   $-6x + 4y = -20$
   $0 = 0$

   We get an obviously true equation, so there are infinitely many solutions.

30. No solution

31. $0.06x + 0.05y = 0.07,$
    $0.04x - 0.03y = 0.11$

   We first multiply each equation by 100 to clear the decimals.

   $6x + 5y = 7,$ (1)
   $4x - 3y = 11$ (2)

   We use the multiplication principle with both equations of the resulting system.

   $18x + 15y = 21$ Multiplying (1) by 3
   $20x - 15y = 55$ Multiplying (2) by 5
   $38x = 76$ Adding
   $x = 2$

   Substitute 2 for x in equation (1) and solve for y.

   $6x + 5y = 7$
   $6 \cdot 2 + 5y = 7$
   $12 + 5y = 7$
   $5y = -5$
   $y = -1$

31. (continued)

   Check:

$0.06x + 0.05y = 0.07$	
$0.06(2) + 0.05(-1)$	0.07
$0.12 - 0.05$	
0.07	

$0.04x - 0.03y = 0.11$	
$0.04(2) - 0.03(-1)$	0.11
$0.08 + 0.03$	
0.11	

   Since $(2, -1)$ checks, it is the solution.

32. $(10, -2)$

33. $\frac{1}{3}x + \frac{3}{2}y = \frac{5}{4},$ (1)
    $\frac{3}{4}x - \frac{5}{6}y = \frac{3}{8}$ (2)

   First we clear the fractions. We multiply on both sides of the first equation by 12 and on both sides of the second equation by 24.

   $12\left(\frac{1}{3}x + \frac{3}{2}y\right) = 12 \cdot \frac{5}{4}$

   $12 \cdot \frac{1}{3}x + 12 \cdot \frac{3}{2}y = 15$

   $4x + 18y = 15$

   $24\left(\frac{3}{4}x - \frac{5}{6}y\right) = 24 \cdot \frac{3}{8}$

   $24 \cdot \frac{3}{4}x - 24 \cdot \frac{5}{6}y = 9$

   $18x - 20y = 9$

   The resulting system is
   $4x + 18y = 15,$ (1)
   $18x - 20y = 9.$ (2)

   We use the multiplication principle with both equations.

   $72x + 324y = 270$ Multiplying (1) by 18
   $-72x + 80y = -36$ Multiplying (2) by -4
   $404y = 234$

   $y = \frac{234}{404},$ or $\frac{117}{202}$

   Substitute $\frac{117}{202}$ for y in (1) and solve for x.

   $4x + 18\left(\frac{117}{202}\right) = 15$

   $4x + \frac{1053}{101} = 15$

   $4x = \frac{462}{101}$

   $x = \frac{1}{4} \cdot \frac{462}{101}$

   $x = \frac{231}{202}$

   The ordered pair $\left(\frac{231}{202}, \frac{117}{202}\right)$ checks in both equations. It is the solution.

34. $\left(\dfrac{231}{202}, \dfrac{117}{202}\right)$

35. The Familiarize and Translate steps were done in Exercise 5 of Exercise Set 8.1. The resulting system is

$$53.95 + 0.30m = c,$$
$$54.95 + 0.20m = c,$$

where m represents the mileage and c the cost.

Carry out. We solve the system of equations. We clear the decimals by multiplying both sides of each equation by 100.

$$5395 + 30m = 100c, \quad (1)$$
$$5495 + 20m = 100c \quad (2)$$

We multiply (1) by −1 and then add.

$$-5395 - 30m = -100c$$
$$\underline{5495 + 20m = 100c}$$
$$100 - 10m = 0 \quad \text{Adding}$$
$$100 = 10m$$
$$10 = m$$

Check. For 10 mi, the cost of the Avis car is 53.95 + 0.30(10), or 53.95 + 3, or $56.95. For 10 mi, the cost of the other car is 54.95 + 0.20(10), or 54.95 + 2, or $56.95, so the costs are the same when the mileage is 10.

State. When the cars are driven 10 miles, the cost will be the same.

36. 5 miles

37. The Familiarize and Translate steps were done in Exercise 10 of Exercise Set 8.1. The resulting system is

$$x + y = 180,$$
$$y = 2x + 30,$$

where x represents the smaller angle and y the larger angle.

Carry out. We solve the system. We will use the elimination method although we could also easily use the substitution method. First we get the second equation in the form Ax + By = C.

$$x + y = 180, \quad (1)$$
$$-2x + y = 30 \quad (2) \quad \text{Adding } -2x$$

Now we multiply equation (2) by −1 and add.

$$x + y = 180$$
$$\underline{2x - y = -30}$$
$$3x = 150$$
$$x = 50$$

Then we substitute 50 for x in equation (1) and solve for y.

$$x + y = 180 \quad (1)$$
$$50 + y = 180 \quad \text{Substituting}$$
$$y = 130$$

37. (continued)

Check. The sum of the angles is 50° + 130°, or 180°, so the angles are supplementary. Also, 30° more than two times the 50° angle is 30° + 2·50°, or 30° + 100°, 130°, the other angle. These numbers check.

State. The angles are 50° and 130°.

38. 47° and 133°

39. The Familiarize and Translate steps were done in Exercise 11 of Exercise Set 8.1. The resulting system is

$$x + y = 90,$$
$$x - y = 34,$$

where x represents the larger angle and y the smaller angle.

Carry out. We solve the system.

$$x + y = 90, \quad (1)$$
$$\underline{x - y = 34} \quad (2)$$
$$2x = 124 \quad \text{Adding}$$
$$x = 62$$

Now we substitute 62 for x in equation (1) and solve for y.

$$x + y = 90 \quad (1)$$
$$62 + y = 90 \quad \text{Substituting}$$
$$y = 28$$

Check. The sum of the angles is 62° + 28°, or 90°, so the angles are complementary. The difference of the angles is 62° − 28°, or 34°. These numbers check.

State. The angles are 62° and 28°.

40. 32° and 58°

41. The Familiarize and Translate steps were done in Exercise 13 of Exercise Set 8.1. The resulting system is

$$x + y = 820,$$
$$y = x + 140,$$

where x represents the number of hectares of Riesling grapes that should be planted and y represents the number of hectares of Chardonnay grapes.

Carry out. We solve the system. We will use the elimination method, although we could also easily use the substitution method. First we get the second equation in the form Ax + By = C. Then we add the equations.

$$x + y = 820, \quad (1)$$
$$\underline{-x + y = 140} \quad (2) \quad \text{Adding } -x$$
$$2y = 960 \quad \text{Adding}$$
$$y = 480$$

Now we substitute 480 for y in equation (1) and solve for x.

$$x + y = 820 \quad (1)$$
$$x + 480 = 820 \quad \text{Substituting}$$
$$x = 340$$

Chapter 8   (8.4)

41. (continued)

    Check. The total number of hectares is 340 + 480, or 820. The number of hectares of Chardonnay grapes, 480, is 140 more than the number of hectares of Riesling grapes, 340. These numbers check.

    State. The vintner should plant 340 hectares of Riesling grapes and 480 hectares of Chardonnay grapes.

42. 415 hectares of hay, 235 hectares of oats

43. $\dfrac{(a^2b^{-3})^4}{a^5b^{-6}} = \dfrac{a^{2(4)}b^{-3(4)}}{a^5b^{-6}} = \dfrac{a^8b^{-12}}{a^5b^{-6}} = a^{8-5}b^{-12-(-6)} = a^3b^{-6}$

44. $\dfrac{9a^4}{4b^6}$

45. $\dfrac{x+7}{x^2-1} = \dfrac{3}{x+1}$, LCM is $(x+1)(x-1)$

    $(x+1)(x-1) \cdot \dfrac{x+7}{(x+1)(x-1)} = (x+1)(x-1) \cdot \dfrac{3}{x+1}$

                            $x + 7 = 3(x - 1)$

                            $x + 7 = 3x - 3$

                            $10 = 2x$

                            $5 = x$

The number 5 checks, so it is the solution.

46. $\dfrac{-2x + 10}{(x+1)(x-1)}$

47. Familiarize. Let x represent the number of rabbits and y the number of pheasants in the cage. Each rabbit has one head and four feet. Thus, there are x rabbit heads and 4x rabbit feet in the cage. Each pheasant has one head and two feet. Thus, there are y pheasant heads and 2y pheasant feet in the cage.

    Translate. We reword the problem.

    <u>Rabbit heads</u>, plus <u>pheasant heads</u>, is 35.
           x   +   y   = 35

    <u>Rabbit feet</u>, plus <u>pheasant feet</u>, is 94.
         4x   +   2y   = 94

The resulting system is

     x + y = 35,   (1)

     4x + 2y = 94.   (2)

Carry out. We solve the system of equations. We multiply equation (1) by -2 and then add.

     -2x - 2y = -70
      4x + 2y = 94
       2x   = 24   Adding
        x   = 12

Substitute 12 for x in one of the original equations and solve for y.

     x + y = 35   (1)
     12 + y = 35   Substituting
          y = 23

47. (continued)

    Check. If there are 12 rabbits and 23 pheasants, the total number of heads in the cage is 12 + 23, or 35. The total number of feet in the cage is 4·12 + 2·23, or 48 + 46, or 94. The numbers check.

    State. There are 12 rabbits and 23 pheasants.

48. Patrick: 6, father: 30

49. Carry out. Solve the system

    $\dfrac{x+5}{5} = y$,

     x - 5 = 8(y - 5),

where x = the man's age and y = his daughter's age.

Multiply the first equation by 5 to clear the fraction.

     x + 5 = 5y

     x - 5y = -5

Simplify the second equation.

     x - 5 = 8(y - 5)

     x - 5 = 8y - 40

     x - 8y = -35

The resulting system is

     x - 5y = -5,   (1)

     x - 8y = -35.   (2)

Multiply equation (2) by -1 and add.

     x - 5y = -5
     -x + 8y = 35   Multiplying by -1
        3y = 30   Adding
         y = 10

Substitute 10 for y in equation (1) and solve for x.

     x - 5y = -5
     x - 5·10 = -5   Substituting
     x - 50 = -5
     x = 45

Possible solution: Man is 45, daughter is 10.

Check. If 5 is added to the man's age, 5 + 45, the result is 50. If 50 is divided by 5, the result is 10, the daughter's age. Five years ago the father and daughter were 40 and 5, respectively, and 40 = 8·5. The numbers check.

State. The man is 45 years old; his daughter is 10 years old.

50. Base: 10 ft, height: 5 ft

Chapter 8   (8.4)

51. $3(x - y) = 9,$
    $x + y = 7$

First we remove parentheses in the first equation.
    $3x - 3y = 9,$ (1)
    $x + y = 7$ (2)

Then we multiply equation (2) by 3 and add.
    $3x - 3y = 9$
    $\underline{3x + 3y = 21}$
    $6x = 30$
    $x = 5$

Now we substitute 5 for x in equation (2) and solve for y.
    $x + y = 7$
    $5 + y = 7$
    $y = 2$

The ordered pair (5,2) checks and is the solution.

52. $(2,0)$

53. $2(x - y) = 3 + x,$
    $x = 3y + 4$

Simplify the first equation.
    $2x - 2y = 3 + x$
    $x - 2y = 3$

Now we have the system
    $x - 2y = 3,$ (1)
    $x = 3y + 4.$ (2)

Substitute $3y + 4$ for x in equation (1) and solve for y.
    $x - 2y = 3$
    $(3y + 4) - 2y = 3$
    $y + 4 = 3$
    $y = -1$

Substitute -1 for y in equation (2) and solve for x.
    $x = 3y + 4$ (2)
    $x = 3(-1) + 4$
    $x = -3 + 4$
    $x = 1$

The ordered pair (1,-1) checks and is the solution.

54. $(0,-1)$

55. $1.5x + 0.85y = 1637.5,$
    $0.01(x + y) = 15.25,$ or $0.01x + 0.01y = 15.25$

Multiply both equations by 100 to clear the decimals.
    $150x + 85y = 163{,}750,$ (1)
    $x + y = 1525$ (2)

Multiply equation (2) by -85 and then add.
    $150x + 85y = 163{,}750$
    $\underline{-85x - 85y = 129{,}625}$
    $65x = 34{,}125$
    $x = 525$

Substitute 525 for x in equation (2).
    $x + y = 1525$
    $525 + y = 1525$
    $y = 1000$

The ordered pair (525,1000) checks and is the solution.

56. $(4,0)$

57. $y = ax + b,$ (1)
    $y = x + c$ (2)

Substitute $x + c$ for y in equation (1) and solve for x.
    $y = ax + b$
    $x + c = ax + b$ Substituting
    $x - ax = b - c$
    $(1 - a)x = b - c$
    $x = \dfrac{b - c}{1 - a}$

Substitute $\dfrac{b - c}{1 - a}$ for x in equation (2) and simplify to find y.
    $y = x + c$
    $y = \dfrac{b - c}{1 - a} + c$
    $y = \dfrac{b - c}{1 - a} + c \cdot \dfrac{1 - a}{1 - a}$
    $y = \dfrac{b - c + c - ac}{1 - a}$
    $y = \dfrac{b - ac}{1 - a}$

The ordered pair $\left(\dfrac{b - c}{1 - a}, \dfrac{b - ac}{1 - a}\right)$ checks and is the solution.

58. $\left(\dfrac{-b - c}{a}, 1\right)$

Chapter 8 (8.5)

59. Simplify the first equation.
$$3(7 - a) - 2(1 + 2b) + 5 = 0$$
$$21 - 3a - 2 - 4b + 5 = 0$$
$$-3a - 4b + 24 = 0$$
$$24 = 3a + 4b$$

Add 18 to both sides of the second equation.
$$3a + 2b - 18 = 0$$
$$3a + 2b = 18$$

The resulting system is
$$3a + 4b = 24, \quad (1)$$
$$3a + 2b = 18. \quad (2)$$

Multiply equation (2) by -1 and then add.
$$3a + 4b = 24$$
$$\underline{-3a - 2b = -18}$$
$$2b = 6$$
$$b = 3$$

Substitute 3 for b in equation (1) and solve for a.
$$3a + 4b = 24$$
$$3a + 4(3) = 24$$
$$3a + 12 = 24$$
$$3a = 12$$
$$a = 4$$

The ordered pair (4,3) checks and is the solution.

60. (4,3)

Exercise Set 8.5

1. **Familiarize.** Let x = the number of cars and y = the number of trucks.

   Translate. We reword the problem.

   The number of cars plus the number of trucks is 510.
   $$x + y = 510$$

   The number of cars is the number of trucks plus 190.
   $$x = y + 190$$

   Carry out. We solve the system of equations.
   $$x + y = 510, \quad (1)$$
   $$x = y + 190 \quad (2)$$

   We substitute y + 190 for x in equation (1) and solve for y.
   $$x + y = 510$$
   $$(y + 190) + y = 510 \quad \text{Substituting}$$
   $$2y + 190 = 510$$
   $$2y = 320$$
   $$y = 160$$

1. (continued)

   Next we substitute 160 for y in one of the original equations and solve for x.
   $$x = y + 190 \quad (2)$$
   $$x = 160 + 190 \quad \text{Substituting}$$
   $$x = 350$$

   Check. If there are 350 cars and 160 trucks, then the total number of vehicles is 350 + 160, or 510. Since the number of trucks plus 190 is 160 + 190, or 350, we know that it is true that the number of cars is 190 more than the number of trucks.

   State. The firm should have 350 cars and 160 trucks.

2. 11 km

3. **Familiarize.** We make a table to organize the information. We let x represent Sammy's age now and y his daughter's age now.

	Age now	Age in 4 years	Age 6 years ago
Sammy	x	x + 4	x - 6
Daughter	y	y + 4	y - 6

   Translate.

   Sammy's age now is twice his daughter's age now.
   $$x = 2 \cdot y$$

   Sammy's age in 4 years will be three times his daughter's age 6 years ago.
   $$x + 4 = 3 \cdot (y - 6)$$

   The resulting system is
   $$x = 2y, \quad (1)$$
   $$x + 4 = 3(y - 6). \quad (2)$$

   Carry out. We use the substitution method. We substitute 2y for x in equation (2) and solve for y.
   $$x + 4 = 3(y - 6)$$
   $$2y + 4 = 3(y - 6) \quad \text{Substituting}$$
   $$2y + 4 = 3y - 18$$
   $$22 = y$$

   Next we substitute 22 for y in one of the original equations and solve for x.
   $$x = 2y \quad (1)$$
   $$x = 2 \cdot 22 \quad \text{Substituting}$$
   $$x = 44$$

   Check. Sammy's age is 44, which is twice his daughter's age, 22. In four years Sammy will be 44 + 4, or 48. Six years ago his daughter was 22 - 6, or 16. Sammy's age four years from now, 48, is three times his daughter's age six years ago, 16.

   State. Sammy is now 44 and his daughter is 22.

Chapter 8   (8.5)

4. Ann: 28, son: 10

5. <u>Familiarize</u>. We make a table to organize the information. We let x represent Marge's age now and y Consuelo's age now.

	Age now	Age seven years ago
Marge	x	x - 7
Consuelo	y	y - 7

<u>Translate</u>.

Marge's age now, is twice Consuelo's age now.
       x   =   2y

Marge's age seven years ago, plus Consuelo's age seven years ago, was 13.
      (x - 7)   +   (y - 7)   =   13

The resulting system is

$x = 2y$,   (1)
$(x - 7) + (y - 7) = 13$. (2)

<u>Carry out</u>. We use the substitution method. We substitute 2y for x in equation (2) and solve for y.

$(x - 7) + (y - 7) = 13$
$(2y - 7) + (y - 7) = 13$ Substituting
$3y - 14 = 13$
$3y = 27$
$y = 9$

Next we substitute 9 for y in one of the original equations and solve for x.

$x = 2y$ (1)
$x = 2 \cdot 9$ Substituting
$x = 18$

<u>Check</u>. Marge's age is 18, which is twice Consuelo's age, 9. Seven years ago when Marge was 11 and Consuelo 2, the sum of their ages was 11 + 2, or 13.

<u>State</u>. Marge is 18 and Consuelo is 9.

6. Andy: 24, Wendy: 6

7. <u>Familiarize</u>. Let d represent the number of dimes and q the number of quarters. Then, 10d represents the value of the dimes in cents, and 25q represents the value of the quarters in cents. The total value is $15.25, or 1525¢. The total number of coins is 103.

<u>Translate</u>.

Number of dimes, plus number of quarters, is 103.
       d   +   q   = 103

Value of dimes, plus value of quarters, is $15.25
      10d   +   25q   = 1525

7. (continued)
The resulting system is
  $d + q = 103$,   (1)
  $10d + 25q = 1525$.   (2)

<u>Carry out</u>. We use the addition method. We multiply equation (1) by -10 and then add.

$-10d - 10q = -1030$ Multiplying by -10
$\underline{10d + 25q = 1525}$
$15q = 495$ Adding
$q = 33$

Next we substitute 33 for q in one of the original equations and solve for d.

$d + q = 103$ (1)
$d + 33 = 103$ Substituting
$d = 70$

<u>Check</u>. The number of dimes plus the number of quarters is 70 + 33, or 103. The total value in cents is 10·70 + 25·33, or 700 + 825, or 1525. This is equal to $15.25. This checks.

<u>State</u>. There are 70 dimes and 33 quarters.

8. 10 nickels, 3 quarters

9. <u>Familiarize</u>. Let n represent the number of nickels and d the number of dimes. Then, 5n represents the value of the nickels in cents, and 10d represents the value of the dimes in cents. The total value is $25, or 2500 cents.

<u>Translate</u>.

Value of nickels, plus value of dimes, is $25.
      5n   +   10d   = 2500

Number of nickels is three times the number of dimes.
      n   =   3   ·   d

The resulting system is
  $5n + 10d = 2500$,   (1)
  $n = 3d$.   (2)

<u>Carry out</u>. We use the substitution method. We substitute 3d for n in equation (1) and solve for d.

$5n + 10d = 2500$
$5 \cdot 3d + 10d = 2500$ Substituting
$15d + 10d = 2500$
$25d = 2500$
$d = 100$

We now substitute 100 for d in one of the original equations and solve for n.

$n = 3d$ (2)
$n = 3 \cdot 100$ Substituting
$n = 300$

<u>Check</u>. The number of nickels is three times the number of dimes (300 = 3·100). The total value in cents is 5·300 + 10·100, or 1500 + 1000, or 2500. This is equal to $25. This checks.

<u>State</u>. There are 300 nickels and 100 dim

Chapter 8   (8.5)

10. 32 nickels, 13 dimes

11. Familiarize. We list the information in a table. Let x = the number of adults and y = the number of children.

	Adults	Children	Totals
Paid	$1.00	$0.75	
Number attending	x	y	429
Money taken in	1.00x	0.75y	$372.50

Translate. The last two rows of the table give us two equations.
The total number of people attending was 429, so
$$x + y = 429.$$
The total amount taken in was $372.50, so
$$1.00x + 0.75y = 372.50,$$
or $100x + 75y = 37,250$ Multiplying by 100

Carry out. We use the addition method.
$x + y = 429,$ (1)
$100x + 75y = 37,250$ (2)

We multiply on both sides of the first equation by -75 and then add.
$-75x - 75y = -32,175$ Multiplying by -75
$\underline{100x + 75y = 37,250}$
$25x = 5075$ Adding
$x = 203$

Next we substitute 203 for x in one of the original equations and solve for y.
$x + y = 429$ (1)
$203 + y = 429$ Substituting
$y = 226$

Check. The total attending was 203 adults plus 226 children, or 429. The total receipts were $1.00(203) + 0.75(226)$. This is $203 + $169.50, or $372.50. The numbers check.
State. 203 adults and 226 children attended the play.

12. 236 adults, 342 children

13. Familiarize. We list the information in a table. Let x = the number of student tickets sold and y = the number of adult tickets sold.

	Students	Adults	Totals
Paid	$0.50	$0.75	
Number attending	x	y	200
Money taken in	0.50x	0.75y	$132.50

13. (continued)
Translate. The last two rows of the table give us two equations.
The total number of tickets sold was 200, so
$$x + y = 200.$$
The total amount collected was $132.50, so
$$0.50x + 0.75y = 132.50,$$
or $50x + 75y = 13,250$ Multiplying by 100

Carry out. We use the addition method.
$x + y = 200,$ (1)
$50x + 75y = 13,250$ (2)

We multiply on both sides of equation (1) by -50 and then add.
$-50x - 50y = -10,000$ Multiplying by -50
$\underline{50x + 75y = 13,250}$
$25y = 3250$
$y = 130$

Next we substitute 130 for y in one of the original equations and solve for x.
$x + y = 200$ (1)
$x + 130 = 200$ Substituting
$x = 70$

Check. The total number of tickets sold was 70 students plus 130 adults, or 200. The total receipts were $0.50(70) + 0.75(130)$. This amount is $35 + $97.50, or $132.50. The numbers check.
State. Thus, 70 student tickets and 130 adult tickets were sold.

14. 128 card holders, 75 noncard holders

15. Familiarize. We complete the table in the text. Note that x represents the number of liters of solution A to be used and y represents the number of liters of solution B.

Type of solution	A	B	Mixture
Amount of solution	x	y	100 liters
Percent of acid	50%	80%	68%
Amount of acid in solution	0.5x	0.8y	0.68 × 100, or 68 liters

$0.5x + 0.8y = 68$
$x + y = 100$

Translate. Since the total amount of solution is 100 liters, we have
$$x + y = 100.$$
The amount of acid in the mixture is to be 68% of 100, or 68 liters. The amounts of acid from the two solutions are 50%x and 80%y. Thus
$50\%x + 80\%y = 68,$
or $0.5x + 0.8y = 68,$
or $5x + 8y = 680$ Clearing decimals

15. (continued)

   Carry out. We use the addition method.
   $$x + y = 100, \quad (1)$$
   $$5x + 8y = 680 \quad (2)$$

   We multiply equation (1) by -5 and then add.
   $$-5x - 5y = -500 \quad \text{Multiplying by -5}$$
   $$\underline{5x + 8y = 680}$$
   $$3y = 180$$
   $$y = 60$$

   Next we substitute 60 for y in one of the original equations and solve for x.
   $$x + y = 100 \quad (1)$$
   $$x + 60 = 100 \quad \text{Substituting}$$
   $$x = 40$$

   Check. We consider $x = 40$ and $y = 60$. The sum is 100. Now 50% of 40 is 20 and 80% of 60 is 48. These add up to 68. The numbers check.

   State. 40 liters of solution A and 60 liters of solution B should be used.

16. $55\frac{5}{9}$ L of A, $44\frac{4}{9}$ L of B

17. Familiarize. We can arrange the information in a table. We let x represent the amount of 30% solution and y represent the amount of 50% solution.

Type of insecticide	30% solution	50% solution	Mixture
Amount of solution	x	y	200 L
Percent of insecticide	30%	50%	42%
Amount of insecticide in solution	0.3x	0.5y	0.42 × 200, or 84 L

   Translate. Since the total amount of solution is 200 liters, we have
   $$x + y = 200.$$

   The amount of insecticide in the mixture is to be 42% of 200, or 84 liters. The amounts of insecticide from the two solutions are 30%x and 50%y. Thus
   $$30\%x + 50\%y = 84,$$
   $$\text{or} \quad 0.3x + 0.5y = 84,$$
   $$\text{or} \quad 3x + 5y = 840 \quad \text{Clearing decimals}$$

17. (continued)

   Carry out. We use the addition method.
   $$x + y = 200, \quad (1)$$
   $$3x + 5y = 840 \quad (2)$$

   We multiply equation (1) by -3 and then add.
   $$-3x - 3y = -600 \quad \text{Multiplying by -3}$$
   $$\underline{3x + 5y = 840}$$
   $$2y = 240$$
   $$y = 120$$

   Next we substitute 120 for y in one of the original equations and solve for x.
   $$x + y = 200 \quad (1)$$
   $$x + 120 = 200 \quad \text{Substituting}$$
   $$x = 80$$

   Check. We consider $x = 80$ and $y = 120$. The sum is 200. Now 30% of 80 is 24 and 50% of 120 is 60. These add up to 84. The numbers check.

   State. 80 L of the 30% solution and 120 L of the 50% solution should be used.

18. 100 L of 28%, 200 L of 40%

19. Familiarize. We organize the information in a table. Let x = the number of kilograms of cashews and y = the number of kilograms of pecans.

	Cashews	Pecans	Mixture
Cost of nuts	$8.00	$9.00	$8.40
Amount (in kg)	x	y	10
Mixture	8x	9y	$8.40 × 10, or $84

   Translate. The last two rows of the table give us two equations.

   Since the total number of kilograms is 10, we have
   $$x + y = 10.$$

   The value of the cashews is 8x (x pounds at $8 per pound). The value of pecans is 9y (y pounds at $9 per pound). The value of the mixture is 8.40 × 10, or $84. Thus we have
   $$8x + 9y = 84.$$

   Carry out. We use the addition method.
   $$x + y = 10, \quad (1)$$
   $$8x + 9y = 84 \quad (2)$$

   We multiply equation (1) by -8 and then add.
   $$-8x - 8y = -80 \quad \text{Multiplying by -8}$$
   $$\underline{8x + 9y = 84}$$
   $$y = 4$$

   Next we substitute 4 for y in one of the original equations and solve for x.
   $$x + y = 10 \quad (1)$$
   $$x + 4 = 10 \quad \text{Substituting}$$
   $$x = 6$$

Chapter 8 (8.5)

19. (continued)

    Check. We consider $x = 6$ kg and $y = 4$ kg. The sum is 10 kg. The value of the mixture of nuts is $8·6 + $9·4, or $48 + $36, or $84. These values check.

    State. The mixture consists of 6 kg of cashews and 4 kg of pecans.

20. 100 kg of Brazilian, 200 kg of Turkish

21. Familiarize. We organize the information in a table. Let x = the amount of seed A and y = the amount of seed B to be used.

Type of seed	A	B	Mixture
Cost of seed	$1.00	$1.35	$1.14
Amount (in pounds)	x	y	50
Mixture	1.00x	1.35y	$1.14(50), or $57

    Translate. The last two rows of the table give us two equations.

    Since the total amount of grass seed is 50 lb, we have

    $x + y = 50$.

    The value of seed A is 1.00x (x lb at $1.00 per pound), and the value of seed B is 1.35y (y lb at $1.35 per pound). The value of the mixture is $1.14(50), or $57, so we have

    $1.00x + 1.35y = 57$, or

    $100x + 135y = 5700$ Clearing decimals

    Carry out. We use the addition method.

    $x + y = 50$,   (1)
    $100x + 135y = 5700$ (2)

    We multiply equation (1) by -100 and then add.

    $-100x - 100y = -5000$
    $\underline{100x + 135y = 5700}$
    $35y = 700$
    $y = 20$

    Next we substitute 20 for y in one of the original equations and solve for x.

    $x + y = 50$ (1)
    $x + 20 = 50$
    $x = 30$

    Check. We consider $x = 30$ lb and $y = 20$ lb. The sum is 50 lb. The value of the mixture is $1.00(30) + $1.35(20), or $30 + $27, or $57. These values check.

    State. 30 lb of seed A and 20 lb of seed B should be used.

22. 135 lb of the less expensive nuts, 345 lb of the more expensive nuts

23. Familiarize. We organize the information in a table. Let a = the number of type A questions and b = the number of type B questions.

Type of question	A	B	Mixture (Test)
Number	a	b	16
Time	3 min	6 min	
Value	10 points	15 points	
Mixture (Test)	3a min, 10a points	6b min, 15b points	60 min, 180 points

    Translate. The table actually gives us three equations. Since the total number of questions is 16, we have

    $a + b = 16$.

    The total time is 60 min, so we have

    $3a + 6b = 60$.

    The total number of points is 180, so we have

    $10a + 15b = 180$.

    The resulting system is

    $a + b = 16$,   (1)
    $3a + 6b = 60$,   (2)
    $10a + 15b = 180$. (3)

    Carry out. We will solve the system composed of equations (1) and (2) and then check to see that this solution also satisfies equation (3). We multiply equation (1) by -3 and add.

    $-3a - 3b = -48$
    $\underline{3a + 6b = 60}$
    $3b = 12$
    $b = 4$

    Now we substitute 4 for b in equation (1) and solve for a.

    $a + b = 16$
    $a + 4 = 16$
    $a = 12$

    Check. We consider $a = 12$ questions and $b = 4$ questions. The total number of questions is 16. The time required is $3·12 + 6·4$, or $36 + 24$, or 60 min. The total points are $10·12 + 15·4$, or $120 + 60$, or 180. These values check.

    State. 12 questions of type A and 4 questions of type B were answered correctly.

24. 45 ounces of three-fourths gold, 15 ounces of five-twelfths gold

218

25. <u>Familiarize</u>. We organize the information in a table. Let x = the price per gallon of the inexpensive paint and y = the price per gallon of the expensive paint.

	Inexpensive paint	Expensive paint	Mixture
Price per gallon	x	y	
First mixture amount	9 gal	7 gal	16 gal
First mixture cost	9x	7y	$19.70(16) or $315.20
Second mixture amount	3	5	8
Second mixture cost	3x	5y	$19.825(8), or $158.60

<u>Translate</u>. The third and fifth rows of the table give us two equations. The total cost of the first mixture is:

$9x + 7y = 315.20$, or
$90x + 70y = 3152$ Clearing the decimal

The total cost of the second mixture is:

$3x + 5y = 158.60$, or
$30x + 50y = 1586$ Clearing the decimal

<u>Carry out</u>. We use the addition method.

$90x + 70y = 3152$,   (1)
$30x + 50y = 1586$,   (2)

We multiply equation (2) by -3 and add.

$90x + 70y = 3152$
$-90x - 150y = -4758$
$-80y = -1606$
$y = 20.075$

Now we substitute 20.075 for y in equation (2) and solve for x.

$30x + 50y = 1586$
$30x + 50(20.075) = 1586$
$30x + 1003.75 = 1586$
$30x = 582.25$
$x \approx 19.408$

<u>Check</u>. We check x ≈ $19.408 and y = $20.075. The cost of the first mixture is $19.408(9) + $20.075(7), or about $315.20. The cost of the second mixture is $19.408(3) + $20.075(5), or about $158.60. These values check.

<u>State</u>. The inexpensive paint costs $19.408 per gallon, and the expensive paint costs $20.075 per gallon.

26. 6.68 pages in large type, 5.32 pages in small type

27. <u>Familiarize</u>. Let x represent the part invested at 12% and y represent the part invested at 13%. The interest earned from the 12% investment is 12%·x. The interest earned from the 13% investment is 13%·y. The total investment is $27,000, and the total interest earned is $3385.

<u>Translate</u>.

$x + y = 27,000$,
$12\%x + 13\%y = \$3385$ or $12x + 13y = 338,500$

<u>Carry out</u>. Multiply the first equation by -12 and add.

$-12x - 12y = -324,000$
$12x + 13y = 338,500$
$y = 14,500$

Substitute 14,500 for y in the first equation and solve for x.

$x + y = 27,000$
$x + 14,500 = 27,000$
$x = 12,500$

<u>Check</u>. We consider $12,500 invested at 12% and $14,500 invested at 13%. The sum of the investments is $27,000. The interest earned is 12%·12,500 + 13%·14,500, or 1500 + 1885, or $3385. These numbers check.

<u>State</u>. $12,500 was invested at 12%, and $14,500 was invested at 13%.

28. 9%, 10.5%

29. <u>Familiarize</u>. Let x represent the ten's digit and y the one's digit. Then the number is 10x + y.

<u>Translate</u>.

<u>The number</u>, is 6 times <u>the sum of its digits</u>.
$10x + y$ = 6 · $(x + y)$

<u>The ten's digit</u>, is 1 <u>more than</u>, <u>the one's digit</u>.
x = 1 + y

We simplify the first equation.

$10x + y = 6(x + y)$
$10x + y = 6x + 6y$
$4x - 5y = 0$

The system of equations is

$4x - 5y = 0$,   (1)
$x = 1 + y$.   (2)

<u>Carry out</u>. We use the substitution method. We substitute 1 + y for x in equation (1) and solve for y.

$4(1 + y) - 5y = 0$
$4 + 4y - 5y = 0$
$4 - y = 0$
$4 = y$

Then we substitute 4 for y in equation (2) and compute x.

$x = 1 + y = 1 + 4 = 5$

219

Chapter 8   (8.5)

29. (continued)

   Check. We consider the number 54. The number is 6 times the sum of the digits, 9. The ten's digit is 1 more than the one's digit. This number checks.

   State. The number is 54.

30. 75

31. Familiarize. We organize the information in a table. Let x = the number of liters of skim milk and y = the number of liters of 3.2% milk.

Type of milk	4.6%	Skim	3.2% (Mixture)
Amount of milk	100 L	x	y
Percent of butterfat	4.6%	0%	3.2%
Amount of butterfat in milk	4.6% × 100, or 4.6 L	0%·x, or 0 L	3.2%y

Translate. The first and third rows of the table give us two equations.

Liters of 4.6% milk + Liters of skim milk = Liters of 3.2% milk.
$$100 + x = y$$

Amt. of butterfat in 4.6% milk + Amt. of butterfat in skim milk = Amt. of butterfat in 3.2% milk
$$4.6 + 0 = 3.2\%y$$

The resulting system is

$$100 + x = y,$$
$$4.6 = 3.2\%y, \text{ or}$$

$$100 + x = y,$$
$$4.6 = 0.032y.$$

Carry out. We solve the second equation for y.

$$4.6 = 3.2\%y$$
$$4.6 = 0.032y$$
$$\frac{4.6}{0.032} = y$$
$$143.75 = y$$

We substitute 143.75 for y in the first equation and solve for x.

$$100 + x = y$$
$$100 + x = 143.75$$
$$x = 43.75$$

Check. We consider x = 43.75 L and y = 143.75 L. The difference between 143.75 and 43.75 is 100 L. There is no butterfat in the skim milk. There are 4.6 liters of butterfat in the 100 liters of the 4.6% milk. Thus there are 4.6 liters of butterfat in the mixture. This checks because 3.2% of 143.75 is 4.6.

State. 43.75 L of skim milk should be used.

32. $2666\frac{2}{3}$ L

33. Familiarize. In a table we organize the information regarding the solution after some of the 30% solution is drained and replaced with pure antifreeze. We let x represent the amount of the original (30%) solution remaining, and we let y represent the amount of the 30% mixture that is drained and replaced with pure antifreeze.

Type of solution	Original (30%)	Pure antifreeze	Mixture
Amount of solution	x	y	16
Percent of antifreeze	30%	100%	50%
Amount of antifreeze in solution	0.3x	1·y, or y	0.5(16), or 8

Translate. The table gives us two equations.

Amount of solution: $x + y = 16$

Amount of antifreeze in solution: $0.3x + y = 8$, or $3x + 10y = 80$

The resulting system is

$$x + y = 16, \quad (1)$$
$$3x + 10y = 80. \quad (2)$$

Carry out. We multiply the first equation by -3 and solve for y.

$$-3x - 3y = -48$$
$$\underline{3x + 10y = 80}$$
$$7y = 32$$
$$y = \frac{32}{7}, \text{ or } 4\frac{4}{7}$$

Then we substitute $4\frac{4}{7}$ for y in equation (1) and solve for x.

$$x + y = 16$$
$$x + 4\frac{4}{7} = 16$$
$$x = 11\frac{3}{7}$$

Check. When $x = 11\frac{3}{7}$ L and $y = 4\frac{4}{7}$ L, the total is 16 L. The amount of antifreeze in the mixture is $0.3\left(11\frac{3}{7}\right) + 4\frac{4}{7}$, or $\frac{3}{10} \cdot \frac{80}{7} + \frac{32}{7}$, or $\frac{24}{7} + \frac{32}{7} = \frac{56}{7}$, or 8 L. This is 50% of 16 L, so the numbers check.

State. $4\frac{4}{7}$ of the original mixture should be drained and replaced with pure antifreeze.

34. 10 $20 workers, 5 $25 workers

Chapter 8   (8.6)

35. Familiarize. Let x represent the ten's digit and y represent the one's digit. Then the number is $10x + y$.

Translate.

One's digit plus the number is 43 more than 5 times ten's digit

$y + 10x + y = 43 + 5 \cdot x$

Sum of the digits is 11.

$x + y = 11$

We simplify the first equation.

$y + 10x + y = 43 + 5x$
$10x + 2y = 43 + 5x$
$5x + 2y = 43$

The resulting system is

$5x + 2y = 43$, (1)
$x + y = 11$ (2)

Carry out. We multiply equation (2) by -2 and add.

$5x + 2y = 43$
$-2x - 2y = -22$
$\overline{3x = 21}$
$x = 7$

Then we substitute 7 for x in equation (2) and solve for y.

$x + y = 11$
$7 + y = 11$
$y = 4$

Check. We check the number 74. The sum of the one's digit and the number is $4 + 74$, or 78. This is 43 more than 5 times the ten's digit, or 35. The sum of the digits is 11. The number checks.

State. The number is 74.

36. 126

37. Familiarize. Let x = the cost of the bat, y = the cost of the ball, and z = the cost of the glove.

Translate.

Cost of bat + Cost of ball + Cost of glove = $99.

$x + y + z = 99$

Cost of bat = $9.95 + Cost of ball

$x = 9.95 + y$

Cost of glove = $65.45 + Cost of bat

$z = 65.45 + x$

37. (continued)

Carry out. We solve the system.

$x + y + z = 99$,   (1)
$x = 9.95 + y$,   (2)
$z = 65.45 + x$.   (3)

Solve the second equation for y.

$x = 9.95 + y$
$x - 9.95 = y$ (4)

Substitute $x - 9.95$ for y and $65.45 + x$ for z in the first equation and solve for x.

$x + y + z = 99$
$x + (x - 9.95) + (65.45 + x) = 99$
$3x + 55.5 = 99$
$3x = 43.5$
$x = 14.5$

Then substitute 14.5 for x in equations (3) and (4) and compute z and y.

$z = 65.45 + x = 65.45 + 14.5 = 79.95$
$y = x - 9.95 = 14.5 - 9.95 = 4.55$

Check. The total cost is $14.50 + $4.55 + $79.95, or $99. The cost of the bat, $14.50, is $9.95 more than $4.55, the cost of the ball. Also, the cost of the glove $79.95 is $65.45 more than $14.50, the cost of the bat. These numbers check.

State. The bat costs $14.50, the ball costs $4.55, and the glove costs $79.95.

38. Tweedledum: 120 pounds, Tweedledee: 121 pounds

Exercise Set 8.6

1. Familiarize. First make a drawing.

```
  Distance of      Distance of
   slow car         fast car
  ←── 48 mph ──→ ←── 55 mph ──→
  ├──────── 206 mi ────────┤
```

The sum of the distances is 206 mi. The times the cars travel are the same. We complete the table in the text.

$d = r \cdot t$

	Distance	Speed	Time
Fast car	Distance of fast car	55	t
Slow car	Distance of slow car	48	t

Translate. From the drawing we see that

(Slow car distance) + (Fast car distance) = 206

Then using $d = rt$ in each row of the chart, we get

$48t + 55t = 206$.

Chapter 8   (8.6)

1. (continued)

   Carry out. We solve this equation for t.
   $$48t + 55t = 206$$
   $$103t = 206$$
   $$t = 2$$

   Check. If the time is 2 hr, then the distance the slow car travels is $48 \cdot 2$, or 96 mi. The fast car travels $55 \cdot 2$, or 110 mi. Since the sum of the distances, $96 + 110$, is 206 mi, the answer checks.

   State. In 2 hours, the cars will be 206 miles apart.

2. 3 hr

3. Familiarize. First make a drawing.

   ```
   Town      30 mph
   Slow car    t hours    d miles

   Town      46 mph
   Fast car    t hours    (d + 72) miles
   ```

   We see that the distances are not the same, but the times the cars travel are the same. We organize the information in a chart.

   $$d = r \cdot t$$

	Distance	Speed	Time
Slow car	d	30	t
Fast car	d + 72	46	t

   Translate. Using $d = rt$ in each row of the chart, we get the following system of equations:
   $$d = 30t,$$
   $$d + 72 = 46t$$

   Carry out. Substitute 30t for d in the second equation and solve for t.
   $$d + 72 = 46t$$
   $$30t + 72 = 46t \quad \text{Substituting}$$
   $$72 = 16t$$
   $$\frac{72}{16} = t$$
   $$t = \frac{9}{2}, \text{ or } 4.5$$

   Check. In 4.5 hours the slow car will travel $30(4.5)$, or 135 mi. In 4.5 hours the fast car will travel $46(4.5)$, or 207 miles. The difference between 207 and 135 is 72.

   State. In 4.5 hours the cars will be 72 miles apart.

4. 4.6 hr

5. Familiarize. First make a drawing.

   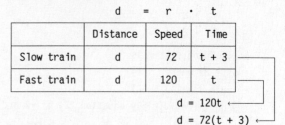

   ```
   Station     72 km/h
   Slow train    t + 3 hours    d kilometers

   Station     120 km/h
   Fast train    t hours    d kilometers
                                    Trains
                                    meet here
   ```

   From the drawing we see that the distances are the same. Let's call the distance d. Let t represent the time for the faster train and $t + 3$ represent the time for the slower train. We complete the table in the text.

   $$d = r \cdot t$$

	Distance	Speed	Time
Slow train	d	72	t + 3
Fast train	d	120	t

   $$d = 120t$$
   $$d = 72(t + 3)$$

   Translate. Using $d = rt$ in each row of the chart, we get the following system of equations:
   $$d = 72(t + 3),$$
   $$d = 120t$$

   Carry out. Substitute 120t for d in the first equation and solve for t.
   $$d = 72(t + 3)$$
   $$120t = 72(t + 3) \quad \text{Substituting}$$
   $$120t = 72t + 216$$
   $$48t = 216$$
   $$t = \frac{216}{48}$$
   $$t = 4.5$$

   Check. When $t = 4.5$ hours, the faster train will travel $120(4.5)$, or 540 km, and the slower train will travel $72(7.5)$, or 540 km. In both cases we get the distance 540 km.

   State. In 4.5 hours after the second train leaves, the second train will overtake the first train. We can also state the answer as 7.5 hours after the first train leaves.

6. $2\frac{1}{2}$ hr after the private plane leaves

Chapter 8   (8.6)

7. Familiarize. We first make a drawing.

   With the current   r + 6
   ←——— 4 hours ———→   d kilometers
   Against the current   r - 6
   ←——— 10 hours ———→ d kilometers

   From the drawing we see that the distances are the same. Let d represent the distance. Let r represent the speed of the canoe in still water. Then, when the canoe is traveling with the current, its speed is r + 6. When it is traveling against the current, its speed is r - 6. We complete the table in the text.

   $d = r \cdot t$

	Distance	Speed	Time
With current	d	r + 6	4
Against current	d	r - 6	10

   $d = (r - 6)10$
   $d = (r + 6)4$

   Translate. Using d = rt in each row of the chart, we get the following system of equations:
   $d = (r + 6)4$,
   $d = (r - 6)10$

   Carry out. Substitute (r + 6)4 for d in the second equation and solve for r.
   $d = (r - 6)10$
   $(r + 6)4 = (r - 6)10$ Substituting
   $4r + 24 = 10r - 60$
   $84 = 6r$
   $14 = r$

   Check. When r = 14, r + 6 = 20 and 20·4 = 80, the distance. When r = 14, r - 6 = 8 and 8·10 = 80. In both cases, we get the same distance.

   State. The speed of the canoe in still water is 14 km/h.

8. 180 km/h

9. Familiarize. First make a drawing.

   Passenger   96 km/h
   ←——— t - 2 hours ———→ d kilometers
   Freight   64 km/h
   ←——— t hours ———→ d kilometers
   Central City   Clear Creek

   From the drawing we see that the distances are the same. Let d represent the distance. Let t represent the time for the slower train (freight). Then the time for the faster train (passenger) is t - 2. We organize the information in a chart.

9. (continued)

   $d = r \cdot t$

	Distance	Speed	Time
Passenger	d	96	t - 2
Freight	d	64	t

   Translate. From each row of the chart we get an equation.
   $d = 96(t - 2)$,
   $d = 64t$

   Carry out. Substitute 64t for d in the first equation and solve for t.
   $d = 96(t - 2)$
   $64t = 96(t - 2)$ Substituting
   $64t = 96t - 192$
   $192 = 32t$
   $6 = t$

   Next we substitute 6 for t in one of the original equations and solve for d.
   $d = 64t$
   $d = 64 \cdot 6$ Substituting
   $d = 384$

   Check. If the time is 6 hr, then the distance the passenger train travels is 96(6 - 2), or 384 km. The freight train travels 64(6), or 384 km. The distances are the same.

   State. It is 384 km from Central City to Clear Creek.

10. 1911 km

11. Familiarize. We first make a drawing.

    Against the wind   r - w
    ←——— 2 hours ———→   600 kilometers
    With the wind   r + w
    ←——— 1⅔ hours ———→ 600 kilometers

    We let r represent the speed of the airplane in still air and w represent the speed of the wind. Then when flying against a head wind, the rate is r - w, and when flying with the wind, the rate is r + w. We organize the information in a chart.

    $d = r \cdot t$

	Distance	Speed	Time
Against	600	r - w	2
With	600	r + w	$1\frac{2}{3}$, or $\frac{5}{3}$

    Translate. Using d = rt in each row of the chart, we get the following system of equations:
    $600 = (r - w)2$ $300 = r - w$ Multiplying by $\frac{1}{2}$
    or
    $600 = (r + w)\frac{5}{3}$ $360 = r + w$ Multiplying by $\frac{3}{5}$

223

Chapter 8   (8.6)

11. (continued)

    Carry out. We use the addition method with the resulting system.

    $300 = r - w$
    $360 = r + w$
    ─────────────
    $660 = 2r$
    $330 = r$

    Next we substitute 330 for r and solve for w.

    $360 = r + w$
    $360 = 330 + w$ Substituting
    $30 = w$

    Check. If $r = 330$ and $w = 30$, then $r - w = 300$, and $r + w = 360$. If the plane flies 2 hours against the wind, it travels $300 \cdot 2$ or 600 km. If the plane flies $1\frac{2}{3}$ hours with the wind, it travels $360 \cdot \frac{5}{3}$, or 600 km. All values check.

    State. The speed of the plane in still air is 330 km/h.

12. 9 km/h

13. Familiarize. First make a drawing.

    Home   t hr   45 mph | (2-t) hr   6 mph   Work
    |Motorcycle distance | Walking distance|
    ←──────────── 25 miles ────────────→

    Let t represent the time the motorcycle was driven. Then $2 - t$ represents the time the rider walked. We organize the information in a chart.

    $d = r \cdot t$

	Distance	Speed	Time
Motorcycling	Motorcycle distance	45	t
Walking	Walking distance	6	2 - t

    Translate. From the drawing we see that

    Motorcycle distance + Walking distance = 25

    Then using $d = rt$ in each row of the chart we get

    $45t + 6(2 - t) = 25$

    Carry out. We solve this equation for t.

    $45t + 12 - 6t = 25$
    $39t + 12 = 25$
    $39t = 13$
    $t = \frac{13}{39}$
    $t = \frac{1}{3}$

13. (continued)

    Check. The problem asks us to find how far the motorcycle went before it broke down. If $t = \frac{1}{3}$, then $45t$ (the distance the motorcycle traveled) $= 45 \cdot \frac{1}{3}$, or 15 and $6(2 - t)$ (the distance walked) $= 6\left(2 - \frac{1}{3}\right) = 6 \cdot \frac{5}{3}$, or 10. The total of these distances is 25, so $\frac{1}{3}$ checks.

    State. The motorcycle went 15 miles before it broke down.

14. $6\frac{3}{4}$ km

15. Familiarize. We first make a drawing.

    With the wind $r + 25.5$
    ──────────────────────────
    4.23 hours   d kilometers

    Against the wind $r - 25.5$
    ──────────────────────────
    4.97 hours   d kilometers

    We let r represent the speed of the airplane in still air. When flying with a tail wind, the rate is $r + 25.5$. When flying against a head wind, the rate is $r - 25.5$. The distance, d, is the same for both directions. We organize the information in a chart.

    $d = r \cdot t$

	Distance	Speed	Time
With	d	r + 25.5	4.23
Against	d	r - 25.5	4.97

    Translate. Using $d = rt$ in each row of the chart, we get the following system of equations:

    $d = (r + 25.5)4.23$,
    $d = (r - 25.5)4.97$

    Carry out. Substitute $(r + 25.5)4.23$ for d in the second equation and solve for r.

    $d = (r - 25.5)4.97$
    $(r + 25.5)4.23 = (r - 25.5)4.97$
    $4.23r + 107.865 = 4.97r - 126.735$
    $234.6 = 0.74r$
    $317.03 \approx r$

    Check. When $r = 317.03$, $r + 25.5 = 342.53$ and $342.53(4.23) \approx 1448.9$, the distance. When $r = 317.03$, $r - 25.5 = 291.53$ and $291.53(4.97) \approx 1448.9$. In both cases, we get the same distance.

    State. The speed of the plane in still air is approximately 317.03 km/h.

16. Wind: 50 mph, plane: 200 mph

Chapter 8   (8.6)

17. Familiarize. First we make a drawing.

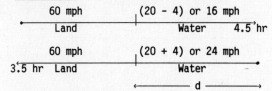

Let t = time spent on land on each part of the trip. Then 4.5 - t represents the time spent in water before delivering the package and 3.5 - t represents the time spent in water after delivering the package. Let d = distance traveled in water on each part of the trip. In a chart we organize the information concerning the part of the trip in water.

	Distance	Speed	Time
Before delivery	d	20 - 4, or 16	4.5 - t
After delivery	d	20 + 4, or 24	3.5 - t

Translate. Using d = rt in each row of the chart, we get the following system of equations:
   d = 16(4.5 - t)
   d = 24(3.5 - t)

Carry out. Substitute 16(4.5 - t) for d in the second equation and solve for t.
   d = 24(3.5 - t)
   16(4.5 - t) = 24(3.5 - t)   Substituting
   72 - 16t = 84 - 24t
   8t = 12
   $t = \frac{12}{8}$
   t = 1.5

Substitute 1.5 for t and solve for d.
   d = 16(4.5 - t)
   d = 16(4.5 - 1.5)
   d = 16(3)
   d = 48

If d = 48, then the total distance traveled by water is 2·48, or 96 mi.

If the time spent on land on each part of the trip is 1.5 hr, then the distance traveled on land, one way, is 60(1.5), or 90 mi. Then the round-trip land distance is 2·90, or 180 mi.

Check. If t = 1.5, then 4.5 - t = 3 and the distance in water before delivery is 3·16, or 48 mi. The distance in water after delivery is 24(3.5 - 1.5), or 24·2, or 48 mi. The distances by water are the same. The values check.

State. For the round trip, the messenger traveled 180 mi on land and 96 mi by water.

18. Wind: 40 mph, plane: 620 mph

19. Familiarize. First we draw a picture.

Slow trip   32 mph
t hr   d miles

Fast trip   32 + 4, or 36 mph
$t - \frac{1}{2}$ hr   d miles

Let d represent the distance between the towns. Let t represent the time for the slower trip. Then $t - \frac{1}{2}$ represents the time for the faster trip. We organize the information in a chart.

          d   =   r   ·   t

	Distance	Speed	Time
Slow trip	d	32	t
Fast trip	d	32 + 4, or 36	$t - \frac{1}{2}$

Translate. Using d = rt in each row of the chart, we get the following system of equations:
   d = 32t,
   $d = 36\left(t - \frac{1}{2}\right)$

Carry out. We substitute 32t for d in the second equation and solve for t.
   $d = 36\left(t - \frac{1}{2}\right)$
   $32t = 36\left(t - \frac{1}{2}\right)$
   32t = 36t - 18
   18 = 4t
   $\frac{18}{4} = t$
   4.5 = t

Substitute 4.5 for t and solve for d.
   d = 32t
   d = 32(4.5)   Substituting
   d = 144

Check. If t = 4.5, then the distance of the slow trip is 32(4.5), or 144. If t = 4.5, then the distance of the fast trip is $36\left(4.5 - \frac{1}{2}\right)$, or 36(4), or 144. The distances are the same. The values check.

State. The towns are 144 miles apart.

20. 1 hr 20 min

Chapter 8   (8.6)

21. <u>Familiarize</u>.

```
               55 mph
Truck •————————————————————→
        t hours    d + 10 miles

               40 mph
Car   •————————————————→
        t hours    d miles
```

Let t represent the time it takes before they lose contact. Since the speed of the truck is faster then the speed of the car, the truck will travel more miles than the car in the same amount of time. Let d represent the distance the car travels before they lose contact. Then d + 10 represents the distance the truck travels. We organize the information in a chart.

$$d = r \cdot t$$

	Distance	Speed	Time
Truck	d + 10	55	t
Car	d	40	t

<u>Translate</u>. Using d = rt in each row of the chart, we get the following system of equations:

    d + 10 = 55t,

    d = 40t

<u>Carry out</u>. We substitute 40t for d in the first equation and solve for t.

       d + 10 = 55t

   40t + 10 = 55t

         10 = 15t

        $\frac{10}{15}$ = t

         $\frac{2}{3}$ = t

<u>Check</u>. If $t = \frac{2}{3}$, the truck travels $55 \cdot \frac{2}{3}$, or $36\frac{2}{3}$ miles. If $t = \frac{2}{3}$, the car travels $40 \cdot \frac{2}{3}$, or $26\frac{2}{3}$ miles. The truck and car are 10 miles apart. The values check.

<u>State</u>. The truck and car will lose contact in $\frac{2}{3}$ hr, or 40 minutes.

22. ≈ 3603 mi

CHAPTER 9   INEQUALITIES

Exercise Set 9.1

1. x > 4
   a) Since 4 > 4 is false, 4 is not a solution.
   b) Since 0 > 4 is false, 0 is not a solution.
   c) Since -4 > 4 is false, -4 is not a solution.
   d) Since 6 > 4 is true, 6 is a solution.
   e) Since 5.6 > 4 is true, 5.6 is a solution.

2. a) Yes, b) No, c) Yes, d) Yes, e) No

3. x ≥ 6
   a) Since -6 ≥ 6 is false, -6 is not a solution.
   b) Since 0 ≥ 6 is false, 0 is not a solution.
   c) Since 6 ≥ 6 is true, 6 is a solution.
   d) Since 8 ≥ 6 is true, 8 is a solution.
   e) Since $-3\frac{1}{2}$ ≥ 6 is false, $-3\frac{1}{2}$ is not a solution.

4. a) Yes, b) Yes, c) Yes, d) No, e) Yes

5. x < -8
   a) Since 0 < -8 is false, 0 is not a solution.
   b) Since -8 < -8 is false, -8 is not a solution.
   c) Since -9 < -8 is true, -9 is a solution.
   d) Since -7 < -8 is false, -7 is not a solution.
   e) Since $-9\frac{1}{4}$ < -8 is true, $-9\frac{1}{4}$ is a solution.

6. a) Yes, b) No, c) Yes, d) Yes, e) Yes

7. y ≥ -5
   a) Since 0 ≥ -5 is true, 0 is a solution.
   b) Since -4 ≥ -5 is true, -4 is a solution.
   c) Since -5 ≥ -5 is true, -5 is a solution.
   d) Since -6 ≥ -5 is false, -6 is not a solution.
   e) Since 213.4 ≥ -5 is true, 213.4 is a solution.

8. a) Yes, b) Yes, c) No, d) Yes, e) Yes

9. The solutions of x > 4 are those numbers greater than 4. They are shown on the graph by shading all points to the right of 4. The open circle at 4 indicates that 4 is not part of the graph.

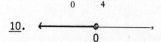

10.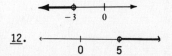

11. The solutions of t < -3 are those numbers less than -3. They are shown on the graph by shading all points to the left of -3. The open circle at -3 indicates that -3 is not part of the graph.

12.

13. The solutions of x ≥ 6 are shown by shading the point for 6 and all points to the right of 6. The closed circle at 6 indicates that 6 is part of the graph.

14.

15. The solutions of m ≤ -1 are shown by shading the point for -1 and all points to the left of -1. The closed circle at -1 indicates that -1 is part of the graph.

16.

17. In order to be a solution of the inequality -3 < x ≤ 4, a number must be a solution of both -3 < x and x ≤ 4. The solution set is graphed as follows:

    The open circle at -3 means that -3 is not part of the graph. The closed circle at 4 means that 4 is part of the graph.

18.

19. In order to be a solution of the inequality 0 < x < 3, a number must be a solution of both 0 < x and x < 3. The solution set is graphed as follows:

    The open circles at 0 and 3 mean that 0 and 3 are not part of the graph.

20.

21. x + 7 > 2
     x + 7 - 7 > 2 - 7   Adding -7
            x > -5   Simplifying

    The solution set is {x|x > -5}.
    The graph is as follows:

22. {x|x > -3}

23. $y + 5 > 8$
    $y + 5 - 5 > 8 - 5$ Adding -5
          $y > 3$

The solution set is $\{y \mid y > 3\}$.
The graph is as follows:

24. $\{y \mid y > 2\}$

25. $x + 8 \leqslant -10$
   $x + 8 - 8 \leqslant -10 - 8$
        $x \leqslant -18$

The solution set is $\{x \mid x \leqslant -18\}$.
The graph is as follows:

26. $\{x \mid x \leqslant -21\}$

27. $a + 12 < 6$
   $a + 12 - 12 < 6 - 12$
        $a < -6$

The solution set is $\{a \mid a < -6\}$.
The graph is as follows:

28. $\{a \mid a < -12\}$

29. $x - 7 \leqslant 9$
   $x - 7 + 7 \leqslant 9 + 7$
        $x \leqslant 16$

The solution set is $\{x \mid x \leqslant 16\}$.
The graph is as follows:

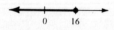

30. $\{x \mid x \leqslant 17\}$

31. $x - 6 > 2$
   $x - 6 + 6 > 2 + 6$
        $x > 8$

The solution set is $\{x \mid x > 8\}$.
The graph is as follows:

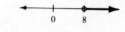

32. $\{x \mid x > 13\}$

33. $y - 7 > -12$
   $y - 7 + 7 > -12 + 7$
        $y > -5$

The solution set is $\{y \mid y > -5\}$.
The graph is as follows:

34. $\{y \mid y > -6\}$

35. $2x + 3 > x + 5$
   $2x + 3 - 3 > x + 5 - 3$ Adding -3
        $2x > x + 2$ Simplifying
   $2x - x > x + 2 - x$ Adding $-x$
        $x > 2$ Simplifying

The solution set is $\{x \mid x > 2\}$.
The graph is as follows:

36. $\{x \mid x > 3\}$

37. $3x + 9 \leqslant 2x + 6$
   $3x + 9 - 9 \leqslant 2x + 6 - 9$ Adding -9
        $3x \leqslant 2x - 3$
   $3x - 2x \leqslant 2x - 3 - 2x$ Adding $-2x$
        $x \leqslant -3$

The solution set is $\{x \mid x \leqslant -3\}$.

38. $\{x \mid x \leqslant -2\}$

Chapter 9   (9.1)

39. $3x - 6 \geq 2x + 7$
     $3x - 6 + 6 \geq 2x + 7 + 6$ Adding 6
     $3x \geq 2x + 13$
     $3x - 2x \geq 2x + 13 - 2x$ Adding $-2x$
     $x \geq 13$

The solution set is $\{x | x \geq 13\}$.

40. $\{x | x \geq 20\}$

41. $5x - 6 < 4x - 2$
     $5x - 6 + 6 < 4x - 2 + 6$
     $5x < 4x + 4$
     $5x - 4x < 4x + 4 - 4x$
     $x < 4$

The solution set is $\{x | x < 4\}$.

42. $\{x | x < -1\}$

43. $3y + 4 \geq 2y - 7$
     $3y + 4 - 4 \geq 2y - 7 - 4$
     $3y \geq 2y - 11$
     $3y - 2y \geq 2y - 11 - 2y$
     $y \geq -11$

The solution set is $\{y | y \geq -11\}$.

44. $\{y | y \leq -13\}$

45. $7 + c > 7$
     $-7 + 7 + c > -7 + 7$
     $c > 0$

The solution set is $\{c | c > 0\}$.

46. $\{c | c > 18\}$

47. $y + \frac{1}{4} \leq \frac{1}{2}$
     $y + \frac{1}{4} - \frac{1}{4} \leq \frac{1}{2} - \frac{1}{4}$
     $y \leq \frac{2}{4} - \frac{1}{4}$ Obtaining a common denominator
     $y \leq \frac{1}{4}$

The solution set is $\{y | y \leq \frac{1}{4}\}$.

48. $\{y | y \leq \frac{1}{2}\}$

49. $x - \frac{1}{3} > \frac{1}{4}$
     $x - \frac{1}{3} + \frac{1}{3} > \frac{1}{4} + \frac{1}{3}$
     $x > \frac{3}{12} + \frac{4}{12}$ Obtaining a common denominator
     $x > \frac{7}{12}$

The solution set is $\{x | x > \frac{7}{12}\}$.

50. $\{x | x > \frac{5}{8}\}$

51. $-14x + 21 > 21 - 15x$
     $-14x + 21 + 15x > 21 - 15x + 15x$
     $x + 21 > 21$
     $x + 21 - 21 > 21 - 21$
     $x > 0$

The solution set is $\{x | x > 0\}$.

52. $\{x | x > 3\}$

53. Familiarize. We draw a picture.

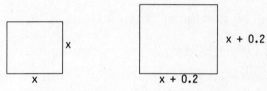

Let x represent the length of a side of the original square. Then $x + 0.2$ represents a side of the new square. Recall that the area of a square is the length of a side squared.

Translate.

Length of new side squared is area of new square.
$(x + 0.2)^2 = 0.64$

Carry out. We solve the equation.
$(x + 0.2)^2 = 0.64$
$x^2 + 0.4x + 0.04 = 0.64$
$x^2 + 0.4x - 0.6 = 0$
$10x^2 + 4x - 6 = 0$ Multiplying by 10 to clear decimals
$2(5x^2 + 2x - 3) = 0$
$2(5x - 3)(x + 1) = 0$
$5x - 3 = 0$ or $x + 1 = 0$
$5x = 3$ or $x = -1$
$x = \frac{3}{5}$ or $x = -1$

Check. Since the length of a side cannot be negative, -1 is not a solution. If $x = \frac{3}{5}$, or 0.6, then $x + 0.2$ is $0.6 + 0.2$, or 0.8 and the area is $(0.8)^2$, or 0.64. The value checks.

State. The length of a side of the original square is 0.6 km.

54. 19,000 hunting licenses, 24,500 fishing licenses

55. $F = \frac{9}{5}C + 32$
     $F - 32 = \frac{9}{5}C$ Adding -32
     $\frac{5}{9}(F - 32) = C$ Multiplying by $\frac{5}{9}$

Chapter 9 (9.2)

56. $x = \dfrac{Q + by + 4}{a}$

57. $3(r + 2) < 2r + 4$
    $3r + 6 < 2r + 4$ Removing parentheses
    $3r - 2r < 4 - 6$ Adding $-2r$ and -6
    $r < -2$

    The solution set is $\{r | r < -2\}$.

58. $\{r | r \geqslant -13\}$

59. $0.8x + 5 \geqslant 6 - 0.2x$
    $0.8x + 0.2x \geqslant 6 - 5$ Adding $0.2x$ and -5
    $x \geqslant 1$

    The solution set is $\{x | x \geqslant 1\}$.

60. $\{x | x \leqslant 1\}$

61. $2x + 2.4 > x - 9.4$
    $2x - x > -9.4 - 2.4$
    $x > -11.8$

    The solution set is $\{x | x > -11.8\}$.

62. $\{x | x > -4\}$

63. $12x + 1.2 \leqslant 11x$
    $12x - 11x \leqslant -1.2$
    $x \leqslant -1.2$

    The solution set is $\{x | x \leqslant -1.2\}$.

64. $\{x | x \leqslant 1\}$

65. $17x + 9,479,756 \leqslant 16x - 8,579,243$
    $17x - 16x \leqslant -8,579,243 - 9,479,756$
    $x \leqslant -18,058,999$

    The solution set is $\{x | x \leqslant -18,058,999\}$.

66. Yes

67. $|x| < 3$

    a) Since $|0| = 0$, and $0 < 3$ is true, 0 is a solution.
    b) Since $|-2| = 2$ and $2 < 3$ is true, -2 is a solution.
    c) Since $|-3| = 3$ and $3 < 3$ is false, -3 is not a solution.
    d) Since $|4| = 4$ and $4 < 3$ is false, 4 is not a solution.
    e) Since $|3| = 3$ and $3 < 3$ is false, 3 is not a solution.
    f) Since $|1.7| = 1.7$ and $1.7 < 3$ is true, 1.7 is a solution.
    g) Since $|-2.8| = 2.8$ and $2.8 < 3$ is true, -2.8 is a solution.

68.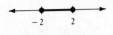

69. $|x| \geqslant 4$

    a) Since $|0| = 0$ and $0 \geqslant 4$ is false, 0 is not a solution.
    b) Since $|-5| = 5$ and $5 \geqslant 4$ is true, 5 is a solution.
    c) Since $|6| = 6$ and $6 \geqslant 4$ is true, 6 is a solution.
    d) Since $|-3| = 3$ and $3 \geqslant 4$ is false, -3 is not a solution.
    e) Since $|3| = 3$ and $3 \geqslant 4$ is false, 3 is not a solution.
    f) Since $|-8| = 8$ and $8 \geqslant 4$ is true, -8 is a solution.
    g) Since $|9.7| = 9.7$ and $9.7 \geqslant 4$ is true, 9.7 is a solution.

70.

71. The solutions of $|x| \leqslant 2$ are all numbers that are 2 units or less from 0 on the number line. The graph is as follows:

    ![number line with closed dots at -2 and 2]

72. ![number line with open dots at -1 and 1]

73. The solutions of $|x| > 3$ are all numbers that are more than 3 units from 0 on the number line. The graph is as follows:

    ![number line with open dots at -3 and 3]

74. ![number line with closed dots at -2 and 2]

75. $x + 3 \leqslant 3 + x$
    $x - x \leqslant 3 - 3$ Adding $-x$ and -3
    $0 \leqslant 0$

    We get an inequality that is true for all values of x, so the inequality is true for all real numbers.

76. No solution

Exercise Set 9.2

1. $5x < 35$
    $\dfrac{1}{5} \cdot 5x < \dfrac{1}{5} \cdot 35$ Multiplying by $\dfrac{1}{5}$
    $x < 7$

    The solution set is $\{x | x < 7\}$. The graph is as follows:

    ![number line with open dot at 7]

Chapter 9   (9.2)

2. {x|x ≥ 4}

3.   9y ≤ 81

   $\frac{1}{9} \cdot 9y \le \frac{1}{9} \cdot 81$ Multiplying by $\frac{1}{9}$

         y ≤ 9

   The solution set is {y|y ≤ 9}. The graph is as follows:

4. {x|x > 24}

5.   7x < 13

   $\frac{1}{7} \cdot 7x < \frac{1}{7} \cdot 13$

         x < $\frac{13}{7}$

   The solution set is $\{x | x < \frac{13}{7}\}$. The graph is as follows:

6. $\{y | y < \frac{17}{8}\}$

7.   12x > -36

   $\frac{1}{12} \cdot 12x > \frac{1}{12} \cdot (-36)$

         x > -3

   The solution set is {x|x > -3}. The graph is as follows:

8. {x|x < -4}
   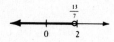

9.   5y ≥ -2

   $\frac{1}{5} \cdot 5y \ge \frac{1}{5} \cdot (-2)$

         y ≥ $-\frac{2}{5}$

   The solution set is $\{y | y \ge -\frac{2}{5}\}$.

10. $\{x | x > -\frac{4}{7}\}$

11.   -2x ≤ 12

    $-\frac{1}{2} \cdot (-2x) \ge -\frac{1}{2} \cdot 12$ Multiplying by $-\frac{1}{2}$

             The symbol has to be reversed.

          x ≥ -6   Simplifying

    The solution set is {x|x ≥ -6}.

12. {y|y ≥ -5}

13.   -4y ≥ -16

    $-\frac{1}{4} \cdot (-4y) \le -\frac{1}{4} \cdot (-16)$ Multiplying by $-\frac{1}{4}$

             The symbol has to be reversed.

          y ≤ 4

    The solution set is {y|y ≤ 4}.

14. {x|x > 3}

15.   -3x < -17

    $-\frac{1}{3} \cdot (-3x) > -\frac{1}{3} \cdot (-17)$ Multiplying by $-\frac{1}{3}$

             The symbol has to be reversed.

          x > $\frac{17}{3}$

    The solution set is $\{x | x > \frac{17}{3}\}$.

16. $\{y | y < \frac{23}{5}\}$

17.   -2y > $\frac{1}{7}$

    $-\frac{1}{2} \cdot (-2y) < -\frac{1}{2} \cdot \frac{1}{7}$

             The symbol has to be reversed.

          y < $-\frac{1}{14}$

    The solution set is $\{y | y < -\frac{1}{14}\}$.

18. $\{x | x \ge -\frac{1}{36}\}$

19. $-\frac{6}{5}$ ≤ -4x

    $-\frac{1}{4} \cdot \left(-\frac{6}{5}\right) \ge -\frac{1}{4} \cdot (-4x)$

          $\frac{6}{20}$ ≥ x

          $\frac{3}{10}$ ≥ x, or x ≤ $\frac{3}{10}$

    The solution set is $\{x | \frac{3}{10} \ge x\}$, or $\{x | x \le \frac{3}{10}\}$.

20. $\{t | t > \frac{1}{64}\}$

Chapter 9   (9.2)

21.
$$4 + 3x < 28$$
$$-4 + 4 + 3x < -4 + 28 \quad \text{Adding } -4$$
$$3x < 24 \quad \text{Simplifying}$$
$$\tfrac{1}{3} \cdot 3x < \tfrac{1}{3} \cdot 24 \quad \text{Multiplying by } \tfrac{1}{3}$$
$$x < 8 \quad \text{Simplifying}$$
The solution set is $\{x \mid x < 8\}$.

22. $\{y \mid y < 8\}$

23.
$$6 + 5y \geq 36$$
$$-6 + 6 + 5y \geq -6 + 36 \quad \text{Adding } -6$$
$$5y \geq 30$$
$$\tfrac{1}{5} \cdot 5y \geq \tfrac{1}{5} \cdot 30 \quad \text{Multiplying by } \tfrac{1}{5}$$
$$y \geq 6$$
The solution set is $\{y \mid y \geq 6\}$.

24. $\{x \mid x \geq 8\}$

25.
$$3x - 5 \leq 13$$
$$3x - 5 + 5 \leq 13 + 5 \quad \text{Adding } 5$$
$$3x \leq 18$$
$$\tfrac{1}{3} \cdot 3x \leq \tfrac{1}{3} \cdot 18 \quad \text{Multiplying by } \tfrac{1}{3}$$
$$x \leq 6$$
The solution set is $\{x \mid x \leq 6\}$.

26. $\{y \mid y \leq 6\}$

27.
$$10y - 9 > 31$$
$$10y - 9 + 9 > 31 + 9$$
$$10y > 40$$
$$\tfrac{1}{10} \cdot 10y > \tfrac{1}{10} \cdot 40$$
$$y > 4$$
The solution set is $\{y \mid y > 4\}$.

28. $\{y \mid y > 4\}$

29.
$$13x - 7 < -46$$
$$13x - 7 + 7 < -46 + 7$$
$$13x < -39$$
$$\tfrac{1}{13} \cdot 13x < \tfrac{1}{13} \cdot (-39)$$
$$x < -3$$
The solution set is $\{x \mid x < -3\}$.

30. $\{y \mid y < -6\}$

31.
$$5x + 3 \geq -7$$
$$5x + 3 - 3 \geq -7 - 3$$
$$5x \geq -10$$
$$\tfrac{1}{5} \cdot 5x \geq \tfrac{1}{5} \cdot (-10)$$
$$x \geq -2$$
The solution set is $\{x \mid x \geq -2\}$.

32. $\{y \mid y \geq -2\}$

33.
$$13 < 4 - 3y$$
$$13 - 4 < 4 - 3y - 4 \quad \text{Adding } -4$$
$$9 < -3y$$
$$-\tfrac{1}{3} \cdot 9 > -\tfrac{1}{3} \cdot (-3y) \quad \text{Multiplying by } -\tfrac{1}{3}$$
$$\text{The symbol has to reversed.}$$
$$-3 > y$$
The solution set is $\{y \mid -3 > y\}$, or $\{y \mid y < -3\}$.

34. $\{x \mid x < -2\}$

35.
$$30 > 3 - 9x$$
$$30 - 3 > 3 - 9x - 3 \quad \text{Adding } -3$$
$$27 > -9x$$
$$-\tfrac{1}{9} \cdot 27 < -\tfrac{1}{9} \cdot (-9x) \quad \text{Multiplying by } -\tfrac{1}{9}$$
$$\text{The symbol as to be reversed.}$$
$$-3 < x$$
The solution set is $\{x \mid -3 < x\}$, or $\{x \mid x > -3\}$.

36. $\{y \mid y > -5\}$

37.
$$3 - 6y > 23$$
$$-3 + 3 - 6y > -3 + 23$$
$$-6y > 20$$
$$-\tfrac{1}{6} \cdot (-6y) < -\tfrac{1}{6} \cdot 20$$
$$\text{The symbol has to be reversed.}$$
$$y < -\tfrac{20}{6}$$
$$y < -\tfrac{10}{3}$$
The solution set is $\left\{y \mid y < -\tfrac{10}{3}\right\}$.

38. $\{y \mid y < -3\}$

39.
$$4x + 2 - 3x \leq 9$$
$$x + 2 \leq 9 \quad \text{Collecting like terms}$$
$$x + 2 - 2 \leq 9 - 2$$
$$x \leq 7$$
The solution set is $\{x \mid x \leq 7\}$.

Chapter 9   (9.2)

40. $\{x | x \leq 4\}$

41. $-3 < 8x + 7 - 7x$
    $-3 < x + 7$ Collecting like terms
$-3 - 7 < x + 7 - 7$
    $-10 < x$

The solution set is $\{x | -10 < x\}$, or $\{x | x > -10\}$.

42. $\{x | x > -13\}$

43. $6 - 4y > 4 - 3y$
$6 - 4y + 4y > 4 - 3y + 4y$ Adding $4y$
       $6 > 4 + y$
$-4 + 6 > -4 + 4 + y$ Adding -4
      $2 > y$, or $y < 2$

The solution set is $\{y | 2 > y\}$, or $\{y | y < 2\}$.

44. $\{y | y < 2\}$

45. $5 - 9y \leq 2 - 8y$
$5 - 9y + 9y \leq 2 - 8y + 9y$
      $5 \leq 2 + y$
  $-2 + 5 \leq -2 + 2 + y$
      $3 \leq y$, or $y \geq 3$

The solution set is $\{y | 3 \leq y\}$, or $\{y | y \geq 3\}$.

46. $\{y | y \geq 2\}$

47. $19 - 7y - 3y < 39$
    $19 - 10y < 39$ Collecting like terms
$-19 + 19 - 10y < -19 + 39$
      $-10y < 20$
  $-\frac{1}{10} \cdot (-10y) > -\frac{1}{10} \cdot 20$
            The symbol has to be reversed.
      $y > -2$

The solution set is $\{y | y > -2\}$.

48. $\{y | y > -3\}$

49. $21 - 8y < 6y + 49$
$21 - 8y + 8y < 6y + 49 + 8y$
      $21 < 14y + 49$
$21 - 49 < 14y + 49 - 49$
      $-28 < 14y$
  $\frac{1}{14} \cdot -28 < \frac{1}{14} \cdot 14y$
      $-2 < y$, or $y > -2$

The solution set is $\{y | -2 < y\}$, or $\{y | y > -2\}$.

50. $\{x | x > -4\}$

51. $14 - 5y - 2y \geq -19$
    $14 - 7y \geq -19$
$-14 + 14 - 7y \geq -14 - 19$
    $-7y \geq -33$
  $-\frac{1}{7} \cdot (-7y) \leq -\frac{1}{7} \cdot (-33)$
           The symbol has to be reversed.
    $y \leq \frac{33}{7}$

The solution set is $\{y | y \leq \frac{33}{7}\}$.

52. $\{y | y \geq \frac{30}{13}\}$

53. $27 - 11x > 14x - 18$
$27 - 11x + 11x > 14x - 18 + 11x$
    $27 > 25x - 18$
$27 + 18 > 25x - 18 + 18$
    $45 > 25x$
  $\frac{1}{25} \cdot 45 > \frac{1}{25} \cdot 25x$
    $\frac{45}{25} > x$
    $\frac{9}{5} > x$, or $x < \frac{9}{5}$

The solution set is $\{x | \frac{9}{5} > x\}$, or $\{x | x < \frac{9}{5}\}$.

54. $\{y | y < \frac{61}{28}\}$

55. $2.1x + 45.2 > 3.2 - 8.4x$
$10(2.1x + 45.2) > 10(3.2 - 8.4x)$ Multiplying by 10 to clear decimals
    $21x + 452 > 32 - 84x$
    $21x + 84x > 32 - 452$ Adding $84x$ and -452
    $105x > -420$
    $x > -4$ Multiplying by $\frac{1}{105}$

The solution set is $\{x | x > -4\}$.

56. $\{y | y \leq \frac{5}{3}\}$

57. $0.7n - 15 + n \geq 2n - 8 - 0.4n$
    $1.7n - 15 \geq 1.6n - 8$ Collecting like terms
$10(1.7n - 15) \geq 10(1.6n - 8)$ Multiplying by 10
    $17n - 150 \geq 16n - 80$
    $17n - 16n \geq -80 + 150$ Adding $-16n$ and 150
    $n \geq 70$

The solution set is $\{n | n \geq 70\}$.

58. $\{t | t > 1\}$

Chapter 9   (9.2)

59. $\frac{x}{3} - 2 \leq 1$

$3\left[\frac{x}{3} - 2\right] \leq 3 \cdot 1$ Multiplying by 3 to clear the fraction

$x - 6 \leq 3$ Simplifying

$x \leq 9$ Adding 6

The solution set is $\{x | x \leq 9\}$.

60. $\{x | x > 2\}$

61. $\frac{y}{5} + 1 \leq \frac{2}{5}$

$5\left[\frac{y}{5} + 1\right] \leq 5 \cdot \frac{2}{5}$ Clearing fractions

$y + 5 \leq 2$

$y \leq -3$ Adding -5

The solution set is $\{y | y \leq -3\}$.

62. $\{x | x \geq -25\}$

63. $\frac{-x}{4} - \frac{3x}{8} + 2 > 3 - x$

$8\left[\frac{-x}{4} - \frac{3x}{8} + 2\right] > 8(3 - x)$ Clearing fractions

$-2x - 3x + 16 > 24 - 8x$

$-5x + 16 > 24 - 8x$ Collecting like terms

$-5x + 8x > 24 - 16$ Adding 8x and -16

$3x > 8$

$x > \frac{8}{3}$ Multiplying by $\frac{1}{3}$

The solution set is $\left\{x \middle| x > \frac{8}{3}\right\}$.

64. $\left\{x \middle| x < \frac{30}{7}\right\}$

65. $3(2y - 3) < 27$

$6y - 9 < 27$ Removing parentheses

$6y < 36$ Adding 9

$y < 6$ Multiplying by $\frac{1}{6}$

The solution set is $\{y | y < 6\}$.

66. $\{y | y > 5\}$

67. $2(3 + 4m) - 9 \geq 45$

$6 + 8m - 9 \geq 45$ Removing parentheses

$8m - 3 \geq 45$ Collecting like terms

$8m \geq 48$ Adding 3

$m \geq 6$ Multiplying by $\frac{1}{8}$

The solution set is $\{m | m \geq 6\}$.

68. $\{m | m \leq 9\}$

69. $5(d + 4) \leq 7(d - 2)$

$5d + 20 \leq 7d - 14$ Removing parentheses

$5d - 7d \leq -14 - 20$ Adding -7d and -20

$-2d \leq -34$

$d \geq 17$ Multiplying by $-\frac{1}{2}$

         The symbol has to be reversed.

The solution set is $\{d | d \geq 17\}$.

70. $\{t | t \leq -4\}$

71. $8(2t + 1) > 4(7t + 7)$

$16t + 8 > 28t + 28$

$16t - 28t > 28 - 8$

$-12t > 20$

$t < -\frac{20}{12}$ Multiplying by $-\frac{1}{12}$ and reversing the symbol

$t < -\frac{5}{3}$

The solution set is $\left\{t \middle| t < -\frac{5}{3}\right\}$.

72. $\{x | x > -8\}$

73. $3(r - 6) + 2 < 4(r + 2) - 21$

$3r - 18 + 2 < 4r + 8 - 21$

$3r - 16 < 4r - 13$

$-16 + 13 < 4r - 3r$

$-3 < r$, or $r > -3$

The solution set is $\{r | r > -3\}$.

74. $\{t | t > -12\}$

75. $19 - (2x + 3) \leq 2(x + 3) + x$

$19 - 2x - 3 \leq 2x + 6 + x$

$16 - 2x \leq 3x + 6$

$-2x - 3x \leq 6 - 16$

$-5x \leq -10$

$x \geq 2$

The solution set is $\{x | x \geq 2\}$.

76. $\{c | c \leq 1\}$

77. $\frac{1}{4}(8y + 4) - 17 > -\frac{1}{2}(4y - 8)$

$2y + 1 - 17 > -2y + 4$ Removing parentheses

$2y - 16 > -2y + 4$ Collecting like terms

$2y + 2y > 4 + 16$

$4y > 20$

$y > 5$

The solution set is $\{y | y > 5\}$.

78. $\{x | x < 6\}$

Chapter 9 (9.2)

79. $\frac{2}{3}(2x - 1) \geq 10$

$\frac{3}{2} \cdot \frac{2}{3}(2x - 1) \geq \frac{3}{2} \cdot 10$ Multiplying by $\frac{3}{2}$

$2x - 1 \geq 15$

$2x \geq 16$

$x \geq 8$

The solution set is $\{x | x \geq 8\}$.

80. $\{x | x \leq 7\}$

81. $\frac{3}{4}(3 + 2x) + 1 < 13$

$4\left[\frac{3}{4}(3 + 2x) + 1\right] < 4 \cdot 13$ Clearing the fraction

$4 \cdot \frac{3}{4}(3 + 2x) + 4 \cdot 1 < 52$

$3(3 + 2x) + 4 < 52$

$9 + 6x + 4 < 52$

$6x + 13 < 52$

$6x < 39$

$x < \frac{39}{6}$

$x < \frac{13}{2}$

The solution set is $\left\{x \big| x < \frac{13}{2}\right\}$.

82. $\left\{x \big| x < -\frac{405}{28}\right\}$

83. $\frac{3}{4}\left(3x - \frac{1}{2}\right) - \frac{2}{3} < \frac{1}{3}$

$\frac{3}{4}\left(3x - \frac{1}{2}\right) < 1$ Adding $\frac{2}{3}$

$\frac{9}{4}x - \frac{3}{8} < 1$ Removing parentheses

$8 \cdot \left(\frac{9}{4}x - \frac{3}{8}\right) < 8 \cdot 1$ Clearing fractions

$18x - 3 < 8$

$18x < 11$

$x < \frac{11}{18}$

The solution set is $\left\{x \big| x < \frac{11}{18}\right\}$.

84. $\left\{x \big| x > -\frac{5}{32}\right\}$

85. $0.7(3x + 6) \geq 1.1 - (x + 2)$

$2.1x + 4.2 \geq 1.1 - x - 2$

$10(2.1x + 4.2) \geq 10(1.1 - x - 2)$ Clearing decimals

$21x + 42 \geq 11 - 10x - 20$

$21x + 42 \geq -10x - 9$ Collecting like terms

$21x + 10x \geq -9 - 42$

$31x \geq -51$

$x \geq -\frac{51}{31}$

The solution set is $\left\{x \big| x \geq -\frac{51}{31}\right\}$.

86. $\left\{x \big| x \geq \frac{39}{14}\right\}$

87. $a + (a - 3) < (a + 2) - (a + 1)$

$2a - 3 < a + 2 - a - 1$

$2a - 3 < 1$

$2a < 4$

$a < 2$

The solution set is $\{a | a < 2\}$.

88. $\{b | b < -7.4\}$

89. <u>Familiarize</u>. First make a drawing. Let x represent the measure of the second angle. Then 5x represents the first angle and x - 16 represents the third angle. The sum of the three angles is 180°.

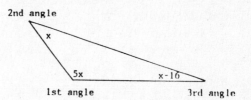

<u>Translate</u>.

$\underline{\text{First angle}} + \underline{\text{Second angle}} + \underline{\text{Third angle}} = 180°$

$5x + x + (x - 16) = 180$

<u>Carry out</u>.

$5x + x + x - 16 = 180$

$7x - 16 = 180$

$7x = 196$

$x = 28$

Possible angle measures:

First angle: $5x = 5 \cdot 28 = 140°$
Second angle: $x = 28°$
Third angle: $x - 16 = 28 - 16 = 12°$

<u>Check</u>. The sum of the measures, 140° + 28° + 12°, is 180°. The measure of the first angle is five times the second angle, 140 = 5·28. The measure of the third angle is sixteen less than the second, 28 - 16 = 12. The values check.

89. (continued)

    State. The measures of the angles are 140°, 28°, and 12°.

90. $-\frac{3}{2}$

91. $2x - 3y = 9$ (1)
    $\underline{5x + 3y = 5}$ (2)
    $7x = 14$ Adding
    $x = 2$

    Substitute 2 for x in equation (2) and solve for y.
    $$5x + 3y = 5$$
    $$5 \cdot 2 + 3y = 5$$
    $$10 + 3y = 5$$
    $$3y = -5$$
    $$y = -\frac{5}{3}$$

    The ordered pair $\left(2, -\frac{5}{3}\right)$ checks. It is the solution.

92. $\left(\frac{92}{29}, -\frac{85}{29}\right)$

93. $2[4 - 2(3 - x)] - 1 \geq 4[2(4x - 3) + 7] - 25$
    $2[4 - 6 + 2x] - 1 \geq 4[8x - 6 + 7] - 25$
    $2[-2 + 2x] - 1 \geq 4[8x + 1] - 25$
    $-4 + 4x - 1 \geq 32x + 4 - 25$
    $-5 + 4x \geq 32x - 21$
    $4x - 32x \geq -21 + 5$
    $-28x \geq -16$
    $x \leq \frac{-16}{-28}$
    $x \leq \frac{4}{7}$

    The solution set is $\left\{x \mid x \leq \frac{4}{7}\right\}$.

94. $\left\{t \mid t > -\frac{27}{19}\right\}$

95. $-(x + 5) \geq 4a - 5$
    $-x - 5 \geq 4a - 5$
    $-x \geq 4a - 5 + 5$
    $-x \geq 4a$
    $-1(-x) \leq -1 \cdot 4a$
    $x \leq -4a$

    The solution set is $\{x \mid x \leq -4a\}$.

96. $\{x \mid x > 7\}$

97. $-6(x + 3) \leq -9(y + 2)$
    $-6x - 18 \leq -9y - 18$
    $-6x \leq -9y - 18 + 18$
    $-6x \leq -9y$
    $-\frac{1}{6} \cdot (-6x) \geq -\frac{1}{6} \cdot (-9y)$
    $x \geq \frac{9}{6}y$
    $x \geq \frac{3}{2}y$

    The solution set is $\left\{x \mid x \geq \frac{3}{2}y\right\}$.

98. $\left\{x \mid x > \frac{y - b}{a}, a > 0\right\}$

99. $x^2 > 0$

    The square of any nonzero real number is positive. All positive numbers are greater than 0. Thus the solution is the set of all nonzero real numbers.

100. All real numbers.

Exercise Set 9.3

1. $x > 4$

2. $x < 7$

3. $x \leq -6$

4. $y \geq 13$

5. $p \geq 1200$

6. $c \leq \$3457.95$

7. $y \leq 500$

8. $c \geq \$0.94$

9. $3x + 2 < 13$

10. $\frac{1}{2}n - 5 > 17$

11. Familiarize. The average of the five scores is their sum divided by the number of quizzes, 5. We let s represent the student's score on the last quiz.

    Translate. The average of the five scores is given by
    $$\frac{73 + 75 + 89 + 91 + s}{5}.$$

    Since this average must be at least 85, this means that it must be greater than or equal to 85. Thus, we can translate the problem to the inequality
    $$\frac{73 + 75 + 89 + 91 + s}{5} \geq 85.$$

Chapter 9 (9.3)

11. (continued)

    Carry out. We first multiply by 5 to clear of fractions.

    $$5\left[\frac{73 + 75 + 89 + 91 + s}{5}\right] \geq 5 \cdot 85$$
    $$73 + 75 + 89 + 91 + s \geq 425$$
    $$328 + s \geq 425$$
    $$s \geq 425 - 328$$
    $$s \geq 97$$

    Check. Suppose s is a score greater than or equal to 97. Then by successively adding 73, 75, 89, and 91 on both sides of the inequality we get

    $$73 + 75 + 89 + 91 + s \geq 425$$

    so

    $$\frac{73 + 75 + 89 + 91 + s}{5} \geq \frac{425}{5}, \text{ or } 85$$

    State. Any score which is at least 97 will give an average quiz grade of 85. The solution set is $\{s | s \geq 97\}$.

12. $\{C | C > 37°\}$

13. Familiarize.

    $R = -0.075t + 3.85$

    In the formula R represents the world record and t represents the years since 1930. When $t = 0$ (1930), the record was $-0.075 \cdot 0 + 3.85$, or 3.85 minutes. When $t = 2$ (1932), the record was $-0.075(2) + 3.85$, or 3.7. For what values of t will $-0.075t + 3.85$ be less than 3.5?

    Translate. The record is to be less than 3.5. We have the inequality

    $R < 3.5$.

    To find the t values which satisfy this condition we substitute $-0.075t + 3.85$ for R.

    $-0.075 + 3.85 < 3.5$

    Carry out.

    $$-0.075t + 3.85 < 3.5$$
    $$-0.075t < 3.5 - 3.85$$
    $$-0.075t < -0.35$$
    $$t > \frac{-0.35}{-0.075}$$
    $$t > 4\frac{2}{3}$$

    Check. With inequalities it is impossible to check each solution. But we can check to see if the solution set we obtained seems reasonable.

    When $t = 4\frac{1}{2}$, $R = -0.075(4.5) + 3.85$, or 3.5125.

    When $t = 4\frac{2}{3}$, $R = -0.075\left(\frac{14}{3}\right) + 3.85$, or 3.5.

    When $t = 4\frac{3}{4}$, $R = -0.075(4.75) + 3.85$, or 3.49375.

    Since $R = 3.5$ when $t = 4\frac{2}{3}$ and R decreases as t increases, R will be less than 3.5 when t is greater than $4\frac{2}{3}$.

13. (continued)

    State. Thus, the world record will be less than 3.5 minutes when t is greater than $4\frac{2}{3}$ years (more than $4\frac{2}{3}$ years since 1930). The solution set is $\{t | t > 4\frac{2}{3}\}$.

14. $\{t | t > 64\frac{2}{7}\}$

15. Familiarize. Let n represent the number.

    Translate.

The number	plus	15	is less than	4 times	the number.
n	+	15	<	4 ·	n

    Carry out.

    $$n + 15 < 4n$$
    $$15 < 3n$$
    $$5 < n, \text{ or } n > 5$$

    Check. With inequalities it is impossible to check each solution. But we can check to see if the solution set we obtained seems reasonable.

    When $n = 4$, we have $4 + 15 < 4 \cdot 4$, or $19 < 16$. This is false.

    When $n = 5$, we have $5 + 15 < 4 \cdot 5$, or $20 < 20$. This is false.

    When $n = 6$, we have $6 + 15 < 4 \cdot 6$, or $21 < 24$. This is true.

    Since the inequality is false for the numbers less than or equal to 5 that we tried and true for the number greater than 5, it would appear that $n > 5$ is correct.

    State. All numbers greater than 5 are solutions. The solution set is $\{n | n > 5\}$.

16. $\{n | n \leq 0\}$

17. Familiarize. Let m represent the number of miles per day. Then the cost per day for those miles is $0.46m. The total cost is the daily rate plus the daily mileage cost. The total cost cannot exceed $200. In other words the total cost must be less than or equal to $200, the daily budget.

    Translate.

Daily rate	+	Mileage cost	≤	Budget
42.95	+	0.46 m	≤	200

    Carry out.

    $$42.95 + 0.46m \leq 200$$
    $$4295 + 46m \leq 20,000 \quad \text{Clearing decimals}$$
    $$46m \leq 15,705$$
    $$m \leq \frac{15,705}{46}$$
    $$m \leq 341.4 \quad \text{Rounding to the nearest tenth}$$

17. (continued)

    Check. We can check to see if the solution set seems reasonable.

    When m = 342, the total cost is
    42.95 + 0.46(342), or $200.27.

    When m = 341.4, the total cost is
    42.95 + 0.46(341.4), or $199.99.

    When m = 341, the total cost is
    42.95 + 0.46(341), or $199.81.

    From these calculations it would appear that m ≤ 341.4 is the correct solution.

    State. To stay within the budget, the number of miles the family drives must not exceed 341.4. The solution set is {m|m ≤ 341.4 mi}.

18. {m|m ≤ 525.8 mi}

19. Familiarize. We first make a drawing. We let ℓ represent the length.

    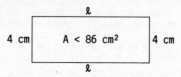

    The area is the length times the width, or 4ℓ.

    Translate.

    Area is less than 86 cm².
    4ℓ   <   86

    Carry out.
       4ℓ < 86
        ℓ < 21.5

    Check. We check to see if the solution seems reasonable.

    When ℓ = 22, the area is 22·4, or 88 cm².
    When ℓ = 21.5, the area is 21.5(4), or 86 cm².
    When ℓ = 21, the area is 21·4, or 84 cm².

    From these calculations, it would appear that the solution is correct.

    State. The area will be less than 86 cm² for lengths less than 21.5 cm. The solution set is {ℓ|ℓ < 21.5 cm}.

20. {ℓ|ℓ ≥ 16.5 yd}

21. Familiarize. We first make a drawing. We let w represent the width.

    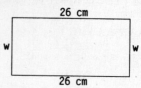

    The perimeter is P = 2ℓ + 2w, or 2·26 + 2w, or 52 + 2w.

    Translate.

    The perimeter is greater than 80 cm.
      52 + 2w   >   80

    Carry out.
       52 + 2w > 80
           2w > 28
            w > 14

    Check. We check to see if the solution seems reasonable.

    When w = 13, P = 2·26 + 2·13, or 78 cm.
    When w = 14, P = 2·26 + 2·14, or 80 cm.
    When w = 15, P = 2·26 + 2·15, or 82 cm.

    From these calculations, it appears that the solution is correct.

    State. Widths greater than 14 cm will make the perimeter greater than 80 cm. The solution set is {w|w > 14 cm}.

22. {ℓ|ℓ ≤ 92 ft}

23. Familiarize. We first make a drawing. Let b represent the length of the base. Then the lengths of the other sides are b - 2 and b + 3.

    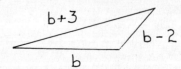

    The perimeter is the sum of the lengths of the sides or b + b - 2 + b + 3, or 3b + 1.

    Translate.

    The perimter is greater than 19 cm.
      3b + 1   >   19

    Carry out.
       3b + 1 > 19
           3b > 18
            b > 6

    Check. We check to see if the solution seems reasonable.

    When b = 5, the perimeter is 3·5 + 1, or 16 cm.
    When b = 6, the perimeter is 3·6 + 1, or 19 cm.
    When b = 7, the perimeter is 3·7 + 1, or 22 cm.

Chapter 9 (9.3)

23. (continued)

    From these calculations, it would appear that the solution is correct.

    State. For lengths of the base greater than 6 cm the perimeter will be greater than 19 cm. The solution set is $\{b \mid b > 6 \text{ cm}\}$.

24. $\{w \mid w \leq \frac{35}{3} \text{ ft}\}$

25. Familiarize. The average number of calls per week is the sum of the calls for the three weeks divided by the number of weeks, 3. We let c represent the number of calls made during the third week.

    Translate. The average of the three weeks is given by
    $$\frac{18 + 22 + c}{3}.$$
    Since the average must be at least 20, this means that it must be greater than or equal to 20. Thus, we can translate the problem to the inequality
    $$\frac{18 + 22 + c}{3} \geq 20.$$

    Carry out. We first multiply by 3 to clear of fractions.
    $$3\left[\frac{18 + 22 + c}{3}\right] \geq 3 \cdot 20$$
    $$18 + 22 + c \geq 60$$
    $$40 + c \geq 60$$
    $$c \geq 20$$

    Check. Suppose c is a number greater than or equal to 20. Then by adding 18 and 22 on both sides of the inequality we get
    $$18 + 22 + c \geq 18 + 22 + 20$$
    $$18 + 22 + c \geq 60$$
    so
    $$\frac{18 + 22 + c}{3} \geq \frac{60}{3}, \text{ or } 20$$

    State. Any number of calls which is at least 20 will maintain an average of at least 20 for the three-week period. The solution set is $\{c \mid c \geq 20\}$.

26. George worked more than 12 hours, and Joan worked more than 15 hours.

27. Familiarize. Let s represent the amount the student can spend on each sweater. Then the total amount the student can spend is represented by $21.95 + 2s$.

    Translate.

Total spent	is less than or equal to	$120.00
$21.95 + 2s$	$\leq$	120

    Carry out.
    $$21.95 + 2s \leq 120$$
    $$2195 + 200s \leq 12{,}000 \quad \text{Clearing decimals}$$
    $$200s \leq 9805$$
    $$s \leq 49.02 \quad \text{Rounding}$$

    Check. We check to see if the solution seems reasonable.

    When s = $49.01, the student spends
    $21.95 + 2(49.01)$, or $119.97.

    When s = $49.02, the student spends
    $21.95 + 2(49.02)$, or $119.99.

    When s = $49.03, the student spends
    $21.95 + 2(49.03)$, or $120.01.

    From these calculations, it would appear that the solution is correct.

    State. The student can spend at most $49.02 for each sweater. The solution set is $\{s \mid s \leq \$49.02\}$.

28. $\{s \mid s \leq 9 \text{ lb}\}$

29. Familiarize. We first make a drawing. Let ℓ represent the length.

    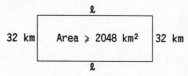

    The area is length times width, or 32ℓ.

    Translate.

The area	is at least	2048 km².
32ℓ	$\geq$	2048

    Carry out.
    $$32\ell \geq 2048$$
    $$\ell \geq 64$$

    Check. We check to see if the solution seems reasonable.

    When ℓ = 63, the area is $32 \cdot 63$, or 2016 km².
    When ℓ = 64, the area is $32 \cdot 64$, or 2048 km².
    When ℓ = 65, the area is $32 \cdot 65$, or 2080 km².

    From these calculations, it would appear that the solution is correct.

    State. Lengths of 64 km or more will make the area at least 2048 km². The solution set is $\{\ell \mid \ell \geq 64 \text{ km}\}$.

Chapter 9 (9.4)

30. $\{b | b \leqslant 4 \text{ cm}\}$

31. <u>Familiarize</u>. We will use the formula
    $$P = 0.1522Y - 298.592.$$
    <u>Translate</u>. We have the inequality $P \geqslant 6$. To find the years that satisfy this condition we substitute $0.1522Y - 298.592$ for P:
    $$0.1522Y - 298.592 \geqslant 6$$
    <u>Carry out</u>.
    $$0.1522Y - 298.592 \geqslant 6$$
    $$0.1522Y \geqslant 304.592$$
    $$Y \geqslant 2001.2615 \quad \text{Rounding}$$
    <u>Check</u>. We check to see if the solution seems reasonable.

    When $Y = 2001.26$, $P = 0.1522(2001.26) - 298.592$, or 5.9998.

    When $Y = 2001.2615$, $P = 0.1522(2001.2615) - 298.592$, or 6.00.

    When $Y = 2001.27$, $P = 0.1522(2001.27) - 298.592$, or 6.0013.

    From these calculations, it would appear that the solution is correct.

    <u>State</u>. From about 2001.2615 on, the average price of a movie ticket will be at least $6. The solution set is $\{Y | Y \geqslant 2001.2615\}$.

32. $\left\{ x \Big| x \leqslant 215\frac{5}{27} \text{ mi} \right\}$

33. <u>Familiarize</u>. We make a drawing. Let s represent the length of a side of the square.

    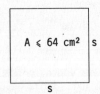

    The area s is the square of the length of a side, or s^2.

    <u>Translate</u>.

    <u>The area</u> is no more than 64 cm².
    $s^2 \qquad \leqslant \qquad 64$

    <u>Carry out</u>.
    $$s^2 \leqslant 64$$
    $$s^2 - 64 \leqslant 0$$
    $$(s + 8)(s - 8) \leqslant 0$$

    We know that $(s + 8)(s - 8) = 0$ for $s = -8$ or $s = 8$. Now $(s + 8)(s - 8) < 0$ when the two factors have opposite signs. That is:

    $s+8 > 0$ <u>and</u> $s-8 < 0$ or $s+8 < 0$ <u>and</u> $s-8 > 0$
    $s > -8$ and $s < 8$ or $s < -8$ and $s > 8$
    This can be expressed as $-8 < s < 8$. This is not possible.

    Then $(s + 8)(s - 8) \leqslant 0$ for $-8 \leqslant s \leqslant 8$.

33. (continued)

    <u>Check</u>. Since the length of a side cannot be negative we only consider positive values of s, or $0 < s \leqslant 8$. We check to see if this solution seems reasonable.

    When $s = 7$, the area is 7^2, or 49 cm².
    When $s = 8$, the area is 8^2, or 64 cm².
    When $s = 9$, the area is 9^2, or 81 cm².

    From these calculations, it appears that the solution is correct.

    <u>State</u>. Sides of length 8 or less will allow an area of no more than 64 cm². The solution set is $\{s | s \leqslant 8 \text{ and } s \text{ is positive}\}$, or $\{s | 0 < s \leqslant 8\}$.

34. 47 and 49

35. <u>Familiarize</u>. Let x = the smallest odd integer. Then $x + 2$ = the next odd integer and $(x + 2) + 2$, or $x + 4$ = the third odd integer.

    <u>Translate</u>.

    First integer plus second integer plus third integer is less than 30.
    $x \quad + \quad (x + 2) \quad + \quad (x + 4) \quad < \quad 30$

    <u>Carry out</u>.
    $$x + (x + 2) + (x + 4) < 30$$
    $$3x + 6 < 30$$
    $$3x < 24$$
    $$x < 8$$

    The largest odd integer less than 8 is 7, so the possible solution is:

    First integer: 7
    Second integer: 7 + 2, or 9
    Third integer: 7 + 4, or 11

    <u>Check</u>. The sum of 7, 9, and 11 is 27 which is less than 30. To determine if these are the largest integers possible, we check to see if $x = 9$ would work. When $x = 9$, the three odd integers would be 9, 11, and 13, and their sum is 33. Thus, our solution seems correct.

    <u>State</u>. The integers are 7, 9, and 11.

36. $\{s | s < \$20,000\}$

Exercise Set 9.4

1. We use alphabetical order of variables. We replace x by -3 and y by -5.

$-x - 3y < 18$
$-(-3) - 3(-5)$ | 18
3 + 15
18

    Since $18 < 18$ is false, $(-3,-5)$ is not a solution.

2. Yes

Chapter 9   (9.4)

3. We use alphabetical order of variables. We substitute $\frac{1}{2}$ for x and $-\frac{1}{4}$ for y.

$$\begin{array}{c|c} 7y - 9x > -3 \\ \hline 7\left[-\frac{1}{4}\right] - 9 \cdot \frac{1}{2} & -3 \\ -\frac{7}{4} - \frac{9}{2} & \\ -\frac{7}{4} - \frac{18}{4} & \\ -\frac{25}{4} & \\ -6\frac{1}{4} & \end{array}$$

Since $-6\frac{1}{4} > -3$ is false, $\left(\frac{1}{2}, -\frac{1}{4}\right)$ is not a solution.

4. Yes

5. Graph $x > 2y$.

First graph the line $x = 2y$, or $y = \frac{1}{2}x$. Two points on the line are (0,0) and (4,2). We draw a dashed line since the inequality symbol is >. Then we pick a test point that is not on the line. We try (-2,1).

$$\begin{array}{c|c} x > 2y \\ \hline -2 & 2 \cdot 1 \\ & 2 \end{array}$$

We see that (-2,1) is not a solution of the inequality, so we shade the points in the half-plane that does not contain (-2,1).

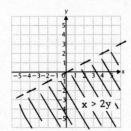

6.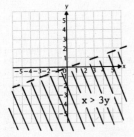

7. Graph $y \leq x - 3$.

First graph the line $y = x - 3$. The intercepts are (0,-3) and (3,0). We draw a solid line since the inequality symbol is $\leq$. Then we pick a test point that is not on the line. We try (0,0).

$$\begin{array}{c|c} y \leq x - 3 \\ \hline 0 & 0 - 3 \\ & -3 \end{array}$$

We see that (0,0) is not a solution of the inequality, so we shade the half-plane that does not contain (0,0).

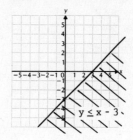

8.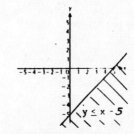

9. Graph $y < x + 1$.

First graph the line $y = x + 1$. The intercepts are (0,1) and (-1,0). We draw a dashed line since the inequality symbol is <. Then we pick a test point that is not on the line. We try (0,0).

$$\begin{array}{c|c} y < x + 1 \\ \hline 0 & 0 + 1 \\ & 1 \end{array}$$

Since (0,0) is a solution of the inequality, we shade the half-plane that contains (0,0).

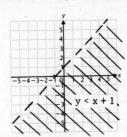

10.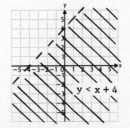

Chapter 9   (9.4)

11. Graph $y \geq x - 2$.

    First graph the line $y = x - 2$. The intercepts are $(0,-2)$ and $(2,0)$. We draw a solid line since the inequality symbol is $\geq$. Then we test the point $(0,0)$.

    $$\begin{array}{c|c} y & \geq x - 2 \\ \hline 0 & 0 - 2 \\ & -2 \end{array}$$

    Since $(0,0)$ is a solution of the inequality, we shade the half-plane containing $(0,0)$.

    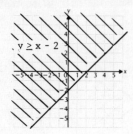

12.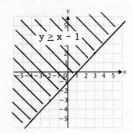

13. Graph $y \leq 2x - 1$.

    First graph the line $y = 2x - 1$. The intercepts are $(0,-1)$ and $\left(\frac{1}{2},0\right)$. We draw a solid line since the inequality symbol is $\leq$. Then we test the point $(0,0)$.

    $$\begin{array}{c|c} y & \leq 2x - 1 \\ \hline 0 & 2 \cdot 0 - 1 \\ & -1 \end{array}$$

    Since $(0,0)$ is not a solution of the inequality, we shade the half-plane that does not contain $(0,0)$.

    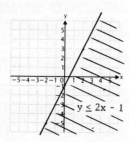

14.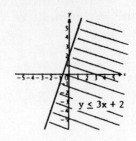

15. Graph $x + y \leq 3$.

    First graph the line $x + y = 3$. The intercepts are $(0,3)$ and $(3,0)$. We draw a solid line since the inequality symbol is $\leq$. Then we test the point $(0,0)$.

    $$\begin{array}{c|c} x + y & \leq 3 \\ \hline 0 + 0 & 3 \\ 0 & \end{array}$$

    Since $(0,0)$ is a solution of the inequality, we shade the half-plane that contains $(0,0)$.

    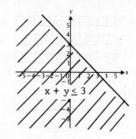

16.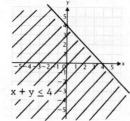

17. Graph $x - y > 7$.

    First graph the line $x - y = 7$. The intercepts are $(0,-7)$ and $(7,0)$. We draw a dashed line since the inequality symbol is $>$. Then we test the point $(0,0)$.

    $$\begin{array}{c|c} x - y & > 7 \\ \hline 0 - 0 & 7 \\ 0 & \end{array}$$

    Since $(0,0)$ is not a solution of the inequality, we shade the half-plane that does not contain $(0,0)$.

    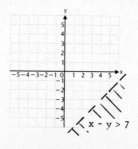

18.

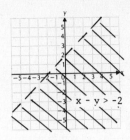

19. Graph $x - 3y < 6$.

First graph the line $x - 3y = 6$. The intercepts are $(0,-2)$ and $(6,0)$. We draw a dashed line since the inequality symbol is $<$. Then we test the point $(0,0)$.

$$\begin{array}{c|c} x - 3y < 6 \\ \hline 0 - 3\cdot 0 & 6 \\ 0 & \end{array}$$

Since $(0,0)$ is a solution of the inequality, we shade the half-plane containing $(0,0)$.

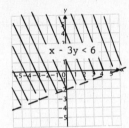

20.

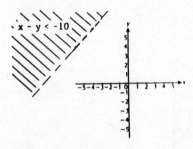

21. Graph $2x + 3y \leq 12$.

First graph the line $2x + 3y = 12$. The intercepts are $(0,4)$ and $(6,0)$. We draw a solid line since the inequality symbol is $\leq$. Then we test the point $(0,0)$.

$$\begin{array}{c|c} 2x + 3y \leq 12 \\ \hline 2\cdot 0 + 3\cdot 0 & 12 \\ 0 & \end{array}$$

Since $(0,0)$ is a solution of the inequality, we shade the half-plane containing $(0,0)$.

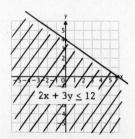

22.

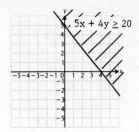

23. Graph $y \geq 1 - 2x$.

First graph the line $y = 1 - 2x$. The intercepts are $(0,1)$ and $\left(\frac{1}{2},0\right)$. We draw a solid line since the inequality symbol is $\geq$. Then we test the point $(0,0)$.

$$\begin{array}{c|c} y \geq 1 - 2x \\ \hline 0 & 1 - 2\cdot 0 \\ & 1 \end{array}$$

Since $(0,0)$ is not a solution of the inequality, we shade the half-plane that does not contain $(0,0)$.

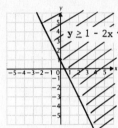

24.

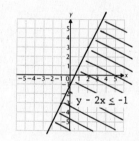

Chapter 9   (9.4)

25. Graph $y + 4x > 0$.

First graph the line $y + 4x = 0$, or $y = -4x$. Two points on the line are (0,0) and (1,-4). We draw a dashed line, since the inequality symbol is >. Then we test the point (2,-3), which is not a point on the line.

$$\begin{array}{c|c} y + 4x & > 0 \\ \hline -3 + 4 \cdot 2 & 0 \\ -3 + 8 & \\ 5 & \end{array}$$

Since (2,-3) is a solution of the inequality, we shade the half-plane containing (2,-3).

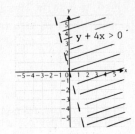

26.

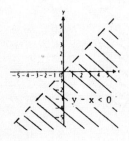

27. Graph $y > -3x$.

First graph the line $y = -3x$. Two points on the line are (0,0) and (-1,3). We draw a dashed line since the inequality symbol is >. Then we test the point (2,4), which is not a point on the line.

$$\begin{array}{c|c} y & > -3x \\ \hline 4 & -3 \cdot 2 \\ & -6 \end{array}$$

Since (2,4) is a solution of the inequality, we shade the half-plane containing (2,4).

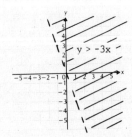

28.

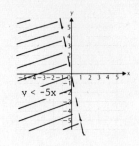

29. Graph $x < 4$ on a number line:

Graph $x < 4$ on a plane:

Graph the line $x = 4$ using a dashed line since the inequality symbol is <. Then use (-1,2) as a test point. We can write the inequality as $x + 0y < 4$.

$$\begin{array}{c|c} x + 0y & < 4 \\ \hline -1 + 0 \cdot 2 & 4 \\ -1 & \end{array}$$

Since (-1,2) is a solution of the inequality, we shade the half-plane containing (-1,2).

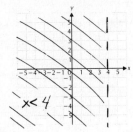

30.

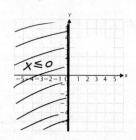

Chapter 9   (9.4)

31. Graph x ⩾ 3 on a number line:

Graph x ⩾ 3 on a plane:

Graph the line x = 3 using a solid line since the inequality symbol is ⩾. Then use (4,-3) as a test point. We can write the inequality as x + 0y ⩾ 3.

$$\begin{array}{c|c} x + 0y \geqslant 3 \\ \hline 4 + 0(-3) & 3 \\ 4 & \end{array}$$

Since (4,-3) is a solution of the inequality, we shade the half-plane containing (4,-3).

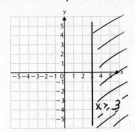

32.

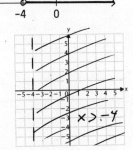

33. Graph y ⩽ 3.

Graph the line y = 3 using a solid line since the inequality symbol is ⩽. Then use (1,-2) as a test point. We can write the inequality as 0x + y ⩽ 3.

$$\begin{array}{c|c} 0x + y \leqslant 3 \\ \hline 0 \cdot 1 + (-2) & 3 \\ -2 & \end{array}$$

Since (1,-2) is a solution of the inequality, we shade the half-plane containing (1,-2).

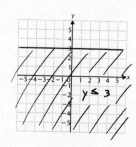

34.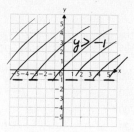

35. Graph y ⩾ -5.

Graph the line y = -5 using a solid line since the inequality symbol is ⩾. Then use (2,3) as a test point. We can write the inequality as 0x + y ⩾ -5.

$$\begin{array}{c|c} 0x + y \geqslant -5 \\ \hline 0 \cdot 2 + 3 & -5 \\ 3 & \end{array}$$

Since (2,3) is a solution of the inequality, we shade the half-plane containing (2,3).

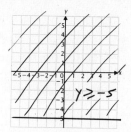

36.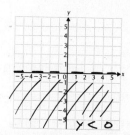

37. Familiarize. We organize the information in a table. Let a = the amount of A and b = the amount of B to be used.

Type of seed	A	B	Mixture
Amount of seed	a	b	50
Price per pound	$1	$1.35	$1.14
Value	1·a, or a	1.35b	50($1.14), or $57

Translate. The first and third rows of the table give us two equations.

Total amount of mixture: a + b = 50

Total value of mixture: a + 1.35b = 57, or
                        100a + 135b = 5700

245

37. (continued)

Carry out. We solve the system using the addition method. We multiply the first equation by -100 and add.

$$-100a - 100b = -5000$$
$$100a + 135b = 5700$$
$$35b = 700$$
$$b = 20$$

Substitute 20 for b in the first equation and solve for a.

$$a + b = 50$$
$$a + 20 = 50$$
$$a = 30$$

Check. If 30 lb of A and 20 lb of B are used, the total weight of the mixture is 50 lb. The value of the mixture is 30($1) + 20($1.35), or $30 + $27, or $57. These values check.

State. 30 lb of A and 20 lb of B should be used.

38. 4

39. $S = \dfrac{4a + b}{3a - b}$

$(3a - b)S = 4a + b$ Clearing the fraction

$3aS - bS = 4a + b$

$3aS - 4a = b + bS$ Adding bS and -4a

$3aS - 4a = b(1 + S)$

$\dfrac{3aS - 4a}{1 + S} = b$

40. $t = \dfrac{F}{a - 3}$

41. First find the equation of the line containing the points (2,0) and (0,-2). The slope is

$\dfrac{-2 - 0}{0 - 2} = \dfrac{-2}{-2} = 1.$

We know that the y-intercept is (0,-2), so we write the equation using the slope-intercept form.

$y = mx + b$
$y = 1 \cdot x + (-2)$
$y = x - 2$

Since the line is dashed, the inequality symbol will be < or >. To determine which, we substitute the coordinates of a point in the shaded region. We will use (0,0).

y	x - 2
0	0 - 2
	-2

Since 0 > -2 is true, the correct symbol is >. The inequality is y > x - 2.

42. $x \geq -2$

43. Graph $xy \leq 0$.

From the principle of zero products, we know that xy = 0 when x = 0 or y = 0. Therefore, the graph contains the lines x = 0 and y = 0, or the y- and x-axes. Also, xy < 0 when x and y have different signs. This is the case for all points in the second quadrant (x is negative and y is positive) and in the fourth quadrants (x is positive and y is negative). Thus, we shade the second and fourth quadrants.

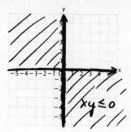

44.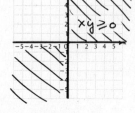

Exercise Set 9.5

1. $x + y \leq 1,$
   $x - y \leq 5$

We graph the lines x + y = 1 and x - y = 5 using solid lines. We indicate the region for each inequality by the arrows at the ends of the lines. We shade the area where the regions overlap.

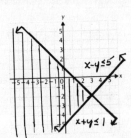

2.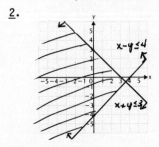

Chapter 9   (9.5)

3. $y - 2x > 1$,
   $y - 2x < 3$

   We graph the lines $y - 2x = 1$ and $y - 2x = 3$ using dashed lines. We indicate the region for each inequality by the arrows at the ends of the lines. We shade the area where the regions overlap.

   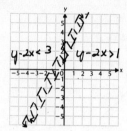

4.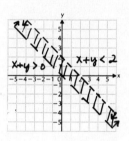

5. $y \geq 1$,
   $x > 2 + y$

   We graph the line $y = 1$ using a solid line and the line $x = 2 + y$ using a dashed line. We indicate the region for each inequality by the arrows at the ends of the lines. We shade the area where the regions overlap.

   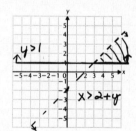

6.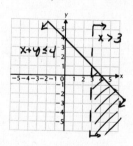

7. $x + y \leq 6$,
   $x \geq 0$,
   $y \geq 0$,
   $y \leq 5$

   We graph the lines $x + y = 6$, $x = 0$, $y = 0$, and $y = 5$ using solid lines. We indicate the region for each inequality by the arrows at the ends of the lines. We shade the area where the regions overlap.

   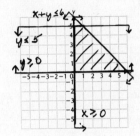

8.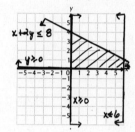

9. $2x - 3y \geq 9$,
   $2y + x > 6$

   We graph the line $2x - 3y = 9$ using a solid line and the line $2y + x = 6$ using a dashed line. We indicate the region for each inequality by the arrows at the ends of the lines. We shade the area where the regions overlap.

   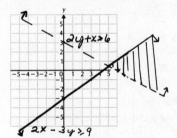

10.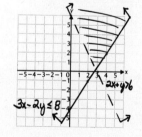

247

11. $y > 4x - 1$,
    $y < -2x + 3$

    We graph the lines $y = 4x - 1$ and $y = -2x + 3$ using dashed lines. We indicate the region for each inequality by the arrows at the ends of the lines. We shade the area where the regions overlap.

    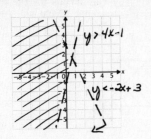

12.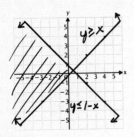

13. $y - x \geq 1$,
    $y - x \leq 3$,
    $x \leq 5$,
    $x \geq 2$

    We graph the lines $y - x = 1$, $y - x = 3$, $x = 5$, and $x = 2$ using solid lines. We indicate the region for each inequality by the arrows at the ends of the lines. We shade the area where the regions overlap.

    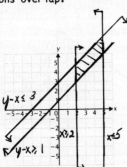

14.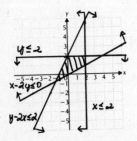

15. $y > 5x + 2$,
    $y \leq 1 - x$

    We graph the line $y = 5x + 2$ using a dashed line and the line $y = 1 - x$ using a solid line. We indicate the region for each inequality by the arrows at the ends of the lines. We shade the area where the regions overlap.

    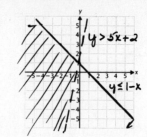

16.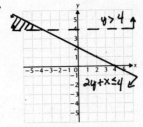

17. $x \leq 3$,
    $y \leq 4$

    We graph the lines $x = 3$ and $y = 4$ using solid lines. We indicate the region for each inequality by the arrows at the ends of the lines. We shade the area where the regions overlap.

    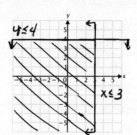

18.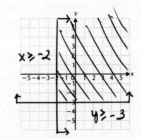

Chapter 9   (9.5)

19. $x \leq 0$,
    $y \leq 0$

    We graph the lines $x = 0$ and $y = 0$ using solid lines. We indicate the region for each inequality by the arrows at the ends of the lines. We shade the area where the regions overlap.

    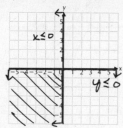

20.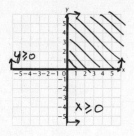

21. $x \leq 4$,
    $x + y \leq 3$,
    $y \leq 3$

    We graph the lines $x = 4$, $x + y = 3$, and $y = 3$ using solid lines. We indicate the region for each inequality by the arrows at the ends of the lines. We shade the area where the regions overlap.

    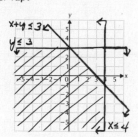

22.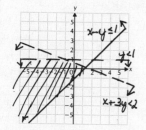

23. $y \leq x$,
    $x \geq -2$,
    $x \leq -y$

    We graph the lines $y = x$, $x = -2$ and $x = -y$ using solid lines. We indicate the region for each inequality by the arrows at the ends of the lines. We shade the area where the regions overlap.

    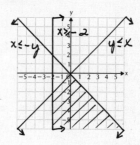

24.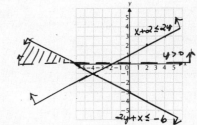

25. $y - x = 1$,
    $y - x = 3$

    Multiply the first equation by -1 and add.
    $$-y + x = -1$$
    $$\underline{y - x = 3}$$
    $$0 = 2$$

    We get a false equation, so there is no solution.

26. $\left(-\dfrac{1}{6}, \dfrac{7}{6}\right)$

27. $x - 2y = 0$, (1)
    $y - 2x = 2$ (2)

Solve equation (1) for x.
$$x - 2y = 0$$
$$x = 2y \quad (3)$$

Substitute 2y for x in equation (2) and solve for y.
$$y - 2(2y) = 2$$
$$y - 4y = 2$$
$$-3y = 2$$
$$y = -\frac{2}{3}$$

Substitute $-\frac{2}{3}$ for y in equation (3) and compute x.
$$x = 2y = 2\left(-\frac{2}{3}\right) = -\frac{4}{3}$$

The ordered pair $\left(-\frac{4}{3}, -\frac{2}{3}\right)$ checks. It is the solution.

28. No solution

29. $5a + 3b \geq 30$,
    $2a + 3b \geq 21$,
    $3a + 6b \geq 36$,
    $a \geq 0$,
    $b \geq 0$

We graph the related equations, find the region for each inequality, and shade the area where the regions overlap.

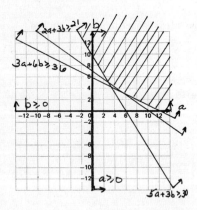

30.

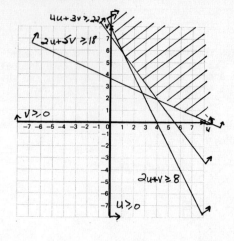

CHAPTER 10 RADICAL EXPRESSIONS AND EQUATIONS

Exercise Set 10.1

1. The square roots of 1 are 1 and -1, because $1^2 = 1$ and $(-1)^2 = 1$.

2. 2, -2

3. The square roots of 16 are 4 and -4, because $4^2 = 16$ and $(-4)^2 = 16$.

4. 3, -3

5. The square roots of 100 are 10 and -10, because $10^2 = 100$ and $(-10)^2 = 100$.

6. 11, -11

7. The square roots of 169 are 13 and -13, because $13^2 = 169$ and $(-13)^2 = 169$.

8. 12, -12

9. $\sqrt{4} = 2$, taking the principal square root.

10. 1

11. $\sqrt{9} = 3$, so $-\sqrt{9} = -3$.

12. -5

13. $\sqrt{64} = 8$, so $-\sqrt{64} = -8$.

14. -9

15. $\sqrt{225} = 15$, so $-\sqrt{225} = -15$.

16. 20

17. $\sqrt{361} = 19$, taking the principal square root.

18. 21

19. $\sqrt{2}$ is irrational, since 2 is not a perfect square.

20. Irrational

21. $\sqrt{8}$ is irrational, since 8 is not a perfect square.

22. Irrational

23. $\sqrt{49}$ is rational, since 49 is a perfect square.

24. Rational

25. $\sqrt{98}$ is irrational, since 98 is not a perfect square.

26. Irrational

27. $-\sqrt{4}$ is rational, since 4 is a perfect square.

28. Rational

29. $-\sqrt{12}$ is irrational, since 12 is not a perfect square.

30. Irrational

31. 4.23 is rational, since the decimal notation ends.

32. Rational

33. 23 is rational since it can be expressed as the ratio of two integers $\left(23 = \frac{23}{1}\right)$.

34. Rational

35. $-\frac{2}{3}$ is rational since it can be expressed as the ratio of two integers $\left(-\frac{2}{3} = \frac{-2}{3}\right)$.

36. Rational

37. $\frac{2.3}{0.01}$ is rational since it can be expressed as the ratio of two integers $\left(\frac{2.3}{0.01} \times \frac{100}{100} = \frac{230}{1}\right)$.

38. Rational

39. 0.156156156... is rational, since the digits 156 repeat.

40. Irrational

41. 7.767767776... is irrational, since the decimal notation neither ends nor repeats.

42. Irrational

43. -63.030030003... is irrational, since the decimal notation neither ends nor repeats.

44. Rational

45. $9.\overline{23}$ is rational, since the digits 23 repeat.

46. Rational

47. $-9.6\overline{5}$ is rational, since the digit 5 repeats.

48. 2.236

49. 2.449

50. 4.123

51. 4.359

52. 9.644

53. 6.557

Chapter 10   (10.2)

54. 6.325

55. a) We substitute 25 into the formula:
    $N = 2.5\sqrt{25} = 2.5(5) = 12.5 \approx 13$

    b) We substitute 89 into the formula and use Table 2 or a calculator to find an approximation.
    $N = 2.5\sqrt{89} \approx 2.5(9.434) = 23.585 \approx 24$

56. a) 20

    b) 25

57. $m = \dfrac{\text{change in y}}{\text{change in x}} = \dfrac{-6 - 4}{5 - (-3)} = \dfrac{-10}{8} = -\dfrac{5}{4}$

58. $\dfrac{3}{5}$

59. Use the point-slope equation.
    $y - y_1 = m(x - x_1)$
    $y - 4 = 2[x - (-3)]$
    $y - 4 = 2(x + 3)$
    $y - 4 = 2x + 6$
    $y = 2x + 10$

60. $y = -\dfrac{5}{4}x + \dfrac{1}{4}$

61. We find the inner square root first.
    $\sqrt{\sqrt{16}} = \sqrt{4} = 2$

62. 5

63. $-\sqrt{36} < -\sqrt{33} < -\sqrt{25}$, or $-6 < -\sqrt{33} < -5$
    $-\sqrt{33}$ is between -6 and -5.

64. 3.578

65. 14.071

66. 30.496

67. 70.228

68. 32.309

69. Consider 10 and 20. The number halfway between 10 and 20 is 15, or $\dfrac{10 + 20}{2}$. Then the number halfway between x and y is $\dfrac{x + y}{2}$.

70. 64

71. $2\dfrac{3}{4} = \dfrac{11}{4} = \dfrac{22}{8}$, $4\dfrac{5}{8} = \dfrac{37}{8}$

    The difference between $\dfrac{37}{8}$ and $\dfrac{22}{8}$ is $\dfrac{15}{8}$.

    $\dfrac{1}{3}$ of $\dfrac{15}{8}$ is $\dfrac{1}{3} \cdot \dfrac{15}{8}$, or $\dfrac{5}{8}$.

    The number one third of the way from $2\dfrac{3}{4}$ to $4\dfrac{5}{8}$ is

    $2\dfrac{3}{4} + \dfrac{5}{8} = \dfrac{22}{8} + \dfrac{5}{8} = \dfrac{27}{8}$, or $3\dfrac{3}{8}$.

72. $-1\dfrac{1}{15}$

Exercise Set 10.2

1. a - 4

2. t + 3

3. The radicand is the expression under the radical, $t^2 + 1$.

4. $x^2 + 5$

5. The radicand is the expression under the radical, $\dfrac{3}{x + 2}$.

6. $\dfrac{a}{a - b}$

7. No, because the radicand is negative

8. No

9. Yes, because the radicand is nonnegative

10. Yes

11. Yes, because the radicand will be nonnegative

12. No

13. We replace t by -11:
    $\sqrt{t - 5} = \sqrt{-11 - 5} = \sqrt{-16}$
    No, -11 is not a meaningful replacement since it yields a negative radicand.

14. Yes

15. The radicand must be greater than or equal to 0. We solve: $5x \geq 0$
    $x \geq 0$
    Any number greater than or equal to 0 is meaningful.

16. $y \geq 0$

Chapter 10   (10.2)

17. The radicand must be greater than or equal to 0.
    We solve: $t - 5 \geq 0$
    $t \geq 5$
    Any number greater than or equal to 5 is meaningful.

18. $y \geq 8$

19. The radicand must be greater than or equal to 0.
    We solve: $y + 8 \geq 0$
    $y \geq -8$
    Any number greater than or equal to -8 is meaningful.

20. $x \geq -6$

21. The radicand must be greater than or equal to 0.
    We solve: $x + 20 \geq 0$
    $x \geq -20$
    Any number greater than or equal to -20 is meaningful.

22. $m \geq 18$

23. The radicand must be greater than or equal to 0.
    We solve: $2y - 7 \geq 0$
    $2y \geq 7$
    $y \geq \frac{7}{2}$
    Any number greater than or equal to $\frac{7}{2}$ is meaningful.

24. $x \geq -\frac{8}{3}$

25. Since t^2 is never negative, then $t^2 + 5$ is never negative. All real-number replacements are meaningful.

26. All real numbers

27. $\sqrt{t^2} = t$ Since t is assumed to be nonnegative

28. x

29. $\sqrt{9x^2} = \sqrt{(3x)^2} = 3x$ Since $3x$ is assumed to be nonnegative

30. $2a$

31. $\sqrt{(ab)^2} = ab$

32. $6y$

33. $\sqrt{(34d)^2} = 34d$

34. $53b$

35. $\sqrt{(x + 3)^2} = x + 3$

36. $x - 7$

37. $\sqrt{a^2 - 10a + 25} = \sqrt{(a - 5)^2} = a - 5$

38. $x + 1$

39. $\sqrt{4x^2 - 20x + 25} = \sqrt{(2x - 5)^2} = 2x - 5$

40. $3p + 2$

41. Familiarize. This problem states that we have direct variation between F and I. Thus, an equation $F = kI$, $k > 0$, applies. As the income increases, the amount spent on food increases.
    Translate. We write an equation of variation.
    Amount spent on food varies directly as the income.
    This translates to $F = kI$.
    Carry out.
    a) First find an equation of variation.
       $F = kI$
       $5096 = k \cdot 19{,}600$ Substituting 5096 for F and 19,600 for I
       $\frac{5096}{19{,}600} = k$
       $0.26 = k$
       The equation of variation is $F = 0.26I$.
    b) We use the equation to find how much a family spends on food when their income is $20,500.
       $F = 0.26I$
       $F = 0.26(\$20{,}500)$ Substituting $20,500 for I
       $F = \$5330$
    Check. Let us do some reasoning about the answer. The income increased from $19,600 to $20,500. Similarly, the amount spent on food increased from $5096 to $5330. This is what we would expect with direct variation.
    State. The amount spent on food is $5330.

42. $y = 0.625x$

43. $-2x + 5x - 4 > 19 - 7x + 2$
    $3x - 4 > 21 - 7x$ Collecting like terms
    $3x + 7x > 21 + 4$ Adding 7x and 4
    $10x > 25$
    $x > \frac{25}{10}$
    $x > \frac{5}{2}$ Simplifying
    The solution set is $\{x \mid x > \frac{5}{2}\}$.

44. $\{x \mid x \geq -3\}$

45. If $\sqrt{x^2} = 6$, then $x^2 = (6)^2$, or 36. Thus $x = 6$ or $x = -6$.

46. No solution

253

Chapter 10   (10.3)

47. If $-\sqrt{x^2} = -3$, then $\sqrt{x^2} = 3$ and $x^2 = 3^2$, or 9. Thus, $x = 3$ or $x = -3$.

48. $-7, 7$

49. $\sqrt{(3a)^2} = 3a$ Since $3a$ is assumed to be nonnegative

50. $3a$

51. $\sqrt{\dfrac{144x^8}{36y^6}} = \sqrt{\dfrac{4x^8}{y^6}} = \sqrt{\left(\dfrac{2x^4}{y^3}\right)^2} = \dfrac{2x^4}{y^3}$

52. $\dfrac{y^6}{90}$

53. $\sqrt{\dfrac{169}{m^{16}}} = \sqrt{\left(\dfrac{13}{m^8}\right)^2} = \dfrac{13}{m^8}$

54. $\dfrac{p}{60}$

Exercise Set 10.3

1. $\sqrt{2}\,\sqrt{3} = \sqrt{2\cdot 3} = \sqrt{6}$

2. $\sqrt{15}$

3. $\sqrt{4}\,\sqrt{3} = \sqrt{12}$, or
   $\sqrt{4}\,\sqrt{3} = 2\sqrt{3}$ Taking the square root of 4

4. $\sqrt{18}$, or $3\sqrt{2}$

5. $\sqrt{\dfrac{2}{5}}\,\sqrt{\dfrac{3}{4}} = \sqrt{\dfrac{2\cdot 3}{5\cdot 4}} = \sqrt{\dfrac{3}{10}}$

6. $\sqrt{\dfrac{3}{40}}$, or $\dfrac{1}{2}\sqrt{\dfrac{3}{10}}$

7. $\sqrt{17}\,\sqrt{17} = \sqrt{17\cdot 17} = \sqrt{17^2} = 17$

8. 18

9. $\sqrt{25}\,\sqrt{3} = \sqrt{75}$, or
   $\sqrt{25}\,\sqrt{3} = 5\sqrt{3}$ Taking the square root of 25

10. $\sqrt{72}$, or $6\sqrt{2}$

11. $\sqrt{2}\,\sqrt{x} = \sqrt{2\cdot x} = \sqrt{2x}$

12. $\sqrt{3a}$

13. $\sqrt{0.24}\,\sqrt{3} = \sqrt{0.24\cdot 3} = \sqrt{0.72}$

14. $\sqrt{1.12}$

15. $\sqrt{x}\,\sqrt{t} = \sqrt{x\cdot t} = \sqrt{xt}$

16. $\sqrt{ay}$

17. $\sqrt{x}\,\sqrt{x-3} = \sqrt{x(x-3)} = \sqrt{x^2 - 3x}$

18. $\sqrt{x^2 + x}$

19. $\sqrt{5}\,\sqrt{2x-1} = \sqrt{5(2x-1)} = \sqrt{10x - 5}$

20. $\sqrt{12x + 6}$

21. $\sqrt{x+2}\,\sqrt{x+1} = \sqrt{(x+2)(x+1)} = \sqrt{x^2 + 3x + 2}$

22. $\sqrt{x^2 + x - 12}$

23. $\sqrt{x-3}\,\sqrt{2x+4} = \sqrt{(x-3)(2x+4)} = \sqrt{2x^2 - 2x - 12}$

24. $\sqrt{2x^2 - 3x - 20}$

25. $\sqrt{x+4}\,\sqrt{x-4} = \sqrt{(x+4)(x-4)} = \sqrt{x^2 - 16}$

26. $\sqrt{x^2 - 4}$

27. $\sqrt{x+y}\,\sqrt{x-y} = \sqrt{(x+y)(x-y)} = \sqrt{x^2 - y^2}$

28. $\sqrt{a^2 - b^2}$

29. $\sqrt{3}\,\sqrt{2x} = \sqrt{3\cdot 2x} = \sqrt{6x}$

30. $\sqrt{20x}$, or $2\sqrt{5x}$

31. $\sqrt{12} = \sqrt{4\cdot 3}$ 4 is a perfect square
    $= \sqrt{4}\,\sqrt{3}$ Factoring into a product of radicals
    $= 2\sqrt{3}$ Taking the square root

32. $2\sqrt{2}$

33. $\sqrt{75} = \sqrt{25\cdot 3}$ 25 is a perfect square
    $= \sqrt{25}\,\sqrt{3}$ Factoring into a product of radicals
    $= 5\sqrt{3}$

34. $5\sqrt{2}$

35. $\sqrt{20} = \sqrt{4\cdot 5}$ 4 is a perfect square
    $= \sqrt{4}\,\sqrt{5}$
    $= 2\sqrt{5}$ Taking the square root

36. $3\sqrt{5}$

37. $\sqrt{200} = \sqrt{100\cdot 2}$ 100 is a perfect square
    $= \sqrt{100}\,\sqrt{2}$
    $= 10\sqrt{2}$

38. $10\sqrt{3}$

39. $\sqrt{9x} = \sqrt{9\cdot x} = \sqrt{9}\,\sqrt{x} = 3\sqrt{x}$

Chapter 10   (10.3)

40. $2\sqrt{y}$

41. $\sqrt{48x} = \sqrt{16 \cdot 3x} = \sqrt{16}\sqrt{3x} = 4\sqrt{3x}$

42. $2\sqrt{10m}$

43. $\sqrt{16a} = \sqrt{16 \cdot a} = \sqrt{16}\sqrt{a} = 4\sqrt{a}$

44. $7\sqrt{b}$

45. $\sqrt{64y^2} = \sqrt{64}\sqrt{y^2} = 8y$, or
    $\sqrt{64y^2} = \sqrt{(8y)^2} = 8y$

46. $3x$

47. $\sqrt{13x^2} = \sqrt{13}\sqrt{x^2} = \sqrt{13} \cdot x$, or $x\sqrt{13}$

48. $t\sqrt{29}$

49. $\sqrt{8t^2} = \sqrt{4 \cdot t^2 \cdot 2} = \sqrt{4}\sqrt{t^2}\sqrt{2} = 2t\sqrt{2}$

50. $5a\sqrt{5}$

51. $\sqrt{180} = \sqrt{36 \cdot 5} = 6\sqrt{5}$

52. $8\sqrt{7}$

53. $\sqrt{288y} = \sqrt{144 \cdot 2y} = \sqrt{144}\sqrt{2y} = 12\sqrt{2y}$

54. $11\sqrt{3p}$

55. $\sqrt{20x^2} = \sqrt{4 \cdot x^2 \cdot 5} = \sqrt{4}\sqrt{x^2}\sqrt{5} = 2x\sqrt{5}$

56. $2x\sqrt{7}$

57. $\sqrt{8x^2 + 8x + 2} = \sqrt{2(4x^2 + 4x + 1)} = \sqrt{2(2x+1)^2} = \sqrt{2}\sqrt{(2x+1)^2} = \sqrt{2}\,(2x+1)$

58. $\sqrt{3}\,(3x - 2)$

59. $\sqrt{36y + 12y^2 + y^3} = \sqrt{y(36 + 12y + y^2)} = \sqrt{y(6+y)^2} = \sqrt{y}\sqrt{(6+y)^2} = \sqrt{y}\,(6+y)$

60. $\sqrt{x}\,(1 - x)$

61. $\sqrt{125} = \sqrt{25 \cdot 5}$ Factoring the radicand
    $= \sqrt{25}\sqrt{5}$ Factoring the radical expression
    $= 5\sqrt{5}$
    $\approx 5(2.236)$ From Table 2, $\sqrt{5} \approx 2.236$
    ≈ 11.180

62. 13.416

63. $\sqrt{360} = \sqrt{36 \cdot 10}$
    $= \sqrt{36}\sqrt{10}$
    $= 6\sqrt{10}$
    $\approx 6(3.162)$ From Table 2
    ≈ 18.972

64. 10.248

65. $\sqrt{300} = \sqrt{100 \cdot 3}$
    $= \sqrt{100}\sqrt{3}$
    $= 10\sqrt{3}$
    $\approx 10(1.732)$ From Table 2
    ≈ 17.320

66. 11.961

67. $\sqrt{124} = \sqrt{2 \cdot 62}$ Factoring using numbers shown in Table 2
    $= \sqrt{2}\sqrt{62}$
    $\approx 1.414 \times 7.874$ From Table 2
    ≈ 11.134 Rounding to 3 decimal places

68. 44.720

69. First we substitute 20 for L in the formula:
    $r = 2\sqrt{5L} = 2\sqrt{5 \cdot 20} = 2\sqrt{100} = 2 \cdot 10 = 20$ mph
    Then we substitute 150 for L:
    $r = 2\sqrt{5 \cdot 150} = 2\sqrt{750} = 2\sqrt{25 \cdot 30} = 2\sqrt{25}\sqrt{30} = 2 \cdot 5\sqrt{30} = 10\sqrt{30} \approx 10(5.477) \approx 54.77$ mph, or 54.8 mph (rounded to the nearest tenth)

70. 24.5 mph, 37.4 mph

71. $\sqrt{3x - 3} = \sqrt{3(x-1)} = \sqrt{3}\sqrt{x-1}$

72. $\sqrt{x-2}\sqrt{x+1}$

73. $\sqrt{x^2 - 4} = \sqrt{(x+2)(x-2)} = \sqrt{x+2}\sqrt{x-2}$

74. $\sqrt{2x+3}\sqrt{x-4}$

75. $\sqrt{x^3 - 2x^2} = \sqrt{x^2(x-2)} = \sqrt{x^2}\sqrt{x-2} = x\sqrt{x-2}$

76. $\sqrt{a+b}\sqrt{a-b}$

77. $\sqrt{0.01} = \sqrt{(0.1)^2} = 0.1$

78. 0.5

79. $\sqrt{x^8} = \sqrt{(x^4)^2} = x^4$

80. $3a^3$

255

Chapter 10   (10.4)

81. $\sqrt{49} = 7$

    $\sqrt{490} = \sqrt{49 \cdot 10} = 7\sqrt{10}$

    $\sqrt{4900} = 70$

    $\sqrt{49{,}000} = \sqrt{4900 \cdot 10} = 70\sqrt{10}$

    $\sqrt{490{,}000} = 700$

    Each is $\sqrt{10}$ times the last.

82. >

83. $\sqrt{450} = \sqrt{225 \cdot 2} = 15\sqrt{2}$, so $15\sqrt{2} = \sqrt{450}$.

84. >

85. $3\sqrt{11} = \sqrt{9}\,\sqrt{11} = \sqrt{99}$ and

    $7\sqrt{2} = \sqrt{49}\,\sqrt{2} = \sqrt{98}$, so

    $3\sqrt{11} > 7\sqrt{2}$.

86. <

87. Using a calculator or Table 2, we find $\sqrt{15} \approx 3.873$ and $\sqrt{17} \approx 4.123$, so $\sqrt{15} + \sqrt{17} \approx 3.873 + 4.123 \approx 7.996$. Then $8 > \sqrt{15} + \sqrt{17}$.

Exercise Set 10.4

1. $\sqrt{x^6} = \sqrt{(x^3)^2} = x^3$

2. x^5

3. $\sqrt{x^{12}} = \sqrt{(x^6)^2} = x^6$

4. x^8

5. $\sqrt{x^5} = \sqrt{x^4 x}$ One factor is a perfect square

    $= \sqrt{x^4}\,\sqrt{x}$

    $= \sqrt{(x^2)^2}\,\sqrt{x}$

    $= x^2\sqrt{x}$

6. $x\sqrt{x}$

7. $\sqrt{t^{19}} = \sqrt{t^{18}t} = \sqrt{t^{18}}\,\sqrt{t} = \sqrt{(t^9)^2}\,\sqrt{t} = t^9\sqrt{t}$

8. $p^8\sqrt{p}$

9. $\sqrt{(y-2)^8} = \sqrt{[(y-2)^4]^2} = (y-2)^4$

10. $(x+3)^3$

11. $\sqrt{4(x+5)^{10}} = \sqrt{4[(x+5)^5]^2} =$

    $\sqrt{4}\,\sqrt{[(x+5)^5]^2} = 2(x+5)^5$

12. $4(a-7)^2$

13. $\sqrt{36m^3} = \sqrt{36 \cdot m^2 \cdot m} = \sqrt{36}\,\sqrt{m^2}\,\sqrt{m} = 6m\sqrt{m}$

14. $5y\sqrt{10y}$

15. $\sqrt{8a^5} = \sqrt{4a^4(2a)} = \sqrt{4(a^2)^2(2a)} =$

    $\sqrt{4}\,\sqrt{(a^2)^2}\,\sqrt{2a} = 2a^2\sqrt{2a}$

16. $2b^3\sqrt{3b}$

17. $\sqrt{104p^{17}} = \sqrt{4p^{16}(26p)} = \sqrt{4(p^8)^2(26p)} =$

    $\sqrt{4}\,\sqrt{(p^8)^2}\,\sqrt{26p} = 2p^8\sqrt{26p}$

18. $2m^{11}\sqrt{71m}$

19. $\sqrt{448x^6y^3} = \sqrt{64x^6y^2(7y)} = \sqrt{64(x^3)^2y^2(7y)} =$

    $\sqrt{64}\,\sqrt{(x^3)^2}\,\sqrt{y^2}\,\sqrt{7y} = 8x^3y\sqrt{7y}$

20. $9x^2y^2\sqrt{3x}$

21. $\sqrt{3}\,\sqrt{18} = \sqrt{3 \cdot 18}$ Multiplying

    $\quad\quad\quad\;\; = \sqrt{3 \cdot 3 \cdot 6}$ Looking for perfect-square factors or pairs of factors

    $\quad\quad\quad\;\; = \sqrt{3 \cdot 3}\,\sqrt{6}$

    $\quad\quad\quad\;\; = 3\sqrt{6}$

22. $5\sqrt{2}$

23. $\sqrt{15}\,\sqrt{6} = \sqrt{15 \cdot 6}$ Multiplying

    $\quad\quad\quad\;\; = \sqrt{5 \cdot 3 \cdot 3 \cdot 2}$ Looking for perfect-square factors or pairs of factors

    $\quad\quad\quad\;\; = \sqrt{3 \cdot 3}\,\sqrt{5 \cdot 2}$

    $\quad\quad\quad\;\; = 3\sqrt{10}$

24. 9

25. $\sqrt{18}\,\sqrt{14x} = \sqrt{18 \cdot 14x} = \sqrt{3 \cdot 3 \cdot 2 \cdot 2 \cdot 7 \cdot x} =$

    $\sqrt{3 \cdot 3}\,\sqrt{2 \cdot 2}\,\sqrt{7x} = 3 \cdot 2\sqrt{7x} = 6\sqrt{7x}$

26. $6\sqrt{6x}$

27. $\sqrt{3x}\,\sqrt{12y} = \sqrt{3x \cdot 12y} = \sqrt{3 \cdot x \cdot 3 \cdot 4 \cdot y} =$

    $\sqrt{3 \cdot 3 \cdot 4 \cdot x \cdot y} = \sqrt{3 \cdot 3}\,\sqrt{4}\,\sqrt{x \cdot y} = 3 \cdot 2\sqrt{xy} =$

    $6\sqrt{xy}$

28. $7\sqrt{3xy}$

29. $\sqrt{10}\,\sqrt{10} = \sqrt{10 \cdot 10} = 10$

30. $11\sqrt{x}$

31. $\sqrt{5b}\,\sqrt{15b} = \sqrt{5b \cdot 15b} = \sqrt{5 \cdot b \cdot 5 \cdot 3 \cdot b} =$

    $\sqrt{5 \cdot 5 \cdot b \cdot b \cdot 3} = \sqrt{5 \cdot 5}\,\sqrt{b \cdot b}\,\sqrt{3} = 5b\sqrt{3}$

Chapter 10   (10.4)

32. $6a\sqrt{3}$

33. $\sqrt{2t}\ \sqrt{2t} = \sqrt{2t \cdot 2t} = 2t$

34. $3a$

35. $\sqrt{ab}\ \sqrt{ac} = \sqrt{ab \cdot ac} = \sqrt{a \cdot a \cdot b \cdot c} = \sqrt{a \cdot a}\ \sqrt{b \cdot c} = a\sqrt{bc}$

36. $x\sqrt{yz}$

37. $\sqrt{2x^2y}\ \sqrt{4xy^2} = \sqrt{2x^2y \cdot 4xy^2} = \sqrt{4 \cdot x^2 \cdot y^2 \cdot 2 \cdot x \cdot y} = \sqrt{4}\ \sqrt{x^2}\ \sqrt{y^2}\ \sqrt{2xy} = 2xy\sqrt{2xy}$

38. $5mn\sqrt{3mn}$

39. $\sqrt{18x^2y^3}\ \sqrt{6xy^4} = \sqrt{18x^2y^3 \cdot 6xy^4} = \sqrt{3 \cdot 6 \cdot x^2 \cdot y^2 \cdot y \cdot 6 \cdot x \cdot y^4} = \sqrt{6 \cdot 6 \cdot x^2 \cdot y^6 \cdot 3 \cdot x \cdot y} = \sqrt{6 \cdot 6}\ \sqrt{x^2}\ \sqrt{y^6}\ \sqrt{3xy} = 6xy^3\sqrt{3xy}$

40. $4x^2y\sqrt{6y}$

41. $\sqrt{50ab}\ \sqrt{10a^2b^4} = \sqrt{50ab \cdot 10a^2b^4} = \sqrt{5 \cdot 10 \cdot a \cdot b \cdot 10 \cdot a^2 \cdot b^4} = \sqrt{10 \cdot 10 \cdot a^2 \cdot b^4 \cdot 5 \cdot a \cdot b} = \sqrt{10 \cdot 10}\ \sqrt{a^2}\ \sqrt{b^4}\ \sqrt{5ab} = 10ab^2\sqrt{5ab}$

42. $5xy^2\sqrt{2xy}$

43. Familiarize. We organize the information in a table.

	Distance	Speed	Time
First car	d	56	t
Second car	d	84	t - 1

$d = r \cdot t$

Translate. From the rows of the table we get two equations:
$$d = 56t,$$
$$d = 84(t - 1).$$

Carry out. We use the substitution method.
$56t = 84(t - 1)$ Substituting $56t$ for d
$56t = 84t - 84$
$-28t = -84$
$t = 3$

The problem asks how far from Hereford the second car will overtake the first, so we need to find d. Substitute 3 for t in the first equation.
$d = 56t$
$d = 56 \cdot 3$
$d = 168$

43. (continued)

Check. If $t = 3$, then the first car travels $56 \cdot 3$, or 168 km, and the second car travels $84(3 - 1)$, or $84 \cdot 2$, or 168 km. Since the distances are the same, our answer checks.

State. The second car overtakes the first 168 km from Hereford.

44. 275 km/h

45. $\sqrt{2y}\ \sqrt{3}\ \sqrt{8y} = \sqrt{2y \cdot 3 \cdot 8y} = \sqrt{2 \cdot y \cdot 3 \cdot 2 \cdot 4 \cdot y} = \sqrt{2 \cdot 2 \cdot 4 \cdot y \cdot y \cdot 3} = \sqrt{2 \cdot 2}\ \sqrt{4}\ \sqrt{y \cdot y}\ \sqrt{3} = 2 \cdot 2 \cdot y\sqrt{3} = 4y\sqrt{3}$

46. $a^2 - 5\sqrt{a}$

47. $\sqrt{27(x+1)}\ \sqrt{12y(x+1)^2} = \sqrt{27(x+1) \cdot 12y(x+1)^2} = \sqrt{9 \cdot 3 \cdot (x+1) \cdot 4 \cdot 3 \cdot y(x+1)^2} = \sqrt{9 \cdot 3 \cdot 3 \cdot 4 \cdot (x+1)^2 \cdot (x+1)y} = \sqrt{9}\ \sqrt{3 \cdot 3}\ \sqrt{4}\ \sqrt{(x+1)^2}\ \sqrt{(x+1)y} = 3 \cdot 3 \cdot 2(x+1)\sqrt{(x+1)y} = 18(x+1)\sqrt{(x+1)y}$

48. $6(x - 2)^2\sqrt{10}$

49. $\sqrt{x}\ \sqrt{2x}\ \sqrt{10x^5} = \sqrt{x \cdot 2x \cdot 10x^5} = \sqrt{x \cdot 2 \cdot x \cdot 2 \cdot 5 \cdot x^4 \cdot x} = \sqrt{x \cdot x \cdot 2 \cdot 2 \cdot x^4 \cdot 5 \cdot x} = \sqrt{x \cdot x}\ \sqrt{2 \cdot 2}\ \sqrt{x^4}\ \sqrt{5x} = x \cdot 2 \cdot x^2\sqrt{5x} = 2x^3\sqrt{5x}$

50. $2^{54}x^{158}\sqrt{2x}$

51. $\sqrt{x^{8n}} = \sqrt{(x^{4n})^2} = x^{4n}$

52. $0.2x^{2n}$

53. Let $A = 25$ and $B = 16$. Then
$\sqrt{A} - \sqrt{B} = \sqrt{25} - \sqrt{16} = 5 - 4 = 1$, but
$\sqrt{A - B} = \sqrt{25 - 16} = \sqrt{9} = 3$.

Thus, $\sqrt{A} - \sqrt{B} \neq \sqrt{A - B}$.
However, if $B = 0$, then
$\sqrt{A} - \sqrt{B} = \sqrt{A} - \sqrt{0} = \sqrt{A} - 0 = \sqrt{A}$ and
$\sqrt{A - B} = \sqrt{A - 0} = \sqrt{A}$.
Thus, $\sqrt{A} - \sqrt{B} = \sqrt{A - B}$ if $B = 0$.

54. $y^{\frac{n}{2}}$

55. If n is an odd whole number ≥ 3, then $n - 1$ is an even whole number ≥ 2 and $\frac{n-1}{2}$ is a whole number.

$\sqrt{y^n} = \sqrt{y^{n-1}y^1} = \sqrt{\left(y^{\frac{n-1}{2}}\right)^2 y} = y^{\frac{n-1}{2}}\sqrt{y}$

Chapter 10   (10.5)

56. $x^4 + y^4$; $(x^4 + \sqrt{2}x^2y^2 + y^4)(x^4 - \sqrt{2}x^2y^2 + y^4)$

Exercise Set 10.5

1. $\sqrt{\frac{9}{49}} = \frac{\sqrt{9}}{\sqrt{49}} = \frac{3}{7}$

2. $\frac{4}{5}$

3. $\sqrt{\frac{1}{36}} = \frac{\sqrt{1}}{\sqrt{36}} = \frac{1}{6}$

4. $\frac{1}{2}$

5. $-\sqrt{\frac{16}{81}} = -\frac{\sqrt{16}}{\sqrt{81}} = -\frac{4}{9}$

6. $-\frac{5}{7}$

7. $\sqrt{\frac{64}{289}} = \frac{\sqrt{64}}{\sqrt{289}} = \frac{8}{17}$

8. $\frac{9}{19}$

9. $\sqrt{\frac{1690}{1960}} = \sqrt{\frac{169 \cdot 10}{196 \cdot 10}} = \sqrt{\frac{169}{196} \cdot \frac{10}{10}} = \sqrt{\frac{169}{196} \cdot 1} = \sqrt{\frac{169}{196}} = \frac{\sqrt{169}}{\sqrt{196}} = \frac{13}{14}$

10. $\frac{12}{25}$

11. $\sqrt{\frac{36}{a^2}} = \frac{\sqrt{36}}{\sqrt{a^2}} = \frac{6}{a}$

12. $\frac{5}{x}$

13. $\sqrt{\frac{9a^2}{625}} = \frac{\sqrt{9a^2}}{\sqrt{625}} = \frac{3a}{25}$

14. $\frac{xy}{16}$

15. $\sqrt{\frac{2}{5}} = \sqrt{\frac{2}{5} \cdot \frac{5}{5}} = \sqrt{\frac{10}{25}} = \frac{\sqrt{10}}{\sqrt{25}} = \frac{\sqrt{10}}{5}$

16. $\frac{\sqrt{14}}{7}$

17. $\sqrt{\frac{3}{8}} = \sqrt{\frac{3}{8} \cdot \frac{2}{2}} = \sqrt{\frac{6}{16}} = \frac{\sqrt{6}}{\sqrt{16}} = \frac{\sqrt{6}}{4}$

18. $\frac{\sqrt{14}}{4}$

19. $\sqrt{\frac{7}{12}} = \sqrt{\frac{7}{12} \cdot \frac{3}{3}} = \sqrt{\frac{21}{36}} = \frac{\sqrt{21}}{\sqrt{36}} = \frac{\sqrt{21}}{6}$

20. $\frac{\sqrt{3}}{6}$

21. $\sqrt{\frac{1}{18}} = \sqrt{\frac{1}{18} \cdot \frac{2}{2}} = \sqrt{\frac{2}{36}} = \frac{\sqrt{2}}{\sqrt{36}} = \frac{\sqrt{2}}{6}$

22. $\frac{\sqrt{10}}{6}$

23. $\frac{3}{\sqrt{5}} = \frac{3}{\sqrt{5}} \cdot \frac{\sqrt{5}}{\sqrt{5}} = \frac{3\sqrt{5}}{5}$

24. $\frac{4\sqrt{3}}{3}$

25. $\sqrt{\frac{8}{3}} = \sqrt{\frac{8}{3} \cdot \frac{3}{3}} = \sqrt{\frac{24}{9}} = \frac{\sqrt{4 \cdot 6}}{\sqrt{9}} = \frac{\sqrt{4}\sqrt{6}}{\sqrt{9}} = \frac{2\sqrt{6}}{3}$

26. $\frac{2\sqrt{15}}{5}$

27. $\sqrt{\frac{3}{x}} = \sqrt{\frac{3}{x} \cdot \frac{x}{x}} = \sqrt{\frac{3x}{x^2}} = \frac{\sqrt{3x}}{\sqrt{x^2}} = \frac{\sqrt{3x}}{x}$

28. $\frac{\sqrt{2x}}{x}$

29. $\sqrt{\frac{x}{y}} = \sqrt{\frac{x}{y} \cdot \frac{y}{y}} = \sqrt{\frac{xy}{y^2}} = \frac{\sqrt{xy}}{\sqrt{y^2}} = \frac{\sqrt{xy}}{y}$

30. $\frac{\sqrt{ab}}{b}$

31. $\sqrt{\frac{x^2}{18}} = \sqrt{\frac{x^2}{18} \cdot \frac{2}{2}} = \sqrt{\frac{2x^2}{36}} = \frac{\sqrt{x^2}\sqrt{2}}{\sqrt{36}} = \frac{x\sqrt{2}}{6}$

32. $\frac{x\sqrt{5}}{10}$

33. $\frac{\sqrt{7}}{\sqrt{3}} = \frac{\sqrt{7}}{\sqrt{3}} \cdot \frac{\sqrt{3}}{\sqrt{3}} = \frac{\sqrt{21}}{3}$

34. $\frac{\sqrt{77}}{7}$

35. $\frac{\sqrt{9}}{\sqrt{8}} = \frac{\sqrt{9}}{\sqrt{8}} \cdot \frac{\sqrt{2}}{\sqrt{2}} = \frac{\sqrt{9 \cdot 2}}{\sqrt{16}} = \frac{3\sqrt{2}}{4}$

36. $\frac{2\sqrt{3}}{9}$

37. $\frac{\sqrt{2}}{\sqrt{5}} = \frac{\sqrt{2}}{\sqrt{5}} \cdot \frac{\sqrt{5}}{\sqrt{5}} = \frac{\sqrt{10}}{5}$

38. $\frac{\sqrt{6}}{2}$

39. $\frac{2}{\sqrt{2}} = \frac{2}{\sqrt{2}} \cdot \frac{\sqrt{2}}{\sqrt{2}} = \frac{2\sqrt{2}}{2} = \sqrt{2}$

40. $\sqrt{3}$

Chapter 10   (10.5)

41. $\dfrac{\sqrt{5}}{\sqrt{11}} = \dfrac{\sqrt{5}}{\sqrt{11}} \cdot \dfrac{\sqrt{11}}{\sqrt{11}} = \dfrac{\sqrt{55}}{11}$

42. $\dfrac{\sqrt{21}}{9}$

43. $\dfrac{\sqrt{7}}{\sqrt{12}} = \dfrac{\sqrt{7}}{\sqrt{12}} \cdot \dfrac{\sqrt{3}}{\sqrt{3}} = \dfrac{\sqrt{21}}{\sqrt{36}} = \dfrac{\sqrt{21}}{6}$

44. $\dfrac{\sqrt{10}}{6}$

45. $\dfrac{\sqrt{48}}{\sqrt{32}} = \sqrt{\dfrac{48}{32}} = \sqrt{\dfrac{3}{2}} = \sqrt{\dfrac{3}{2} \cdot \dfrac{2}{2}} = \sqrt{\dfrac{6}{4}} = \dfrac{\sqrt{6}}{\sqrt{4}} = \dfrac{\sqrt{6}}{2}$

46. $\dfrac{\sqrt{35}}{5}$

47. $\dfrac{\sqrt{450}}{\sqrt{18}} = \sqrt{\dfrac{450}{18}} = \sqrt{25} = 5$

48. 4

49. $\dfrac{\sqrt{3}}{\sqrt{x}} = \dfrac{\sqrt{3}}{\sqrt{x}} \cdot \dfrac{\sqrt{x}}{\sqrt{x}} = \dfrac{\sqrt{3x}}{x}$

50. $\dfrac{\sqrt{2y}}{y}$

51. $\dfrac{4y}{\sqrt{3}} = \dfrac{4y}{\sqrt{3}} \cdot \dfrac{\sqrt{3}}{\sqrt{3}} = \dfrac{4y\sqrt{3}}{3}$

52. $\dfrac{8x\sqrt{5}}{5}$

53. $\dfrac{\sqrt{a^3}}{\sqrt{8}} = \dfrac{\sqrt{a^3}}{\sqrt{8}} \cdot \dfrac{\sqrt{2}}{\sqrt{2}} = \dfrac{\sqrt{2a^3}}{\sqrt{16}} = \dfrac{\sqrt{a^2 \cdot 2a}}{\sqrt{16}} = \dfrac{a\sqrt{2a}}{4}$

54. $\dfrac{x\sqrt{3x}}{9}$

55. $\dfrac{\sqrt{56}}{\sqrt{12x}} = \sqrt{\dfrac{56}{12x}} = \sqrt{\dfrac{14}{3x}} = \sqrt{\dfrac{14}{3x} \cdot \dfrac{3x}{3x}} = \sqrt{\dfrac{42x}{3x \cdot 3x}} = \dfrac{\sqrt{42x}}{3x}$

56. $\dfrac{3\sqrt{10a}}{4a}$

57. $\dfrac{\sqrt{27c}}{\sqrt{32c^3}} = \sqrt{\dfrac{27c}{32c^3}} = \sqrt{\dfrac{27}{32c^2}} = \sqrt{\dfrac{27}{32c^2} \cdot \dfrac{2}{2}} = \sqrt{\dfrac{54}{64c^2}} =$

$\sqrt{\dfrac{9 \cdot 6}{64c^2}} = \dfrac{3\sqrt{6}}{8c}$

58. $\dfrac{x\sqrt{21}}{6}$

59. $\dfrac{\sqrt{y^5}}{\sqrt{xy^2}} = \sqrt{\dfrac{y^5}{xy^2}} = \sqrt{\dfrac{y^3}{x}} = \sqrt{\dfrac{y^3}{x} \cdot \dfrac{x}{x}} = \sqrt{\dfrac{xy^3}{x^2}} = \sqrt{\dfrac{y^2 \cdot xy}{x^2}} =$

$\dfrac{y\sqrt{xy}}{x}$

60. $\dfrac{x\sqrt{y}}{y}$

61. $\dfrac{\sqrt{16a^4b^6}}{\sqrt{128a^6b^6}} = \sqrt{\dfrac{16a^4b^6}{128a^6b^6}} = \sqrt{\dfrac{1}{8a^2}} = \sqrt{\dfrac{1}{8a^2} \cdot \dfrac{2}{2}} =$

$\sqrt{\dfrac{2}{16a^2}} = \dfrac{\sqrt{2}}{4a}$

62. $\dfrac{3n\sqrt{10}}{8}$

63. Method 1. Using a calculator.

$\sqrt{\dfrac{1}{3}} \approx \sqrt{0.333333} \approx 0.577$

Method 2. Using Table 2.

$\sqrt{\dfrac{1}{3}} = \sqrt{\dfrac{1}{3} \cdot \dfrac{3}{3}} = \sqrt{\dfrac{3}{9}} = \dfrac{\sqrt{3}}{\sqrt{9}} = \dfrac{\sqrt{3}}{3} \approx \dfrac{1.732}{3} \approx 0.577$

64. 1.225

65. Method 1. Using a calculator.

$\sqrt{\dfrac{7}{8}} = \sqrt{0.875} \approx 0.935$

Method 2. Using Table 2.

$\sqrt{\dfrac{7}{8}} = \sqrt{\dfrac{7}{8} \cdot \dfrac{2}{2}} = \sqrt{\dfrac{14}{16}} = \dfrac{\sqrt{14}}{4} \approx \dfrac{3.742}{4} \approx 0.936$

66. 0.612

67. Method 1. Using a calculator.

$\sqrt{\dfrac{1}{12}} \approx \sqrt{0.0833333} \approx 0.289$

Method 2. Using Table 2.

$\sqrt{\dfrac{1}{12}} = \sqrt{\dfrac{1}{12} \cdot \dfrac{3}{3}} = \sqrt{\dfrac{3}{36}} = \dfrac{\sqrt{3}}{\sqrt{36}} = \dfrac{\sqrt{3}}{6} \approx \dfrac{1.732}{6} \approx 0.289$

68. 0.645

69. Method 1. Using a calculator.

$\sqrt{\dfrac{1}{2}} = \sqrt{0.5} \approx 0.707$

Method 2. Using Table 2.

$\sqrt{\dfrac{1}{2}} = \sqrt{\dfrac{1}{2} \cdot \dfrac{2}{2}} = \sqrt{\dfrac{2}{4}} = \dfrac{\sqrt{2}}{\sqrt{4}} = \dfrac{\sqrt{2}}{2} \approx \dfrac{1.414}{2} \approx 0.707$

70. 0.378

Chapter 10   (10.5)

71. Method 1. Using a calculator.
$\dfrac{17}{\sqrt{20}} \approx \dfrac{17}{4.472135955} \approx 3.801$

Method 2. Using Table 2.
$\dfrac{17}{\sqrt{20}} = \dfrac{17}{\sqrt{20}} \cdot \dfrac{\sqrt{5}}{\sqrt{5}} = \dfrac{17\sqrt{5}}{\sqrt{100}} = \dfrac{17\sqrt{5}}{10} \approx \dfrac{17(2.236)}{10} \approx \dfrac{38.012}{10} \approx 3.801$

72. 7.766

73. Method 1. Using a calculator.
$\dfrac{\sqrt{13}}{\sqrt{18}} \approx \dfrac{3.605551275}{4.242640687} \approx 0.850$

Method 2. Using Table 2.
$\dfrac{\sqrt{13}}{\sqrt{18}} = \dfrac{\sqrt{13}}{\sqrt{18}} \cdot \dfrac{\sqrt{2}}{\sqrt{2}} = \dfrac{\sqrt{26}}{\sqrt{36}} = \dfrac{\sqrt{26}}{6} \approx \dfrac{5.099}{6} \approx 0.850$

74. 0.782

75. $\dfrac{\sqrt{18}}{\sqrt{2}} = \sqrt{\dfrac{18}{2}} = \sqrt{9} = 3$

76. 2

77. $\dfrac{\sqrt{60}}{\sqrt{15}} = \sqrt{\dfrac{60}{15}} = \sqrt{4} = 2$

78. 6

79. $\dfrac{\sqrt{75}}{\sqrt{15}} = \sqrt{\dfrac{75}{15}} = \sqrt{5}$

80. $\sqrt{6}$

81. $\dfrac{\sqrt{3}}{\sqrt{75}} = \sqrt{\dfrac{3}{75}} = \sqrt{\dfrac{1}{25}} = \dfrac{1}{5}$

82. $\dfrac{1}{4}$

83. $\dfrac{\sqrt{12}}{\sqrt{75}} = \sqrt{\dfrac{12}{75}} = \sqrt{\dfrac{4}{25}} = \dfrac{2}{5}$

84. $\dfrac{3}{4}$

85. $\dfrac{\sqrt{8x}}{\sqrt{2x}} = \sqrt{\dfrac{8x}{2x}} = \sqrt{4} = 2$

86. 3

87. $\dfrac{\sqrt{63y^3}}{\sqrt{7y}} = \sqrt{\dfrac{63y^3}{7y}} = \sqrt{9y^2} = 3y$

88. $4x$

89. $\dfrac{\sqrt{15x^5}}{\sqrt{3x}} = \sqrt{\dfrac{15x^5}{3x}} = \sqrt{5x^4} = x^2\sqrt{5}$

90. 1.57 sec, 3.14 sec

91. First convert 2/3 in. to ft.
$\dfrac{2}{3}$ in. $\times \dfrac{1 \text{ ft}}{12 \text{ in.}} = \dfrac{1}{18}$ ft

Then substitute $\dfrac{1}{18}$ for L and 3.14 for π in the formula.

$T = 2\pi\sqrt{\dfrac{L}{32}} \approx 2(3.14)\sqrt{\dfrac{\frac{1}{18}}{32}} \approx 6.28\sqrt{\dfrac{1}{18} \cdot \dfrac{1}{32}} \approx 6.28\sqrt{\dfrac{1}{576}} \approx 6.28\left(\dfrac{1}{24}\right) \approx 0.262$ sec

92. 1 sec

93. Substitute $\dfrac{45}{\pi^2}$ for L in the formula.

$T = 2\pi\sqrt{\dfrac{L}{32}} = 2\pi\sqrt{\dfrac{\frac{45}{\pi^2}}{32}} = 2\pi\sqrt{\dfrac{45}{\pi^2} \cdot \dfrac{1}{32}} = 2\pi\sqrt{\dfrac{45}{32\pi^2}} =$
$2\pi\sqrt{\dfrac{45}{32\pi^2} \cdot \dfrac{2}{2}} = 2\pi\sqrt{\dfrac{90}{64\pi^2}} = 2\pi\sqrt{\dfrac{9 \cdot 10}{64\pi^2}} =$
$2\pi \cdot \dfrac{3\sqrt{10}}{8\pi} = \dfrac{3\sqrt{10}}{4}$ sec.

It takes $\dfrac{3\sqrt{10}}{4}$ sec to move from one side to the other and back. Thus it takes $\dfrac{1}{2} \cdot \dfrac{3\sqrt{10}}{4} = \dfrac{3\sqrt{10}}{8}$ sec, or approximately 1.186 sec, to swing from one side to the other.

94. $\dfrac{\sqrt{5}}{40}$

95. $\sqrt{\dfrac{3}{1000}} = \sqrt{\dfrac{3}{1000} \cdot \dfrac{10}{10}} = \sqrt{\dfrac{30}{10,000}} = \dfrac{\sqrt{30}}{\sqrt{10,000}} = \dfrac{\sqrt{30}}{100}$

96. $\dfrac{\sqrt{5x}}{5x^2}$

97. $\sqrt{\dfrac{3x^2y}{a^2x^5}} = \sqrt{\dfrac{3y}{a^2x^3}} = \sqrt{\dfrac{3y}{a^2x^3} \cdot \dfrac{x}{x}} = \sqrt{\dfrac{3xy}{a^2x^4}} = \dfrac{\sqrt{3xy}}{\sqrt{a^2x^4}} = \dfrac{\sqrt{3xy}}{ax^2}$

98. $\dfrac{\sqrt{3ab}}{b}$

99. $\sqrt{\dfrac{1}{5zw^2}} = \sqrt{\dfrac{1}{5zw^2} \cdot \dfrac{5z}{5z}} = \sqrt{\dfrac{5z}{25z^2w^2}} = \dfrac{\sqrt{5z}}{\sqrt{25z^2w^2}} = \dfrac{\sqrt{5z}}{5zw}$

100. $\dfrac{\sqrt{70}}{100}$

101. $\sqrt{0.012} = \sqrt{\dfrac{12}{1000}} = \sqrt{\dfrac{3}{250}} = \sqrt{\dfrac{3}{250} \cdot \dfrac{10}{10}} = \sqrt{\dfrac{30}{2500}} = \dfrac{\sqrt{30}}{\sqrt{2500}} = \dfrac{\sqrt{30}}{50}$

Chapter 10   (10.6)

Exercise Set 10.6

1. $3\sqrt{2} + 4\sqrt{2} = (3 + 4)\sqrt{2}$
   $= 7\sqrt{2}$

2. $11\sqrt{3}$

3. $7\sqrt{5} - 3\sqrt{5} = (7 - 3)\sqrt{5}$
   $= 4\sqrt{5}$

4. $3\sqrt{2}$

5. $6\sqrt{x} + 7\sqrt{x} = (6 + 7)\sqrt{x}$
   $= 13\sqrt{x}$

6. $12\sqrt{y}$

7. $9\sqrt{x} - 11\sqrt{x} = (9 - 11)\sqrt{x}$
   $= -2\sqrt{x}$

8. $-8\sqrt{a}$

9. $5\sqrt{8} + 15\sqrt{2} = 5\sqrt{4\cdot 2} + 15\sqrt{2}$
   $= 5\cdot 2\sqrt{2} + 15\sqrt{2}$
   $= 10\sqrt{2} + 15\sqrt{2}$
   $= 25\sqrt{2}$

10. $8\sqrt{3}$

11. $\sqrt{27} - 2\sqrt{3} = \sqrt{9\cdot 3} - 2\sqrt{3}$
    $= 3\sqrt{3} - 2\sqrt{3}$
    $= (3 - 2)\sqrt{3}$
    $= 1\sqrt{3}$
    $= \sqrt{3}$

12. $32\sqrt{2}$

13. $\sqrt{45} - \sqrt{20} = \sqrt{9\cdot 5} - \sqrt{4\cdot 5}$
    $= 3\sqrt{5} - 2\sqrt{5}$
    $= (3 - 2)\sqrt{5}$
    $= 1\sqrt{5}$
    $= \sqrt{5}$

14. $\sqrt{3}$

15. $\sqrt{72} + \sqrt{98} = \sqrt{36\cdot 2} + \sqrt{49\cdot 2}$
    $= 6\sqrt{2} + 7\sqrt{2}$
    $= (6 + 7)\sqrt{2}$
    $= 13\sqrt{2}$

16. $7\sqrt{5}$

17. $2\sqrt{12} + \sqrt{27} - \sqrt{48} = 2\sqrt{4\cdot 3} + \sqrt{9\cdot 3} - \sqrt{16\cdot 3}$
    $= 2\cdot 2\sqrt{3} + 3\sqrt{3} - 4\sqrt{3}$
    $= (4 + 3 - 4)\sqrt{3}$
    $= 3\sqrt{3}$

18. $19\sqrt{2}$

19. $3\sqrt{18} - 2\sqrt{32} - 5\sqrt{50} = 3\sqrt{9\cdot 2} - 2\sqrt{16\cdot 2} - 5\sqrt{25\cdot 2}$
    $= 3\cdot 3\sqrt{2} - 2\cdot 4\sqrt{2} - 5\cdot 5\sqrt{2}$
    $= (9 - 8 - 25)\sqrt{2}$
    $= -24\sqrt{2}$

20. $2\sqrt{2}$

21. $2\sqrt{27} - 3\sqrt{48} + 2\sqrt{18} = 2\sqrt{9\cdot 3} - 3\sqrt{16\cdot 3} + 2\sqrt{9\cdot 2}$
    $= 2\cdot 3\sqrt{3} - 3\cdot 4\sqrt{3} + 2\cdot 3\sqrt{2}$
    $= (6 - 12)\sqrt{3} + 6\sqrt{2}$
    $= -6\sqrt{3} + 6\sqrt{2}$, or
    $6\sqrt{2} - 6\sqrt{3}$

22. $6\sqrt{3} - 6\sqrt{2}$

23. $\sqrt{4x} + \sqrt{81x^3} = \sqrt{4\cdot x} + \sqrt{81\cdot x^2\cdot x}$
    $= 2\sqrt{x} + 9x\sqrt{x}$
    $= (2 + 9x)\sqrt{x}$

24. $(2x + 3)\sqrt{3}$

25. $\sqrt{27} - \sqrt{12x^2} = \sqrt{9\cdot 3} - \sqrt{4\cdot 3\cdot x^2}$
    $= 3\sqrt{3} - 2x\sqrt{3}$
    $= (3 - 2x)\sqrt{3}$

26. $(9x - 2)\sqrt{x}$

27. $\sqrt{8x + 8} + \sqrt{2x + 2} = \sqrt{4(2x + 2)} + \sqrt{2x + 2}$
    $= 2\sqrt{2x + 2} + 1\sqrt{2x + 2}$
    $= (2 + 1)\sqrt{2x + 2}$
    $= 3\sqrt{2x + 2}$

28. $3\sqrt{3x + 3}$

29. $\sqrt{x^5 - x^2} + \sqrt{9x^3 - 9} = \sqrt{x^2(x^3 - 1)} + \sqrt{9(x^3 - 1)}$
    $= x\sqrt{x^3 - 1} + 3\sqrt{x^3 - 1}$
    $= (x + 3)\sqrt{x^3 - 1}$

30. $(4 + 5x)\sqrt{x - 1}$

Chapter 10 (10.6)

31. $3x\sqrt{y^3x} - x\sqrt{yx^3} + y\sqrt{y^3x}$
$= 3x\sqrt{y^2 \cdot y \cdot x} - x\sqrt{y \cdot x^2 \cdot x} + y\sqrt{y^2 \cdot y \cdot x}$
$= 3x \cdot y\sqrt{yx} - x \cdot x\sqrt{yx} + y \cdot y\sqrt{yx}$
$= (3xy - x^2 + y^2)\sqrt{yx}$

32. $(4a^2 + a^2b - 5b)\sqrt{b}$

33. $\sqrt{8(a+b)^3} - \sqrt{32(a+b)^3}$
$= \sqrt{4(a+b)^2 \cdot 2(a+b)} - \sqrt{16(a+b)^2 \cdot 2(a+b)}$
$= 2(a+b)\sqrt{2(a+b)} - 4(a+b)\sqrt{2(a+b)}$
$= (2 - 4)[(a+b)\sqrt{2(a+b)}]$
$= -2(a+b)\sqrt{2(a+b)}$

34. $(x + 3 + y)\sqrt{y}$

35. $\sqrt{3} - \sqrt{\frac{1}{3}} = \sqrt{3} - \sqrt{\frac{1}{3} \cdot \frac{3}{3}}$
$= \sqrt{3} - \frac{\sqrt{3}}{3}$
$= \left(1 - \frac{1}{3}\right)\sqrt{3}$
$= \frac{2}{3}\sqrt{3}, \text{ or } \frac{2\sqrt{3}}{3}$

36. $\frac{\sqrt{2}}{2}$

37. $5\sqrt{2} + 3\sqrt{\frac{1}{2}} = 5\sqrt{2} + 3\sqrt{\frac{1}{2} \cdot \frac{2}{2}}$
$= 5\sqrt{2} + \frac{3}{2}\sqrt{2}$
$= \left(5 + \frac{3}{2}\right)\sqrt{2}$
$= \frac{13}{2}\sqrt{2}, \text{ or } \frac{13\sqrt{2}}{2}$

38. $\frac{14\sqrt{3}}{3}$

39. $\sqrt{\frac{2}{3}} - \sqrt{\frac{1}{6}} = \sqrt{\frac{2}{3} \cdot \frac{3}{3}} - \sqrt{\frac{1}{6} \cdot \frac{6}{6}}$
$= \frac{\sqrt{6}}{3} - \frac{\sqrt{6}}{6}$
$= \left(\frac{1}{3} - \frac{1}{6}\right)\sqrt{6}$
$= \frac{1}{6}\sqrt{6}, \text{ or } \frac{\sqrt{6}}{6}$

40. $\frac{\sqrt{2}}{4}$

41. $\sqrt{\frac{1}{12}} - \sqrt{\frac{1}{27}} = \sqrt{\frac{1}{12} \cdot \frac{3}{3}} - \sqrt{\frac{1}{27} \cdot \frac{3}{3}}$
$= \frac{\sqrt{3}}{\sqrt{36}} - \frac{\sqrt{3}}{\sqrt{81}}$
$= \frac{\sqrt{3}}{6} - \frac{\sqrt{3}}{9}$
$= \left(\frac{1}{6} - \frac{1}{9}\right)\sqrt{3}$
$= \frac{1}{18}\sqrt{3}, \text{ or } \frac{\sqrt{3}}{18}$

42. $-\frac{\sqrt{30}}{30}$

43. $(\sqrt{5} + 7)(\sqrt{5} - 7) = (\sqrt{5})^2 - 7^2$
  (Using $(A + B)(A - B) = A^2 - B^2$)
$= 5 - 49$
$= -44$

44. -4

45. $(\sqrt{6} - \sqrt{3})(\sqrt{6} + \sqrt{3}) = (\sqrt{6})^2 - (\sqrt{3})^2$
  (Using $(A - B)(A + B) = A^2 - B^2$)
$= 6 - 3$
$= 3$

46. -4

47. $(3\sqrt{5} - 2)(\sqrt{5} + 1) = 3\sqrt{5}\sqrt{5} + 3\sqrt{5} - 2\sqrt{5} - 2$
  Using FOIL
$= 3 \cdot 5 + 3\sqrt{5} - 2\sqrt{5} - 2$
$= 15 + \sqrt{5} - 2$
$= 13 + \sqrt{5}$

48. $7\sqrt{2} - 5\sqrt{5}$

49. $(\sqrt{x} - \sqrt{y})^2 = (\sqrt{x})^2 - 2\sqrt{x}\sqrt{y} + (\sqrt{y})^2$
  Using $(A - B)^2 = A^2 - 2AB + B^2$
$= x - 2\sqrt{xy} + y$

50. $w + 22\sqrt{w} + 121$

51. $\frac{2}{\sqrt{3} - \sqrt{5}} = \frac{2}{\sqrt{3} - \sqrt{5}} \cdot \frac{\sqrt{3} + \sqrt{5}}{\sqrt{3} + \sqrt{5}}$ Multiplying by 1
$= \frac{2(\sqrt{3} + \sqrt{5})}{(\sqrt{3} - \sqrt{5})(\sqrt{3} + \sqrt{5})}$ Multiplying
$= \frac{2\sqrt{3} + 2\sqrt{5}}{(\sqrt{3})^2 - (\sqrt{5})^2} = \frac{2\sqrt{3} + 2\sqrt{5}}{3 - 5}$
$= \frac{2\sqrt{3} + 2\sqrt{5}}{-2} = \frac{2(\sqrt{3} + \sqrt{5})}{-2}$
$= -(\sqrt{3} + \sqrt{5}) = -\sqrt{3} - \sqrt{5}$

Chapter 10   (10.6)

52. $\dfrac{15 - 5\sqrt{7}}{2}$

53. $\dfrac{\sqrt{3} - \sqrt{2}}{\sqrt{3} + \sqrt{2}} = \dfrac{\sqrt{3} - \sqrt{2}}{\sqrt{3} + \sqrt{2}} \cdot \dfrac{\sqrt{3} - \sqrt{2}}{\sqrt{3} - \sqrt{2}}$ Multiplying by 1

$= \dfrac{(\sqrt{3} - \sqrt{2})^2}{(\sqrt{3} + \sqrt{2})(\sqrt{3} - \sqrt{2})}$

$= \dfrac{(\sqrt{3})^2 - 2\sqrt{3}\sqrt{2} + (\sqrt{2})^2}{(\sqrt{3})^2 - (\sqrt{2})^2}$

$= \dfrac{3 - 2\sqrt{6} + 2}{3 - 2} = \dfrac{5 - 2\sqrt{6}}{1}$

$= 5 - 2\sqrt{6}$

54. $2\sqrt{3} + 2\sqrt{2} - \sqrt{21} - \sqrt{14}$

55. $\dfrac{4}{\sqrt{10} + 1} = \dfrac{4}{\sqrt{10} + 1} \cdot \dfrac{\sqrt{10} - 1}{\sqrt{10} - 1}$

$= \dfrac{4(\sqrt{10} - 1)}{(\sqrt{10} + 1)(\sqrt{10} - 1)}$

$= \dfrac{4\sqrt{10} - 4}{(\sqrt{10})^2 - 1^2} = \dfrac{4\sqrt{10} - 4}{10 - 1}$

$= \dfrac{4\sqrt{10} - 4}{9}$

56. $3\sqrt{11} + 9$

57. $\dfrac{1 - \sqrt{7}}{3 + \sqrt{7}} = \dfrac{1 - \sqrt{7}}{3 + \sqrt{7}} \cdot \dfrac{3 - \sqrt{7}}{3 - \sqrt{7}}$

$= \dfrac{(1 - \sqrt{7})(3 - \sqrt{7})}{(3 + \sqrt{7})(3 - \sqrt{7})}$

$= \dfrac{3 - \sqrt{7} - 3\sqrt{7} + \sqrt{7}\sqrt{7}}{3^2 - (\sqrt{7})^2}$

$= \dfrac{3 - \sqrt{7} - 3\sqrt{7} + 7}{9 - 7} = \dfrac{10 - 4\sqrt{7}}{2}$

$= \dfrac{2(5 - 2\sqrt{7})}{2} = 5 - 2\sqrt{7}$

58. $\dfrac{-1 + \sqrt{5} + \sqrt{2} + \sqrt{10}}{2}$

59. $\sqrt{10} + \sqrt{50} = \sqrt{10} + \sqrt{10}\sqrt{5} = \sqrt{10}(1 + \sqrt{5})$

$\sqrt{10} + \sqrt{50} = \sqrt{10} + \sqrt{25 \cdot 2} = \sqrt{10} + 5\sqrt{2}$

$\sqrt{10} + \sqrt{50} = \sqrt{2}\sqrt{5} + \sqrt{2}\sqrt{25} = \sqrt{2}(\sqrt{5} + \sqrt{25}) = \sqrt{2}(\sqrt{5} + 5)$, or $\sqrt{2}(5 + \sqrt{5})$

All three are correct.

60. $\sqrt{10} + 5\sqrt{2}$

61. $\sqrt{125} - \sqrt{45} + 2\sqrt{5} = \sqrt{25 \cdot 5} - \sqrt{9 \cdot 5} + 2\sqrt{5}$

$= 5\sqrt{5} - 3\sqrt{5} + 2\sqrt{5}$

$= (5 - 3 + 2)\sqrt{5}$

$= 4\sqrt{5}$

62. $16\sqrt{2}$

63. $\dfrac{3}{5}\sqrt{24} + \dfrac{2}{5}\sqrt{150} - \sqrt{96} = \dfrac{3}{5}\sqrt{4 \cdot 6} + \dfrac{2}{5}\sqrt{25 \cdot 6} - \sqrt{16 \cdot 6}$

$= \dfrac{3}{5} \cdot 2\sqrt{6} + \dfrac{2}{5} \cdot 5\sqrt{6} - 4\sqrt{6}$

$= \dfrac{6}{5}\sqrt{6} + 2\sqrt{6} - 4\sqrt{6}$

$= \left(\dfrac{6}{5} + 2 - 4\right)\sqrt{6}$

$= -\dfrac{4}{5}\sqrt{6}$, or $-\dfrac{4\sqrt{6}}{5}$

64. $11\sqrt{3} - 10\sqrt{2}$

65. $\sqrt{ab^6} + b\sqrt{a^3} + a\sqrt{a} = \sqrt{b^6 \cdot a} + b\sqrt{a^2 \cdot a} + a\sqrt{a}$

$= b^3\sqrt{a} + ab\sqrt{a} + a\sqrt{a}$

$= (b^3 + ab + a)\sqrt{a}$

66. 0

67. $7x\sqrt{12xy^2} - 9y\sqrt{27x^3} + 5\sqrt{300x^3y^2}$

$= 7x\sqrt{4y^2 \cdot 3x} - 9y\sqrt{9x^2 \cdot 3x} + 5\sqrt{100x^2y^2 \cdot 3x}$

$= 7x \cdot 2y\sqrt{3x} - 9y \cdot 3x\sqrt{3x} + 5 \cdot 10xy\sqrt{3x}$

$= 14xy\sqrt{3x} - 27xy\sqrt{3x} + 50xy\sqrt{3x}$

$= (14xy - 27xy + 50xy)\sqrt{3x}$

$= 37xy\sqrt{3x}$

68. $\dfrac{x + 1}{x}\sqrt{x}$

69. $5\sqrt{\dfrac{3}{10}} + 2\sqrt{\dfrac{5}{6}} - 6\sqrt{\dfrac{15}{32}}$

$= 5\sqrt{\dfrac{3}{10} \cdot \dfrac{10}{10}} + 2\sqrt{\dfrac{5}{6} \cdot \dfrac{6}{6}} - 6\sqrt{\dfrac{15}{32} \cdot \dfrac{2}{2}}$

$= \dfrac{5\sqrt{30}}{10} + \dfrac{2\sqrt{30}}{6} - \dfrac{6\sqrt{30}}{8}$

$= \dfrac{\sqrt{30}}{2} + \dfrac{\sqrt{30}}{3} - \dfrac{3\sqrt{30}}{4}$ Simplifying

$= \dfrac{6\sqrt{30}}{12} + \dfrac{4\sqrt{30}}{12} - \dfrac{9\sqrt{30}}{12}$

$= \left(\dfrac{6}{12} + \dfrac{4}{12} - \dfrac{9}{12}\right)\sqrt{30}$

$= \dfrac{1}{12}\sqrt{30}$, or $\dfrac{\sqrt{30}}{12}$

70. $\left(\dfrac{2}{b} - \dfrac{2}{a^2} + \dfrac{5a}{4}\right)\sqrt{2ab}$

71. $\sqrt{1+x^2} + \dfrac{1}{\sqrt{1+x^2}} = \sqrt{1+x^2} + \dfrac{1}{\sqrt{1+x^2}} \cdot \dfrac{\sqrt{1+x^2}}{\sqrt{1+x^2}}$

$= \sqrt{1+x^2} + \dfrac{\sqrt{1+x^2}}{1+x^2}$

$= \left[1 + \dfrac{1}{1+x^2}\right]\sqrt{1+x^2}$

$= \left[\dfrac{1+x^2+1}{1+x^2}\right]\sqrt{1+x^2}$

$= \left[\dfrac{2+x^2}{1+x^2}\right]\sqrt{1+x^2}$, or

$\dfrac{(2+x^2)\sqrt{1+x^2}}{1+x^2}$

72. Any pairs of numbers a, b such that $a = 0, b \geq 0$ or $a \geq 0, b = 0$.

73. a) $\sqrt{a^2 + c^2} = \sqrt{1^2 + 2^2} = \sqrt{1+4} = \sqrt{5}$
    $\sqrt{a^2} + \sqrt{c^2} = \sqrt{1^2} + \sqrt{2^2} = 1 + 2 = 3$

  b) $\sqrt{b^2 + c^2} = \sqrt{3^2 + 2^2} = \sqrt{9+4} = \sqrt{13}$
    $\sqrt{b^2} + \sqrt{c^2} = \sqrt{3^2} + \sqrt{2^2} = 3 + 2 = 5$

  c) $\sqrt{a^2 + d^2} = \sqrt{1^2 + 4^2} = \sqrt{1+16} = \sqrt{17}$
    $\sqrt{a^2} + \sqrt{d^2} = \sqrt{1^2} + \sqrt{4^2} = 1 + 4 = 5$

  d) $\sqrt{b^2 + d^2} = \sqrt{3^2 + 4^2} = \sqrt{9+16} = \sqrt{25} = 5$
    $\sqrt{b^2} + \sqrt{d^2} = \sqrt{3^2} + \sqrt{4^2} = 3 + 4 = 7$

  e) $\sqrt{a^2 + b^2} = \sqrt{1^2 + 3^2} = \sqrt{1+9} = \sqrt{10}$
    $\sqrt{a^2} + \sqrt{b^2} = \sqrt{1^2} + \sqrt{3^2} = 1 + 3 = 4$

  f) $\sqrt{c^2 + d^2} = \sqrt{2^2 + 4^2} = \sqrt{4+16} = \sqrt{20} = \sqrt{4 \cdot 5} = 2\sqrt{5}$
    $\sqrt{c^2} + \sqrt{d^2} = \sqrt{2^2} + \sqrt{4^2} = 2 + 4 = 6$

74. Any pair of numbers x, y such that $x = 0$ and y is any real number or $y = 0$ and x is any real number.

75. Substitute 30 for T and 25 for v.

$T_w = 91.4 - \dfrac{(10.45 + 6.68\sqrt{25} - 0.447 \cdot 25)(457 - 5 \cdot 30)}{110}$

$= 91.4 - \dfrac{(10.45 + 33.4 - 11.175)(457 - 150)}{110}$

$= 91.4 - \dfrac{(32.675)(307)}{110} \approx 91.4 - 91.2 \approx 0.2$

$\approx 0°$ To the nearest degree

76. $-10°$

77. Substitute 20 for T and 40 for v.

$T_w = 91.4 - \dfrac{(10.45 + 6.68\sqrt{40} - 0.447 \cdot 40)(457 - 5 \cdot 20)}{110}$

$\approx 91.4 - \dfrac{(10.45 + 42.25 - 17.88)(457 - 100)}{110}$

$\approx 91.4 - \dfrac{(34.82)(357)}{110} \approx 91.4 - 113.01 \approx -21.61$

$\approx -22°$ To the nearest degree

78. $-64°$

Exercise Set 10.7

1. $\sqrt{x} = 5$
   $(\sqrt{x})^2 = 5^2$
   $x = 25$

   Check: $\sqrt{x} = 5$
   $\sqrt{25} \mid 5$
   5

2. 49

3. $\sqrt{x} = 6.2$
   $(\sqrt{x})^2 = (6.2)^2$
   $x = 38.44$

   Check: $\sqrt{x} = 6.2$
   $\sqrt{38.44} \mid 6.2$
   6.2

4. 18.49

5. $\sqrt{x + 3} = 20$
   $(\sqrt{x+3})^2 = 20^2$
   $x + 3 = 400$
   $x = 397$

   Check: $\sqrt{x+3} = 20$
   $\sqrt{397 + 3} \mid 20$
   $\sqrt{400}$
   20

6. 117

7. $\sqrt{2x + 4} = 25$
   $(\sqrt{2x+4})^2 = 25^2$
   $2x + 4 = 625$
   $2x = 621$
   $x = \dfrac{621}{2}$

   Check: $\sqrt{2x+4} = 25$
   $\sqrt{2 \cdot \dfrac{621}{2} + 4} \mid 25$
   $\sqrt{621 + 4}$
   $\sqrt{625}$
   25

8. 84

9. $3 + \sqrt{x-1} = 5$
   $\sqrt{x-1} = 2$
   $(\sqrt{x-1})^2 = 2^2$
   $x - 1 = 4$
   $x = 5$

   Check: $3 + \sqrt{x-1} = 5$
   $3 + \sqrt{5-1} \mid 5$
   $3 + \sqrt{4}$
   $3 + 2$
   5

10. 52

Chapter 10   (10.7)

11. $6 - 2\sqrt{3n} = 0$ Check: $\begin{array}{c} 6 - 2\sqrt{3n} = 0 \\ \hline 6 - 2\sqrt{3\cdot 3} \\ 6 - 2\cdot 3 \\ 6 - 6 \\ 0 \end{array}$

$6 = 2\sqrt{3n}$
$6^2 = (2\sqrt{3n})^2$
$36 = 4\cdot 3n$
$36 = 12n$
$3 = n$

12. $\dfrac{4}{5}$

13. $\sqrt{5x - 7} = \sqrt{x + 10}$ Check:
$(\sqrt{5x - 7})^2 = (\sqrt{x + 10})^2$
$5x - 7 = x + 10$
$4x = 17$
$x = \dfrac{17}{4}$

$\begin{array}{c|c} \sqrt{5x - 7} = \sqrt{x + 10} \\ \hline \sqrt{5\cdot\dfrac{17}{4} - 7} & \sqrt{\dfrac{17}{4} + 10} \\ \sqrt{\dfrac{85}{4} - \dfrac{28}{4}} & \sqrt{\dfrac{57}{4}} \\ \sqrt{\dfrac{57}{4}} \end{array}$

14. $\dfrac{14}{3}$

15. $\sqrt{x} = -7$
There is no solution. The principal square root of x cannot be negative.

16. No solution

17. $\sqrt{2y + 6} = \sqrt{2y - 5}$
$(\sqrt{2y + 6})^2 = (\sqrt{2y - 5})^2$
$2y + 6 = 2y - 5$
$6 = -5$
The equation $6 = -5$ is false; there is no solution.

18. No solution

19. $x - 7 = \sqrt{x - 5}$
$(x - 7)^2 = (\sqrt{x - 5})^2$
$x^2 - 14x + 49 = x - 5$
$x^2 - 15x + 54 = 0$
$(x - 9)(x - 6) = 0$
$x - 9 = 0$ or $x - 6 = 0$
$x = 9$ or $x = 6$

Check: $\begin{array}{c|c} x - 7 = \sqrt{x - 5} \\ \hline 9 - 7 & \sqrt{9 - 5} \\ 2 & \sqrt{4} \\ & 2 \end{array}$ $\begin{array}{c|c} x - 7 = \sqrt{x - 5} \\ \hline 6 - 7 & \sqrt{6 - 5} \\ -1 & \sqrt{1} \\ & 1 \end{array}$

The number 9 checks, but 6 does not. The solution is 9.

20. 9

21. $\sqrt{x + 18} = x - 2$
$(\sqrt{x + 18})^2 = (x - 2)^2$
$x + 18 = x^2 - 4x + 4$
$0 = x^2 - 5x - 14$
$0 = (x - 7)(x + 2)$
$x - 7 = 0$ or $x + 2 = 0$
$x = 7$ or $x = -2$

Check: $\begin{array}{c|c} \sqrt{x + 18} = x - 2 \\ \hline \sqrt{7 + 18} & 7 - 2 \\ \sqrt{25} & 5 \\ 5 \end{array}$ $\begin{array}{c|c} \sqrt{x + 18} = x - 2 \\ \hline \sqrt{-2 + 18} & -2 - 2 \\ \sqrt{16} & -4 \\ 4 \end{array}$

The number 7 checks, but -2 does not. The solution is 7.

22. 12

23. $2\sqrt{x - 1} = x - 1$
$(2\sqrt{x - 1})^2 = (x - 1)^2$
$4(x - 1) = x^2 - 2x + 1$
$4x - 4 = x^2 - 2x + 1$
$0 = x^2 - 6x + 5$
$0 = (x - 5)(x - 1)$
$x - 5 = 0$ or $x - 1 = 0$
$x = 5$ or $x = 1$
Both numbers check. The solutions are 5 and 1.

24. 0, 8

25. $\sqrt{5x + 21} = x + 3$
$(\sqrt{5x + 21})^2 = (x + 3)^2$
$5x + 21 = x^2 + 6x + 9$
$0 = x^2 + x - 12$
$0 = (x + 4)(x - 3)$
$x + 4 = 0$ or $x - 3 = 0$
$x = -4$ or $x = 3$

Check: $\begin{array}{c|c} \sqrt{5x + 21} = x + 3 \\ \hline \sqrt{5(-4) + 21} & -4 + 3 \\ \sqrt{1} & -1 \\ 1 \end{array}$ $\begin{array}{c|c} \sqrt{5x + 21} = x + 3 \\ \hline \sqrt{5\cdot 3 + 21} & 3 + 3 \\ \sqrt{36} & 6 \\ 6 \end{array}$

The number 3 checks, but -4 does not. The solution is 3.

26. 6

Chapter 10 (10.7)

27.
$$x = 1 + 6\sqrt{x - 9}$$
$$x - 1 = 6\sqrt{x - 9} \quad \text{Isolating the radical}$$
$$(x - 1)^2 = (6\sqrt{x - 9})^2$$
$$x^2 - 2x + 1 = 36(x - 9)$$
$$x^2 - 2x + 1 = 36x - 324$$
$$x^2 - 38x + 325 = 0$$
$$(x - 13)(x - 25) = 0$$
$$x - 13 = 0 \quad \text{or} \quad x - 25 = 0$$
$$x = 13 \quad \text{or} \quad x = 25$$
Both numbers check.

28. 5

29. $\sqrt{x^2 + 6} - x + 3 = 0$
$$\sqrt{x^2 + 6} = x - 3 \quad \text{Isolating the radical}$$
$$(\sqrt{x^2 + 6})^2 = (x - 3)^2$$
$$x^2 + 6 = x^2 - 6x + 9$$
$$-3 = -6x \quad \text{Adding } -x^2 \text{ and } -9$$
$$\frac{1}{2} = x$$

Check:
$$\frac{\sqrt{x^2 + 6} - x + 3 = 0}{\sqrt{\left(\frac{1}{2}\right)^2 + 6} - \frac{1}{2} + 3 \;\bigg|\; 0}$$
$$\sqrt{\frac{25}{4}} - \frac{1}{2} + 3$$
$$\frac{5}{2} - \frac{1}{2} + 3$$
$$5$$

The number $\frac{1}{2}$ does not check. There are no real-number solutions.

30. No solution

31. $\sqrt{(p + 6)(p + 1)} - 2 = p + 1$
$$\sqrt{(p + 6)(p + 1)} = p + 3 \quad \text{Isolating the radical}$$
$$\left[\sqrt{(p + 6)(p + 1)}\right]^2 = (p + 3)^2$$
$$(p + 6)(p + 1) = p^2 + 6p + 9$$
$$p^2 + 7p + 6 = p^2 + 6p + 9$$
$$p = 3$$

The number 3 checks. It is the solution.

32. 5

33.
$$\sqrt{2 - x} = \sqrt{3x - 7}$$
$$(\sqrt{2 - x})^2 = (\sqrt{3x - 7})^2$$
$$2 - x = 3x - 7$$
$$9 = 4x$$
$$\frac{9}{4} = x$$

Check:
$$\frac{\sqrt{2 - x} = \sqrt{3x - 7}}{\sqrt{2 - \frac{9}{4}} \;\bigg|\; \sqrt{3 \cdot \frac{9}{4} - 7}}$$
$$\sqrt{-\frac{1}{4}} \;\bigg|\; \sqrt{-\frac{1}{4}}$$

Since $\sqrt{-\frac{1}{4}}$ is not a real number, there are no real-number solutions.

34. No solution

35.
$$\sqrt{x + 9} = 1 + \sqrt{x}$$
$$(\sqrt{x + 9})^2 = (1 + \sqrt{x})^2$$
$$x + 9 = 1 + 2 \cdot 1 \cdot \sqrt{x} + (\sqrt{x})^2$$
$$x + 9 = 1 + 2\sqrt{x} + x$$
$$8 = 2\sqrt{x} \quad \text{Isolating the radical}$$
$$4 = \sqrt{x} \quad \text{Multiplying by } \frac{1}{2}$$
$$4^2 = (\sqrt{x})^2$$
$$16 = x$$

The number 16 checks. It is the solution.

36. 9

37.
$$\sqrt{3x + 1} = 1 - \sqrt{x + 4}$$
$$(\sqrt{3x + 1})^2 = (1 - \sqrt{x + 4})^2$$
$$3x + 1 = 1 - 2 \cdot 1 \cdot \sqrt{x + 4} + (\sqrt{x + 4})^2$$
$$3x + 1 = 1 - 2\sqrt{x + 4} + x + 4$$
$$3x + 1 = 5 - 2\sqrt{x + 4} + x$$
$$2x - 4 = -2\sqrt{x + 4} \quad \text{Isolating the radical}$$
$$2(x - 2) = -2\sqrt{x + 4}$$
$$x - 2 = -\sqrt{x + 4} \quad \text{Multiplying by } \frac{1}{2}$$
$$(x - 2)^2 = (-\sqrt{x + 4})^2$$
$$x^2 - 4x + 4 = x + 4$$
$$x^2 - 5x = 0$$
$$x(x - 5) = 0$$
$$x = 0 \quad \text{or} \quad x - 5 = 0$$
$$x = 0 \quad \text{or} \quad x = 5$$

Check:
$$\frac{\sqrt{3x + 1} = 1 - \sqrt{x + 4}}{\sqrt{3 \cdot 0 + 1} \;\bigg|\; 1 - \sqrt{0 + 4}}$$
$$\sqrt{1} \;\bigg|\; 1 - 2$$
$$1 \;\bigg|\; -1$$

$$\frac{\sqrt{3x + 1} = 1 - \sqrt{x + 4}}{\sqrt{3 \cdot 5 + 1} \;\bigg|\; 1 - \sqrt{5 + 4}}$$
$$\sqrt{16} \;\bigg|\; 1 - 3$$
$$4 \;\bigg|\; -2$$

Neither number checks. There is no solution.

Chapter 10 (10.7)

38. 1

39. $V = 3.5\sqrt{h}$

$V = 3.5\sqrt{9800}$ Substituting 9800 for h

$= 3.5\sqrt{4900 \cdot 2}$

$= 3.5(70)\sqrt{2}$

$= 245\sqrt{2}$

≈ 346.48

You can see about 346.48 km to the horizon.

40. 17.15 km

41. $V = 3.5\sqrt{h}$

$371 = 3.5\sqrt{h}$ Substituting 371 for V

$\frac{371}{3.5} = \sqrt{h}$

$106 = \sqrt{h}$

$106^2 = (\sqrt{h})^2$

$11,236 = h$

The airplane is 11,236 m high.

42. 806.56 m

43. $r = 2\sqrt{5L}$

$50 = 2\sqrt{5L}$ Substituting 50 for r

$25 = \sqrt{5L}$

$25^2 = (\sqrt{5L})^2$

$625 = 5L$

$125 = L$

The car will skid 125 ft at 50 mph.

$70 = 2\sqrt{5L}$ Substituting 70 for r

$35 = \sqrt{5L}$

$35^2 = (\sqrt{5L})^2$

$1225 = 5L$

$245 = L$

The car will skid 245 ft at 70 mph.

44. 180 ft, 500 ft

45. <u>Familiarize</u>. Let x represent the number.
<u>Translate</u>. We reword the problem.

Two times   the square root   is 14.
              of a number

2   · $\sqrt{x}$ = 14

<u>Carry out</u>.

$2\sqrt{x} = 14$

$\sqrt{x} = 7$

$(\sqrt{x})^2 = 7^2$

$x = 49$

<u>Check</u>. The principal square root of 49 is 7 and twice 7 is 14.
<u>State</u>. The number is 49.

46. 121

47. <u>Familiarize</u>. Let x represent the number. Then 4 more than 5 times the number is represented by 5x + 4.
<u>Translate</u>.

The square root   is 8.
of 5x + 4

$\sqrt{5x + 4}$ = 8

<u>Carry out</u>.

$\sqrt{5x + 4} = 8$

$(\sqrt{5x + 4})^2 = 8^2$

$5x + 4 = 64$

$5x = 60$

$x = 12$

<u>Check</u>. Four more than five times 12 is 5·12 + 4, or 64. The principal square root of 64 is 8.
<u>State</u>. The number is 12.

48. About 2.08 ft

49. $T = 2\pi\sqrt{\frac{L}{32}}$

$3 = 2(3.14)\sqrt{\frac{L}{32}}$ Substituting 3 for T and 3.14 for π

$3 = 6.28\sqrt{\frac{L}{32}}$

$(3)^2 = \left[6.28\sqrt{\frac{L}{32}}\right]^2$

$9 = \frac{39.4384L}{32}$

$\frac{32 \cdot 9}{39.4384} = L$

$7.30 \approx L$

The pendulum is about 7.30 ft long.

50. 2, -2

267

Chapter 10 (10.7)

51.
$$\sqrt{x} = -x$$
$$(\sqrt{x})^2 = (-x)^2$$
$$x = x^2$$
$$0 = x^2 - x$$
$$0 = x(x - 1)$$
$x = 0$ or $x - 1 = 0$
$x = 0$ or $x = 1$

Check:
$\sqrt{x} = -x$		$\sqrt{x} = -x$	
$\sqrt{0}$	-0	$\sqrt{1}$	-1
0	0	1	

Only 0 checks. It is the solution.

52. $\frac{1}{2}$

53.
$$4 + \sqrt{10 - x} = 6 + \sqrt{4 - x}$$
$$\sqrt{10 - x} = 2 + \sqrt{4 - x} \quad \text{Isolating one radical}$$
$$(\sqrt{10 - x})^2 = (2 + \sqrt{4 - x})^2$$
$$10 - x = 4 + 4\sqrt{4 - x} + (4 - x)$$
$$10 - x = 4\sqrt{4 - x} + 8 - x$$
$$2 = 4\sqrt{4 - x}$$
$$1 = 2\sqrt{4 - x}$$
$$1^2 = (2\sqrt{4 - x})^2$$
$$1 = 4(4 - x)$$
$$1 = 16 - 4x$$
$$-15 = -4x$$
$$\frac{15}{4} = x$$

$\frac{15}{4}$ checks, so it is the solution.

54. 0, 4

55.
$$\sqrt{x + 3} = \frac{8}{\sqrt{x - 9}}$$
$$(\sqrt{x + 3})^2 = \left[\frac{8}{\sqrt{x - 9}}\right]^2$$
$$x + 3 = \frac{64}{x - 9}$$
$$(x - 9)(x + 3) = 64 \quad \text{Multiplying by } x - 9$$
$$x^2 - 6x - 27 = 64$$
$$x^2 - 6x - 91 = 0$$
$$(x - 13)(x + 7) = 0$$
$x - 13 = 0$ or $x + 7 = 0$
$x = 13$ or $x = -7$

The number 13 checks, but -7 does not. The solution is 13.

56. 2

57.
$$2\sqrt{x - 1} - \sqrt{3x - 5} = \sqrt{x - 9}$$
$$(2\sqrt{x - 1} - \sqrt{3x - 5})^2 = (\sqrt{x - 9})^2$$
$$4(x - 1) - 4\sqrt{(x - 1)(3x - 5)} + (3x - 5) = x - 9$$
$$4x - 4 - 4\sqrt{3x^2 - 8x + 5} + 3x - 5 = x - 9$$
$$7x - 9 - 4\sqrt{3x^2 - 8x + 5} = x - 9$$
$$-4\sqrt{3x^2 - 8x + 5} = -6x$$
$$2\sqrt{3x^2 - 8x + 5} = 3x$$
$$(2\sqrt{3x^2 - 8x + 5})^2 = (3x)^2$$
$$4(3x^2 - 8x + 5) = 9x^2$$
$$12x^2 - 32x + 20 = 9x^2$$
$$3x^2 - 32x + 20 = 0$$
$$(3x - 2)(x - 10) = 0$$
$3x - 2 = 0$ or $x - 10 = 0$
$3x = 2$ or $x = 10$
$x = \frac{2}{3}$ or $x = 10$

The number 10 checks, but $\frac{2}{3}$ does not. The solution is 10.

58. 3

59.
$$A = \sqrt{1 + \frac{a^2}{b^2}}$$
$$A = \sqrt{\frac{b^2 + a^2}{b^2}}$$
$$A^2 = \left[\sqrt{\frac{b^2 + a^2}{b^2}}\right]^2$$
$$A^2 = \frac{b^2 + a^2}{b^2}$$
$$A^2 b^2 = b^2 + a^2$$
$$A^2 b^2 - b^2 = a^2$$
$$b^2(A^2 - 1) = a^2$$
$$b^2 = \frac{a^2}{A^2 - 1}$$
$$b = \sqrt{\frac{a^2}{A^2 - 1}} \quad \text{Taking the principal square root}$$

60. $s = \frac{gt^2}{2}$

61. Using the result of Exercise 60, we substitute 32.2 for g and 5 for t.
$$\frac{gt^2}{2} = s$$
$$\frac{32.2(5)^2}{2} = s$$
$$\frac{32.2(25)}{2} = s$$
$$402.5 = s$$

The object falls 402.5 ft.

62. 490 m

Chapter 10 (10.7)

63. <u>Familiarize</u>. We will use the formula $V = 3.5\sqrt{h}$. We organize the information in a table.

	Height	Distance to the horizon
First sighting	h	V
Second sighting	h + 100	V + 20

<u>Translate</u>. The rows of the table give us two equations.

$V = 3.5\sqrt{h}$,   (1)

$V + 20 = 3.5\sqrt{h + 100}$ (2)

<u>Carry out</u>. We substitute $3.5\sqrt{h}$ for V in equation (2) and solve for h.

$3.5\sqrt{h} + 20 = 3.5\sqrt{h + 100}$

$(3.5\sqrt{h} + 20)^2 = (3.5\sqrt{h + 100})^2$

$12.25h + 140\sqrt{h} + 400 = 12.25(h + 100)$

$12.25h + 140\sqrt{h} + 400 = 12.25h + 1225$

$140\sqrt{h} = 825$

$28\sqrt{h} = 165$ Multiplying by $\frac{1}{5}$

$(28\sqrt{h})^2 = (165)^2$

$784h = 27{,}225$

$h \approx 34.726$

<u>Check</u>. When $h \approx 34.726$, then $V \approx 3.5\sqrt{34.726} \approx 20.625$ km. When $h \approx 100 + 34.726$, or 134.726, then $V \approx 3.5\sqrt{134.726} \approx 40.625$ km. This is 20 km more than 20.625. The answer checks.

<u>State</u>. The climber was at a height of about 34.726 m when the first computation was made.

64.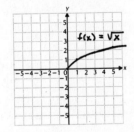

65. Graph $f(x) = \sqrt{x - 5}$.

We make a table of values. Note that we must choose values for x that are greater than or equal to 5 in order to have a nonnegative radicand.

x	f(x)
5	0
6	1
9	2

We plot these points and connect them with a smooth curve.

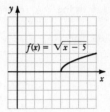

66.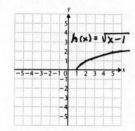

67. Graph $f(x) = \sqrt{x + 1}$.

We make a table of values. Note that we must choose values for x that are greater than or equal to -1 in order to have a nonnegative radicand.

x	f(x)
-1	0
0	1
3	2

We plot these points and connect them with a smooth curve.

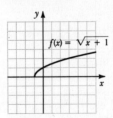

68.

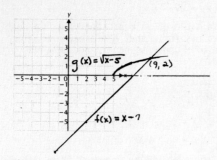

The solution is 9.

69. We can graph $f(x) = x - 5$ using the intercepts, $(0,-5)$ and $(5,0)$.

We make a table of values for $f(x) = \sqrt{x + 7}$.

x	f(x)
-7	0
-6	1
-3	2
2	3
9	4

We plot these points and connect them with a smooth curve.

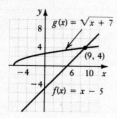

The graphs intersect at (9,4), so the solution of $x - 5 = \sqrt{x + 7}$ is 9.

Exercise Set 10.8

1. $a^2 + b^2 = c^2$
   $8^2 + 15^2 = c^2$ Substituting
   $64 + 225 = c^2$
   $289 = c^2$
   $\sqrt{289} = c$
   $17 = c$

2. $c = \sqrt{34} \approx 5.831$

3. $a^2 + b^2 = c^2$
   $4^2 + 4^2 = c^2$ Substituting
   $16 + 16 = c^2$
   $32 = c^2$
   $\sqrt{32} = c$ Exact answer
   $5.657 \approx c$ Approximation

4. $c = \sqrt{98} \approx 9.899$

5. $a^2 + b^2 = c^2$
   $5^2 + b^2 = 13^2$
   $25 + b^2 = 169$
   $b^2 = 144$
   $b = 12$

6. $a = 5$

7. $a^2 + b^2 = c^2$
   $(4\sqrt{3})^2 + b^2 = 8^2$
   $16 \cdot 3 + b^2 = 64$
   $48 + b^2 = 64$
   $b^2 = 16$
   $b = 4$

8. $b = \sqrt{31} \approx 5.568$

9. $a^2 + b^2 = c^2$
   $10^2 + 24^2 = c^2$
   $100 + 576 = c^2$
   $676 = c^2$
   $26 = c$

10. $c = 13$

11. $a^2 + b^2 = c^2$
    $9^2 + b^2 = 15^2$
    $81 + b^2 = 225$
    $b^2 = 144$
    $b = 12$

12. $b = 24$

13. $a^2 + b^2 = c^2$
    $a^2 + 1^2 = (\sqrt{5})^2$
    $a^2 + 1 = 5$
    $a^2 = 4$
    $a = 2$

14. $a = 1$

15. $a^2 + b^2 = c^2$
    $1^2 + b^2 = (\sqrt{3})^2$
    $1 + b^2 = 3$
    $b^2 = 2$
    $b = \sqrt{2}$ Exact answer
    $b \approx 1.414$ Approximation

16. $c = \sqrt{8} \approx 2.828$

Chapter 10   (10.8)

17. $a^2 + b^2 = c^2$
     $a^2 + (5\sqrt{3})^2 = 10^2$
     $a^2 + 25 \cdot 3 = 100$
     $a^2 + 75 = 100$
     $a^2 = 25$
     $a = 5$

18. $c = \sqrt{50} \approx 7.071$

19. Familiarize. We first make a drawing. We label the unknown height h.

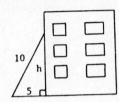

Translate. We use the Pythagorean property.
$a^2 + b^2 = c^2$
$5^2 + h^2 = 10^2$ Substituting 5 for a, h for b, and 10 for c

Carry out. We solve the equation.
$5^2 + h^2 = 10^2$
$25 + h^2 = 100$
$h^2 = 75$
$h = \sqrt{75}$ Exact answer
$h \approx 8.660$ Approximation

Check. We check by substituting 5, $\sqrt{75}$, and 10 into the Pythagorean equation:

$a^2 + b^2 = c^2$	
$5^2 + (\sqrt{75})^2$	10^2
$25 + 75$	100
100	

State. The top of the ladder is $\sqrt{75}$ or about 8.660 m from the ground.

20. $\sqrt{18} \approx 4.243$ cm

21. Familiarize. We first make a drawing. We label the unknown length w.

Translate. We use the Pythagorean property.
$a^2 + b^2 = c^2$
$8^2 + 12^2 = w^2$ Substituting 8 for a, 12 for b, and w for c.

21. (continued)
Carry out. We solve the equation.
$8^2 + 12^2 = w^2$
$64 + 144 = w^2$
$208 = w^2$
$\sqrt{208} = w$
$4\sqrt{13} = w$ Exact answer
$14.422 \approx w$ Approximation

Check. We check by substituting 8, 12, and $\sqrt{208}$ into the Pythagorean equation:

$a^2 + b^2 = c^2$	
$8^2 + 12^2$	$(\sqrt{208})^2$
$64 + 144$	208
208	

State. The guy wire is $4\sqrt{13}$ or about 14.422 feet long.

22. $5\sqrt{10} \approx 15.811$ m

23. Familiarize. We first make a drawing. We label the length from home to second base d.

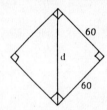

Translate. We use the Pythagorean property.
$a^2 + b^2 = c^2$
$60^2 + 60^2 = d^2$ Substituting 60 for a, 60 for b and d for c

Carry out. We solve the equation.
$60^2 + 60^2 = d^2$
$3600 + 3600 = d^2$
$7200 = d^2$
$\sqrt{7200} = d$
$60\sqrt{2} = d$ Exact answer
$84.853 \approx d$ Approximation

Check. We check by substituting 60, 60, and $\sqrt{7200}$ into the Pythagorean equation:

$a^2 + b^2 = c^2$	
$60^2 + 60^2$	$(\sqrt{7200})^2$
$3600 + 3600$	7200
7200	

State. It is $60\sqrt{2}$ or about 84.853 feet from home to second base.

24. $90\sqrt{2} \approx 127.279$ ft

25. Familiarize. Referring to the drawing in the text, we let d represent the distance from P to R.
Translate. We use the Pythagorean property.
$$a^2 + b^2 = c^2$$
$$25^2 + 35^2 = d^2 \quad \text{Substituting 25 for a, 35 for b, and d for c}$$
Carry out. We solve the equation.
$$25^2 + 35^2 = d^2$$
$$625 + 1225 = d^2$$
$$1850 = d^2$$
$$\sqrt{1850} = d$$
$$43 \approx d$$
Check. We check by substituting 25, 35, and $\sqrt{1850}$ into the Pythagorean equation:

$a^2 + b^2 = c^2$	
$25^2 + 35^2$	$(\sqrt{1850})^2$
$625 + 1225$	1850
1850	

State. It is about 43 yd from P to R.

26. $\approx 14{,}533$ ft

27. Write the equation in the slope-intercept form.
$$4 - x = 3y$$
$$\tfrac{1}{3}(4 - x) = y$$
$$\tfrac{4}{3} - \tfrac{1}{3}x = y, \text{ or}$$
$$y = -\tfrac{1}{3}x + \tfrac{4}{3}$$
The slope is $-\tfrac{1}{3}$.

28. $\tfrac{5}{8}$

29. $-\tfrac{3}{5}x < 15$
$$x > -\tfrac{5}{3} \cdot 15 \quad \text{Multiplying by } -\tfrac{5}{3} \text{ and reversing the inequality}$$
$$x > -25$$
The solution set is $\{x | x > -25\}$.

30. $\{x | x \leq 1\}$

31. Using the Pythagorean property we get:
$$h^2 + \left(\tfrac{a}{2}\right)^2 = a^2$$
$$h^2 + \tfrac{a^2}{4} = a^2$$
$$h^2 = a^2 - \tfrac{a^2}{4}$$
$$h^2 = \tfrac{3a^2}{4}$$
$$h = \sqrt{\tfrac{3a^2}{4}}$$
$$h = \tfrac{a\sqrt{3}}{2}$$

32. $A = \dfrac{a^2\sqrt{3}}{4}$

33. Familiarize. We first make a drawing. We let x represent the width; then x + 1 represents the length.

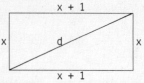

Using an area formula we can determine the value of x. Then we can use the Pythagorean property to determine the length of the diagonal labeled d.
Translate.
The area of the rectangle is 90 cm².
This translates to $(x + 1)x = 90$.
Carry out. We solve the equation.
$$x^2 + x = 90$$
$$x^2 + x - 90 = 0$$
$$(x + 10)(x - 9) = 0$$
$$x + 10 = 0 \quad \text{or} \quad x - 9 = 0$$
$$x = -10 \quad \text{or} \quad x = 9$$
Check. Since the width of the rectangle cannot be negative, we only check $x = 9$. If the width is 9 cm, then the length is 10 cm and the area is 9·10, or 90 cm².
We repeat the previous three steps to determine the length of the diagonal.
Translate. We use the Pythagorean property.

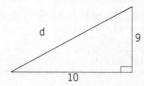

$$a^2 + b^2 = c^2$$
$$10^2 + 9^2 = d^2 \quad \text{Substituting 10 for a, 9 for b, and d for c}$$
Carry out.
$$100 + 81 = d^2$$
$$181 = d^2$$
$$\sqrt{181} = d$$
$$13.454 \approx d \quad \text{Using a calculator}$$

Check. We check by substituting 10, 9, and $\sqrt{181}$ into the Pythagorean equation:

$a^2 + b^2 = c^2$	
$10^2 + 9^2$	$(\sqrt{181})^2$
$100 + 81$	181
181	

State. The length of the diagonal of the rectangle is $\sqrt{181}$, or about 13.454 cm.

34. ≈ 39.1 mi

Chapter 10 (10.8)

35. <u>Familiarize</u>. After 3.5 hr the car traveling south has gone 3.5(55), or 192.5 mi, and the car traveling west has gone 3.5(65), or 227.5 mi. We make a drawing. We label the distance between the cars d.

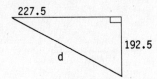

<u>Translate</u>. We use the Pythagorean property.
$$a^2 + b^2 = c^2$$
$(227.5)^2 + (192.5)^2 = d^2$ Substituting 227.5 for a, 192.5 for b, and d for c

<u>Carry out</u>. We solve the equation.
$$(227.5)^2 + (192.5)^2 = d^2$$
$$51{,}756.25 + 37{,}056.25 = d^2$$
$$88{,}812.5 = d^2$$
$$\sqrt{88{,}812.5} = d$$
$$298 \approx d$$

<u>Check</u>. We check by substituting 227.5, 192.5, and $\sqrt{88{,}812.5}$ into the Pythagorean equation:

$a^2 + b^2$	$= c^2$
$(227.5)^2 + (192.5)^2$	$(\sqrt{88{,}812.5})^2$
$51{,}756.25 + 37{,}056.25$	$88{,}812.5$
$88{,}812.5$	

<u>State</u>. The cars will be about 298 mi apart.

36. $\frac{2}{3}$

37. <u>Familiarize</u>. We make a drawing. The outfielder is 40 ft + 90 ft, or 130 ft, from home plate. We label the unknown distance d.

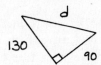

<u>Translate</u>. We use the Pythagorean property.
$$a^2 + b^2 = c^2$$
$130^2 + 90^2 = d^2$ Substituting 130 for a, 90 for b, and d for c

<u>Carry out</u>. We solve the equation.
$$130^2 + 90^2 = d^2$$
$$16{,}900 + 8100 = d^2$$
$$25{,}000 = d^2$$
$$\sqrt{25{,}000} = d$$
$$50\sqrt{10} = d$$
$$158 \approx d$$

37. (continued)

<u>Check</u>. We check by substituting 130, 90, and $\sqrt{25{,}000}$ into the Pythagorean equation.

$a^2 + b^2$	$= c^2$
$130^2 + 90^2$	$(\sqrt{25{,}000})^2$
$16{,}900 + 8100$	$25{,}000$
$25{,}000$	

<u>State</u>. The outfielder will have to throw the ball $50\sqrt{10}$ or about 158 ft.

38. 8 ft

39. <u>Familiarize</u>. We make a drawing. We label the length of the diagonal d.

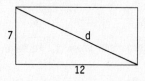

<u>Translate</u>. We use the Pythagorean property.
$$a^2 + b^2 = c^2$$
$7^2 + 12^2 = d^2$ Substituting 7 for a, 12 for b, and d for c

<u>Carry out</u>. We solve the equation.
$$7^2 + 12^2 = d^2$$
$$49 + 144 = d^2$$
$$193 = d^2$$
$$\sqrt{193} = d$$
$$13.89 \approx d$$

<u>Check</u>. We check by substituting 7, 12, and $\sqrt{193}$ into the Pythagorean equation:

$a^2 + b^2$	$= c^2$
$7^2 + 12^2$	$(\sqrt{193})^2$
$49 + 144$	193
193	

<u>State</u>. The length of the diagonal is $\sqrt{193}$ or about 13.89 in.

40. 3, 4, and 5

41.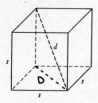

From the drawing we see that the diagonal of the cube is the hypotenuse of a right triangle with one leg of length s, where s is the length of a side of the cube, and the other leg of length D, where D is the length of the diagonal of the base of the cube. First we find D:

Using the Pythagorean property we have:
$s^2 + s^2 = D^2$
$2s^2 = D^2$
$\sqrt{2s^2} = D$
$s\sqrt{2} = D$

Then we find d:

Using the Pythagorean property again we have:
$s^2 + (s\sqrt{2})^2 = d^2$
$s^2 + 2s^2 = d^2$
$3s^2 = d^2$
$\sqrt{3s^2} = d$
$s\sqrt{3} = d$

42. $12 - 2\sqrt{6}$

43. Using the Pythagorean property we can label the figure with additional information.

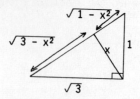

Next we use the Pythagorean property with the largest right triangle and solve for x.

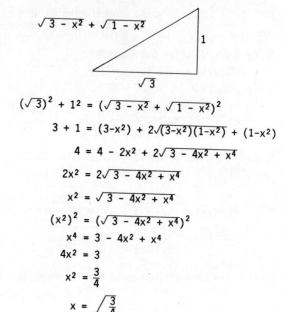

$(\sqrt{3})^2 + 1^2 = (\sqrt{3-x^2} + \sqrt{1-x^2})^2$
$3 + 1 = (3-x^2) + 2\sqrt{(3-x^2)(1-x^2)} + (1-x^2)$
$4 = 4 - 2x^2 + 2\sqrt{3 - 4x^2 + x^4}$
$2x^2 = 2\sqrt{3 - 4x^2 + x^4}$
$x^2 = \sqrt{3 - 4x^2 + x^4}$
$(x^2)^2 = (\sqrt{3 - 4x^2 + x^4})^2$
$x^4 = 3 - 4x^2 + x^4$
$4x^2 = 3$
$x^2 = \frac{3}{4}$
$x = \sqrt{\frac{3}{4}}$
$x = \frac{\sqrt{3}}{2}$

44. 6

45. The perimeter of the smaller square plot is 2 mi, so each side is $\frac{1}{4} \cdot 2$, or $\frac{1}{2}$ mi, and the area is $\left(\frac{1}{2} \text{ mi}\right)^2$, or $\frac{1}{4}$ mi². This tells us that $\frac{1}{4}$ mi² is equivalent to 160 acres. The perimeter of the larger square plot is 4 mi, so each side is $\frac{1}{4} \cdot 4$, or 1 mi, and the area is (1 mi)², or 1 mi². Since 1 mi² = $4 \cdot \frac{1}{4}$ mi², then 1 mi² is equivalent to 4·160, or 640 acres. Thus, 4 mi of fencing will enclose a square whose area is 640 acres.

46. 50 ft²

Chapter 10 (10.9)

Exercise Set 10.9

1. $\sqrt[3]{-8} = -2$ $(-2)^3 = (-2)(-2)(-2) = -8$
2. 4
3. $\sqrt[3]{1000} = 10$ $10^3 = 10 \cdot 10 \cdot 10 = 1000$
4. -3
5. $-\sqrt[3]{125} = -5$ $\sqrt[3]{125} = 5$, so $-\sqrt[3]{125} = -5$
6. -2
7. $\sqrt[3]{216} = 6$ $6^3 = 6 \cdot 6 \cdot 6 = 216$
8. -7
9. $\sqrt[4]{81} = 3$ $3^4 = 3 \cdot 3 \cdot 3 \cdot 3 = 81$
10. 5
11. $\sqrt[5]{0} = 0$ $0^5 = 0 \cdot 0 \cdot 0 \cdot 0 \cdot 0 = 0$
12. 1
13. $\sqrt[5]{-1} = -1$ $(-1)^5 = (-1)(-1)(-1)(-1)(-1) = -1$
14. -3
15. $\sqrt[4]{-81}$ does not represent a real number, because it is an even root of a negative number.
16. Not defined
17. $\sqrt[4]{10{,}000} = 10$ $10^4 = 10 \cdot 10 \cdot 10 \cdot 10 = 10{,}000$
18. 10
19. $\sqrt[3]{5^3} = 5$ $5^3 = 5 \cdot 5 \cdot 5$
20. 7
21. $\sqrt[6]{64} = 2$ $2^6 = 2 \cdot 2 \cdot 2 \cdot 2 \cdot 2 \cdot 2 = 64$
22. 1
23. $\sqrt[3]{x^3} = x$ $x^3 = x \cdot x \cdot x$
24. t
25. $\sqrt[3]{32} = \sqrt[3]{8 \cdot 4} = \sqrt[3]{8}\sqrt[3]{4} = 2\sqrt[3]{4}$
26. $3\sqrt[3]{2}$
27. $\sqrt[4]{48} = \sqrt[4]{16 \cdot 3} = \sqrt[4]{16}\sqrt[4]{3} = 2\sqrt[4]{3}$
28. $2\sqrt[5]{5}$
29. $\sqrt[3]{\frac{27}{64}} = \frac{\sqrt[3]{27}}{\sqrt[3]{64}} = \frac{3}{4}$
30. $\frac{5}{4}$
31. $\sqrt[4]{\frac{256}{625}} = \frac{\sqrt[4]{256}}{\sqrt[4]{625}} = \frac{4}{5}$
32. $\frac{3}{2}$
33. $\sqrt[3]{\frac{17}{8}} = \frac{\sqrt[3]{17}}{\sqrt[3]{8}} = \frac{\sqrt[3]{17}}{2}$
34. $\frac{\sqrt[5]{11}}{2}$
35. $\sqrt[4]{\frac{13}{81}} = \frac{\sqrt[4]{13}}{\sqrt[4]{81}} = \frac{\sqrt[4]{13}}{3}$
36. $\frac{\sqrt[3]{10}}{3}$
37. $25^{1/2} = \sqrt{25} = 5$
38. 3
39. $1000^{1/3} = \sqrt[3]{1000} = 10$
40. 5
41. $16^{1/4} = \sqrt[4]{16} = 2$
42. 2
43. $81^{3/4} = (81^{1/4})^3 = (\sqrt[4]{81})^3 = 3^3 = 27$
44. 16
45. $4^{5/2} = (4^{1/2})^5 = (\sqrt{4})^5 = 2^5 = 32$
46. 27
47. $64^{2/3} = (64^{1/3})^2 = (\sqrt[3]{64})^2 = 4^2 = 16$
48. 4
49. $8^{2/3} = (8^{1/3})^2 = (\sqrt[3]{8})^2 = 2^2 = 4$
50. 8
51. $25^{5/2} = (25^{1/2})^5 = (\sqrt{25})^5 = 5^5 = 3125$
52. $\frac{1}{2}$
53. $36^{-1/2} = \frac{1}{36^{1/2}} = \frac{1}{\sqrt{36}} = \frac{1}{6}$
54. $\frac{1}{2}$
55. $256^{-1/4} = \frac{1}{256^{1/4}} = \frac{1}{\sqrt[4]{256}} = \frac{1}{4}$

Chapter 10 (10.9)

56. $\dfrac{1}{1000}$

57. $81^{-3/4} = \dfrac{1}{81^{3/4}} = \dfrac{1}{(\sqrt[4]{81})^3} = \dfrac{1}{3^3} = \dfrac{1}{27}$

58. $\dfrac{1}{11}$

59. $144^{-1/2} = \dfrac{1}{144^{1/2}} = \dfrac{1}{\sqrt{144}} = \dfrac{1}{12}$

60. $\dfrac{1}{8}$

61. $81^{-5/4} = \dfrac{1}{81^{5/4}} = \dfrac{1}{(\sqrt[4]{81})^5} = \dfrac{1}{3^5} = \dfrac{1}{243}$

62. $\dfrac{1}{4}$

63. $8^{-2/3} = \dfrac{1}{8^{2/3}} = \dfrac{1}{(\sqrt[3]{8})^2} = \dfrac{1}{2^2} = \dfrac{1}{4}$

64. $\dfrac{1}{125}$

65. $122^{1.3} \approx 515.546$ Using a calculator with a $\boxed{y^x}$ key

66. 64,366,759.72

67. $2^{0.34} \approx 1.266$ Using a calculator with a $\boxed{y^x}$ key

68. 2.776

69. $(x^{2/3})^{5/3} = x^{\frac{2}{3} \cdot \frac{5}{3}}$ Multiplying exponents

            $= x^{\frac{10}{9}}$

70. $a^{\frac{3}{4}}$

71. $\dfrac{p^{4/5}}{p^{2/3}} = p^{\frac{4}{5} - \frac{2}{3}}$ Subtracting exponents

       $= p^{\frac{12}{15} - \frac{10}{15}} = p^{\frac{2}{15}}$

72. $m^{\frac{7}{12}}$

73. Graph $f(x) = \sqrt[3]{x}$.

We make a table of values.

x	f(x)
-8	-2
-1	-1
0	0
1	1
8	2

We plot these points and connect them with a smooth curve.

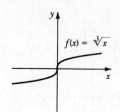

74.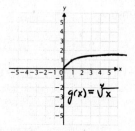

CHAPTER 11   QUADRATIC EQUATIONS

Exercise Set 11.1

1. $x^2 - 3x + 2 = 0$
   This equation is already in standard form.
   $a = 1, b = -3, c = 2$

2. $a = 1, b = -8, c = -5; x^2 - 8x - 5 = 0$

3. $\quad 2x^2 = 3$
   $2x^2 - 3 = 0 \quad$ Adding -3. This is standard form.
   Think: $2x^2 + 0x - 3 = 0$
   $a = 2, b = 0, c = -3$

4. $a = 5, b = 0, c = -9; 5x^2 - 9 = 0$

5. $\quad 7x^2 = 4x - 3$
   $7x^2 - 4x + 3 = 0 \quad$ Standard form
   $a = 7, b = -4, c = 3$

6. $a = 9, b = -1, c = -5; 9x^2 - x - 5 = 0$

7. $\quad 5 = -2x^2 + 3x$
   $2x^2 - 3x + 5 = 0 \quad$ Standard form
   $a = 2, b = -3, c = 5$

8. $a = 3, b = -2, c = 8; 3x^2 - 2x + 8 = 0$

9. $x^2 + 7x = 0$
   $x(x + 7) = 0$
   $x = 0 \quad$ or $\quad x + 7 = 0$
   $x = 0 \quad$ or $\quad x = -7$
   The solutions are 0 and -7.

10. $0, 5$

11. $3x^2 + 6x = 0$
    $3x(x + 2) = 0$
    $3x = 0 \quad$ or $\quad x + 2 = 0$
    $x = 0 \quad$ or $\quad x = -2$
    The solutions are 0 and -2.

12. $0, -2$

13. $\quad 5x^2 = 2x$
    $5x^2 - 2x = 0$
    $x(5x - 2) = 0$
    $x = 0 \quad$ or $\quad 5x - 2 = 0$
    $x = 0 \quad$ or $\quad 5x = 2$
    $x = 0 \quad$ or $\quad x = \frac{2}{5}$
    The solutions are 0 and $\frac{2}{5}$.

14. $0, \frac{7}{3}$

15. $4x^2 + 4x = 0$
    $4x(x + 1) = 0$
    $4x = 0 \quad$ or $\quad x + 1 = 0$
    $x = 0 \quad$ or $\quad x = -1$
    The solutions are 0 and -1.

16. $0, 1$

17. $0 = 10x^2 - 30x$
    $0 = 10x(x - 3)$
    $10x = 0 \quad$ or $\quad x - 3 = 0$
    $x = 0 \quad$ or $\quad x = 3$
    The solutions are 0 and 3.

18. $0, 5$

19. $11x = 55x^2$
    $0 = 55x^2 - 11x$
    $0 = 11x(5x - 1)$
    $11x = 0 \quad$ or $\quad 5x - 1 = 0$
    $x = 0 \quad$ or $\quad 5x = 1$
    $x = 0 \quad$ or $\quad x = \frac{1}{5}$
    The solutions are 0 and $\frac{1}{5}$.

20. $0, -\frac{1}{3}$

21. $\quad 14t^2 = 3t$
    $14t^2 - 3t = 0$
    $t(14t - 3) = 0$
    $t = 0 \quad$ or $\quad 14t - 3 = 0$
    $t = 0 \quad$ or $\quad 14t = 3$
    $t = 0 \quad$ or $\quad t = \frac{3}{14}$
    The solutions are 0 and $\frac{3}{14}$.

22. $0, \frac{8}{17}$

23. $5y^2 - 3y^2 = 72y + 9y$
    $2y^2 = 81y$
    $2y^2 - 81y = 0$
    $y(2y - 81) = 0$
    $y = 0 \quad$ or $\quad 2y - 81 = 0$
    $y = 0 \quad$ or $\quad 2y = 81$
    $y = 0 \quad$ or $\quad y = \frac{81}{2}$
    The solutions are 0 and $\frac{81}{2}$.

24. $0, \frac{23}{37}$

Chapter 11   (11.1)

25. $x^2 + 8x - 48 = 0$
     $(x + 12)(x - 4) = 0$
     $x + 12 = 0$ or $x - 4 = 0$
          $x = -12$ or $x = 4$
     The solutions are -12 and 4.

26. $12, 4$

27. $5 + 6x + x^2 = 0$
     $(5 + x)(1 + x) = 0$
     $5 + x = 0$ or $1 + x = 0$
        $x = -5$ or $x = -1$
     The solutions are -5 and -1.

28. $-6, -1$

29. $18 = 7p + p^2$
     $0 = p^2 + 7p - 18$
     $0 = (p + 9)(p - 2)$
     $p + 9 = 0$ or $p - 2 = 0$
        $p = -9$ or $p = 2$
     The solutions are -9 and 2.

30. $-7, 3$

31. $-15 = -8y + y^2$
     $0 = y^2 - 8y + 15$
     $0 = (y - 5)(y - 3)$
     $y - 5 = 0$ or $y - 3 = 0$
        $y = 5$ or $y = 3$
     The solutions are 5 and 3.

32. $7, 2$

33. $x^2 + 6x + 9 = 0$
     $(x + 3)(x + 3) = 0$
     $x + 3 = 0$ or $x + 3 = 0$
        $x = -3$ or $x = -3$
     The solution is -3.

34. -5

35. $r^2 = 8r - 16$
     $r^2 - 8r + 16 = 0$
     $(r - 4)(r - 4) = 0$
     $r - 4 = 0$ or $r - 4 = 0$
        $r = 4$ or $r = 4$
     The solution is 4.

36. 1

37. $6x^2 + x - 2 = 0$
     $(3x + 2)(2x - 1) = 0$
     $3x + 2 = 0$ or $2x - 1 = 0$
     $3x = -2$ or $2x = 1$
     $x = -\frac{2}{3}$ or $x = \frac{1}{2}$
     The solutions are $-\frac{2}{3}$ and $\frac{1}{2}$.

38. $\frac{3}{2}, 5$

39. $15b - 9b^2 = 4$
     $0 = 9b^2 - 15b + 4$
     $0 = (3b - 1)(3b - 4)$
     $3b - 1 = 0$ or $3b - 4 = 0$
     $3b = 1$ or $3b = 4$
     $b = \frac{1}{3}$ or $b = \frac{4}{3}$
     The solutions are $\frac{1}{3}$ and $\frac{4}{3}$.

40. $-\frac{2}{3}, 4$

41. $6x^2 - 4x = 10$
     $6x^2 - 4x - 10 = 0$
     $2(3x^2 - 2x - 5) = 0$
     $2(3x - 5)(x + 1) = 0$
     $3x - 5 = 0$ or $x + 1 = 0$
     $3x = 5$ or $x = -1$
     $x = \frac{5}{3}$ or $x = -1$
     The solutions are $\frac{5}{3}$ and -1.

42. $-\frac{5}{3}, 4$

43. $12w^2 - 5w = 2$
     $12w^2 - 5w - 2 = 0$
     $(4w + 1)(3w - 2) = 0$
     $4w + 1 = 0$ or $3w - 2 = 0$
     $4w = -1$ or $3w = 2$
     $w = -\frac{1}{4}$ or $w = \frac{2}{3}$
     The solutions are $-\frac{1}{4}$ and $\frac{2}{3}$.

44. $-5, -1$

Chapter 11   (11.1)

45. $t(t - 5) = 14$
         $t^2 - 5t = 14$
         $t^2 - 5t - 14 = 0$
         $(t + 2)(t - 7) = 0$
     $t + 2 = 0$ or $t - 7 = 0$
         $t = -2$ or $t = 7$
The solutions are -2 and 7.

46. $\frac{1}{3}, -\frac{1}{2}$

47. $t(9 + t) = 4(2t + 5)$
         $9t + t^2 = 8t + 20$
         $t^2 + t - 20 = 0$
         $(t + 5)(t - 4) = 0$
     $t + 5 = 0$ or $t - 4 = 0$
         $t = -5$ or $t = 4$
The solutions are -5 and 4.

48. $-\frac{5}{3}, 3$

49. $16(p - 1) = p(p + 8)$
       $16p - 16 = p^2 + 8p$
           $0 = p^2 - 8p + 16$
           $0 = (p - 4)(p - 4)$
     $p - 4 = 0$ or $p - 4 = 0$
         $p = 4$ or $p = 4$
The solution is 4.

50. $\frac{3}{2}, 3$

51. $(x - 2)(x + 2) = x + 2$
         $x^2 - 4 = x + 2$
         $x^2 - x - 6 = 0$
         $(x - 3)(x + 2) = 0$
     $x - 3 = 0$ or $x + 2 = 0$
         $x = 3$ or $x = -2$
The solutions are 3 and -2.

52. $-2, 1$

53. $\frac{8}{x + 2} + \frac{8}{x - 2} = 3$
The LCM is $(x + 2)(x - 2)$.
$(x+2)(x-2)\left[\frac{8}{x + 2} + \frac{8}{x - 2}\right] = (x+2)(x-2)(3)$
$(x+2)(x-2) \cdot \frac{8}{x+2} + (x+2)(x-2) \cdot \frac{8}{x-2} = 3(x+2)(x-2)$
     $8(x - 2) + 8(x + 2) = 3(x^2 - 4)$
     $8x - 16 + 8x + 16 = 3x^2 - 12$
         $16x = 3x^2 - 12$
         $0 = 3x^2 - 16x - 12$
         $0 = (3x+2)(x-6)$
$3x + 2 = 0$ or $x - 6 = 0$
   $3x = -2$ or $x = 6$
    $x = -\frac{2}{3}$ or $x = 6$
Both numbers check. The solutions are $-\frac{2}{3}$ and 6.

54. $-\frac{2}{5}, 10$

55. $\frac{1}{x} + \frac{1}{x + 6} = \frac{1}{4}$
The LCM is $4x(x + 6)$.
$4x(x + 6)\left[\frac{1}{x} + \frac{1}{x + 6}\right] = 4x(x + 6) \cdot \frac{1}{4}$
$4x(x + 6) \cdot \frac{1}{x} + 4x(x + 6) \cdot \frac{1}{x + 6} = x(x + 6)$
     $4(x + 6) + 4x = x(x + 6)$
     $4x + 24 + 4x = x^2 + 6x$
        $8x + 24 = x^2 + 6x$
         $0 = x^2 - 2x - 24$
         $0 = (x - 6)(x + 4)$
$x - 6 = 0$ or $x + 4 = 0$
    $x = 6$ or $x = -4$
Both numbers check. The solutions are 6 and -4.

56. $36, -5$

57. $1 + \frac{12}{x^2 - 4} = \frac{3}{x - 2}$
The LCM is $(x + 2)(x - 2)$.
$(x+2)(x-2)\left[1 + \frac{12}{(x+2)(x-2)}\right] =$
                           $(x+2)(x-2) \cdot \frac{3}{x-2}$
$(x+2)(x-2) \cdot 1 + (x+2)(x-2) \cdot \frac{12}{(x+2)(x-2)} = 3(x+2)$
         $x^2 - 4 + 12 = 3x + 6$
            $x^2 + 8 = 3x + 6$
         $x^2 - 3x + 2 = 0$
         $(x - 2)(x - 1) = 0$
$x - 2 = 0$ or $x - 1 = 0$
    $x = 2$ or $x = 1$
The number 1 checks, but 2 does not. (It makes the denominators $x^2 - 4$ and $x - 2$ zero.) The solution is 1.

Chapter 11   (11.1)

58. 2

59.
$$\frac{r}{r-1} + \frac{2}{r^2-1} = \frac{8}{r+1}$$
The LCM is $(r-1)(r+1)$.
$$(r-1)(r+1)\left[\frac{r}{r-1} + \frac{2}{r^2-1}\right] = (r-1)(r+1)\cdot\frac{8}{r+1}$$
$$(r-1)(r+1)\cdot\frac{r}{r-1} + (r-1)(r+1)\cdot\frac{2}{(r-1)(r+1)} = 8(r-1)$$
$$r(r+1) + 2 = 8(r-1)$$
$$r^2 + r + 2 = 8r - 8$$
$$r^2 - 7r + 10 = 0$$
$$(r-5)(r-2) = 0$$
$r - 5 = 0$ or $r - 2 = 0$
$r = 5$ or $r = 2$

Both numbers check. The solutions are 5 and 2.

60. $-\frac{5}{3}$, 2

61.
$$\frac{4-x}{x-4} + \frac{x+3}{x-3} = 0$$
The LCM is $(x-4)(x-3)$.
$$(x-4)(x-3)\left[\frac{4-x}{x-4} + \frac{x+3}{x-3}\right] = (x-4)(x-3)\cdot 0$$
$$(x-4)(x-3)\cdot\frac{4-x}{x-4} + (x-4)(x-3)\cdot\frac{x+3}{x-3} = 0$$
$$(x-3)(4-x) + (x-4)(x+3) = 0$$
$$4x - x^2 - 12 + 3x + x^2 - x - 12 = 0$$
$$6x - 24 = 0$$
$$6x = 24$$
$$x = 4$$

The number 4 does not check. (It makes the denominator $x - 4$ zero.) There is no solution.

62. $-4, -1$

63. Familiarize. We will use the formula
$$d = \frac{n^2 - 3n}{2},$$
where d is the number of diagonals and n is the number of sides.

Translate. We substitute 6 for n.
$$d = \frac{6^2 - 3\cdot 6}{2}$$

Carry out. We do the computation.
$$d = \frac{6^2 - 3\cdot 6}{2} = \frac{36 - 18}{2} = \frac{18}{2} = 9$$

Check. We can substitute 9 for d in the original formula and see if this yields n = 6. This is left to the student.

State. A hexagon has 9 diagonals.

64. 35

65. Familiarize. We will use the formula
$$d = \frac{n^2 - 3n}{2},$$
where d is the number of diagonals and n is the number of sides.

Translate. We substitute 14 for d.
$$14 = \frac{n^2 - 3n}{2}$$

Carry out. We solve the equation.
$$\frac{n^2 - 3n}{2} = 14$$
$$n^2 - 3n = 28 \quad \text{Multiplying by 2}$$
$$n^2 - 3n - 28 = 0$$
$$(n-7)(n+4) = 0$$
$n - 7 = 0$ or $n + 4 = 0$
$n = 7$ or $n = -4$

Check. Since the number of sides cannot be negative, -4 cannot be a solution. To check 7, we substitute 7 for n in the original formula and determine if this yields d = 14. This is left to the student.

State. The polygon has 7 sides.

66. 6

67. $\sqrt{20} = \sqrt{4\cdot 5} = \sqrt{4}\sqrt{5} = 2\sqrt{5}$

68. $\frac{17}{16}$

69. $\sqrt{\frac{3240}{2560}} = \sqrt{\frac{81\cdot 40}{64\cdot 40}} = \sqrt{\frac{81}{64}\cdot\frac{40}{40}} = \sqrt{\frac{81}{64}\cdot 1} = \frac{\sqrt{81}}{\sqrt{64}} = \frac{9}{8}$

70. $2\sqrt{22}$

71.
$$4m^2 - (m+1)^2 = 0$$
$$4m^2 - (m^2 + 2m + 1) = 0$$
$$4m^2 - m^2 - 2m - 1 = 0$$
$$3m^2 - 2m - 1 = 0$$
$$(3m+1)(m-1) = 0$$
$3m + 1 = 0$ or $m - 1 = 0$
$3m = -1$ or $m = 1$
$m = -\frac{1}{3}$ or $m = 1$

The solutions are $-\frac{1}{3}$ and 1.

72. $0, -\sqrt{3}$

Chapter 11   (11.2)

73. $\sqrt{5}x^2 - x = 0$

   $x(\sqrt{5}x - 1) = 0$

   $x = 0$ or $\sqrt{5}x - 1 = 0$

   $x = 0$ or $\sqrt{5}x = 1$

   $x = 0$ or $x = \dfrac{1}{\sqrt{5}}$, or $\dfrac{\sqrt{5}}{5}$

   The solutions are 0 and $\dfrac{\sqrt{5}}{5}$.

74. $0, -\dfrac{\sqrt{3}}{\sqrt{7}}$, or $-\dfrac{\sqrt{21}}{7}$

75. $\dfrac{5}{y+4} - \dfrac{3}{y-2} = 4$,   LCM is $(y+4)(y-2)$

   $(y+4)(y-2)\left[\dfrac{5}{y+4} - \dfrac{3}{y-2}\right] = (y+4)(y-2)\cdot 4$

   $5(y-2) - 3(y+4) = 4(y^2 + 2y - 8)$

   $5y - 10 - 3y - 12 = 4y^2 + 8y - 32$

   $2y - 22 = 4y^2 + 8y - 32$

   $0 = 4y^2 + 6y - 10$

   $0 = 2(2y^2 + 3y - 5)$

   $0 = 2(2y + 5)(y - 1)$

   $2y + 5 = 0$ or $y - 1 = 0$

   $2y = -5$ or $y = 1$

   $y = -\dfrac{5}{2}$ or $y = 1$

   The solutions are $-\dfrac{5}{2}$ and 1.

76. $-\dfrac{1}{4}, -3$

77. $ax^2 + bx = 0$

   $x(ax + b) = 0$

   $x = 0$ or $ax + b = 0$

   $x = 0$ or $ax = -b$

   $x = 0$ or $x = -\dfrac{b}{a}$

   The solutions are 0 and $-\dfrac{b}{a}$.

78. $0, -28,160,000$

79. $y^4 - 4y^2 + 4 = 0$

   Let $x = y^2$. Then $x^2 = y^4$. We substitute x for y^2 and x^2 for y^4.

   $x^2 - 4x + 4 = 0$

   $(x - 2)(x - 2) = 0$

   $x - 2 = 0$ or $x - 2 = 0$

   $x = 2$ or $x = 2$

   If $x = 2$ and $x = y^2$, then $y^2 = 2$. We solve $y^2 = 2$ for y.

   $y^2 = 2$

   $y = \sqrt{2}$ or $y = -\sqrt{2}$

   The solutions are $\sqrt{2}$ and $-\sqrt{2}$.

80. 81, 1

81. $(x - 2)^2 + 3(x - 2) = 4$

   We can substitute y for x - 2 and y^2 for $(x - 2)^2$, obtaining a quadratic equation in y, or we can simplify the given equation and solve directly. We will choose the latter method.

   $(x - 2)^2 + 3(x - 2) = 4$

   $x^2 - 4x + 4 + 3x - 6 = 4$

   $x^2 - x - 2 = 4$

   $x^2 - x - 6 = 0$

   $(x - 3)(x + 2) = 0$

   $x - 3 = 0$ or $x + 2 = 0$

   $x = 3$ or $x = -2$

   The solutions are 3 and -2.

82. $2\sqrt{2}, -2\sqrt{2}, 1, -1$

Exercise Set 11.2

1. $x^2 = 121$

   $x = 11$ or $x = -11$ Principle of square roots

   The solutions are 11 and -11.

2. $\sqrt{10}, -\sqrt{10}$

3. $5x^2 = 35$

   $x^2 = 7$ Multiplying by $\dfrac{1}{5}$

   $x = \sqrt{7}$ or $x = -\sqrt{7}$ Principle of square roots

   The solutions are $\sqrt{7}$ and $-\sqrt{7}$.

4. $\sqrt{10}, -\sqrt{10}$

5. $5x^2 = 3$

   $x^2 = \dfrac{3}{5}$

   $x = \sqrt{\dfrac{3}{5}}$ or $x = -\sqrt{\dfrac{3}{5}}$ Principle of square roots

   $x = \sqrt{\dfrac{3}{5} \cdot \dfrac{5}{5}}$ or $x = -\sqrt{\dfrac{3}{5} \cdot \dfrac{5}{5}}$ Rationalizing denominators

   $x = \dfrac{\sqrt{15}}{5}$ or $x = -\dfrac{\sqrt{15}}{5}$

   The solutions are $\dfrac{\sqrt{15}}{5}$ and $-\dfrac{\sqrt{15}}{5}$.

6. $\dfrac{\sqrt{10}}{2}, -\dfrac{\sqrt{10}}{2}$

Chapter 11   (11.2)

7. $4x^2 - 25 = 0$
$$4x^2 = 25$$
$$x^2 = \frac{25}{4}$$
$$x = \frac{5}{2} \text{ or } x = -\frac{5}{2}$$
The solutions are $\frac{5}{2}$ and $-\frac{5}{2}$.

8. $\frac{2}{3}, -\frac{2}{3}$

9. $3x^2 - 49 = 0$
$$3x^2 = 49$$
$$x^2 = \frac{49}{3}$$
$$x = \frac{7}{\sqrt{3}} \text{ or } x = -\frac{7}{\sqrt{3}}$$
$$x = \frac{7}{\sqrt{3}} \cdot \frac{\sqrt{3}}{\sqrt{3}} \text{ or } x = -\frac{7}{\sqrt{3}} \cdot \frac{\sqrt{3}}{\sqrt{3}}$$
$$x = \frac{7\sqrt{3}}{3} \text{ or } x = -\frac{7\sqrt{3}}{3}$$
The solutions are $\frac{7\sqrt{3}}{3}$ and $-\frac{7\sqrt{3}}{3}$.

10. $\frac{4\sqrt{5}}{5}, -\frac{4\sqrt{5}}{5}$

11. $4y^2 - 3 = 9$
$$4y^2 = 12$$
$$y^2 = 3$$
$$y = \sqrt{3} \text{ or } y = -\sqrt{3}$$
The solutions are $\sqrt{3}$ and $-\sqrt{3}$.

12. $\frac{4}{7}, -\frac{4}{7}$

13. $25y^2 - 36 = 0$
$$25y^2 = 36$$
$$y^2 = \frac{36}{25}$$
$$y = \frac{6}{5} \text{ or } y = -\frac{6}{5}$$
The solutions are $\frac{6}{5}$ and $-\frac{6}{5}$.

14. $2\sqrt{5}, -2\sqrt{5}$

15. $(x - 2)^2 = 49$
$$x - 2 = 7 \text{ or } x - 2 = -7 \quad \text{Principle of square roots}$$
$$x = 9 \text{ or } x = -5$$
The solutions are 9 and -5.

16. $-1 \pm \sqrt{6}$

17. $(x + 3)^2 = 21$
$$x + 3 = \sqrt{21} \text{ or } x + 3 = -\sqrt{21}$$
$$x = -3 + \sqrt{21} \text{ or } x = -3 - \sqrt{21}$$
The solutions are $-3 + \sqrt{21}$ and $-3 - \sqrt{21}$, or $-3 \pm \sqrt{21}$.

18. $3 \pm \sqrt{6}$

19. $(x + 13)^2 = 8$
$$x + 13 = \sqrt{8} \text{ or } x + 3 = -\sqrt{8}$$
$$x + 13 = 2\sqrt{2} \text{ or } x + 13 = -2\sqrt{2}$$
$$x = -13 + 2\sqrt{2} \text{ or } x = -13 - 2\sqrt{2}$$
The solutions are $-13 + 2\sqrt{2}$ and $-13 - 2\sqrt{2}$, or $-13 \pm 2\sqrt{2}$.

20.   21, 5

21. $(x - 7)^2 = 12$
$$x - 7 = \sqrt{12} \text{ or } x - 7 = -\sqrt{12}$$
$$x - 7 = 2\sqrt{3} \text{ or } x - 7 = -2\sqrt{3}$$
$$x = 7 + 2\sqrt{3} \text{ or } x = 7 - 2\sqrt{3}$$
The solutions are $7 + 2\sqrt{3}$ and $7 - 2\sqrt{3}$, or $7 \pm 2\sqrt{3}$.

22. $-1 \pm \sqrt{14}$

23. $(x + 9)^2 = 34$
$$x + 9 = \sqrt{34} \text{ or } x + 9 = -\sqrt{34}$$
$$x = -9 + \sqrt{34} \text{ or } x = -9 - \sqrt{34}$$
The solutions are $-9 + \sqrt{34}$ and $-9 - \sqrt{34}$, or $-9 \pm \sqrt{34}$.

24.   3, -7

25. $\left(x + \frac{3}{2}\right)^2 = \frac{7}{2}$
$$x + \frac{3}{2} = \sqrt{\frac{7}{2}} \text{ or } x + \frac{3}{2} = -\sqrt{\frac{7}{2}}$$
$$x = -\frac{3}{2} + \sqrt{\frac{7}{2}} \text{ or } x = -\frac{3}{2} - \sqrt{\frac{7}{2}}$$
$$x = -\frac{3}{2} + \sqrt{\frac{7}{2} \cdot \frac{2}{2}} \text{ or } x = -\frac{3}{2} - \sqrt{\frac{7}{2} \cdot \frac{2}{2}}$$
$$x = -\frac{3}{2} + \frac{\sqrt{14}}{2} \text{ or } x = -\frac{3}{2} - \frac{\sqrt{14}}{2}$$
$$x = \frac{-3 + \sqrt{14}}{2} \text{ or } x = \frac{-3 - \sqrt{14}}{2}$$
The solutions are $\frac{-3 \pm \sqrt{14}}{2}$.

Chapter 11   (11.2)

26. $\dfrac{3 \pm \sqrt{17}}{4}$

27. $x^2 - 6x + 9 = 64$
$(x - 3)^2 = 64$ Factoring the left side
$x - 3 = 8$ or $x - 3 = -8$ Principle of square roots
$x = 11$ or $x = -5$
The solutions are 11 and -5.

28. 15, -5

29. $y^2 + 14y + 49 = 4$
$(y + 7)^2 = 4$ Factoring the left side
$y + 7 = 2$ or $y + 7 = -2$ Principle of square roots
$y = -5$ or $y = -9$
The solutions are -5 and -9.

30. -3, -5

31. Familiarize. We will use the formula $s = 16t^2$.
Translate. We substitute 1377 for s.
$1377 = 16t^2$
Carry out. We solve the equation.
$1377 = 16t^2$
$\dfrac{1377}{16} = t^2$
$86.0625 = t^2$ Dividing
$\sqrt{86.0625} = t$ or $-\sqrt{86.0625} = t$ Principle of square roots
$9.3 \approx t$ or $-9.3 \approx t$ Using a calculator and rounding to the nearest tenth

Check. The number -9.3 cannot be a solution, because time cannot be negative in this situation. We substitute 9.3 in the original equation.
$s = 16(9.3)^2 = 16(86.49) = 1383.84$
This is close. Remember that we approximated a solution. Thus we have a check.
State. It takes about 9.3 sec for an object to fall to the ground from the top of the World Trade Center.

32. 12.5 sec

33. Familiarize. We will use the formula $s = 16t^2$.
Translate. We substitute 175 for s.
$175 = 16t^2$
Carry out. We solve the equation.
$175 = 16t^2$
$\dfrac{175}{16} = t^2$
$10.9375 = t^2$ Dividing
$\sqrt{10.9375} = t$ or $-\sqrt{10.9375} = t$
$3.3 \approx t$ or $-3.3 \approx t$

Check. The number -3.3 cannot be a solution because time cannot be negative in this situation. We substitute 3.3 in the original equation.
$s = 16(3.3)^2 = 16(10.89) = 174.24$
This is close. Remember that we approximated a solution. Thus we have a check.
State. The fall took about 3.3 sec.

34. 4.4 sec

35. Familiarize. We will use the formula $A = P(1 + r)^t$ where P is $1000, A is $1210 and t is 2.
Translate. We make the substitutions.
$A = P(1 + r)^t$
$1210 = 1000(1 + r)^2$
Carry out. We solve the equation.
$1210 = 1000(1 + r)^2$
$\dfrac{1210}{1000} = (1 + r)^2$
$\dfrac{121}{100} = (1 + r)^2$
$\dfrac{11}{10} = 1 + r$ or $-\dfrac{11}{10} = 1 + r$ Principle of square roots
$\dfrac{11}{10} - \dfrac{10}{10} = r$ or $-\dfrac{11}{10} - \dfrac{10}{10} = r$
$\dfrac{1}{10} = r$ or $-\dfrac{21}{10} = r$

Check. Since the interest rate cannot be negative,
$\dfrac{1}{10} = r$
$0.1 = r$
$10\% = r.$
We check 10%, or 0.1, in the formula.
$1000(1 + 0.1)^2 = 1000(1.1)^2 = 1000(1.21) = 1210$
Our answer checks.
State. The interest rate is 10%.

36. 20%

283

37. **Familiarize.** We will use the formula
$A = P(1 + r)^t$ where P is $2560, A is $3610, and t is 2.

Translate. We make the substitutions.
$$A = P(1 + r)^t$$
$$3610 = 2560(1 + r)^2$$

Carry out. We solve the equation.
$$3610 = 2560(1 + r)^2$$
$$\frac{3610}{2560} = (1 + r)^2$$
$$\frac{361}{256} = (1 + r)^2$$
$$\frac{19}{16} = 1 + r \quad \text{or} \quad -\frac{19}{16} = 1 + r \quad \text{Principle of square roots}$$
$$\frac{19}{16} - \frac{16}{16} = r \quad \text{or} \quad -\frac{19}{16} - \frac{16}{16} = r$$
$$\frac{3}{16} = r \quad \text{or} \quad -\frac{35}{16} = r$$

Check. Since the interest rate cannot be negative,
$$\frac{3}{16} = r$$
$$0.1875 = r$$
$$18.75\% = r.$$
We check 18.75%, or 0.1875, in the formula.
$$2560(1 + 0.1875)^2 = 2560(1.1875)^2 = 3610$$
Our answer checks.

State. The interest rate is 18.75%.

38. 5%

39. **Familiarize.** We will use the formula
$A = P(1 + r)^t$ where P is $6250, A is $7290, and t is 2.

Translate. We make the substitutions.
$$A = P(1 + r)^t$$
$$7290 = 6250(1 + r)^2$$

Carry out. We solve the equation.
$$7290 = 6250(1 + r)^2$$
$$\frac{7290}{6250} = (1 + r)^2$$
$$\frac{729}{625} = (1 + r)^2$$
$$\frac{27}{25} = 1 + r \quad \text{or} \quad -\frac{27}{25} = 1 + r$$
$$\frac{27}{25} - \frac{25}{25} = r \quad \text{or} \quad -\frac{27}{25} - \frac{25}{25} = r$$
$$\frac{2}{25} = r \quad \text{or} \quad -\frac{52}{25} = r$$

39. (continued)

Check. Since the interest rate cannot be negative,
$$\frac{2}{25} = r$$
$$0.08 = r$$
$$8\% = r.$$
We check 8%, or 0.08, in the formula.
$$6250(1 + 0.08)^2 = 6250(1.08)^2 = 6250(1.1664) = 7290$$
Our answer checks.

State. The interest rate is 8%.

40. 4%

41. **Familiarize.** We will use the formula
$A = P(1 + r)^t$ where P is $2500, A is $3600, and t is 2.

Translate. We make the substitutions.
$$A = P(1 + r)^t$$
$$3600 = 2500(1 + r)^2$$

Carry out. We solve the equation.
$$3600 = 2500(1 + r)^2$$
$$\frac{3600}{2500} = (1 + r)^2$$
$$\frac{36}{25} = (1 + r)^2$$
$$\frac{6}{5} = 1 + r \quad \text{or} \quad -\frac{6}{5} = 1 + r$$
$$\frac{6}{5} - \frac{5}{5} = r \quad \text{or} \quad -\frac{6}{5} - \frac{5}{5} = r$$
$$\frac{1}{5} = r \quad \text{or} \quad -\frac{11}{5} = r$$

Check. Since the interest rate cannot be negative,
$$\frac{1}{5} = r$$
$$0.2 = r$$
$$20\% = r.$$
We check 20%, or 0.2, in the formula.
$$2500(1 + 0.2)^2 = 2500(1.2)^2 = 3600$$

State. The interest rate is 20%.

42. 25%

43. $4.82x^2 = 12{,}000$
$$x^2 = \frac{12{,}000}{4.82}$$
$$x = \sqrt{\frac{12{,}000}{4.82}} \quad \text{or} \quad x = -\sqrt{\frac{12{,}000}{4.82}} \quad \text{Principle of square roots}$$
$$x \approx 49.9 \quad \text{or} \quad x \approx -49.9 \quad \text{Using a calculator and rounding to the nearest tenth}$$

The solutions are approximately 49.9 and -49.9.

44. 6, -6

Chapter 11   (11.2)

45. $1 = \frac{1}{3}x^2$
    $3 = x^2$ Multiplying by 3
    $\sqrt{3} = x$ or $-\sqrt{3} = x$ Principle of square roots
    The solutions are $\sqrt{3}$ and $-\sqrt{3}$.

46. 9, -9

47. $\frac{4}{m^2 - 7} = 1$
    $4 = m^2 - 7$ Multiplying by $m^2 - 7$
    $11 = m^2$
    $\sqrt{11} = m$ or $-\sqrt{11} = m$
    The solutions are $\sqrt{11}$ and $-\sqrt{11}$.

48. $\frac{\sqrt{10}}{5}, -\frac{\sqrt{10}}{5}$

49. $\frac{4}{x^2 - 3} = \frac{6}{x^2}$, LCM is $x^2(x^2 - 3)$
    $x^2(x^2 - 3) \cdot \frac{4}{x^2 - 3} = x^2(x^2 - 3) \cdot \frac{6}{x^2}$
    $4x^2 = 6(x^2 - 3)$
    $4x^2 = 6x^2 - 18$
    $18 = 2x^2$
    $9 = x^2$
    $3 = x$ or $-3 = x$
    The solutions are 3 and -3.

50. $\frac{\sqrt{a(b^2 + 3b)}}{a}, -\frac{\sqrt{a(b^2 + 3b)}}{a}$

51. $x^2 + 9a^2 = 9 + ax^2$
    $x^2 - ax^2 = 9 - 9a^2$ Adding $-ax^2$ and $-9a^2$
    $x^2(1 - a) = 9(1 - a^2)$
    $x^2 = \frac{9(1 + a)(1 - a)}{1 - a}$
    $x^2 = 9(1 + a)$
    $x = 3\sqrt{1 + a}$ or $x = -3\sqrt{1 + a}$
    The solutions are $3\sqrt{1 + a}$ and $-3\sqrt{1 + a}$.

52. 5, -3

53. $x^2 + 2x + 1 = 81$
    $(x + 1)^2 = 81$
    $x + 1 = 9$ or $x + 1 = -9$
    $x = 8$ or $x = -10$
    The solutions are 8 and -10.

54. $-8 \pm \sqrt{15}$

55. $x^2 + 4x + 4 = 29$
    $(x + 2)^2 = 29$
    $x + 2 = \sqrt{29}$ or $x + 2 = -\sqrt{29}$
    $x = -2 + \sqrt{29}$ or $x = -2 - \sqrt{29}$
    The solutions are $-2 \pm \sqrt{29}$.

56. $\frac{3 \pm \sqrt{17}}{4}$

57. $t^2 + 3t + \frac{9}{4} = \frac{49}{4}$
    $\left(t + \frac{3}{2}\right)^2 = \frac{49}{4}$
    $t + \frac{3}{2} = \frac{7}{2}$ or $t + \frac{3}{2} = -\frac{7}{2}$
    $t = \frac{4}{2}$ or $t = -\frac{10}{2}$
    $t = 2$ or $t = -5$
    The solutions are 2 and -5.

58. $\frac{22}{9}, -\frac{26}{9}$

59. $9x^2 - 24x + 16 = 2$
    $(3x - 4)^2 = 2$
    $3x - 4 = \sqrt{2}$ or $3x - 4 = -\sqrt{2}$
    $3x = 4 + \sqrt{2}$ or $3x = 4 - \sqrt{2}$
    $x = \frac{4 + \sqrt{2}}{3}$ or $x = \frac{4 - \sqrt{2}}{3}$
    The solutions are $\frac{4 \pm \sqrt{2}}{3}$.

60. $\frac{7}{8}, -\frac{13}{8}$

61. Familiarize. We will use the formula $A = P(1 + r)^t$ where P is $2000, A is 2($2000), or $4000, and t is 2.
    Translate. We make the substitutions.
    $A = P(1 + r)^t$
    $4000 = 2000(1 + r)^2$
    Carry out. We solve the equation.
    $4000 = 2000(1 + r)^2$
    $2 = (1 + r)^2$
    $\sqrt{2} = 1 + r$ or $-\sqrt{2} = 1 + r$
    $\sqrt{2} - 1 = r$ or $-\sqrt{2} - 1 = r$
    $1.4142 - 1 \approx r$ or $-1.4142 - 1 \approx r$
    $0.4142 \approx r$ or $-2.4142 \approx r$

    Check. Since the interest rate cannot be negative we check 0.4142 in the formula.
    $2000(1 + 0.4142)^2 = 2000(1.4142)^2 \approx 4000$
    Our answer checks.
    State. The interest rate would have to be about 0.4142, or 41.42%.

Chapter 11 (11.3)

62. 12.6%

63. <u>Familiarize</u>. We will use the formula
$A = P(1 + r)^t$ where P is $4000, A is $5267.03, and t is 2.

<u>Translate</u>. We make the substitutions.
$$A = P(1 + r)^t$$
$$5267.03 = 4000(1 + r)^2$$

<u>Carry out</u>. We solve the equation.
$$5267.03 = 4000(1 + r)^2$$
$$\frac{5267.03}{4000} = (1 + r)^2$$
$$\sqrt{\frac{5267.03}{4000}} = 1 + r \quad \text{or} \quad -\sqrt{\frac{5267.03}{4000}} = 1 + r$$
$$1.1475 \approx 1 + r \quad \text{or} \quad -1.1475 \approx 1 + r$$
Using a calculator
$$0.1475 \approx r \quad \text{or} \quad -2.1475 \approx r$$

<u>Check</u>. Since the interest rate cannot be negative, we check 0.1475 in the formula.
$$4000(1 + 0.1475)^2 = 4000(1.1475)^2 = 5267.03$$
Our answer checks.

<u>State</u>. The interest rate is about 0.1475, or 14.75%.

64. $2239.13

65.
$$\frac{x - 1}{9} = \frac{1}{x - 1}, \quad \text{LCM is } 9(x - 1)$$
$$9(x - 1) \cdot \frac{x - 1}{9} = 9(x - 1) \cdot \frac{1}{x - 1}$$
$$(x - 1)(x - 1) = 9$$
$$x^2 - 2x + 1 = 9$$
$$x^2 - 2x - 8 = 0$$
$$(x - 4)(x + 2) = 0$$
$$x - 4 = 0 \quad \text{or} \quad x + 2 = 0$$
$$x = 4 \quad \text{or} \quad x = -2$$
The solutions are 4 and -2.

66. 3b, -b

67. $2(3x + 1)^2 = 8$
$$(3x + 1)^2 = 4$$
$$3x + 1 = 2 \quad \text{or} \quad 3x + 1 = -2$$
$$3x = 1 \quad \text{or} \quad 3x = -3$$
$$x = \frac{1}{3} \quad \text{or} \quad x = -1$$
The solutions are $\frac{1}{3}$ and -1.

68. $\frac{4}{5}$, 0

Exercise Set 11.3

1. $x^2 - 2x$
$$\left[\frac{-2}{2}\right]^2 = (-1)^2 = 1 \quad \text{Taking half the x-coefficient and squaring}$$
$$x^2 - 2x + 1$$
The trinomial $x^2 - 2x + 1$ is the square of $x - 1$.

2. $x^2 - 4x + 4$

3. $x^2 + 18x \quad \left[\frac{18}{2}\right]^2 = 9^2 = 81$
The trinomial $x^2 + 18x + 81$ is the square of $x + 9$.

4. $x^2 + 22x + 121$

5. $x^2 - x \quad \left[\frac{-1}{2}\right]^2 = \left[-\frac{1}{2}\right]^2 = \frac{1}{4}$
The trinomial $x^2 - x + \frac{1}{4}$ is the square of $x - \frac{1}{2}$.

6. $x^2 + x + \frac{1}{4}$

7. $t^2 + 5t \quad \left[\frac{5}{2}\right]^2 = \frac{25}{4}$
The trinomial $t^2 + 5t + \frac{25}{4}$ is the square of $t + \frac{5}{2}$.

8. $y^2 - 9y + \frac{81}{4}$

9. $x^2 - \frac{3}{2}x \quad \left[\frac{-\frac{3}{2}}{2}\right]^2 = \left[-\frac{3}{4}\right]^2 = \frac{9}{16}$
The trinomial $x^2 - \frac{3}{2}x + \frac{9}{16}$ is the square of $x - \frac{3}{4}$.

10. $x^2 + \frac{4}{3}x + \frac{4}{9}$

11. $m^2 + \frac{9}{2}m \quad \left[\frac{\frac{9}{2}}{2}\right]^2 = \left[\frac{9}{4}\right]^2 = \frac{81}{16}$
The trinomial $m^2 + \frac{9}{2}m + \frac{81}{16}$ is the square of $m + \frac{9}{4}$.

12. $r^2 - \frac{2}{5}r + \frac{1}{25}$

Chapter 11   (11.3)

13. $x^2 - 6x - 16 = 0$
    $x^2 - 6x = 16$ Adding 16
    $x^2 - 6x + 9 = 16 + 9$ Adding 9: $\left(\frac{-6}{2}\right)^2 = (-3)^2 = 9$
    $(x - 3)^2 = 25$
    $x - 3 = 5$ or $x - 3 = -5$ Principle of square roots
    $x = 8$ or $ x = -2$
    The solutions are 8 and -2.

14. -3, -5

15. $x^2 + 22x + 21 = 0$
    $x^2 + 22x = -21$ Adding -21
    $x^2 + 22x + 121 = -21 + 121$ Adding 121: $\left(\frac{22}{2}\right)^2 = 11^2 = 121$
    $(x + 11)^2 = 100$
    $x + 11 = 10$ or $x + 11 = -10$ Principle of square roots
    $x = -1$ or $ x = -21$
    The solutions are -1 and -21.

16. 1, -15

17. $x^2 - 2x - 5 = 0$
    $x^2 - 2x = 5$
    $x^2 - 2x + 1 = 5 + 1$ Adding 1: $\left(\frac{-2}{2}\right)^2 = (-1)^2 = 1$
    $(x - 1)^2 = 6$
    $x - 1 = \sqrt{6}$ or $x - 1 = -\sqrt{6}$
    $x = 1 + \sqrt{6}$ or $ x = 1 - \sqrt{6}$
    The solutions are $1 \pm \sqrt{6}$.

18. $2 \pm \sqrt{15}$

19. $x^2 - 22x + 102 = 0$
    $x^2 - 22x = -102$
    $x^2 - 22x + 121 = -102 + 121$
    $(x - 11)^2 = 19$
    $x - 11 = \sqrt{19}$ or $x - 11 = -\sqrt{19}$
    $x = 11 + \sqrt{19}$ or $ x = 11 - \sqrt{19}$
    The solutions are $11 \pm \sqrt{19}$.

20. $9 \pm \sqrt{7}$

21. $x^2 + 10x - 4 = 0$
    $x^2 + 10x = 4$
    $x^2 + 10x + 25 = 4 + 25$
    $(x + 5)^2 = 29$
    $x + 5 = \sqrt{29}$ or $x + 5 = -\sqrt{29}$
    $x = -5 + \sqrt{29}$ or $ x = -5 - \sqrt{29}$
    The solutions are $-5 \pm \sqrt{29}$.

22. $5 \pm \sqrt{29}$

23. $x^2 - 7x - 2 = 0$
    $x^2 - 7x = 2$
    $x^2 - 7x + \frac{49}{4} = 2 + \frac{49}{4}$ Adding $\frac{49}{4}$: $\left(\frac{-7}{2}\right)^2 = \frac{49}{4}$
    $\left(x - \frac{7}{2}\right)^2 = \frac{8}{4} + \frac{49}{4} = \frac{57}{4}$
    $x - \frac{7}{2} = \frac{\sqrt{57}}{2}$ or $x - \frac{7}{2} = -\frac{\sqrt{57}}{2}$
    $x = \frac{7}{2} + \frac{\sqrt{57}}{2}$ or $x = \frac{7}{2} - \frac{\sqrt{57}}{2}$
    $x = \frac{7 + \sqrt{57}}{2}$ or $x = \frac{7 - \sqrt{57}}{2}$
    The solutions are $\frac{7 \pm \sqrt{57}}{2}$.

24. $\frac{-7 \pm \sqrt{57}}{2}$

25. $x^2 + 3x - 28 = 0$
    $x^2 + 3x = 28$
    $x^2 + 3x + \frac{9}{4} = 28 + \frac{9}{4}$ Adding $\frac{9}{4}$: $\left(\frac{3}{2}\right)^2 = \frac{9}{4}$
    $\left(x + \frac{3}{2}\right)^2 = \frac{121}{4}$
    $x + \frac{3}{2} = \frac{11}{2}$ or $x + \frac{3}{2} = -\frac{11}{2}$
    $x = \frac{8}{2}$ or $ x = -\frac{14}{2}$
    $x = 4$ or $ x = -7$
    The solutions are 4 and -7.

26. 7, -4

27. $x^2 + \frac{3}{2}x - \frac{1}{2} = 0$
    $x^2 + \frac{3}{2}x = \frac{1}{2}$
    $x^2 + \frac{3}{2}x + \frac{9}{16} = \frac{1}{2} + \frac{9}{16}$ Adding $\frac{9}{16}$: $\left[\frac{\frac{3}{2}}{2}\right]^2 = \left(\frac{3}{4}\right)^2 = \frac{9}{16}$
    $\left(x + \frac{3}{4}\right)^2 = \frac{17}{16}$
    $x + \frac{3}{4} = \frac{\sqrt{17}}{4}$ or $x + \frac{3}{4} = -\frac{\sqrt{17}}{4}$
    $x = -\frac{3}{4} + \frac{\sqrt{17}}{4}$ or $x = -\frac{3}{4} - \frac{\sqrt{17}}{4}$
    $x = \frac{-3 + \sqrt{17}}{4}$ or $x = \frac{-3 - \sqrt{17}}{4}$
    The solutions are $\frac{-3 \pm \sqrt{17}}{4}$.

28. $\frac{3 \pm \sqrt{41}}{4}$

29. $2x^2 + 3x - 17 = 0$

$\frac{1}{2}(2x^2 + 3x - 17) = \frac{1}{2} \cdot 0$ Multiplying by $\frac{1}{2}$ to make the x^2-coefficient 1

$x^2 + \frac{3}{2}x - \frac{17}{2} = 0$

$x^2 + \frac{3}{2}x = \frac{17}{2}$

$x^2 + \frac{3}{2}x + \frac{9}{16} = \frac{17}{2} + \frac{9}{16}$ Adding $\frac{9}{16}$: $\left[\frac{\frac{3}{2}}{2}\right]^2 = \left(\frac{3}{4}\right)^2 = \frac{9}{16}$

$\left(x + \frac{3}{4}\right)^2 = \frac{145}{16}$

$x + \frac{3}{4} = \frac{\sqrt{145}}{4}$ or $x + \frac{3}{4} = -\frac{\sqrt{145}}{4}$

$x = \frac{-3 + \sqrt{145}}{4}$ or $x = \frac{-3 - \sqrt{145}}{4}$

The solutions are $\frac{-3 \pm \sqrt{145}}{4}$.

30. $\frac{3 \pm \sqrt{17}}{4}$

31. $3x^2 + 4x - 1 = 0$

$\frac{1}{3}(3x^2 + 4x - 1) = \frac{1}{3} \cdot 0$

$x^2 + \frac{4}{3}x - \frac{1}{3} = 0$

$x^2 + \frac{4}{3}x = \frac{1}{3}$

$x^2 + \frac{4}{3}x + \frac{4}{9} = \frac{1}{3} + \frac{4}{9}$

$\left(x + \frac{2}{3}\right)^2 = \frac{7}{9}$

$x + \frac{2}{3} = \frac{\sqrt{7}}{3}$ or $x + \frac{2}{3} = -\frac{\sqrt{7}}{3}$

$x = \frac{-2 + \sqrt{7}}{3}$ or $x = \frac{-2 - \sqrt{7}}{3}$

The solutions are $\frac{-2 \pm \sqrt{7}}{3}$.

32. $\frac{2 \pm \sqrt{13}}{3}$

33. $2x^2 = 9x + 5$

$2x^2 - 9x - 5 = 0$ Standard form

$\frac{1}{2}(2x^2 - 9x - 5) = \frac{1}{2} \cdot 0$

$x^2 - \frac{9}{2}x - \frac{5}{2} = 0$

$x^2 - \frac{9}{2}x = \frac{5}{2}$

$x^2 - \frac{9}{2}x + \frac{81}{16} = \frac{5}{2} + \frac{81}{16}$

$\left(x - \frac{9}{4}\right)^2 = \frac{121}{16}$

$x - \frac{9}{4} = \frac{11}{4}$ or $x - \frac{9}{4} = -\frac{11}{4}$

$x = \frac{20}{4}$ or $x = -\frac{2}{4}$

$x = 5$ or $x = -\frac{1}{2}$

The solutions are 5 and $-\frac{1}{2}$.

34. $4, -\frac{3}{2}$

35. $4x^2 + 12x = 7$

$4x^2 + 12x - 7 = 0$ Standard form

$\frac{1}{4}(4x^2 + 12x - 7) = \frac{1}{4} \cdot 0$

$x^2 + 3x - \frac{7}{4} = 0$

$x^2 + 3x = \frac{7}{4}$

$x^2 + 3x + \frac{9}{4} = \frac{7}{4} + \frac{9}{4}$

$\left(x + \frac{3}{2}\right)^2 = \frac{16}{4} = 4$

$x + \frac{3}{2} = 2$ or $x + \frac{3}{2} = -2$

$x = \frac{1}{2}$ or $x = -\frac{7}{2}$

The solutions are $\frac{1}{2}$ and $-\frac{7}{2}$.

36. $\frac{2}{3}, -\frac{5}{2}$

37. $y = \frac{k}{x}$ Inverse variation

$235 = \frac{k}{0.6}$ Substituting 0.6 for x and 235 for y

$141 = k$ Constant of variation

$y = \frac{141}{x}$ Equation of variation

38. $3\frac{1}{3}$ hr

39. $f(x) = 2x - 5$

$f(0) = 2 \cdot 0 - 5 = 0 - 5 = -5$

$f(-1) = 2(-1) - 5 = -2 - 5 = -7$

$f(11) = 2 \cdot 11 - 5 = 22 - 5 = 17$

Chapter 11 (11.4)

40. 1, 0, 9

41. $f(x) = |x| + 1$
$f(-2) = |-2| + 1 = 2 + 1 = 3$
$f(0) = |0| + 1 = 0 + 1 = 1$
$f(12) = |12| + 1 = 12 + 1 = 13$

42. $-1, 4, \frac{1}{2}$

43. $x^2 + bx + 36$

The trinomial is a square if the square of one-half the x-coefficient is equal to 36. Thus we have:
$$\left(\frac{b}{2}\right)^2 = 36$$
$$\frac{b^2}{4} = 36$$
$$b^2 = 144$$
$b = 12$ or $b = -12$ Principle of square roots

44. $2\sqrt{55}, -2\sqrt{55}$

45. $x^2 + bx + 128$

The trinomial is a square if the square of one-half the x-coefficient is equal to 128. Thus we have:
$$\left(\frac{b}{2}\right)^2 = 128$$
$$\frac{b^2}{4} = 128$$
$$b^2 = 512$$
$b = \sqrt{512}$ or $b = -\sqrt{512}$
$b = 16\sqrt{2}$ or $b = -16\sqrt{2}$

46. 16, -16

47. $x^2 + bx + c$

The trinomial is a square if the square of one-half the x-coefficient is equal to c. Thus we have:
$$\left(\frac{b}{2}\right)^2 = c$$
$$\frac{b^2}{4} = c$$
$$b^2 = 4c$$
$b = \sqrt{4c}$ or $b = -\sqrt{4c}$
$b = 2\sqrt{c}$ or $b = -2\sqrt{c}$

48. $2\sqrt{ac}, -2\sqrt{ac}$

Exercise Set 11.4

1. $\quad x^2 - 4x = 21$
$x^2 - 4x - 21 = 0$ Standard form
$a = 1, b = -4, c = -21$

We compute the discriminant:
$b^2 - 4ac = (-4)^2 - 4 \cdot 1 \cdot (-21) = 16 + 84 = 100$

The discriminant is a perfect square, so we can factor.
$x^2 - 4x - 21 = 0$
$(x - 7)(x + 3) = 0$
$x - 7 = 0$ or $x + 3 = 0$
$\quad x = 7$ or $\quad x = -3$
The solutions are 7 and -3.

2. -9, 2

3. $\quad x^2 = 6x - 9$
$x^2 - 6x + 9 = 0$ Standard form
$a = 1, b = -6, c = 9$

We compute the discriminant:
$b^2 - 4ac = (-6)^2 - 4 \cdot 1 \cdot 9 = 36 - 36 = 0$

The discriminant is a perfect square, so we can factor.
$x^2 - 6x + 9 = 0$
$(x - 3)(x - 3) = 0$
$x - 3 = 0$ or $x - 3 = 0$
$\quad x = 3$ or $\quad x = 3$
The solution is 3.

4. 4

5. $3y^2 - 2y - 8 = 0$
$a = 3, b = -2, c = -8$

We compute the discriminant:
$b^2 - 4ac = (-2)^2 - 4 \cdot 3 \cdot (-8) = 4 + 96 = 100$

The discriminant is a perfect square, so we can factor.
$3y^2 - 2y - 8 = 0$
$(3y + 4)(y - 2) = 0$
$3y + 4 = 0$ or $y - 2 = 0$
$\quad 3y = -4$ or $\quad y = 2$
$\quad y = -\frac{4}{3}$ or $\quad y = 2$

The solutions are $-\frac{4}{3}$ and 2.

6. $\frac{4}{3}, 1$

289

Chapter 11 (11.4)

7. $4x^2 + 12x = 7$
$4x^2 + 12x - 7 = 0$
$a = 4, b = 12, c = -7$
We compute the discriminant:
$b^2 - 4ac = 12^2 - 4 \cdot 4 \cdot (-7) = 144 + 112 = 256$
The discriminant is a perfect square, so we can factor.
$4x^2 + 12x - 7 = 0$
$(2x - 1)(2x + 7) = 0$
$2x - 1 = 0$ or $2x + 7 = 0$
$2x = 1$ or $2x = -7$
$x = \frac{1}{2}$ or $x = -\frac{7}{2}$
The solutions are $\frac{1}{2}$ and $-\frac{7}{2}$.

8. $-\frac{5}{2}, \frac{3}{2}$

9. $x^2 - 9 = 0$ Difference of squares
$(x + 3)(x - 3) = 0$
$x + 3 = 0$ or $x - 3 = 0$
$x = -3$ or $x = 3$
The solutions are -3 and 3.

10. $-2, 2$

11. $x^2 - 2x - 2 = 0$
$a = 1, b = -2, c = -2$
We compute the discriminant:
$b^2 - 4ac = (-2)^2 - 4 \cdot 1 \cdot (-2) = 4 + 8 = 12$
The discriminant is positive, so there are real-number solutions. They are given by
$x = \frac{-(-2) \pm \sqrt{12}}{2 \cdot 1}$
$x = \frac{2 \pm \sqrt{12}}{2} = \frac{2 \pm \sqrt{4 \cdot 3}}{2}$
$x = \frac{2 \pm 2\sqrt{3}}{2} = \frac{2(1 \pm \sqrt{3})}{2}$
$x = 1 \pm \sqrt{3}$.

12. $2 \pm \sqrt{11}$

13. $y^2 - 10y + 22 = 0$
$a = 1, b = -10, c = 22$
We compute the discriminant:
$b^2 - 4ac = (-10)^2 - 4 \cdot 1 \cdot 22 = 100 - 88 = 12$
The discriminant is positive, so there are real-number solutions. They are given by
$y = \frac{-(-10) \pm \sqrt{12}}{2 \cdot 1}$
$y = \frac{10 \pm \sqrt{12}}{2} = \frac{10 \pm \sqrt{4 \cdot 3}}{2}$
$y = \frac{10 \pm 2\sqrt{3}}{2} = \frac{2(5 \pm \sqrt{3})}{2}$
$y = 5 \pm \sqrt{3}$.

14. $-3 \pm \sqrt{10}$

15. $x^2 + 4x + 4 = 7$
$x^2 + 4x - 3 = 0$ Adding -7 to get standard form
$a = 1, b = 4, c = -3$
We compute the discriminant:
$b^2 - 4ac = 4^2 - 4 \cdot 1 \cdot (-3) = 16 + 12 = 28$
The discriminant is positive, so there are real-number solutions. They are given by
$x = \frac{-4 \pm \sqrt{28}}{2 \cdot 1}$
$x = \frac{-4 \pm \sqrt{28}}{2} = \frac{-4 \pm \sqrt{4 \cdot 7}}{2}$
$x = \frac{-4 \pm 2\sqrt{7}}{2} = \frac{2(-2 \pm \sqrt{7})}{2}$
$x = -2 \pm \sqrt{7}$.

16. $1 \pm \sqrt{5}$

17. $3x^2 + 8x + 2 = 0$
$a = 3, b = 8, c = 2$
We compute the discriminant:
$b^2 - 4ac = 8^2 - 4 \cdot 3 \cdot 2 = 64 - 24 = 40$
The discriminant is positive, so there are real-number solutions. They are given by
$x = \frac{-8 \pm \sqrt{40}}{2 \cdot 3}$
$x = \frac{-8 \pm \sqrt{40}}{6} = \frac{-8 \pm \sqrt{4 \cdot 10}}{6}$
$x = \frac{-8 \pm 2\sqrt{10}}{6} = \frac{2(-4 \pm \sqrt{10})}{2 \cdot 3}$
$x = \frac{-4 \pm \sqrt{10}}{3}$.

18. $\frac{2 \pm \sqrt{10}}{3}$

Chapter 11 (11.4)

19. $2x^2 - 5x = 1$
    $2x^2 - 5x - 1 = 0$ Adding -1 to get standard form
    $a = 2, b = -5, c = -1$
    We compute the discriminant:
    $b^2 - 4ac = (-5)^2 - 4 \cdot 2 \cdot (-1) = 25 + 8 = 33$
    The discriminant is positive, so there are real-number solutions. They are given by
    $x = \dfrac{-(-5) \pm \sqrt{33}}{2 \cdot 2}$
    $x = \dfrac{5 \pm \sqrt{33}}{4}.$

20. $\dfrac{-1 \pm \sqrt{7}}{2}$

21. $4y^2 - 4y - 1 = 0$
    $a = 4, b = -4, c = -1$
    We compute the discriminant:
    $b^2 - 4ac = (-4)^2 - 4 \cdot 4 \cdot (-1) = 16 + 16 = 32$
    The discriminant is positive, so there are real-number solutions. They are given by
    $y = \dfrac{-(-4) \pm \sqrt{32}}{2 \cdot 4}$
    $y = \dfrac{4 \pm \sqrt{32}}{8} = \dfrac{4 \pm \sqrt{16 \cdot 2}}{8}$
    $y = \dfrac{4 \pm 4\sqrt{2}}{8} = \dfrac{4(1 \pm \sqrt{2})}{4 \cdot 2}$
    $y = \dfrac{1 \pm \sqrt{2}}{2}.$

22. $\dfrac{-1 \pm \sqrt{2}}{2}$

23. $3x^2 + 5x = 0$ We can factor.
    $x(3x + 5) = 0$
    $x = 0$ or $3x + 5 = 0$
    $x = 0$ or $\quad\quad 3x = -5$
    $x = 0$ or $\quad\quad\quad x = -\dfrac{5}{3}$

24. $0, \dfrac{2}{5}$

25. $2t^2 + 6t + 5 = 0$
    $a = 2, b = 6, c = 5$
    We compute the discriminant:
    $b^2 - 4ac = 6^2 - 4 \cdot 2 \cdot 5 = 36 - 40 = -4$
    The discriminant is negative, so there are no real-number solutions.

26. No real-number solutions

27. $4x^2 = 100$
    $4x^2 - 100 = 0$ Difference of squares
    $4(x^2 - 25) = 0$
    $4(x + 5)(x - 5) = 0$
    $x + 5 = 0$ or $x - 5 = 0$
    $x = -5$ or $\quad\quad x = 5$

28. $-4, 4$

29. $3x^2 = 5x + 4$
    $3x^2 - 5x - 4 = 0$
    $a = 3, b = -5, c = -4$
    We compute the discriminant:
    $b^2 - 4ac = (-5)^2 - 4 \cdot 3 \cdot (-4) = 25 + 48 = 73$
    The discriminant is positive, so there are real-number solutions. They are given by
    $x = \dfrac{-(-5) \pm \sqrt{73}}{2 \cdot 3}$
    $x = \dfrac{5 \pm \sqrt{73}}{6}.$

30. $\dfrac{-3 \pm \sqrt{17}}{4}$

31. $2y^2 - 6y = 10$
    $2y^2 - 6y - 10 = 0$
    $y^2 - 3y - 5 = 0$ Multiplying by $\dfrac{1}{2}$ to simplify
    $a = 1, b = -3, c = -5$
    We compute the discriminant:
    $b^2 - 4ac = (-3)^2 - 4 \cdot 1 \cdot (-5) = 9 + 20 = 29$
    The discriminant is positive, so there are real-number solutions. They are given by
    $y = \dfrac{-(-3) \pm \sqrt{29}}{2 \cdot 1}$
    $y = \dfrac{3 \pm \sqrt{29}}{2}.$

32. $\dfrac{11 \pm \sqrt{181}}{10}$

33. $\dfrac{x^2}{x - 4} - \dfrac{7}{x - 4} = 0$, LCM is $x - 4$
    $(x - 4)\left[\dfrac{x^2}{x - 4} - \dfrac{7}{x - 4}\right] = (x - 4) \cdot 0$
    $x^2 - 7 = 0$
    $x^2 = 7$
    $x = \sqrt{7}$ or $x = -\sqrt{7}$ Principle of square roots
    Both numbers check. The solutions are $\sqrt{7}$ and $-\sqrt{7}$.

34. $\sqrt{5}, -\sqrt{5}$

Chapter 11   (11.4)

35. $x + 2 = \dfrac{3}{x+2}$

$(x+2)(x+2) = (x+2) \cdot \dfrac{3}{x+2}$ Clearing the fraction

$x^2 + 4x + 4 = 3$

$x^2 + 4x + 1 = 0$

$a = 1, b = 4, c = 1$

We compute the discriminant:
$b^2 - 4ac = 4^2 - 4 \cdot 1 \cdot 1 = 16 - 4 = 12$

The discriminant is positive, so there are real-number solutions. They are given by

$x = \dfrac{-4 \pm \sqrt{12}}{2 \cdot 1}$

$x = \dfrac{-4 \pm \sqrt{12}}{2} = \dfrac{-4 \pm \sqrt{4 \cdot 3}}{2}$

$x = \dfrac{-4 \pm 2\sqrt{3}}{2} = \dfrac{2(-2 \pm \sqrt{3})}{2}$

$x = -2 \pm \sqrt{3}$.

Both numbers check. The solutions are $-2 + \sqrt{3}$ and $-2 - \sqrt{3}$.

36. $3 \pm \sqrt{5}$

37. $\dfrac{1}{x} + \dfrac{1}{x+6} = \dfrac{1}{5}$,   LCM is $5x(x+6)$

$5x(x+6)\left[\dfrac{1}{x} + \dfrac{1}{x+6}\right] = 5x(x+6) \cdot \dfrac{1}{5}$

$5(x+6) + 5x = x(x+6)$

$5x + 30 + 5x = x^2 + 6x$

$10x + 30 = x^2 + 6x$

$0 = x^2 - 4x - 30$

$a = 1, b = -4, c = -30$

We compute the discriminant:
$b^2 - 4ac = (-4)^2 - 4 \cdot 1 \cdot (-30) = 16 + 120 = 136$

The discriminant is positive, so there are real-number solutions. They are given by

$x = \dfrac{-(-4) \pm \sqrt{136}}{2 \cdot 1}$

$x = \dfrac{4 \pm \sqrt{136}}{2} = \dfrac{4 \pm \sqrt{4 \cdot 34}}{2}$

$x = \dfrac{4 \pm 2\sqrt{34}}{2} = \dfrac{2(2 \pm \sqrt{34})}{2}$

$x = 2 \pm \sqrt{34}$

Both numbers check. The solutions are $2 + \sqrt{34}$ and $2 - \sqrt{34}$.

38. $\dfrac{5 \pm \sqrt{37}}{2}$

39. $x^2 - 4x - 7 = 0$

$a = 1, b = -4, c = -7$

$x = \dfrac{-(-4) \pm \sqrt{(-4)^2 - 4 \cdot 1 \cdot (-7)}}{2 \cdot 1}$

$x = \dfrac{4 \pm \sqrt{16 + 28}}{2} = \dfrac{4 \pm \sqrt{44}}{2}$

$x = \dfrac{4 \pm \sqrt{4 \cdot 11}}{2} = \dfrac{4 \pm 2\sqrt{11}}{2}$

$x = \dfrac{2(2 \pm \sqrt{11})}{2} = 2 \pm \sqrt{11}$

Using a calculator or Table 2, we see that $\sqrt{11} \approx 3.317$:

$2 + \sqrt{11} \approx 2 + 3.317$ or $2 - \sqrt{11} \approx 2 - 3.317$
≈ 5.3 or ≈ -1.3

The approximate solutions, to the nearest tenth, are 5.3 and -1.3.

40. 0.7, -2.7

41. $y^2 - 6y - 1 = 0$

$a = 1, b = -6, c = -1$

$y = \dfrac{-(-6) \pm \sqrt{(-6)^2 - 4 \cdot 1 \cdot (-1)}}{2 \cdot 1}$

$y = \dfrac{6 \pm \sqrt{36 + 4}}{2} = \dfrac{6 \pm \sqrt{40}}{2}$

$y = \dfrac{6 \pm \sqrt{4 \cdot 10}}{2} = \dfrac{6 \pm 2\sqrt{10}}{2}$

$y = \dfrac{2(3 \pm \sqrt{10})}{2} = 3 \pm \sqrt{10}$

Using a calculator or Table 2, we see that $\sqrt{10} \approx 3.162$:

$3 + \sqrt{10} \approx 3 + 3.162$ or $3 - \sqrt{10} \approx 3 - 3.162$
≈ 6.2 or ≈ -0.2

The approximate solutions, to the nearest tenth, are 6.2 and -0.2.

42. -3.3, -6.7

Chapter 11 (11.4)

43. $4x^2 + 4x = 1$

$4x^2 + 4x - 1 = 0$ Standard form

$a = 4, b = 4, c = -1$

$x = \dfrac{-4 \pm \sqrt{4^2 - 4 \cdot 4 \cdot (-1)}}{2 \cdot 4}$

$x = \dfrac{-4 \pm \sqrt{16 + 16}}{8} = \dfrac{-4 \pm \sqrt{32}}{8}$

$x = \dfrac{-4 \pm \sqrt{16 \cdot 2}}{8} = \dfrac{-4 \pm 4\sqrt{2}}{8}$

$x = \dfrac{4(-1 \pm \sqrt{2})}{4 \cdot 2} = \dfrac{-1 \pm \sqrt{2}}{2}$

Using a calculator or Table 2, we see that $\sqrt{2} \approx 1.414$:

$\dfrac{-1 + \sqrt{2}}{2} \approx \dfrac{-1 + 1.414}{2}$ or $\dfrac{-1 - \sqrt{2}}{2} \approx \dfrac{-1 - 1.414}{2}$

$\approx \dfrac{0.414}{2}$ or $\approx \dfrac{-2.414}{2}$

≈ 0.2 or ≈ -1.2

The approximate solutions, to the nearest tenth, are 0.2 and -1.2.

44. 1.2, -0.2

45. $3x^2 + 4x - 2 = 0$

$a = 3, b = 4, c = -2$

$x = \dfrac{-4 \pm \sqrt{4^2 - 4 \cdot 3(-2)}}{2 \cdot 3}$

$x = \dfrac{-4 \pm \sqrt{16 + 24}}{6} = \dfrac{-4 \pm \sqrt{40}}{6}$

$x = \dfrac{-4 \pm \sqrt{4 \cdot 10}}{6} = \dfrac{-4 \pm 2\sqrt{10}}{6}$

$x = \dfrac{2(-2 \pm \sqrt{10})}{2 \cdot 3} = \dfrac{-2 \pm \sqrt{10}}{3}$

Using a calculator or Table 2 we see that $\sqrt{10} \approx 3.162$:

$\dfrac{-2 + \sqrt{10}}{3} \approx \dfrac{-2 + 3.162}{3}$ or $\dfrac{-2 - \sqrt{10}}{3} \approx \dfrac{-2 - 3.162}{3}$

$\approx \dfrac{1.162}{3}$ or $\approx \dfrac{-5.162}{3}$

≈ 0.4 or ≈ -1.7

The approximate solutions, to the nearest tenth, are 0.4 and -1.7.

46. 2.4, 0.3

47. $2y^2 + 2y - 3 = 0$

$a = 2, b = 2, c = -3$

$x = \dfrac{-2 \pm \sqrt{2^2 - 4 \cdot 2 \cdot (-3)}}{2 \cdot 2}$

$x = \dfrac{-2 \pm \sqrt{4 + 24}}{4} = \dfrac{-2 \pm \sqrt{28}}{4}$

$x = \dfrac{-2 \pm \sqrt{4 \cdot 7}}{4} = \dfrac{-2 \pm 2\sqrt{7}}{4}$

$x = \dfrac{2(-1 \pm \sqrt{7})}{2 \cdot 2} = \dfrac{-1 \pm \sqrt{7}}{2}$

Using a calculator or Table 2 we see that $\sqrt{7} \approx 2.646$:

$\dfrac{-1 + \sqrt{2}}{2} \approx \dfrac{-1 + 2.646}{2}$ or $\dfrac{-1 - \sqrt{2}}{2} \approx \dfrac{-1 - 2.646}{2}$

$\approx \dfrac{1.646}{2}$ or $\approx \dfrac{-3.646}{2}$

≈ 0.8 or ≈ -1.8

The approximate solutions, to the nearest tenth, are 0.8 and -1.8.

48. 1.9, -0.3

49. $\sqrt{3x^2}\,\sqrt{9x^3} = \sqrt{27x^5} = \sqrt{9x^4 \cdot 3x} = \sqrt{9x^4}\,\sqrt{3x} = 3x^2\sqrt{3x}$

50. $\sqrt{6}$

51. $\sqrt{80} = \sqrt{16 \cdot 5} = \sqrt{16}\,\sqrt{5} = 4\sqrt{5}$

52. $\dfrac{\sqrt{21}}{3}$

53. $5x + x(x - 7) = 0$

$5x + x^2 - 7x = 0$

$x^2 - 2x = 0$ We can factor.

$x(x - 2) = 0$

$x = 0$ or $x - 2 = 0$

$x = 0$ or $x = 2$

The solutions are 0 and 2.

54. $0, -\dfrac{4}{3}$

55. $3 - x(x - 3) = 4$

$3 - x^2 + 3x = 4$

$0 = x^2 - 3x + 1$ Standard form

$a = 1, b = -3, c = 1$

We compute the discriminant:

$b^2 - 4ac = (-3)^2 - 4 \cdot 1 \cdot 1 = 9 - 4 = 5$

The discriminant is positive, so there are real-number solutions. They are given by

$x = \dfrac{-(-3) \pm \sqrt{5}}{2 \cdot 1} = \dfrac{3 \pm \sqrt{5}}{2}.$

Chapter 11   (11.4)

56. $\dfrac{7 \pm \sqrt{69}}{10}$

57. $(y + 4)(y + 3) = 15$
   $y^2 + 7y + 12 = 15$
   $y^2 + 7y - 3 = 0$ Standard form

$a = 1, b = 7, c = -3$

We compute the discriminant:
$b^2 - 4ac = 7^2 - 4 \cdot 1 \cdot (-3) = 49 + 12 = 61$

The discriminant is positive, so there are real-number solutions. They are given by

$y = \dfrac{-7 \pm \sqrt{61}}{2 \cdot 1}$

$y = \dfrac{-7 \pm \sqrt{61}}{2}$.

58. $-8, 4$

59. $x^2 + (x + 2)^2 = 7$
   $x^2 + x^2 + 4x + 4 = 7$
   $2x^2 + 4x + 4 = 7$
   $2x^2 + 4x - 3 = 0$ Standard form

$a = 2, b = 4, c = -3$

We compute the discriminant:
$b^2 - 4ac = 4^2 - 4 \cdot 2 \cdot (-3) = 16 + 24 = 40$

The discriminant is positive, so there are real-number solutions. They are given by

$x = \dfrac{-4 \pm \sqrt{40}}{2 \cdot 2} = \dfrac{-4 \pm \sqrt{4 \cdot 10}}{4}$

$x = \dfrac{-4 \pm 2\sqrt{10}}{4} = \dfrac{2(-2 \pm \sqrt{10})}{2 \cdot 2}$

$x = \dfrac{-2 \pm \sqrt{10}}{2}$.

60. $-2, 1$

61. $(x + 2)^2 + (x + 1)^2 = 0$
   $x^2 + 4x + 4 + x^2 + 2x + 1 = 0$
   $2x^2 + 6x + 5 = 0$ Standard form

$a = 2, b = 6, c = 5$

We compute the discriminant:
$b^2 - 4ac = 6^2 - 4 \cdot 2 \cdot 5 = 36 - 40 = -4$

The discriminant is negative, so there are no real-number solutions.

62. No real number solutions

63. $ax^2 + 2x = 3$
   $ax^2 + 2x - 3 = 0$ Standard form

$a = a, b = 2, c = -3$

We will assume that the discriminant is nonnegative.

$x = \dfrac{-2 \pm \sqrt{2^2 - 4 \cdot a \cdot (-3)}}{2 \cdot a}$

$x = \dfrac{-2 \pm \sqrt{4 + 12a}}{2a} = \dfrac{-2 \pm \sqrt{4(1 + 3a)}}{2a}$

$x = \dfrac{-2 \pm 2\sqrt{1 + 3a}}{2a} = \dfrac{2(-1 \pm \sqrt{1 + 3a})}{2a}$

$x = \dfrac{-1 \pm \sqrt{1 + 3a}}{a}$

64. $\dfrac{5 \pm \sqrt{25 - 24b^2}}{4b}$

65. $4x^2 - 4cx + c^2 - 3d^2 = 0$

$a = 4, b = -4c, c = c^2 - 3d^2$

We will assume that the discriminant is nonnegative.

$x = \dfrac{-(-4c) \pm \sqrt{(-4c)^2 - 4 \cdot 4 \cdot (c^2 - 3d^2)}}{2 \cdot 4}$

$x = \dfrac{4c \pm \sqrt{16c^2 - 16c^2 + 48d^2}}{8} = \dfrac{4c \pm \sqrt{48d^2}}{8}$

$x = \dfrac{4c \pm \sqrt{16d^2 \cdot 3}}{8} = \dfrac{4c \pm 4d\sqrt{3}}{8}$

$x = \dfrac{4(c \pm d\sqrt{3})}{4 \cdot 2} = \dfrac{c \pm d\sqrt{3}}{2}$

66. $-0.45, 0.25$

67. $bdx^2 + bcx - ac = adx$
   $bdx^2 + bcx - adx - ac = 0$
   $bx(dx + c) - a(dx + c) = 0$ Factoring by grouping
   $(dx + c)(bx - a) = 0$

$dx + c = 0$ or $bx - a = 0$
$dx = -c$ or $bx = a$
$x = -\dfrac{c}{d}$ or $x = \dfrac{a}{b}$

68. $-1, -2b + 1$

69. a) Yes. If $b^2 > 4ac$, then $b^2 - 4ac > 0$, so there will be real-number solutions.
   No. It does not make a difference whether b is positive, negative, or zero.

b) Yes. If $ac < 0$, then $-4ac > 0$ and $b^2 - 4ac > 0$, so there will be real-number solutions.
   No. It does not make a difference whether b is positive, negative, or zero.

c) If a and c are both positive, then $ac > 0$ and $-4ac < 0$. Then $b^2 - 4ac \geq 0$ when $b^2 \geq 4ac$.

Chapter 11   (11.5)

70. Sum of the solutions:

$$\frac{-b + \sqrt{b^2 - 4ac}}{2a} + \frac{-b - \sqrt{b^2 - 4ac}}{2a} = \frac{-2b}{2a} = -\frac{b}{a}$$

Product of solutions:

$$\left(\frac{-b + \sqrt{b^2 - 4ac}}{2a}\right)\left(\frac{-b - \sqrt{b^2 - 4ac}}{2a}\right) =$$

$$\frac{b^2 - (b^2 - 4ac)}{4a^2} = \frac{4ac}{4a^2} = \frac{c}{a}$$

$$-\frac{b}{a} = -\frac{5}{2}, \frac{c}{a} = -\frac{3}{2}$$

Exercise Set 11.5

1. $N = 2.5\sqrt{A}$

   $\frac{N}{2.5} = \sqrt{A}$ Isolating the radical

   $\left(\frac{N}{2.5}\right)^2 = (\sqrt{A})^2$ Principle of squaring

   $\frac{N^2}{6.25} = A$ Simplifying

2. $L = \frac{8T^2}{\pi^2}$

3. $Q = \sqrt{\frac{aT}{c}}$

   $Q^2 = \left(\sqrt{\frac{aT}{c}}\right)^2$ Principle of squaring

   $Q^2 = \frac{aT}{c}$

   $\frac{cQ^2}{a} = T$ Multiplying by $\frac{c}{a}$

4. $E = \frac{mv^2}{2g}$

5. $E = mc^2$

   $\frac{E}{m} = c^2$

   $\sqrt{\frac{E}{m}} = c$ Principle of square roots. Assume c is nonnegative.

6. $r = \frac{1}{2}\sqrt{\frac{S}{\pi}}$

7. $Q = ad^2 - cd$

   $0 = ad^2 - cd - Q$ Standard form

   $a = a, b = -c, c = -Q$

   $d = \frac{-(-c) \pm \sqrt{(-c)^2 - 4 \cdot a \cdot (-Q)}}{2a}$ Using the quadratic formula

   $d = \frac{c \pm \sqrt{c^2 + 4aQ}}{2a}$

8. $A = \frac{-m \pm \sqrt{m^2 + 4kP}}{2k}$

9. $c^2 = a^2 + b^2$

   $c^2 - b^2 = a^2$

   $\sqrt{c^2 - b^2} = a$ Principle of square roots. Assume a is nonnegative.

10. $b = \sqrt{c^2 - a^2}$

11. $s = 16t^2$

    $\frac{s}{16} = t^2$

    $\sqrt{\frac{s}{16}} = t$ Principle of square roots. Assume t is nonnegative.

    $\frac{\sqrt{s}}{4} = t$

12. $r = \sqrt{\frac{V}{\pi h}}$

13. $A = \pi r^2 + 2\pi rh$

    $0 = \pi r^2 + 2\pi hr - A$

    $a = \pi, b = 2\pi h, c = -A$

    $r = \frac{-2\pi h \pm \sqrt{(2\pi h)^2 - 4 \cdot \pi \cdot (-A)}}{2 \cdot \pi}$

    $r = \frac{-2\pi h \pm \sqrt{4\pi^2 h^2 + 4\pi A}}{2\pi} = \frac{-2\pi h \pm \sqrt{4(\pi^2 h^2 + \pi A)}}{2\pi}$

    $r = \frac{-2\pi h \pm 2\sqrt{\pi^2 h^2 + \pi A}}{2\pi}$

    $r = \frac{-\pi h \pm \sqrt{\pi^2 h^2 + \pi A}}{\pi}$

14. $r = \frac{-\pi h \pm \sqrt{\pi^2 h^2 + 2\pi A}}{2\pi}$

15. $A = \frac{\pi r^2 S}{360}$

    $\frac{360A}{\pi S} = r^2$ Multiplying by $\frac{360}{\pi S}$

    $\sqrt{\frac{360A}{\pi S}} = r$ Principle of square roots. Assume r is nonnegative.

    $\sqrt{\frac{36 \cdot 10A}{\pi S}} = r$

    $6\sqrt{\frac{10A}{\pi S}} = r$

16. $D = \sqrt{\frac{2.5H}{N}}$

17. $c = \sqrt{a^2 + b^2}$

    $c^2 = (\sqrt{a^2 + b^2})^2$ Principle of squaring

    $c^2 = a^2 + b^2$

    $c^2 - b^2 = a^2$

    $\sqrt{c^2 - b^2} = a$ Principle of square roots. Assume a is nonnegative.

18. $b = \sqrt{c^2 - a^2}$

Chapter 11   (11.5)

19. $h = \dfrac{a}{2}\sqrt{3}$

   $2h = a\sqrt{3}$

   $\dfrac{2h}{\sqrt{3}} = a$

20. $s = \dfrac{d}{\sqrt{2}}$

21. $n = aT^2 - 4T + m$

   $0 = aT^2 - 4T + m - n$

   $a = a,\ b = -4,\ c = m - n$

   $T = \dfrac{-(-4) \pm \sqrt{(-4)^2 - 4 \cdot a \cdot (m - n)}}{2 \cdot a}$

   $T = \dfrac{4 \pm \sqrt{16 - 4a(m - n)}}{2a} = \dfrac{4 \pm \sqrt{4[4 - a(m - n)]}}{2a}$

   $T = \dfrac{4 \pm 2\sqrt{4 - a(m - n)}}{2a} = \dfrac{2(2 \pm \sqrt{4 - a(m - n)})}{2 \cdot a}$

   $T = \dfrac{2 \pm \sqrt{4 - a(m - n)}}{a}$

22. $x = \dfrac{-b \pm \sqrt{b^2 - 4a(c - y)}}{2a}$

23. $r = 2\sqrt{5L}$

   $r^2 = (2\sqrt{5L})^2$ Principle of squaring

   $r^2 = 4 \cdot 5L$

   $\dfrac{r^2}{20} = L,\ \text{or}\ \dfrac{1}{20}r^2 = L$

24. $L = \dfrac{gT^2}{4\pi^2}$

25. $c = \sqrt{\dfrac{E}{m}}$

   $c^2 = \left(\sqrt{\dfrac{E}{m}}\right)^2$ Principle of squaring

   $c^2 = \dfrac{E}{m}$

   $mc^2 = E$

   $m = \dfrac{E}{c^2}$

26. $r = \sqrt{\dfrac{A}{\pi}}$

27. $N = n^2 - n$

   $0 = n^2 - n - N$

   $a = 1,\ b = -1,\ c = -N$

   $n = \dfrac{-(-1) \pm \sqrt{(-1)^2 - 4 \cdot 1 \cdot (-N)}}{2 \cdot 1}$

   $n = \dfrac{1 \pm \sqrt{1 + 4N}}{2}$

28. $t = \dfrac{-v \pm \sqrt{v^2 + 32h}}{16}$

29. $a^2 + b^2 = c^2$ Pythagorean equation
    $4^2 + 7^2 = c^2$ Substituting
    $16 + 49 = c^2$
    $65 = c^2$
    $\sqrt{65} = c$ Exact answer
    $8.062 \approx c$ Approximate answer

30. $\sqrt{75} \approx 8.660$

31. $a^2 + b^2 = c^2$ Pythagorean equation
    $4^2 + 5^2 = c^2$ Substituting
    $16 + 25 = c^2$
    $41 = c^2$
    $\sqrt{41} = c$ Exact answer
    $6.403 \approx c$ Approximate answer

32. $\sqrt{44} \approx 6.633$

33. a) $C = 2\pi r$

    $\dfrac{C}{2\pi} = r$

    b) $A = \pi r^2$

    $A = \pi \cdot \left(\dfrac{C}{2\pi}\right)^2$ Substituting $\dfrac{C}{2\pi}$ for r

    $A = \pi \cdot \dfrac{C^2}{4\pi^2}$

    $A = \dfrac{C^2}{4\pi}$

34. $C = 2\sqrt{\pi A}$

35. $3ax^2 - x - 3ax + 1 = 0$

    $3ax^2 + (-1 - 3a)x + 1 = 0$

    $a = 3a,\ b = -1 - 3a,\ c = 1$

    $x = \dfrac{-(-1 - 3a) \pm \sqrt{(-1 - 3a)^2 - 4 \cdot 3a \cdot 1}}{2 \cdot 3a}$

    $x = \dfrac{1 + 3a \pm \sqrt{1 + 6a + 9a^2 - 12a}}{6a}$

    $x = \dfrac{1 + 3a \pm \sqrt{9a^2 - 6a + 1}}{6a}$

    $x = \dfrac{1 + 3a \pm \sqrt{(3a - 1)^2}}{6a}$

    $x = \dfrac{1 + 3a \pm (3a - 1)}{6a}$

    $x = \dfrac{1 + 3a + 3a - 1}{6a}$ or $x = \dfrac{1 + 3a - 3a + 1}{6a}$

    $x = \dfrac{6a}{6a}$ or $x = \dfrac{2}{6a}$

    $x = 1$ or $x = \dfrac{1}{3a}$

    The solutions are 1 and $\dfrac{1}{3a}$.

36. $3y,\ \dfrac{2y - 3}{2}$

Chapter 11 (11.6)

Exercise Set 11.6

1. $y = x^2 + 1$

We first find the vertex. The x-coordinate is

$$-\frac{b}{2a} = -\frac{0}{2 \cdot 1} = 0.$$

We substitute into the equation to find the second coordinate of the vertex.

$$y = x^2 + 1 = 0^2 + 1 = 1$$

The vertex is (0,1). The line of symmetry is $x = 0$, the y-axis.

We choose some x-values on both sides of the vertex and graph the parabola.

When $x = 1$, $y = 1^2 + 1 = 1 + 1 = 2$.
When $x = -1$, $y = (-1)^2 + 1 = 1 + 1 = 2$.
When $x = 2$, $y = 2^2 + 1 = 4 + 1 = 5$.
When $x = -2$, $y = (-2)^2 + 1 = 4 + 1 = 5$.

x	y
0	1 ← Vertex
1	2
-1	2
2	5
-2	5

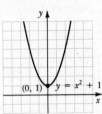

2.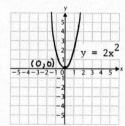

3. $y = -1 \cdot x^2$

Find the vertex. The x-coordinate is

$$-\frac{b}{2a} = -\frac{0}{2(-1)} = 0.$$

The y-coordinate is

$$y = -1 \cdot x^2 = -1 \cdot 0^2 = 0.$$

The vertex is (0,0). The line of symmetry is $x = 0$, the y-axis.

Choose some x-values on both sides of the vertex and graph the parabola.

When $x = -2$, $y = -1 \cdot (-2)^2 = -1 \cdot 4 = -4$.
When $x = -1$, $y = -1 \cdot (-1)^2 = -1 \cdot 1 = -1$.
When $x = 1$, $y = -1 \cdot 1^2 = -1 \cdot 1 = -1$.
When $x = 2$, $y = -1 \cdot 2^2 = -1 \cdot 4 = -4$.

3. (continued)

x	y
0	0 ← Vertex
-2	-4
-1	-1
1	-1
2	-4

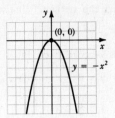

4.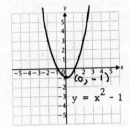

5. $y = -x^2 + 2x$

Find the vertex. The x-coordinate is

$$-\frac{b}{2a} = -\frac{2}{2(-1)} = -\frac{2}{-2} = 1.$$

The y-coordinate is

$$y = -x^2 + 2x = -(1)^2 + 2 \cdot 1 = -1 + 2 = 1.$$

The vertex is (1,1).

We choose some x-values on both sides of the vertex and graph the parabola. We make sure we find y when $x = 0$. This gives us the y-intercept.

x	y
1	1 ← Vertex
0	0 ← y-intercept
-1	-3
2	0
3	-3

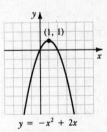

6.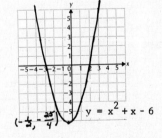

Chapter 11   (11.6)

7. $y = 8 - x - x^2$, or $y = -x^2 - x + 8$

   Find the vertex. The x-coordinate is

   $-\frac{b}{2a} = -\frac{-1}{2(-1)} = -\frac{1}{2}$.

   The y-coordinate is

   $y = 8 - x - x^2 = 8 - \left(-\frac{1}{2}\right) - \left(-\frac{1}{2}\right)^2 =$
   $8 + \frac{1}{2} - \frac{1}{4} = \frac{33}{4}$.

   The vertex is $\left(-\frac{1}{2}, \frac{33}{4}\right)$.

   We choose some x-values on both sides of the vertex and graph the parabola.

x	y	
$-\frac{1}{2}$	$\frac{33}{4}$	← Vertex
0	8	← y-intercept
-1	8	
-2	6	
1	6	

   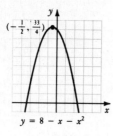

8.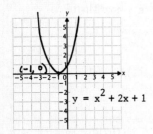

9. $y = x^2 - 2x + 1$

   Find the vertex. The x-coordinate is

   $-\frac{b}{2a} = -\frac{-2}{2 \cdot 1} = -(-1) = 1$.

   The y-coordinate is

   $y = x^2 - 2x + 1 = 1^2 - 2 \cdot 1 + 1 = 1 - 2 + 1 = 0$.

   The vertex is (1,0).

   We choose some x-values on both sides of the vertex and graph the parabola.

x	y	
1	0	← Vertex
0	1	← y-intercept
-1	4	
2	1	
3	4	

   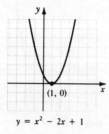

10.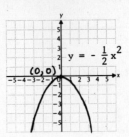

11. $y = -x^2 + 2x + 3$

    Find the vertex. The x-coordinate is

    $-\frac{b}{2a} = -\frac{2}{2(-1)} = -(-1) = 1$.

    The y-coordinate is

    $y = -x^2 + 2x + 3 = -(1)^2 + 2 \cdot 1 + 3 = -1 + 2 + 3 = 4$.

    The vertex is (1,4).

    Choose some x-values on both sides of the vertex and graph the parabola.

x	y	
1	4	← Vertex
0	3	← y-intercept
-1	0	
2	3	
3	0	

    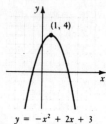

12.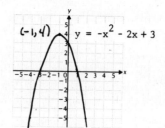

13. $y = -2x^2 - 4x + 1$

    Find the vertex. The x-coordinate is

    $-\frac{b}{2a} = -\frac{-4}{2(-2)} = -1$.

    The y-coordinate is

    $y = -2x^2 - 4x + 1 = -2(-1)^2 - 4(-1) + 1 = -2 + 4 + 1 = 3$.

    The vertex is (-1,3).

    Choose some x-values on both sides of the vertex and graph the parabola.

x	y	
-1	3	← Vertex
0	1	← y-intercept
1	-5	
-2	1	
-3	-5	

    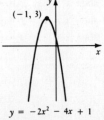

14.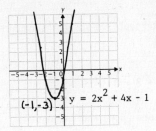

15. $y = \frac{1}{4}x^2$

    Find the vertex. The x-coordinate is

    $-\frac{b}{2a} = -\frac{0}{2\left(\frac{1}{4}\right)} = 0.$

    The y-coordinate is

    $y = \frac{1}{4}x^2 = \frac{1}{4} \cdot 0^2 = 0.$

    The vertex is (0,0).

    Choose some points on both sides of the vertex and graph the parabola.

x	y	
0	0	← Vertex
-2	1	
-4	4	
2	1	
4	4	

    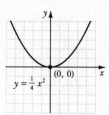

16.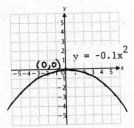

17. $y = 3 - x^2$, or $y = -x^2 + 3$

    Find the vertex. The x-coordinate is

    $-\frac{b}{2a} = -\frac{0}{2(-1)} = 0.$

    The y-coordinate is

    $y = 3 - x^2 = 3 - 0^2 = 3.$

    The vertex is (0,3).

    Choose some x-values on both sides of the vertex and graph the parabola.

x	y	
0	3	← Vertex
-1	2	
-2	-1	
1	2	
2	-1	

    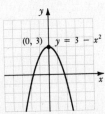

18.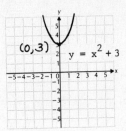

19. $y = -x^2 + x - 1$

    Find the vertex. The x-coordinate is

    $-\frac{b}{2a} = -\frac{1}{2(-1)} = -\left(-\frac{1}{2}\right) = \frac{1}{2}.$

    The y-coordinate is

    $y = -x^2 + x - 1 = -\left(\frac{1}{2}\right)^2 + \frac{1}{2} - 1 = -\frac{1}{4} + \frac{1}{2} - 1 = -\frac{3}{4}.$

    The vertex is $\left(\frac{1}{2}, -\frac{3}{4}\right).$

    Choose some x-values on both sides of the vertex and graph the parabola.

x	y	
$\frac{1}{2}$	$-\frac{3}{4}$	← Vertex
0	-1	← y-intercept
-1	-3	
1	-1	
2	-3	

    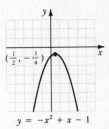

20.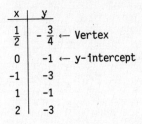

21. $y = -2x^2$

    Find the vertex. The x-coordinate is

    $-\frac{b}{2a} = -\frac{0}{2(-2)} = 0.$

    The y-coordinate is

    $y = -2x^2 = -2 \cdot 0^2 = 0.$

    The vertex is (0,0).

    Choose some x-values on both sides of the vertex and graph the parabola.

x	y	
0	0	← Vertex
-1	-2	
-2	-8	
1	-2	
2	-8	

    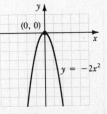

22.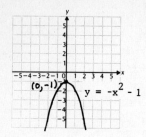

23. $y = x^2 - x - 6$

   Find the vertex. The x-coordinate is
   $$-\frac{b}{2a} = -\frac{-1}{2 \cdot 1} = -\left(-\frac{1}{2}\right) = \frac{1}{2}.$$
   The y-coordinate is
   $$y = x^2 - x - 6 = \left(\frac{1}{2}\right)^2 - \frac{1}{2} - 6 = \frac{1}{4} - \frac{1}{2} - 6 = -\frac{25}{4}.$$
   The vertex is $\left(\frac{1}{2}, -\frac{25}{4}\right)$.

   Choose some x-values on both sides of the vertex and graph the parabola.

x	y	
$\frac{1}{2}$	$-\frac{25}{4}$	← Vertex
0	-6	← y-intercept
-1	-4	
1	-6	
2	-4	

   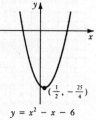

24.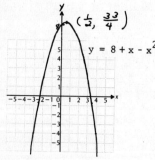

25. $y = x^2 - 5$

   To find the x-intercepts we solve the equation:
   $$x^2 - 5 = 0$$
   $$x^2 = 5$$
   $$x = \sqrt{5} \text{ or } x = -\sqrt{5} \quad \text{Principle of square roots}$$
   The x-intercepts are $(\sqrt{5}, 0)$ and $(-\sqrt{5}, 0)$.

26. $(\sqrt{3}, 0), (-\sqrt{3}, 0)$

27. $y = x^2 + 2x$

   To find the x-intercepts we solve the equation:
   $$x^2 + 2x = 0$$
   $$x(x + 2) = 0$$
   $$x = 0 \text{ or } x + 2 = 0$$
   $$x = 0 \text{ or } \quad x = -2$$
   The x-intercepts are $(0,0)$ and $(-2,0)$.

28. $(0,0), (2,0)$

29. $y = 8 - x - x^2$

   To find the intercepts we solve the equation:
   $$8 - x - x^2 = 0$$
   $$0 = x^2 + x - 8 \quad \text{Standard form}$$
   $$a = 1, b = 1, c = -8$$
   $$x = \frac{-1 \pm \sqrt{1^2 - 4 \cdot 1 \cdot (-8)}}{2 \cdot 1}$$
   $$x = \frac{-1 \pm \sqrt{33}}{2}$$
   The x-intercepts are $\left(\frac{-1 + \sqrt{33}}{2}, 0\right)$ and $\left(\frac{-1 - \sqrt{33}}{2}, 0\right)$.

30. $\left(\frac{1 + \sqrt{33}}{2}, 0\right), \left(\frac{1 - \sqrt{33}}{2}, 0\right)$

31. $y = x^2 + 10x + 25$

   To find the x-intercepts we solve the equation:
   $$x^2 + 10x + 25 = 0$$
   $$(x + 5)(x + 5) = 0$$
   $$x + 5 = 0 \text{ or } x + 5 = 0$$
   $$x = -5 \text{ or } \quad x = -5$$
   The x-intercept is $(-5, 0)$.

32. $(4, 0)$

33. $y = -2x^2 - 4x + 1$

   To find the x-intercepts we solve the equation:
   $$-2x^2 - 4x + 1 = 0$$
   $$2x^2 + 4x - 1 = 0 \quad \text{Standard form}$$
   $$a = 2, b = 4, c = -1$$
   $$x = \frac{-4 \pm \sqrt{4^2 - 4 \cdot 2 \cdot (-1)}}{2 \cdot 2}$$
   $$x = \frac{-4 \pm \sqrt{24}}{4} = \frac{-4 \pm \sqrt{4 \cdot 6}}{4} = \frac{-4 \pm 2\sqrt{6}}{4}$$
   $$x = \frac{2(-2 \pm \sqrt{6})}{2 \cdot 2} = \frac{-2 \pm \sqrt{6}}{2}$$
   The x-intercepts are $\left(\frac{-2 + \sqrt{6}}{2}, 0\right)$ and $\left(\frac{-2 - \sqrt{6}}{2}, 0\right)$.

Chapter 11   (11.6)

34. $\left[\dfrac{-2 + \sqrt{6}}{2}, 0\right]$, $\left[\dfrac{-2 - \sqrt{6}}{2}, 0\right]$

35. $y = x^2 + 5$

To find the x-intercepts we solve the equation:
$$x^2 + 5 = 0$$
$$x^2 = -5$$

The negative number -5 has no real-number square roots. Thus there are no x-intercepts. (The graph does not cross the x-axis.)

36. No x-intercepts

37. $f(x) = x^2 - x + 7$
$f(-1) = (-1)^2 - (-1) + 7 = 1 + 1 + 7 = 9$
$f(0) = 0^2 - 0 + 7 = 0 - 0 + 7 = 7$
$f\left(\dfrac{1}{2}\right) = \left(\dfrac{1}{2}\right)^2 - \dfrac{1}{2} + 7 = \dfrac{1}{4} - \dfrac{1}{2} + 7 = \dfrac{27}{4}$
$f(3) = 3^2 - 3 + 7 = 9 - 3 + 7 = 13$

38. -8, -10, -5, 220

39. $f(x) = -x^2 + 5x$
$f(-3) = -(-3)^2 + 5(-3) = -9 - 15 = -24$
$f(0) = -(0)^2 + 5 \cdot 0 = 0 + 0 = 0$
$f(5) = -(5)^2 + 5 \cdot 5 = -25 + 25 = 0$
$f(10) = -(10)^2 + 5 \cdot 10 = -100 + 50 = -50$

40. -54, 11, -5, -1229

41. We graphed $y = x^2 - x - 6$ in Exercise 23. The graph is shown below.

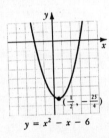

a) We graph $y = 2$ on the same set of axes as $y = x^2 - x - 6$.

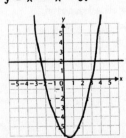

The graph of $y = 2$ intersects the graph of $y = x^2 - x - 6$ at about (-2.4, 2) and (3.4, 2). The solutions of the equation $x^2 - x - 6 = 2$ are about -2.4 and 3.4

41. (continued)
b) We graph $y = -3$ on the same set of axes as $y = x^2 - x - 6$.

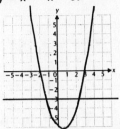

The graph of $y = -3$ intersects the graph of $y = x^2 - x - 6$ at about (-1.3, -3) and (2.3, -3). The solutions of the equation $x^2 - x - 6 = -3$ are about -1.3 and 2.3.

42. (0, c)

43.

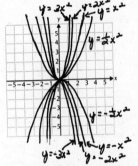

As $|a|$ increases the graph of $y = ax^2$ is stretched vertically.

44.

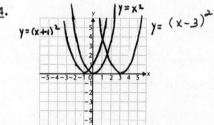

We can move the graph of $y = x^2$ to the right h units if h > 0 or to the left h units if h < 0 to obtain the graph of $y = (x - h)^2$.

45.

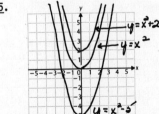

We can move the graph of $y = x^2$ up k units if k > 0 or down k units if k < 0 to obtain the graph of $y = x^2 + k$.

Chapter 11 (11.7)

46. If the coefficient of x^2 is negative, the graph opens downward and has a maximum point. If the coefficient of x^2 is positive, the graph opens upward and has a minimum point.

47. $A = LW$

$A = L(8 - L)$ Substituting $8 - L$ for W

$A = 8L - L^2$

We know the graph of $A = 8L - L^2$ opens downward since the coefficient of the squared term is negative ($-1 < 0$). Thus the vertex is the top point of the curve and the second coordinate of the vertex is the largest value of A. We find the vertex. The first coordinate is

$$-\frac{b}{2a} = -\frac{8}{2(-1)} = -(-4) = 4.$$

The second coordinate is

$A = 8L - L^2 = 8 \cdot 4 - 4^2 = 32 - 16 = 16.$

Thus the largest rectangular area that can be enclosed is 16 ft².

48.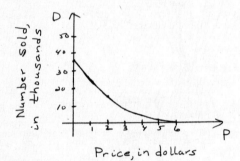

49. $S(p) = p^2 + p + 10$

p	S(p)
0	10
1	12
2	16
3	22
4	30
5	40
6	52

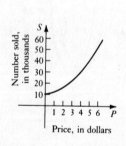

50. $2; 16,000 units

51. a) $D(25) = 25 + 0.05(25)^2 = 56.25$ ft
    $D(40) = 40 + 0.05(40)^2 = 120$ ft
    $D(55) = 55 + 0.05(55)^2 = 206.25$ ft
    $D(65) = 65 + 0.05(65)^2 = 276.25$ ft
    $D(75) = 75 + 0.05(75)^2 = 356.25$ ft
    $D(100) = 100 + 0.05(100)^2 = 600$ ft

51. (continued)

b) Plot the points found in part a) and connect them with a smooth curve.

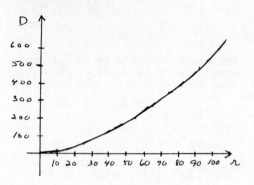

Exercise Set 11.7

1. <u>Familiarize</u>. We first make a drawing and label it with both known and unknown information. We let x represent the width of the frame. The length of the frame is 20 cm and the width is 12 cm. The length of the picture is $20 - 2x$, and the width of the picture is $12 - 2x$.

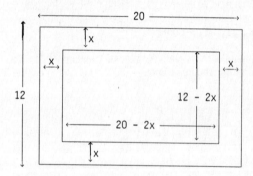

<u>Translate</u>. Recall that area is length × width. Thus, we have two expressions for the area of the picture: $(20 - 2x)(12 - 2x)$ and 84. This gives us a translation.

$(20 - 2x)(12 - 2x) = 84$

<u>Carry out</u>. We solve the equation.

$240 - 64x + 4x^2 = 84$

$4x^2 - 64x + 156 = 0$

$x^2 - 16x + 39 = 0$ Multiplying by $\frac{1}{4}$

$(x - 13)(x - 3) = 0$

$x - 13 = 0$ or $x - 3 = 0$

$x = 13$ or $x = 3$

<u>Check</u>. The number 13 is not a solution because when $x = 13$, $20 - 2x = -6$, and the length of the picture cannot be negative. When $x = 3$, $20 - 2x = 14$. This is the length. When $x = 3$, $12 - 2x = 6$. This is the width. The area is 14×6, or 84. This checks.

<u>State</u>. The width of the frame is 3 cm.

302

Chapter 11   (11.7)

2. 1 cm

3. <u>Familiarize</u>. We first make a drawing and label it. We let x represent the length of one leg. Then x + 1 represents the length of the other leg.

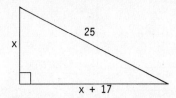

<u>Translate</u>. We use the Pythagorean equation.
$x^2 + (x + 17)^2 = 25^2$

<u>Carry out</u>. We solve the equation.
$x^2 + x^2 + 34x + 289 = 625$
$2x^2 + 34x - 336 = 0$
$x^2 + 17x - 168 = 0$ Multiplying by $\frac{1}{2}$
$(x - 7)(x + 24) = 0$

$x - 7 = 0$ or $x + 24 = 0$
$x = 7$ or $x = -24$

<u>Check</u>. Since the length of a leg cannot be negative, -24 does not check. But 7 does check. If the smaller leg is 7, the other leg is 7 + 17, or 24. Then, $7^2 + 24^2 = 49 + 576 = 625$, and $\sqrt{625} = 25$, the length of the hypotenuse.

<u>State</u>. The legs measure 7 ft and 24 ft.

4. 10 yd, 24 yd

5. <u>Familiarize</u>. We consider the drawing in the text, where w represents the width of the rectangle and w + 2 represents the length.

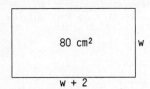

<u>Translate</u>. The area is length × width. Thus, we have two expressions for the area of the rectangle: (w + 2)w and 80. This gives us a translation.
$(w + 2)w = 80$

<u>Carry out</u>. We solve the equation.
$w^2 + 2w = 80$
$w^2 + 2w - 80 = 0$
$(w + 10)(w - 8) = 0$

$w + 10 = 0$ or $w - 8 = 0$
$w = -10$ or $w = 8$

5. (continued)

<u>Check</u>. Since the length of a side cannot be negative, -10 does not check. But 8 does check. If the width is 8, then the length is 8 + 2, or 10. The area is 10 × 8, or 80. This checks.

<u>State</u>. The length is 10 cm, and the width is 8 cm.

6. Length: 10 m, width: 7 m

7. <u>Familiarize</u>. We first make a drawing. We let x represent the length. Then x - 4 represents the width.

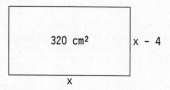

<u>Translate</u>. The area is length × width. Thus, we have two expressions for the area of the rectangle: x(x - 4) and 320. This gives us a translation.
$x(x - 4) = 320$

<u>Carry out</u>. We solve the equation.
$x^2 - 4x = 320$
$x^2 - 4x - 320 = 0$
$(x - 20)(x + 16) = 0$

$x - 20 = 0$ or $x + 16 = 0$
$x = 20$ or $x = -16$

<u>Check</u>. Since the length of a side cannot be negative, -16 does not check. But 20 does check. If the length is 20, then the width is 20 - 4, or 16. The area is 20 × 16, or 320. This checks.

<u>State</u>. The length is 20 cm, and the width is 16 cm.

8. Length: 20 cm, width: 17 cm

9. <u>Familiarize</u>. We first make a drawing. We let x represent the width. Then 2x represents the length.

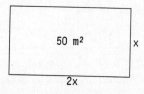

<u>Translate</u>. The area is length × width. Thus, we have two expressions for the area of the rectangle: 2x·x and 50. This gives us a translation.
$2x \cdot x = 50$

9. (continued)

Carry out. We solve the equation.

$2x^2 = 50$

$x^2 = 25$

$x = 5$ or $x = -5$ Principle of square roots

Check. Since the length of a side cannot be negative, -5 does not check. But 5 does check. If the width is 5, then the length is $2 \cdot 5$, or 10. The area is 10×5, or 50. This checks.

State. The length is 10 m, and the width is 5 m.

10. Length: 8 cm, width: 4 cm

11. Familiarize. We first make a drawing. We let x represent the length of one leg. Then x + 2 represents the length of the other leg.

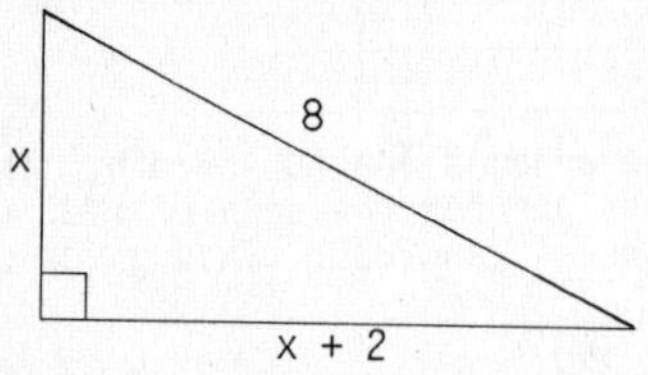

Translate. We use the Pythagorean equation.

$x^2 + (x + 2)^2 = 8^2$

Carry out. We solve the equation.

$x^2 + x^2 + 4x + 4 = 64$

$2x^2 + 4x - 60 = 0$

$x^2 + 2x - 30 = 0$

$a = 1, b = 2, c = -30$

$$x = \frac{-2 \pm \sqrt{2^2 - 4 \cdot 1 \cdot (-30)}}{2 \cdot 1}$$

$$x = \frac{-2 \pm \sqrt{4 + 120}}{2} = \frac{-2 \pm 2\sqrt{124}}{2}$$

$$x = \frac{-2 \pm \sqrt{4 \cdot 31}}{2} = \frac{-2 \pm 2\sqrt{31}}{2}$$

$$x = \frac{2(-1 \pm \sqrt{31})}{2} = -1 \pm \sqrt{31}$$

Using a calculator or Table 2 we find that $\sqrt{31} \approx 5.568$:

$-1 + \sqrt{31} \approx -1 + 5.568$ or $-1 - \sqrt{31} \approx -1 - 5.568$

≈ 4.6 or ≈ -6.6

Check. Since the length of a leg cannot be negative, -6.6 does not check. But 4.6 does check. If the smaller leg is 4.6, the other leg is 4.6 + 2, or 6.6. Then $4.6^2 + 6.6^2 = 21.16 + 43.56 = 64.72$ and using a calculator, $\sqrt{64.72} \approx 8.04 \approx 8$. Note that our check does not come out exact since we are using an approximation.

State. One leg is about 4.6 m, and the other is about 6.6 m long.

12. 2.4 cm, 4.4 cm

13. Familiarize. We first make a drawing. We let x represent the width and x + 2 the length.

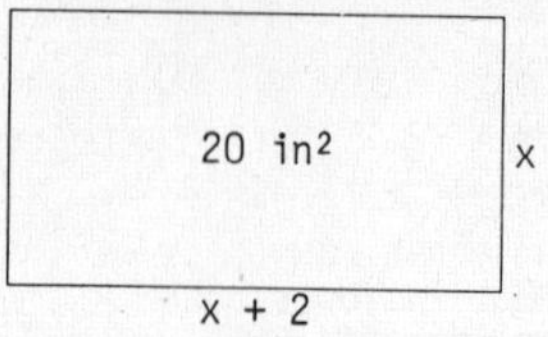

Translate. The area is length × width. We have two expressions for the area of the rectangle: (x + 2)x and 20. This gives us a translation.

$(x + 2)x = 20$

Carry out. We solve the equation.

$x^2 + 2x = 20$

$x^2 + 2x - 20 = 0$

$a = 1, b = 2, c = -20$

$$x = \frac{-2 \pm \sqrt{2^2 - 4 \cdot 1 \cdot (-20)}}{2 \cdot 1}$$

$$x = \frac{-2 \pm \sqrt{4 + 80}}{2} = \frac{-2 \pm \sqrt{84}}{2}$$

$$x = \frac{-2 \pm \sqrt{4 \cdot 21}}{2} = \frac{-2 \pm 2\sqrt{21}}{2}$$

$$x = \frac{2(-1 \pm \sqrt{21})}{2} = -1 \pm \sqrt{21}$$

Using a calculator or Table 2 we find that $\sqrt{21} \approx 4.583$:

$-1 + \sqrt{21} \approx -1 + 4.583$ or $-1 - \sqrt{21} \approx -1 - 4.583$

≈ 3.6 or ≈ -5.6

Check. Since the length of a side cannot be negative, -5.6 does not check. But 3.6 does check. If the width is 3.6, then the length is 3.6 + 2, or 5.6. The area is 5.6(3.6), or $20.16 \approx 20$.

State. The length is about 5.6 in., and the width is about 3.6 in.

14. Length: 5.7 ft, width: 2.7 ft

15. Familiarize. We first make a drawing. We let x represent the width and 2x the length.

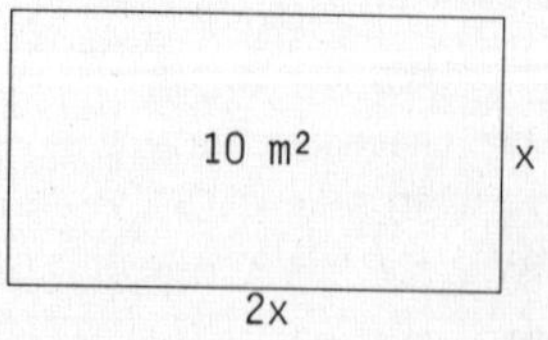

Translate. The area is length × width. We have two expressions for the area of the rectangle: $2x \cdot x$ and 10. This gives us a translation.

$2x \cdot x = 10$

Chapter 11   (11.7)

15. (continued)

   <u>Carry out</u>. We solve the equation.
   $$2x^2 = 10$$
   $$x^2 = 5$$
   $$x = \sqrt{5} \text{ or } x = -\sqrt{5}$$
   $$x \approx 2.2 \text{ or } x \approx -2.2 \quad \text{Using a calculator or Table 2}$$

   <u>Check</u>. Since the length cannot be negative, -2.2 does not check. But 2.2 does check. If the width is 2.2 m, then the length is 2(2.2) or 4.4 m. The area is 4.4(2.2), or 9.68 ≈ 10.

   <u>State</u>. The length is about 4.4 m, and the width is about 2.2 m.

16. Length: 6.4 cm, width: 3.2 cm

17. <u>Familiarize</u>. Referring to the drawing in the text, we complete the table.

	d	r	t
Upstream	40	r - 3	t_1
Downstream	40	r + 3	t_2

   <u>Translate</u>. Using t = d/r and the rows of the table, we have
   $$t_1 = \frac{40}{r - 3} \text{ and } t_2 = \frac{40}{r + 3}.$$
   Since the total time is 14 hr, $t_1 + t_2 = 14$, and we have
   $$\frac{40}{r - 3} + \frac{40}{r + 3} = 14.$$

   <u>Carry out</u>. We solve the equation.
   We multiply by (r - 3)(r + 3), the LCM of the denominators.
   $$(r - 3)(r + 3)\left[\frac{40}{r - 3} + \frac{40}{r + 3}\right] = (r - 3)(r + 3) \cdot 14$$
   $$40(r + 3) + 40(r - 3) = 14(r^2 - 9)$$
   $$40r + 120 + 40r - 120 = 14r^2 - 126$$
   $$80r = 14r^2 - 126$$
   $$0 = 14r^2 - 80r - 126$$
   $$0 = 7r^2 - 40r - 63$$
   $$0 = (7r + 9)(r - 7)$$

   $7r + 9 = 0$ or $r - 7 = 0$
   $7r = -9$ or $r = 7$
   $r = -\frac{9}{7}$ or $r = 7$

   <u>Check</u>. Since speed cannot be negative, $-\frac{9}{7}$ cannot be a solution. If the speed of the boat is 7 km/h, the speed upstream is 7 - 3, or 4 km/h, and the speed downstream is 7 + 3, or 10 km/h. The time upstream is $\frac{40}{4}$, or 10 hr. The time downstream is $\frac{40}{10}$, or 4 hr. The total time is 14 hr. This checks.

   <u>State</u>. The speed of the boat in still water is 7 km/h.

18. 12 km/h

19. <u>Familiarize</u>. We first make a drawing. We let r represent the speed of the boat in still water. Then r - 4 is the speed of the boat traveling upstream and r + 4 is the speed of the boat traveling downstream.

   Upstream
   r - 4 mph
   ───────────────
   4 mi

   Downstream
   r + 4 mph
   ───────────────
   12 mi

   We summarize the information in a table.

	d	r	t
Upstream	4	r - 4	t_1
Downstream	12	r + 4	t_2

   <u>Translate</u>. Using t = d/r and the rows of the table, we have
   $$t_1 = \frac{4}{r - 4} \text{ and } t_2 = \frac{12}{r + 4}.$$
   Since the total time is 2 hr, $t_1 + t_2 = 2$, and we have
   $$\frac{4}{r - 4} + \frac{12}{r + 4} = 2.$$

   <u>Carry out</u>. We solve the equation.
   We multiply by (r - 4)(r + 4), the LCM of the denominators.
   $$(r - 4)(r + 4)\left[\frac{4}{r - 4} + \frac{12}{r + 4}\right] = (r - 4)(r + 4) \cdot 2$$
   $$4(r + 4) + 12(r - 4) = 2(r^2 - 16)$$
   $$4r + 16 + 12r - 48 = 2r^2 - 32$$
   $$16r - 32 = 2r^2 - 32$$
   $$0 = 2r^2 - 16r$$
   $$0 = 2r(r - 8)$$

   $2r = 0$ or $r - 8 = 0$
   $r = 0$ or $r = 8$

   <u>Check</u>. If r = 0, then the speed upstream, 0 - 4, would be negative. Since speed cannot be negative, 0 cannot be a solution. If the speed of the boat is 8 mph, the speed upstream is 8 - 4, or 4 mph and the speed downstream is 8 + 4, or 12 mph. The time upstream is $\frac{4}{4}$, or 1 hr. The time downstream is $\frac{12}{12}$, or 1 hr. The total time is 2 hr. This checks.

   <u>State</u>. The speed of the boat in still water is 8 mph.

20. 9 mph

305

21. __Familiarize__. We first make a drawing. Let let r represent the speed of the current. Then $10 - r$ is the speed of the boat traveling upstream and $10 + r$ is the speed of the boat traveling downstream.

   Upstream
   $10 - r$ km/h
   12 km

   Downstream
   $10 + r$ km/h
   28 km

   We summarize the information in a table.

	d	r	t
Upstream	12	$10 - r$	t_1
Downstream	28	$10 + r$	t_2

   __Translate__. Using $t = d/r$ and the rows of the table, we have

   $$t_1 = \frac{12}{10 - r} \quad \text{and} \quad t_2 = \frac{28}{10 + r}.$$

   Since the total time is 4 hr, $t_1 + t_2 = 4$, and we have

   $$\frac{12}{10 - r} + \frac{28}{10 + r} = 4.$$

   __Carry out__. We solve the equation.

   We multiply by $(10 - r)(10 + r)$, the LCM of the denominators.

   $$(10-r)(10+r)\left[\frac{12}{10-r} + \frac{28}{10+r}\right] = (10-r)(10+r) \cdot 4$$

   $$12(10 + r) + 28(10 - r) = 4(100 - r^2)$$
   $$120 + 12r + 280 - 28r = 400 - 4r^2$$
   $$400 - 16r = 400 - 4r^2$$
   $$4r^2 - 16r = 0$$
   $$r^2 - 4r = 0$$
   $$r(r - 4) = 0$$

   $r = 0$ or $r - 4 = 0$
   $r = 0$ or $r = 4$

   __Check__. Since a stream is defined to be a flow of running water, its rate must be greater than 0. Thus, 0 cannot be a solution.

   If the speed of the current is 4 km/h, the speed upstream is $10 - 4$, or 6 km/h and the speed downstream is $10 + 4$, or 14 km/h. The time upstream is $\frac{12}{6}$, or 2 hours. The time downstream is $\frac{28}{14}$, or 2 hours. The total time is 4 hours.

   __State__. The speed of the stream is 4 km/h.

22. 2 km/h

23. __Familiarize__. First we make a drawing. We let r represent the speed of the wind. Then the speed of the plane flying against the wind is $200 - r$ and the speed of the plane flying with the wind is $200 + r$.

   Against the wind
   $200 - r$ mph
   738 miles

   With the wind
   $200 + r$ mph
   1062 miles

   We summarize the information in a table.

	d	r	t
Against wind	738	$200 - r$	t_1
With wind	1062	$200 + r$	t_2

   __Translate__. Using $t = d/r$ and the rows of the table, we have

   $$t_1 = \frac{738}{200 - r} \quad \text{and} \quad t_2 = \frac{1062}{200 + r}.$$

   Since the total time is 9 hr, $t_1 + t_2 = 9$, and we have

   $$\frac{738}{200 - r} + \frac{1062}{200 + r} = 9.$$

   __Carry out__. We solve the equation.

   We multiply by $(200 - r)(200 + r)$, the LCM of the denominators.

   $$(200-r)(200+r)\left[\frac{738}{200-r} + \frac{1062}{200+r}\right] = (200-r)(200+r) \cdot 9$$

   $$738(200 + r) + 1062(200 - r) = 9(40{,}000 - r^2)$$
   $$147{,}600 + 738r + 212{,}400 - 1062r = 360{,}000 - 9r^2$$
   $$360{,}000 - 324r = 360{,}000 - 9r^2$$
   $$9r^2 - 324r = 0$$
   $$9r(r - 36) = 0$$

   $9r = 0$ or $r - 36 = 0$
   $r = 0$ or $r = 36$

   __Check__. In this problem we assume there is a wind. Thus, the speed of the wind must be greater than 0 and the number 0 cannot be a solution. If the speed of the wind is 36 mph, the speed of the airplane against the wind is $200 - 36$, or 164 mph and the speed with the wind is $200 + 36$, or 236 mph. The time against the wind is $\frac{738}{164}$, or $4\frac{1}{2}$ hr. The time with the wind is $\frac{1062}{236}$, or $4\frac{1}{2}$ hr. The total time is 9 hours. The value checks.

   __State__. The speed of the wind is 36 mph.

24. 40 km/h

25. <u>Familiarize</u>. We first make a drawing. Let r represent the speed of the stream. Then 9 - r represents the speed of the boat traveling upstream and 9 + r represents the speed of the boat traveling downstream.

Upstream
9 - r
|←——————————→|
80 km

Downstream
9 + r
|←——————————→|
80 km

We summarize the information in a table.

	d	r	t
Upstream	80	9 - r	t_1
Downstream	80	9 + r	t_2

<u>Translate</u>. Using t = d/r and the rows of the table, we have

$$t_1 = \frac{80}{9-r} \text{ and } t_2 = \frac{80}{9+r}.$$

Since the total time is 18 hr, $t_1 + t_2 = 18$, and we have

$$\frac{80}{9-r} + \frac{80}{9+r} = 18.$$

<u>Carry out</u>. We solve the equation.
We multiply by (9 - r)(9 + r), the LCM of the denominators.

$$(9-r)(9+r)\left[\frac{80}{9-r} + \frac{80}{9+r}\right] = (9-r)(9+r) \cdot 18$$

$$80(9+r) + 80(9-r) = 18(81 - r^2)$$
$$720 + 80r + 720 - 80r = 1458 - 18r^2$$
$$1440 = 1458 - 18r^2$$
$$18r^2 = 18$$
$$r^2 = 1$$

r = 1 or r = -1

<u>Check</u>. Since speed cannot be negative, -1 cannot be a solution. If the speed of the stream is 1 km/h, the speed upstream is 9 - 1, or 8 km/h and the speed downstream is 9 + 1, or 10 km/h. The time upstream is $\frac{80}{8}$, or 10 hr. The time downstream is $\frac{80}{10}$, or 8 hr. The total time is 18 hr. This checks.

<u>State</u>. The speed of the stream is 1 km/h.

26. 2 km/h

27. <u>Familiarize</u>. First we make a drawing. Let x represent the length of each side. Then x + 1 represents the length of the diagonal.

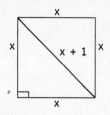

<u>Translate</u>. We use the Pythagorean equation.
$$x^2 + x^2 = (x + 1)^2$$

<u>Carry out</u>. We solve the equation.
$$2x^2 = x^2 + 2x + 1$$
$$x^2 - 2x - 1 = 0$$
$$a = 1, b = -2, c = -1$$

$$x = \frac{-(-2) \pm \sqrt{(-2)^2 - 4 \cdot 1 \cdot (-1)}}{2 \cdot 1}$$

$$x = \frac{2 \pm \sqrt{4+4}}{2} = \frac{2 \pm \sqrt{8}}{2}$$

$$x = \frac{2 \pm \sqrt{4 \cdot 2}}{2} = \frac{2 \pm 2\sqrt{2}}{2}$$

$$x = \frac{2(1 \pm \sqrt{2})}{2} = 1 \pm \sqrt{2}$$

<u>Check</u>. Since $1 - \sqrt{2}$ is negative, it cannot be the length of a side. Thus it cannot be a solution. If the length of a side is $1 + \sqrt{2}$, then the length of a diagonal is $2 + \sqrt{2}$. We check using the Pythagorean equation.

$(1 + \sqrt{2})^2 + (1 + \sqrt{2})^2$	$(2 + \sqrt{2})^2$
$= 2(1 + 2\sqrt{2} + 2)$	$= 4 + 4\sqrt{2} + 2$
$= 2(3 + 2\sqrt{2})$	$= 6 + 4\sqrt{2}$
$= 6 + 4\sqrt{2}$	

Thus, $(1 + \sqrt{2})^2 + (1 + \sqrt{2})^2 = (2 + \sqrt{2})^2$. The value checks.

The area is side × side:
$(1 + \sqrt{2}) \times (1 + \sqrt{2}) = 1 + 2\sqrt{2} + 2 = 3 + 2\sqrt{2}$

<u>State</u>. The area of the square is $3 + 2\sqrt{2}$ (≈ 5.828) square units.

28. 37 and 38 or -38 and -37

Chapter 11   (11.7)

29. From the drawing we see that we have a problem similar to Exercise 27.

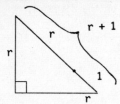

$r^2 + r^2 = (r + 1)^2$

From Exercise 27, we know that $r = 1 + \sqrt{2} \approx 2.41$ cm.

30. 7.5 ft

31. Familiarize. The radius of a 10-in. pizza is $\frac{10}{2}$, or 5 in. The radius of a d-in. pizza is $\frac{d}{2}$ in. The area of a circle is πr^2.

Translate.

$$\underbrace{\text{Area of d-in. pizza}}_{\pi \left(\frac{d}{2}\right)^2} = \underbrace{\text{Area of 10-in. pizza}}_{\pi \cdot 5^2} + \underbrace{\text{Area of 10-in. pizza}}_{\pi \cdot 5^2}$$

Carry out. We solve the equation.

$\frac{d^2}{4}\pi = 25\pi + 25\pi$

$\frac{d^2}{4}\pi = 50\pi$

$\frac{d^2}{4} = 50$ Multiplying by $\frac{1}{\pi}$

$d^2 = 200$

$d = \sqrt{200}$ or $d = -\sqrt{200}$

$d = 10\sqrt{2}$ or $d = -10\sqrt{2}$

$d \approx 14.14$ or $d \approx -14.14$ Using a calculator or Table 2

Check. Since the diameter cannot be negative, -14.14 is not a solution. If $d = 10\sqrt{2}$, or 14.14, then $r = 5\sqrt{2}$ and the area is $\pi(5\sqrt{2})^2$, or 50π. The area of the two 10" pizzas is $2 \cdot \pi \cdot 5^2$, or 50π. The value checks.

State. The diameter of the pizza should be $10\sqrt{2}$, or ≈ 14.14 in.

The area of two 10" pizzas is approximately the same as a 14" pizza. Thus, you get more to eat with two 10" pizzas than with a 13-in. pizza.

32. 5.33 cm

33. Familiarize. First make a drawing.

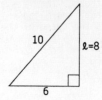

Using the Pythagorean equation, we get 8 for the other leg.

$6^2 + \ell^2 = 10^2$

$36 + \ell^2 = 100$

$\ell^2 = 64$

$\ell = 8$ Taking the principal square root

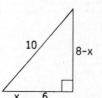

Let x represent how much the lower end would have to be pulled away. Thus, the legs are represented by $x + 6$ and $8 - x$.

Translate. We use the Pythagorean equation.

$(x + 6)^2 + (8 - x)^2 = 10^2$

Carry out. We solve the equation.

$x^2 + 12x + 36 + 64 - 16x + x^2 = 100$

$2x^2 - 4x + 100 = 100$

$2x^2 - 4x = 0$

$2x(x - 2) = 0$

$2x = 0$ or $x - 2 = 0$

$x = 0$ or $x = 2$

Check. If the ladder is moved, x cannot equal 0. We check $x = 2$. If $x = 2$, then the legs are 8 and 6 and $8^2 + 6^2 = 64 + 36 = 100$, the square of the hypotenuse.

State. The lower end of the ladder would have to be pulled away 2 ft.

34. A: 15 mph, B: 20 mph

35. Familiarize. We first make a drawing. Let x represent the length of each side. Then $x + 3$ represents the length of the diagonal.

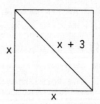

Translate. We use the Pythagorean equation.

$x^2 + x^2 = (x + 3)^2$

308

Chapter 11   (11.8)

35. (continued)

    Carry out. We solve the equation.
    $$2x^2 = x^2 + 6x + 9$$
    $$x^2 - 6x - 9 = 0$$
    $a = 1, b = -6, c = -9$
    $$x = \frac{-(-6) \pm \sqrt{(-6)^2 - 4 \cdot 1 \cdot (-9)}}{2 \cdot 1}$$
    $$x = \frac{6 \pm \sqrt{36 + 36}}{2} = \frac{6 \pm \sqrt{72}}{2}$$
    $$x = \frac{6 \pm \sqrt{36 \cdot 2}}{2} = \frac{6 \pm 6\sqrt{2}}{2}$$
    $$x = \frac{2(3 \pm 3\sqrt{2})}{2} = 3 \pm 3\sqrt{2}$$

    Check. Since $3 - 3\sqrt{2}$ is negative, it cannot be the length of a side. If the length of a side is $3 + 3\sqrt{2}$, then the length of a diagonal is $6 + 3\sqrt{2}$. We check using the Pythagorean equation.

    $(3 + 3\sqrt{2})^2 + (3 + 3\sqrt{2})^2 \qquad (6 + 3\sqrt{2})^2$
    $= 2(9 + 18\sqrt{2} + 18) \qquad\quad = 36 + 36\sqrt{2} + 18$
    $= 2(27 + 18\sqrt{2}) \qquad\qquad\;\; = 54 + 36\sqrt{2}$
    $= 54 + 36\sqrt{2}$

    Thus, $(3 + 3\sqrt{2})^2 + (3 + 3\sqrt{2})^2 = (6 + 3\sqrt{2})^2$. The value checks.

    State. The length of a side of the square is $3 + 3\sqrt{2}$, or about 7.2 cm.

36. 6 hr

Exercise Set 11.8

1. $\sqrt{-1} = i$

2. $3i$

3. $\sqrt{-25} = \sqrt{-1 \cdot 25} = \sqrt{-1} \cdot \sqrt{25} = i \cdot 5 = 5i$

4. $4i$

5. $\sqrt{-8} = \sqrt{-1 \cdot 8} = \sqrt{-1} \cdot \sqrt{8} = i \cdot 2\sqrt{2} = 2i\sqrt{2}$

6. $2i\sqrt{5}$

7. $-\sqrt{-12} = -\sqrt{-1 \cdot 12} = -\sqrt{-1} \cdot \sqrt{12} = -i \cdot 2\sqrt{3} = -2i\sqrt{3}$

8. $-3i\sqrt{5}$

9. $-\sqrt{-27} = -\sqrt{-1 \cdot 27} = -\sqrt{-1} \cdot \sqrt{27} = -i \cdot 3\sqrt{3} = -3i\sqrt{3}$

10. $-2i$

11. $7 + \sqrt{-16} = 7 + \sqrt{-1 \cdot 16} = 7 + \sqrt{-1} \cdot \sqrt{16} = 7 + i \cdot 4 = 7 + 4i$

12. $-8 - 6i$

13. $3 - \sqrt{-98} = 3 - \sqrt{-1 \cdot 98} = 3 - \sqrt{-1} \cdot \sqrt{98} = 3 - i \cdot 7\sqrt{2} = 3 - 7i\sqrt{2}$

14. $-2 + 5i\sqrt{5}$

15. $i^7 = i^6 \cdot i = (i^2)^3 \cdot i = (-1)^3 \cdot i = -1 \cdot i = -i$

16. 1

17. $i^9 = i^8 \cdot i = (i^2)^4 \cdot i = (-1)^4 \cdot i = 1 \cdot i = i$

18. -1

19. $i^{37} = i^{36} \cdot i = (i^2)^{18} \cdot i = (-1)^{18} \cdot i = 1 \cdot i = i$

20. -1

21. $i^{80} = (i^2)^{40} = (-1)^{40} = 1$

22. $-i$

23. $i^3 + i^4 = i^2 \cdot i + (i^2)^2 = -1 \cdot i + (-1)^2 = -i + 1$, or $1 - i$

24. 0

25. $(2 + 3i) + (4 - i) = (2 + 4) + (3 - 1)i$
    $= 6 + 2i$

26. $5 + 9i$

27. $(2 - 2i) + (4 - i) = (2 + 4) + (-2 - 1)i$
    $= 6 - 3i$

28. $-1 - 8i$

29. $(8 - i) + (-3 + 6i) = (8 - 3) + (-1 + 6)i$
    $= 5 + 5i$

30. $11 + i$

31. $(5 - i) - (3 + 2i) = (5 - 3) + (-1 - 2)i$
    $= 2 - 3i$

32. -4

33. $(4 - 4i) - (5 - 3i) = (4 - 5) + [-4 - (-3)]i$
    $= -1 - i$

34. $-3 + 4i$

35. $(7 + 4i) - (-42 - i) = [7 - (-42)] + [4 - (-1)]i$
    $= 49 + 5i$

Chapter 11   (11.8)

36. $-5i$

37. $(2 + 3i)(1 + i) = 2 + 2i + 3i + 3i^2$ FOIL
    $ = 2 + 2i + 3i - 3$ $(i^2 = -1)$
    $ = -1 + 5i$

38. $-14 + 23i$

39. $(3 + 2i)(6 - 2i) = 18 - 6i + 12i - 4i^2$ FOIL
    $ = 18 - 6i + 12i + 4$ $(i^2 = -1)$
    $ = 22 + 6i$

40. $10 + 5i$

41. $(6 - 5i)(3 + 4i) = 18 + 24i - 15i - 20i^2$ FOIL
    $ = 18 + 24i - 15i + 20$ $(i^2 = -1)$
    $ = 38 + 9i$

42. $40 - 13i$

43. $(2 - 7i)(2 + 7i) = 2^2 - (7i)^2$
    $ = 4 - 49i^2$
    $ = 4 - (-49)$ $(i^2 = -1)$
    $ = 53$

44. 25

45. $i^4 = (i^2)^2 = (-1)^2 = 1$

46. -1

47. $(2i)^3 = 2^3 \cdot i^3 = 8 \cdot i^2 \cdot i = 8(-1)i = -8i$

48. $-243i$

49. $(1 - 3i)^2 = 1^2 - 2 \cdot 1 \cdot 3i + (3i)^2$
    $ = 1 - 6i + 9i^2$
    $ = 1 - 6i - 9$
    $ = -8 - 6i$

50. $3 - 4i$

51. $(3 + 5i)^2 = 3^2 + 2 \cdot 3 \cdot 5i + (5i)^2$
    $ = 9 + 30i + 25i^2$
    $ = 9 + 30i - 25$
    $ = -16 + 30i$

52. $32 + 24i$

53. $(-1 + 3i)^2 = (-1)^2 + 2(-1)3i + (3i)^2$
    $ = 1 - 6i + 9i^2$
    $ = 1 - 6i - 9$
    $ = -8 - 6i$

54. $45 + 28i$

55. $\dfrac{1 - i}{1 + i} = \dfrac{1 - i}{1 + i} \cdot \dfrac{1 - i}{1 - i}$ Multiplying by 1
    $\phantom{\dfrac{1 - i}{1 + i}} = \dfrac{(1 - i)(1 - i)}{(1 + i)(1 - i)} = \dfrac{1 - i - i + i^2}{1^2 - i^2}$
    $\phantom{\dfrac{1 - i}{1 + i}} = \dfrac{1 - 2i - 1}{1 + 1} = \dfrac{-2i}{2} = -i$

56. $\dfrac{11}{26} + \dfrac{3}{26}i$

57. $\dfrac{1}{2 - 3i} = \dfrac{1}{2 - 3i} \cdot \dfrac{2 + 3i}{2 + 3i}$ Multiplying by 1
    $\phantom{\dfrac{1}{2 - 3i}} = \dfrac{2 + 3i}{2^2 - (3i)^2} = \dfrac{2 + 3i}{4 - 9i^2} = \dfrac{2 + 3i}{4 + 9}$
    $\phantom{\dfrac{1}{2 - 3i}} = \dfrac{2 + 3i}{13} = \dfrac{2}{13} + \dfrac{3}{13}i$

58. $-\dfrac{4}{25} + \dfrac{3}{25}i$

59. $\dfrac{6 - 5i}{3i} = \dfrac{6 - 5i}{3i} \cdot \dfrac{-3i}{-3i}$ Multiplying by 1
    $\phantom{\dfrac{6 - 5i}{3i}} = \dfrac{-18i + 15i^2}{-9i^2} = \dfrac{-18i - 15}{9}$
    $\phantom{\dfrac{6 - 5i}{3i}} = \dfrac{-15}{9} - \dfrac{18}{9}i = -\dfrac{5}{3} - 2i$

60. $4 - \dfrac{5}{2}i$

61. We substitute $-i$ for x in the equation.
    $$\begin{array}{c|c} x^2 + 1 = 0 & \\ \hline (-i)^2 + 1 & 0 \\ i^2 + 1 & \\ -1 + 1 & \\ 0 & \end{array}$$
    The number $-i$ is a solution.

62. Yes

63. We substitute $-1 + i$ for x in the equation.
    $$\begin{array}{c|c} x^2 + 2x + 2 = 0 & \\ \hline (-1 + i)^2 + 2(-1 + i) + 2 & 0 \\ 1 - 2i + i^2 - 2 + 2i + 2 & \\ 1 - 2i - 1 - 2 + 2i + 2 & \\ (1 - 1 - 2 + 2) + (-2 + 2)i & \\ 0 + 0i & \\ 0 & \end{array}$$
    The number $-1 + i$ is a solution.

64. No

65. We substitute $3 - 2i$ for x in the equation.
    $$\begin{array}{c|c} x^2 + 2x + 1 = 0 & \\ \hline (3 - 2i)^2 + 2(3 - 2i) + 1 & 0 \\ 9 - 12i + 4i^2 + 6 - 4i + 1 & \\ 9 - 12i - 4 + 6 - 4i + 1 & \\ (9 - 4 + 6 + 1) + (-12 - 4)i & \\ 12 - 16i & \end{array}$$
    The number $3 - 2i$ is not a solution.

Chapter 11   (11.8)

66. No

67. $x^2 + 4 = 0$
$x^2 = -4$
$x = \sqrt{-4}$ or $x = -\sqrt{-4}$ Principle of square roots
$x = i\sqrt{4}$ or $x = -i\sqrt{4}$
$x = 2i$ or $x = -2i$
The solutions are $2i$ and $-2i$, or $\pm 2i$.

68. $\pm 3i$

69. $x^2 = -12$
$x = \sqrt{-12}$ or $x = -\sqrt{-12}$ Principle of square roots
$x = i\sqrt{12}$ or $x = -i\sqrt{12}$
$x = 2i\sqrt{3}$ or $x = -2i\sqrt{3}$
The solutions are $2i\sqrt{3}$ and $-2i\sqrt{3}$, or $\pm 2i\sqrt{3}$.

70. $\pm 4i\sqrt{3}$

71. $x^2 - 4x - 5 = 0$
$a = 1, b = -4, c = -5$
We compute the discriminant:
$b^2 - 4ac = (-4)^2 - 4 \cdot 1 \cdot (-5) = 16 + 20 = 36$
The discriminant is a perfect square, so we can factor.
$x^2 - 4x - 5 = 0$
$(x - 5)(x + 1) = 0$
$x - 5 = 0$ or $x + 1 = 0$
$x = 5$ or $x = -1$
The solutions are 5 and -1.

72. $-2 \pm i$

73. $(x - 2)^2 = -16$
$x - 2 = \sqrt{-16}$ or $x - 2 = -\sqrt{-16}$ Principle of square roots
$x - 2 = i\sqrt{16}$ or $x - 2 = -i\sqrt{16}$
$x - 2 = 4i$ or $x - 2 = -4i$
$x = 2 + 4i$ or $x = 2 - 4i$
The solutions are $2 + 4i$ and $2 - 4i$, or $2 \pm 4i$.

74. $-1 \pm 5i$

75. $x^2 + 2x + 2 = 0$
$a = 1, b = 2, c = 2$
We compute the discriminant:
$b^2 - 4ac = 2^2 - 4 \cdot 1 \cdot 2 = 4 - 8 = -4$
The discriminant is negative, so there are no real-number solutions. But there are complex-number solutions given by
$$x = \frac{-b \pm \sqrt{b^2 - 4ac}}{2a}$$
$$x = \frac{-2 \pm \sqrt{-4}}{2 \cdot 1} = \frac{-2 \pm i\sqrt{4}}{2} = \frac{-2 \pm 2i}{2}$$
$$x = \frac{-2}{2} \pm \frac{2i}{2} = -1 \pm i.$$
The solutions are $-1 \pm i$.

76. $1 \pm i$

77. $x^2 + 7 = 4x$
$x^2 - 4x + 7 = 0$ Standard form
$a = 1, b = -4, c = 7$
We compute the discriminant:
$b^2 - 4ac = (-4)^2 - 4 \cdot 1 \cdot 7 = 16 - 28 = -12$
The discriminant is negative, so there are no real-number solutions. But there are complex-number solutions given by
$$x = \frac{-b \pm \sqrt{b^2 - 4ac}}{2a}$$
$$x = \frac{-(-4) \pm \sqrt{-12}}{2 \cdot 1} = \frac{4 \pm \sqrt{-12}}{2} = \frac{4 \pm i\sqrt{12}}{2}$$
$$x = \frac{4 \pm 2i\sqrt{3}}{2} = \frac{4}{2} \pm \frac{2\sqrt{3}}{2}i = 2 \pm \sqrt{3}i$$
The solutions are $2 \pm \sqrt{3}i$.

78. $-2 \pm \sqrt{3}i$

79. $2t^2 + 6t + 5 = 0$
$a = 2, b = 6, t = 5$
We compute the discriminant:
$b^2 - 4ac = 6^2 - 4 \cdot 2 \cdot 5 = 36 - 40 = -4$
The discriminant is negative, so there are no real-number solutions. But there are complex-number solutions given by
$$t = \frac{-b \pm \sqrt{b^2 - 4ac}}{2a}$$
$$t = \frac{-6 \pm \sqrt{-4}}{2 \cdot 2} = \frac{-6 \pm i\sqrt{4}}{4} = \frac{-6 \pm 2i}{4}$$
$$t = \frac{-6}{4} \pm \frac{2}{4}i = -\frac{3}{2} \pm \frac{1}{2}i.$$
The solutions are $-\frac{3}{2} \pm \frac{1}{2}i$.

80. $-\frac{3}{8} \pm \frac{\sqrt{23}}{8}i$

Chapter 11   (11.8)

81. $1 + 2m + 3m^2 = 0$
$3m^2 + 2m + 1 = 0$ Standard form
$a = 3, b = 2, c = 1$

We compute the discriminant:
$b^2 - 4ac = 2^2 - 4 \cdot 3 \cdot 1 = 4 - 12 = -8$

The discriminant is negative, so there are no real-number solutions. But there are complex-number solutions given by

$m = \dfrac{-b \pm \sqrt{b^2 - 4ac}}{2a}$

$m = \dfrac{-2 \pm \sqrt{-8}}{2 \cdot 3} = \dfrac{-2 \pm i\sqrt{8}}{6} = \dfrac{-2 \pm 2i\sqrt{2}}{6}$

$m = \dfrac{-2}{6} \pm \dfrac{2\sqrt{2}}{6}i = -\dfrac{1}{3} \pm \dfrac{\sqrt{2}}{3}i$

The solutions are $-\dfrac{1}{3} \pm \dfrac{\sqrt{2}}{3}i$.

82. $\dfrac{3}{4} \pm \dfrac{\sqrt{3}}{4}i$

83. $y = \dfrac{k}{x}$

$16 = \dfrac{k}{4}$ Substituting

$64 = k$ Constant of variation

$y = \dfrac{64}{x}$ Equation of variation

84. $\sqrt{2}$ ft

85. $\sqrt{72x^6y^8} = \sqrt{36x^6y^8 \cdot 2} = 6x^3y^4\sqrt{2}$

86. -15

87. $1 + i + i^2 + i^3 = 1 + i + (-1) + i^2 \cdot i$
$= 1 + i - 1 + (-1) \cdot i$
$= 1 + i - 1 - i$
$= (1 - 1) + (1 - 1)i$
$= 0$

88. $-1 + i$

89. $\dfrac{(1 + i)^2}{(2 - i)^2} = \dfrac{1 + 2i + i^2}{4 - 4i + i^2} = \dfrac{1 + 2i - 1}{4 - 4i - 1}$

$= \dfrac{2i}{3 - 4i} = \dfrac{2i}{3 - 4i} \cdot \dfrac{3 + 4i}{3 + 4i}$

$= \dfrac{6i + 8i^2}{9 - 16i^2} = \dfrac{6i - 8}{9 + 16}$

$= \dfrac{-8 + 6i}{25} = -\dfrac{8}{25} + \dfrac{6}{25}i$

90. $\dfrac{5}{26} - \dfrac{1}{26}i$

91. $(x + 2)^2 + (x + 1)^2 = 0$
$(x^2 + 4x + 4) + (x^2 + 2x + 1) = 0$
$2x^2 + 6x + 5 = 0$

This is the equation we solved in Exercise 79 (using the variable t instead of x). The solutions are $-\dfrac{3}{2} \pm \dfrac{1}{2}i$.

92. $-2 \pm i$

93. $\dfrac{2x - 1}{5} - \dfrac{2}{x} = \dfrac{x}{2}$

We multiply by 10x, the LCM of the denominators.

$10x\left[\dfrac{2x - 1}{5} - \dfrac{2}{x}\right] = 10x \cdot \dfrac{x}{2}$

$2x(2x - 1) - 10 \cdot 2 = 5x \cdot x$

$4x^2 - 2x - 20 = 5x^2$

$0 = x^2 + 2x + 20$

$a = 1, b = 2, c = 20$

We compute the discriminant:
$b^2 - 4ac = 2^2 - 4 \cdot 1 \cdot 20 = 4 - 80 = -76$

The discriminant is negative, so there are no real-number solutions. But there are complex-number solutions given by

$x = \dfrac{-b \pm \sqrt{b^2 - 4ac}}{2a}$

$x = \dfrac{-2 \pm \sqrt{-76}}{2 \cdot 1} = \dfrac{-2 \pm i\sqrt{76}}{2} = \dfrac{-2 \pm 2i\sqrt{19}}{2}$

$x = \dfrac{-2}{2} \pm \dfrac{2\sqrt{19}}{2}i = -1 \pm \sqrt{19}i$

The solutions are $-1 \pm \sqrt{19}i$.

94. $\dfrac{1}{2} \pm \dfrac{\sqrt{3}}{6}i$

ANSWERS FOR EXERCISES IN THE APPENDIXES

Appendix A

1. $\{3,4,5,6,7,8\}$ 2. $\{101,102,103,104,105,106,107\}$

3. $\{41,43,45,47,49\}$ 4. $\{15,20,25,30,35\}$

5. $\{-3,3\}$ 6. $\{0.008\}$ 7. False   8. True

9. True   10. True   11. True   12. True

13. True   14. True   15. True   16. False

17. False   18. True   19. $\{c,d,e\}$ 20. $\{u,i\}$

21. $\{1,10\}$ 22. $\{0,1\}$ 23. $\emptyset$ 24. $\emptyset$

25. The system has no solution. The lines are parallel. Their intersection is empty.   26. The system has no solution. The lines are parallel. Their intersection is empty.   27. $\{a,e,i,o,u,q,c,k\}$

28. $\{a,b,c,d,e,f,g\}$ 29. $\{0,1,2,5,7,10\}$

30. $\{1,2,5,10,0,7\}$ 31. $\{a,e,i,o,u,m,n,f,g,h\}$

32. $\{1,2,5,10,a,b\}$ 33. The solution set is $\{3,-5\}$. This set is the union of the solution sets of the equations $x - 3 = 0$ and $x + 5 = 0$, which are $\{3\}$ and $\{-5\}$.   34. The solution set is $\{-3,1\}$. This set is the union of the solution sets of the equations $x + 3 = 0$ and $x - 1 = 0$, which are $\{-3\}$ and $\{1\}$.

35. The set of integers   36. $\emptyset$ 37. The set of real numbers   38. The set of positive even integers

39. $\emptyset$ 40. The set of integers   41. a) A, b) A, c) A, d) $\emptyset$ 42. a) Yes, b) No, c) No, d) Yes, e) Yes, f) No   43. True

Appendix B

1. $(t + 3)(t^2 - 3t + 9)$ 2. $(p + 2)(p^2 - 2p + 4)$

3. $(a - 1)(a^2 + a + 1)$ 4. $(w - 4)(w^2 + 4w + 16)$

5. $(z + 5)(z^2 - 5z + 25)$ 6. $(x + 1)(x^2 - x + 1)$

7. $(2a - 1)(4a^2 + 2a + 1)$ 8. $(3x - 1)(9x^2 + 3x + 1)$

9. $(y - 3)(y^2 + 3y + 9)$ 10. $(p - 2)(p^2 + 2p + 4)$

11. $(4 + 5x)(16 - 20x + 25x^2)$

12. $(2 + 3b)(4 - 6b + 9b^2)$

13. $(5p - 1)(25p^2 + 5p + 1)$

14. $(4w - 1)(16w^2 + 4w + 1)$

15. $(3m + 4)(9m^2 - 12m + 16)$

16. $(2t + 3)(4t^2 - 6t + 9)$

17. $(p - q)(p^2 + pq + q^2)$ 18. $(a + b)(a^2 - ab + b^2)$

19. $\left(x + \frac{1}{2}\right)\left(x^2 - \frac{1}{2}x + \frac{1}{4}\right)$ 20. $\left(y + \frac{1}{3}\right)\left(y^2 - \frac{1}{3}y + \frac{1}{9}\right)$

21. $2(y - 4)(y^2 + 4y + 16)$

22. $3(z - 1)(z^2 + z + 1)$

23. $3(2a + 1)(4a^2 - 2a + 1)$

24. $2(3x + 1)(9x^2 - 3x + 1)$

25. $r(s + 4)(s^2 - 4s + 16)$

26. $a(b + 5)(b^2 - 5b + 25)$

27. $5(x - 2z)(x^2 + 2xz + 4z^2)$

28. $2(y - 3z)(y^2 + 3yz + 9z^2)$

29. $(x + 0.1)(x^2 - 0.1x + 0.01)$

30. $(y + 0.5)(y^2 - 0.5y + 0.25)$

31. $8(2x^2 - t^2)(4x^4 + 2x^2t^2 + t^4)$

32. $(5c^2 - 2d^2)(25c^4 + 10c^2d^2 + 4d^4)$

33. $(x - 0.3)(x^2 + 0.3x + 0.09)$

34. $(y + 0.2)(y^2 - 0.2y + 0.04)$

35. $(x^{2a} + y^b)(x^{4a} - x^{2a}y^b + y^{2b})$

36. $(ax - by)(a^2x^2 + abxy + b^2y^2)$

37. $3(x^a + 2y^b)(x^{2a} - 2x^ay^b + 4y^{2b})$

38. $\left(\frac{2}{3}x + \frac{1}{4}y\right)\left(\frac{4}{9}x^2 - \frac{1}{6}xy + \frac{1}{16}y^2\right)$

39. $\frac{1}{3}\left(\frac{1}{2}xy + z\right)\left(\frac{1}{4}x^2y^2 - \frac{1}{2}xyz + z^2\right)$

40. $\frac{1}{2}\left(\frac{1}{2}x^a + y^{2a}z^{3b}\right)\left(\frac{1}{4}x^{2a} - \frac{1}{2}x^ay^{2a}z^{3b} + y^{4a}z^{6b}\right)$

ANSWERS FOR EXERCISES IN THE APPENDIXES

Appendix C

1. {x|x < -5 or x > 3}   2. {x|x < -4 or x > 1}
3. {x|-2 ≤ x ≤ 1}   4. {x|-5 ≤ x ≤ 3}
5. {x|-2 < x < 1}   6. {x|-1 < x < 2}
7. {x|x ≤ -2 or x ≥ 2}   8. {x|x ≤ -3 or x ≥ 3}
9. {p|-5 < p < 5}   10. {m|-6 ≤ m ≤ 6}
11. All reals   12. ∅   13. {x|2 < x < 4}
14. {x|x < -2 or x > 6}   15. {x|x < -2 or 0 < x < 2}
16. {x|-1 < x < 0 or x > 1}
17. {x|x < -3 or -1 < x < 2}
18. {x|-2 < x < 1 or x > 4}
19. {x|x < -3 or -2 < x < 1}
20. {x|x < -1 or 2 < x < 3}
21. {x|-1 < x < 1 or x > 2}
22. {x|-3 ≤ x ≤ 1 or x ≥ 2}
23. {y|y ≤ -2 or -1 ≤ y ≤ 1 or y ≥ 2}
24. {p|-5 ≤ p ≤ -3 or 3 ≤ p ≤ 5}
25. {x|1 - $\sqrt{3}$ < x < 1 + $\sqrt{3}$}
26. {x|x < -1 - $\sqrt{5}$ or x > -1 + $\sqrt{5}$}
27. {x|x < 4}   28. {x|x > -5}
29. {x|x < -1 or x > 3}   30. {x|-5 < x < 2}